INTRODUCTION TO STELLAR DYNAMICS

The study of stellar dynamics is experiencing an exciting new wave of interest thanks to observational campaigns and the ready availability of powerful computers. While its relevance includes many areas of astrophysics, from the structure of the Milky Way to dark matter halos, few texts are suited to advanced students. This volume provides a broad overview of the key concepts beyond the elementary level, bridging the gap between the standard texts and specialist literature. The author reviews Newtonian gravity in depth before examining the dynamical properties of collisional and collisionless stellar-dynamical systems that result from gravitational interactions. Guided examples and exercises ensure a thorough grounding in the mathematics, while discussions of important practical applications give a complete picture of the subject. Readers are given a sound working knowledge of the fundamental ideas and techniques employed in the field and the conceptual background needed to progress to more advanced graduate-level treatises.

LUCA CIOTTI is Professor of Astronomy and Astrophysics at the University of Bologna, where he has also served for many years as Director of the Collegio Superiore. Since 1992, he has been a long-term visitor of the Princeton University Observatory. His work developed in several fields of theoretical astrophysics, with main interests in stellar dynamics, fluid dynamics, and black hole accretion. This is his second book in the field, having previously written *Lecture Notes on Stellar Dynamics* based on the lectures given at the Scuola Normale Superiore in Pisa.

INTRODUCTION TO STELLAR DYNAMICS

LUCA CIOTTI
University of Bologna

CAMBRIDGE
UNIVERSITY PRESS

University Printing House, Cambridge CB2 8BS, United Kingdom

One Liberty Plaza, 20th Floor, New York, NY 10006, USA

477 Williamstown Road, Port Melbourne, VIC 3207, Australia

314-321, 3rd Floor, Plot 3, Splendor Forum, Jasola District Centre, New Delhi - 110025, India

79 Anson Road, #06-04/06, Singapore 079906

Cambridge University Press is part of the University of Cambridge.

It furthers the University's mission by disseminating knowledge in the pursuit of education, learning, and research at the highest international levels of excellence.

www.cambridge.org
Information on this title: www.cambridge.org/9781107001534
DOI: 10.1017/9780511736117

First published 2021

A catalogue record for this publication is available from the British Library.

Library of Congress Cataloging-in-Publication Data
Names: Ciotti, Luca, 1964- author.
Title: Introduction to stellar dynamics / Luca Ciotti, Università degli Studi, Bologna, Italy.
Description: Cambridge, UK ; New York, NY : Cambridge University Press, 2021. | Includes bibliographical references and index.
Identifiers: LCCN 2020046758 (print) | LCCN 2020046759 (ebook) | ISBN 9781107001534 (hardback) | ISBN 9780511736117 (epub)
Subjects: LCSH: Stellar dynamics.
Classification: LCC QB810 .C56 2021 (print) | LCC QB810 (ebook) | DDC 521–dc23
LC record available at https://lccn.loc.gov/2020046758
LC ebook record available at https://lccn.loc.gov/2020046759

ISBN 978-1-107-00153-4 Hardback

Contents

Preface

This book aims to provide an introductory yet solid background in stellar dynamics to astronomy students who later will attend more advanced courses (e.g., at the graduate level) and to researchers in other fields of astrophysics looking for an accessibile introduction to the subject. It should provide the reader with sufficiently broad (yet rigorous) coverage of some of the basic topics of stellar dynamics, opening the way to the study of specialized texts such as *Dynamics of Galaxies* (Bertin 2014), *Galactic Dynamics* (Binney and Tremaine 2008), *Dynamics and Evolution of Galactic Nuclei* (Merritt 2013), and *Dynamical Evolution of Globular Clusters* (Spitzer 1987), and to a fruitful reading of important reviews such as those of Binney (1982a), Cappellari (2016), and de Zeeuw and Franx (1991). However, it is assumed that the student already knows the basics of extragalactic astronomy (e.g., the classification, morphology, kinematics, and phenomenology of stellar systems such as open and globular clusters, galaxies, and clusters of galaxies); from this point of view, *Galactic Astronomy* (Binney and Merrifield 1998) and *Spiral Galaxies* (Bertin and Lin 1996) are excellent complementary works. From the mathematical point of view, a working knowledge of geometry, linear algebra, and calculus in one and several variables is required.

This book does not cover the whole content of my courses (Extragalactic Astrophysics and Dynamics of Stellar Systems for third- and fourth-year undergraduate astronomy students) taught at the University of Bologna. In fact, as any judicious student knows by experience, no book can substitute for the understanding of a subject obtained from direct interaction with a teacher and from doing first-hand exercises (sometimes erroneously!) at the blackboard. I devoted some care to the addition of arguments and illustrative examples that are not usually found in other presentations but that I found quite effective in my classes for clarifying important conceptual points. Informal interactions with colleagues lead me to expect that this book will also be useful to professional astronomers not working directly on stellar dynamics who are looking for a simple but sufficiently rigorous exposition of the most important concepts and techniques used in the field. At the end of each chapter, the student will find worked-out exercises: they should *not* be considered optional, but as complementary material supporting the arguments discussed in the main text. The student will notice that several important topics of stellar dynamics are not discussed in detail, or are not even mentioned at all; examples include instabilities and resonances, the evolution of collisional systems such as globular and open clusters, the kinematics and dynamics of spiral arms, the dynamical evolution of galaxy centers under the effects of single and multiple supermassive black holes, and the field of numerical simulations. The absence of these topics is motivated by the lack of sufficient expertise of the author, or because a proper treatment would be beyond the scope of this introductory book, or, finally (but not less importantly), simply due to personal taste.

I am especially grateful to the many students who attended my classes over the years, both in Bologna and at the Scuola Normale Superiore of Pisa, where I taught in 1998–2003 and 2006–2008, during which time I wrote a little collection of *Lecture Notes on Stellar Dynamics* (Ciotti 2000): their questions often helped to reveal weak points in my understanding of the subject. Discussions and exchanges of opinions with several colleagues have also deeply influenced (directly or indirectly) the writing of this book. Among them, special acknowledgment is due to G. Bertin, J. Binney, A. D'Ercole, T. de Zeeuw, D. Lynden-Bell, J. Ostriker, S. Pellegrini, A. Renzini, R. Sancisi, M. Stiavelli, and T. S. van Albada. I also wish to thank W. Dehnen, W. Evans, O. Gerhard, J. Goodman, L. Hernquist, D. Merritt, L. Ossipkov, D. Pfenniger, E. Remiddi, D. Spergel, A. Toomre, S. Tremaine, and my coauthors and students, who are too numerous to be listed by name. Finally, B. Franchi, A. Parmeggiani, and M. C. Tesi of the Department of Mathematics of the University of Bologna are acknowledged for having been generous with their time spent on clarifying my understanding of particular mathematical issues; of course, I alone am responsible for any error in this book.

V. Higgs, H. Cockburn, L. Edwards, E. Ferns, E. Migueliz, and the entire staff of Cambridge University Press are warmly acknowledged for their untiring patience and support during all stages of the completion of this project. H. Niranjana as project manager and J. S. Marr as copyeditor are thanked for their work.

On a more personal note, I warmly thank G. Bertin for his friendship: without his constant encouragement, this book simply would not exist. Finally, I wish to thank my wife, Silvia, and our daughters, Anna Magdalena and Sofia Elisabetta, for their patience and help over the years of writing: this book is dedicated to them, and to my parents, Marino and Enrica.

La Scienza è ultima perfezione de la nostra anima.
Dante Alighieri (1265–1321), *Convivio*, Trattato Primo, Capitolo Primo

Part I

Potential Theory

1

The Gravitational Field

This chapter is aimed at introducing in an elementary yet rigorous way the mathematical properties of the Newtonian gravitational field – the basic entity – together with the Second Law of Dynamics, upon which stellar dynamics is founded. We begin from the gravitational field of a point mass, and then we move to consider the field of extended mass distributions by using the superposition principle. A direct proof of Newton's first and second theorems for homogeneous shells is worked out, followed by a different derivation based on the Gauss theorem.

1.1 The Gravitational Field of a Point Mass and of Extended Distributions

The understanding of the structure, equilibrium, and dynamical evolution of stellar systems (to be understood in a broad sense, ranging from small galactic open clusters up to giant clusters of galaxies) in terms of the fundamental physical laws, and in particular of classical gravity, is the main subject of stellar dynamics. Stellar systems are immense when compared to the human scale, and so a sense of the astonishing values of the masses, lengths, and times involved is absolutely necessary, especially for students of sister disciplines such as mathematics or physics, who may lack a specific background in observational astronomy.

So, let us start by considering our Sun, a yellow G-dwarf star (but not at all "dwarf" when compared to the masses and sizes of the vast majority of the $N_* \approx 3 \times 10^{11}$ stars present in our galaxy, the Milky Way), with a mass of $M_\odot \simeq 1.98 \times 10^{33}$ g and a radius of $R_\odot \simeq 7 \times 10^{10}$ cm. The shape of our galaxy is that of a quite flattened disk, with a radius of $\approx$26 kpc (1 kpc $= 10^3$ pc $= 3.08 \times 10^{21}$ cm), a central "bulge," and a disk thickness of $\approx$1 kpc. The Sun rotates around the galactic center on an almost circular orbit of radius $\approx$8 kpc within the galactic disk. Detailed information about our galaxy and other stellar systems can be found in books such as Bertin (2014), Bertin and Lin (1996), Binney and Merrifield (1998), Binney and Tremaine (2008), Cimatti et al. (2019), and Sparke and Gallagher (2007). To appreciate these figures in their full glory, it is useful to construct a scaled-down model of the Milky Way, such as one in which the Sun is imagined as a little sphere of $\simeq$0.7 mm radius; in this model, all lengths are reduced by a factor of 10^{-12}.

The Earth is now an invisible grain of dust, the radius of the Moon's orbit is $\simeq$0.4 mm, while the Earth's orbit is an (almost circular) ellipse with a semimajor axis of $\simeq$15 cm. The giant planets Jupiter, Saturn, Uranus, and Neptune rotate around the Sun at average distances of 0.78 m, 1.40 m, 2.87 m, and 4.50 m, respectively. On this scale, the *nearest* star (actually a multiple star, the Alpha Centauri system) is placed at $\approx$41 km from our Sun, the thickness of the Milky Way disk is $\approx 3 \times 10^4$ km, and its diameter is $\approx 1.6 \times 10^6$ km! The Sun would revolve around the galactic center on a roughly circular orbit with a radius of $\approx 2.5 \times 10^5$ km at a velocity of $\approx 2.1 \times 10^{-4}$ mm/s (i.e., $\approx$0.8 mm/hr) and a period[1] of $\approx$230 Myr (corresponding to a physical circular velocity of $\approx$220 km/s).

Therefore, it should not be surprising that, with extremely good approximation, in most (but not all) applications of stellar dynamics, stars can be considered *point masses* (see also Exercise 1.1). We then start our study quite naturally by considering the gravitational field of a point mass. Notice that the relevance of this case goes well beyond the point-mass approximation of stars in galaxies because, as the gravitational field produced by a generic mass distribution is given by the sum of the fields produced by each of its parts, the point masses that will be used in the next two chapters can also be interpreted as atoms or molecules when computing the gravitational field produced by a macroscopic material object!

The starting point of Newtonian gravity is the definition of the gravitational field $\mathbf{g}$, at position $\mathbf{x}$, produced by a material point of mass m at position $\mathbf{y}$ in some reference system S_0. Empirically, it is found that the field is radial, with

$$\mathbf{g}(\mathbf{x}) = -Gm\frac{\mathbf{x}-\mathbf{y}}{||\mathbf{x}-\mathbf{y}||^3}, \tag{1.1}$$

where $G \simeq 6.67 \times 10^{-8}$ cm^3/s^{2}g is the universal gravitational constant, $\|\ldots\| = \sqrt{\langle \ldots, \ldots \rangle}$ is the standard Euclidean norm, and $\langle , \rangle$ is the usual inner product over $\Re^3$ (see Appendix A.1); from now on we indicate vectors with bold letters, if not stated otherwise. Note that the field is not defined at the particle position and that $\mathbf{g}(\mathbf{x})$ depends on time through the position of the particle $\mathbf{y}(t)$: the classical gravitational field "propagates" instantaneously over all of the space. Equation (1.1), formally identical to that describing the electrostatic field produced by a point charge, encapsulates the two most important properties of the gravitational field of a point mass (i.e., the fact that the modulus of the field decreases radially as the inverse of the square distance from the particle and the fact that the force between two particles is attractive). The third fundamental property of classical gravity, confirmed by an enormous body of experimental evidence (a point unfortunately not always sufficiently stressed; however, see e.g. Feynman et al. 1977 among the notable exceptions), is the *superposition principle* (i.e., the fact that the gravitational field produced at the point $\mathbf{x}$ by two point masses of masses m_1 and m_2 placed at $\mathbf{y}_1$ and $\mathbf{y}_2$ is the vector

[1] For order-of-magnitude estimates, it is useful to recall that 1 yr $\simeq \pi \times 10^7$ s, and that a velocity of 1 km/s corresponds to $\simeq$1 pc/Myr.

sum $\mathbf{g}_1(\mathbf{x}) + \mathbf{g}_2(\mathbf{x})$ of the two fields). As a matter of fact, the *whole* theory of classical gravitation can be built from Eq. (1.1) and the superposition principle.

For example, the gravitational field produced at $\mathbf{x}$ by an extended mass distribution of density $\rho(\mathbf{y})$, such as a star, a planet, or a galaxy, can be immediately written as

$$\mathbf{g}(\mathbf{x}) = -G \int_{\Re^3} \frac{\mathbf{x} - \mathbf{y}}{\|\mathbf{x} - \mathbf{y}\|^3} \rho(\mathbf{y}) d^3\mathbf{y}, \tag{1.2}$$

where $\rho(\mathbf{y})d^3\mathbf{y}$ is the mass element in the infinitesimal integration volume $d^3\mathbf{y}$ and the integral (a sum) embodies the superposition principle. Of course, in the special case of N point masses of mass m_i and position $\mathbf{x}_i$, one can define

$$\rho(\mathbf{y}) = \sum_{i=1}^{N} m_i \delta(\mathbf{y} - \mathbf{x}_i), \tag{1.3}$$

where δ is the so-called Dirac δ-function (actually a distribution; see Appendix A.2.5), and so Eq. (1.2) reduces to a standard sum, as expected from Eq. (1.1).

The student should appreciate that, in principle, Eq. (1.2) contains *all* of the information we need to determine the gravitational field of a given density distribution. In other words, Eq. (1.2) is the *general solution* of the problem of the calculation of the Newtonian gravitational field produced by an assigned mass distribution ρ: one could conclude that the only problem to be addressed is just how to calculate (analytically or numerically) the integral (1.2) in all of the cases of interest. However, this conclusion would be very wrong. In fact, the most profound properties of the gravitational field *cannot* be derived by simple evaluation of Eq. (1.2), and (as is common in physics) the defining *equations* of a problem invariably contain much more information than the solution itself. For these reasons, we now start from the "solution" given by Eqs. (1.1) and (1.2), and we look for the differential equations leading to this solution. Significant effort will then be spent in the study of the properties and implications of the obtained equations, and this effort will be repaid by the discovery of powerful mathematical methods that will lead us to a deep understanding of the gravitational field produced by mass distributions and, finally, as a useful by-product, to general techniques for the evaluation of Eq. (1.2).

1.2 Newton's First and Second Theorems

One of Newton's major accomplishments was the discovery of the *first* and *second theorems* concerning the gravitational fields produced by spherical and homogeneous material shells. Usually, the two theorems are proved by using the Gauss divergence theorem, as we will also do in Section 1.3. However, due to their importance, here we prove them from direct integration of Eq. (1.2), in the original spirit (while avoiding the subtleties of Newton's awe-inspiring geometric proof; e.g., see Binney and Tremaine 2008; Chandrasekhar 1995).

The first step is to show that the gravitational field produced by a generic spherical density distribution $\rho(r)$ is radial. This is accomplished by arbitrarily fixing a point of space $\mathbf{x}$, by defining the direction of the z-axis so that[2] $\mathbf{x} = (0, 0, r)$, and finally by changing the integration variables from Cartesian to spherical in Eq. (1.2). Integration over the angles (do it!) proves that the resulting field $\mathbf{g}$ is directed along the z-axis (i.e., along $\mathbf{x}$) and hence is radial. The second step is to specialize the density distribution ρ to the case of a homogeneous and infinitesimally thin shell of radius R and total mass M, so that from Eq. (A.97)

$$\rho(\mathbf{y}) = \frac{M\delta(r - R)}{4\pi r^2}, \qquad r \equiv \|\mathbf{y}\|. \tag{1.4}$$

From the radial symmetry of the field, we can proceed to the explicit integration of the z-component of $\mathbf{g}$ at $\mathbf{x} = (0, 0, r)$, and after some algebra (see also Exercise 1.2), we finally obtain

$$\mathbf{g}(\mathbf{x}) = -\frac{GM}{2r^2}\left(1 + \frac{r - R}{|r - R|}\right)\mathbf{f}_r, \quad r \neq R, \tag{1.5}$$

where $\mathbf{f}_r$ is the radial unit vector in spherical coordinates (see Appendix A.8). It follows that for $r < R$ no field is present (Newton's first theorem), while for $r > R$ the field coincides with that produced by a material point of mass M, placed at the origin (Newton's second theorem).

A somewhat delicate situation arises when considering the field produced by the shell on itself (i.e., at $r = R$). In fact, it is clear from Eq. (1.5) that the function $\mathbf{g}(\mathbf{x})$, evaluated as a limit for $r \to R$, is discontinuous, with different left and right values. A natural question that arises is how to compute the field on the shell itself. Naively, a "reasonable" approach could be to fix $r = R$ before performing the integral over the surface of the shell and to pretend that for symmetry reasons the integral over the azimuthal angle φ is to be performed before the integral over the colatitude ϑ. With this approach, we deduce that a point *on* the shell experiences a *radial* force per unit mass given by

$$\mathbf{g}(\mathbf{x}) = -\frac{GM}{2R^2}\mathbf{f}_r, \tag{1.6}$$

the average value of Eq. (1.5) just inside and just outside the shell. However, a closer inspection of the integral reveals a more delicate situation (i.e., that the *tangential* component of $\mathbf{g}$ to the shell *cannot* be uniquely defined). For example, the student is encouraged to calculate the tangential component of $\mathbf{g}(0, 0, R)$ with a different order of integration from the "natural" one (i.e., by fixing the angle φ) and integrating first over $0 \leq \vartheta \leq \pi$: the tangential component of the field now diverges for $\vartheta \to 0$ (i.e., for the effect of the points of the material meridian line touching the point $\mathbf{x}$). Technically, the problem is due to the fact that, in the present case, the integral in Eq. (1.2) is not absolutely convergent, and the result depends on how the integral is computed. Here, the lesson is that, as a safety

[2] For simplicity, vectors are represented in the text as row vectors. See also Footnote 1 in Appendix A.

rule, symmetry arguments should be used to evaluate integrals only *after* the integrals are known to converge.[3]

1.3 The Gauss Theorem and the Gravitational Field

An alternative and more elegant derivation of Newton's first and second theorems can be obtained by using the Gauss divergence theorem (see Appendix A.4). However, before addressing this problem in Section 1.3.1, we must establish a few fundamental properties of the gravitational field.

We start by considering the application of the Gauss theorem to the gravitational field of a point mass, and so we evaluate first the *divergence* of the gravitational field in Eq. (1.1). It is a simple exercise (do it!) to show that for all $\mathbf{x} \neq \mathbf{y}$, the divergence of the field at $\mathbf{x}$ vanishes (the student may also wish to prove, by using the divergence operator in spherical coordinates, that the radial $1/r^2$ field is the *only* radial field in $\Re^3$ with this property). Therefore, from the Gauss theorem, it follows that the flux of the gravitational field produced by a point mass across a generic closed surface $\partial\Omega$, *not containing* the point mass, is zero. From the superposition principle, this conclusion immediately generalizes to the case of the flux of the field $\mathbf{g}$ produced by an arbitrary mass distribution placed outside the closed surface $\partial\Omega$.

Now, let us ask what happens to the flux if the point mass is *inside* the region Ω. Clearly, being $\operatorname{div}\mathbf{g} = 0$ for $\mathbf{x} \neq \mathbf{y}$, according to Eq. (A.112), the only possible nonzero contribution of the divergence to the flux integral can be due to what happens at the point $\mathbf{x} = \mathbf{y}$. We face two problems here. The first is that we cannot compute $\operatorname{div}\mathbf{g}$ for $\mathbf{x} = \mathbf{y}$, simply because the field is not defined there. The second is even more worrisome because from a naive interpretation of integration theory, one could expect that the volume integral evaluates to zero even if we pretend that the "value" of $\operatorname{div}\mathbf{g}$ at $\mathbf{x} = \mathbf{y}$ is infinite, being a point of a set of zero measure. This would be in stark contrast to the fact that at this stage we can certainly imagine a region containing the point mass with a shape such that the total flux is strictly negative! The only logical conclusion is that the flux is not zero and that integration theory is (obviously) correct, but that $\operatorname{div}\mathbf{g}$ for $\mathbf{x} = \mathbf{y}$ is a more "complicated" object than a function with infinite value: we now show that this is the case. The idea is borrowed from complex analysis, and it substitutes into our problem a different problem that we can solve. So, as is shown in Figure 1.1, we remove from Ω a spherical region of radius R centered on $\mathbf{y}$, and we produce a two-dimensional "cut" (the dashed line c in Figure 1.1) connecting the surface of the inner hole with the external surface $\partial\Omega$. With the exclusion of the spherical hole, our particle is now outside the resulting region, so that

[3] Perhaps the most famous such case is encountered in the computation of the Jeans mass (e.g., see Binney and Tremaine 2008; Shu 1992), when fixing to zero (the so-called Jeans swindle) the gravitational field produced by the unperturbed, infinite, and homogeneous three-dimensional density background. In fact, instead of integrating over spherical shells centered on the point of interest (with the result of a zero field from Newton's first theorem), we could integrate over parallel density slabs of unitary thickness, at the left and right of the point, obtaining a conditionally convergent alternating series (do it!); see also Exercise 1.8 for another example. A full discussion of the conditions for the existence of integral (1.2) can be found in Kellogg (1953).

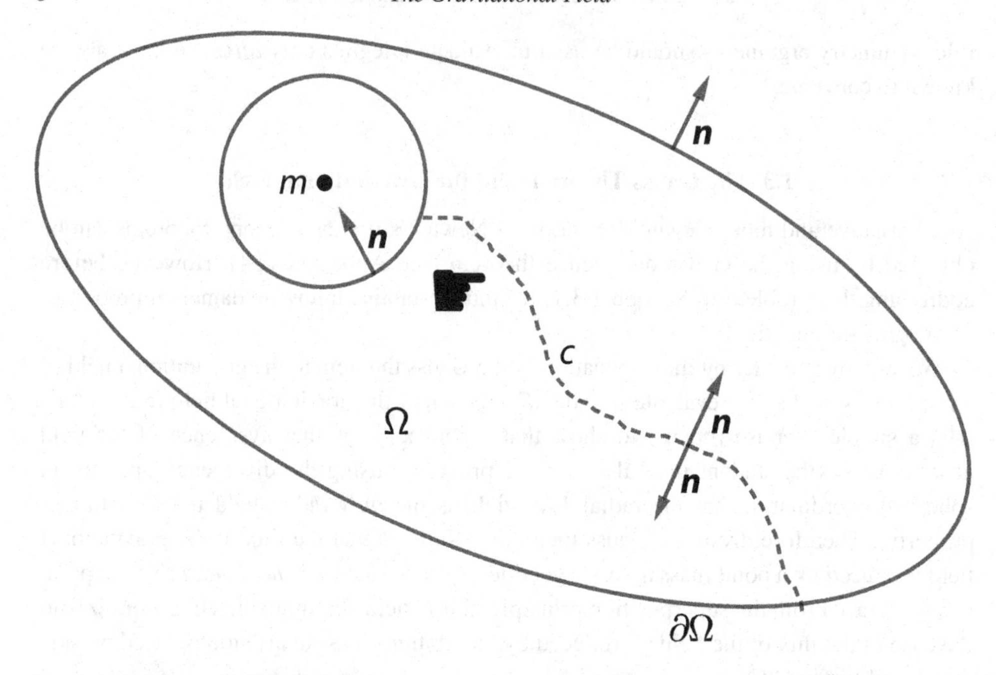

Figure 1.1 Schematic illustration of how, by using the divergence theorem, we can prove that the flux of the gravitational (or electrostatic) field of a particle of mass m (or charge q) contained in a closed and regular (but otherwise arbitrary) region $\Omega \subset \Re^3$ is $-4\pi Gm$ (or q/ϵ_0), independently of the position of the particle inside the volume. The dashed line indicates a generic two-dimensional "cut" with two opposite normals, one for each of the two sides of the cut.

the flux of **g** through the total boundary (made by the original $\partial\Omega$, by the surface of the spherical hole, and by the two geometrically coincident but analytically distinct surfaces of the cut c) evaluates to zero. Therefore, from the divergence theorem, the flux of **g** through $\partial\Omega$ is simply the negative of the flux on the inner spherical hole, because the flux on the two surfaces of the cut cancels out. By reversing the normal at the surface of the hole and performing the surface integral by exploiting the spherical symmetry of **g**, it is now simple to show (do it!) that the flux on the hole,[4] and so onto $\partial\Omega$, is $-4\pi Gm$.

In conclusion, by using the Gauss theorem applied to the special case of the three-dimensional radial $1/r^2$ field, we showed that for a point mass

$$\int_{\Omega} \operatorname{div} \mathbf{g}(\mathbf{x}) d^3\mathbf{x} = \int_{\partial\Omega} \langle \mathbf{g}(\mathbf{x}), \mathbf{n} \rangle d^2\mathbf{x} = -4\pi Gm \times \begin{cases} 0, & \mathbf{y} \notin \Omega, \\ 1, & \mathbf{y} \in \Omega \end{cases}. \tag{1.7}$$

In other words, $\operatorname{div} \mathbf{g}$ behaves as the three-dimensional Dirac δ-function in Eq. (A.95) (technically a *distribution*), and we can formally write

[4] Quite obviously, even before computing it, we knew that the flux on the spherical hole is independent of the adopted value of R, which is consistent with the fact that the flux on $\partial\Omega$ cannot depend on our arbitrary choice of R!

$$\mathrm{div}_{\mathbf{x}} \frac{\mathbf{x}-\mathbf{y}}{||\mathbf{x}-\mathbf{y}||^3} = 4\pi\delta(\mathbf{x}-\mathbf{y}), \tag{1.8}$$

where the subscript indicates the coordinates used to compute the divergence. From the superposition principle, we now conclude that, given an arbitrary mass distribution $\rho(\mathbf{x})$ and an arbitrary closed (and sufficiently regular) region $\Omega \subset \Re^3$, the flux of $\mathbf{g}$ produced by ρ through the boundary $\partial\Omega$ is $-4\pi GM$, where $M = \int_\Omega \rho d^3\mathbf{x}$ is the mass contained in the volume. Highlighting an elementary but (perhaps) not always appreciated point is in order here. In fact, it is clear that a *null* flux over a closed surface $\partial\Omega$ does not imply a *null* field inside the region Ω, as is obvious for the case of a point mass external to a given volume. In turn, this also means that, in general, one *cannot* prove that the gravitational (or electrostatic) field inside an empty cavity is zero simply by considering the null flux through some closed surface contained in the cavity. A naturally related question is then why inside an electrically charged, conducting, and closed surface of arbitrary shape at equilibrium not only the flux but also the field *is* zero (the so-called *Faraday's cage*), while inside a material surface of similar shape the gravitational field *is not* zero, even though the force law in the two cases is mathematically identical. To properly answer this question, we must move to the next chapter, where we encounter the concept of *gravitational potential*, which is fundamental in stellar dynamics.

We will now use Eq. (1.8) to obtain the differential equation relating the gravitational field of a mass distribution to its density. In practice, we compute[5] the divergence under the sign of the integral of Eq. (1.2), and from Eq. (1.8), we obtain the fundamental identity

$$\mathrm{div}\,\mathbf{g}(\mathbf{x}) = -4\pi G\rho(\mathbf{x}). \tag{1.9}$$

If we now insert into Eq. (1.9) the expression for the field of a point mass from Eq. (1.1), we deduce that

$$\rho(\mathbf{x}) = m\delta(\mathbf{x}-\mathbf{y}), \tag{1.10}$$

or, in other words, that *physically* the Dirac δ-function can be imagined as the "density distribution" of a point mass. This shows in the most apparent way that the δ-function is a *dimensional* object with the inverse volume units of the space under consideration, and that, from a mathematical point of view, a point mass is "more" than just a simple point of "infinite density."

1.3.1 Newton's First and Second Theorems Again

We are now in the position to prove again, using a different approach, Newton's first and second theorems, and to show the power of the Gauss theorem in Eq. (1.7). In fact, from the property that the gravitational field of a spherically symmetric mass distribution $\rho(r)$ is

[5] Of course, this step can be rigorously justified (e.g., see Kellogg 1953).

radial (i.e., $\mathbf{g} = g_r(r)\mathbf{f}_r$; see Section 1.2), and using a spherical control volume of radius r centered on the origin, Eq. (1.7) reduces immediately to

$$4\pi r^2 g_r(r) = -4\pi G M(r), \quad M(r) = 4\pi \int_0^r \rho(t)t^2 dt, \tag{1.11}$$

where the (still unknown) component g_r has been "factorized" out of the integral thanks to its geometric properties and to a wise choice for the integration surface. In practice, the gravitational (or electrostatic) field of a spherical mass (charge) distribution at distance r from the center is the same field of a point mass (charge) of mass $M(r)$ or charge $Q(r)$. Newton's two theorems are immediately recovered from Eq. (1.11) for the case of ρ given the homogeneous shell in Eq. (1.4). The student is encouraged to follow the same line of reasoning and deduce the expressions analogous to Eq. (1.11) for cylindrical densities $\rho(R)$ and planar densities $\rho(z)$, taking into account the precautionary notes at the end of Section 1.2.

A final warning to enthusiastic students: unfortunately, the powerful Gauss theorem is not a magic tool that can, elegantly and effortlessly, give the gravitational field of an arbitrary mass distribution while sparing the tedious work of integration. In fact, even though the Gauss theorem holds for arbitrary volumes and mass distributions, it should be clear that the derived expressions for **g** of spherical, cylindrical, and planar distributions are based on the *exceptional* circumstance that the geometry of the surfaces over which the flux integrand can be factorized is known *before* **g** is actually calculated. For generic mass distributions, the geometric properties of **g** are not known in advance, such that the question of what kind of surface one could/should use to repeat the treatment of special geometries is as difficult as the problem of determining **g** itself!

Exercises

1.1 With this exercise, we quantify the possibility of considering stars as point masses in real stellar systems. Consider an idealized stellar system of radius R and homogeneously filled with N identical stars of radius $R_\odot$. We define a *geometric collision* as the situation realized when the centers of two stars are separated by a distance $d \leq 2R_\odot$. By considering the co-volume of N "cylinders" of length λ (a rough estimate of the *mean free path*; e.g., see Born 1969) and radius $2R_\odot$ (why?), argue that an estimate of λ can be obtained by imposing that the total volume of the cylinders equals the volume of the sphere, i.e., formally

$$\frac{\lambda}{2R} = \frac{R^2}{6NR_\odot^2}. \tag{1.12}$$

Estimate λ for an elliptical galaxy and for a globular cluster.

1.2 Let

$$\mathbf{g}(\mathbf{x}) = -Gm\frac{\mathbf{x}-\mathbf{y}}{\|\mathbf{x}-\mathbf{y}\|^{1+\alpha}}, \quad \alpha < 3, \tag{1.13}$$

a natural generalization of the classical gravitational field produced by a particle of mass m, where G is some universal constant (e.g., see Di Cintio and Ciotti 2011). Show that the radial component of the gravitational field produced at radius r by the homogeneous spherical shell of total mass M and radius R given in Eq. (1.4) is

$$g_r(r) = -\frac{GM}{4Rr^2} \times \begin{cases} 2Rr - (r^2 - R^2)\ln\frac{r_-}{r_+}, & \alpha = 1, \\ \frac{r_+^{3-\alpha} - r_-^{3-\alpha}}{3-\alpha} - (r^2 - R^2)\frac{r_+^{1-\alpha} - r_-^{1-\alpha}}{\alpha - 1}, & \alpha \neq 1, \end{cases} \tag{1.14}$$

where $r_+ \equiv r + R$ and $r_- \equiv |r - R|$. Discuss the relevant case of $\alpha = -1$ (the harmonic oscillator) and show that for $\alpha = 2$ we reobtain Eq. (1.5). What happens to the field at $r = R$?

1.3 For $\mathbf{x} \neq \mathbf{y}$, calculate the divergence of the field given in Eqs. (1.13) and (1.14) and discuss the result as a function of α by using the Gauss theorem applied to concentric spherical control surfaces of increasing radius. Explain geometrically why inside a uniform shell a mass point is attracted toward the shell for $\alpha > 2$ and toward the center for $\alpha < 2$.

1.4 Consider a homogeneous sphere of constant density ρ_0, and from Eq. (1.11) show that the field inside the sphere is that of the three-dimensional isotropic harmonic oscillator

$$g_r(r) = -\frac{4\pi G\rho_0}{3} r. \tag{1.15}$$

By using the superposition principle and the Newton's second theorem, determine the field inside a spherical hole – not necessarily concentric – carved in the sphere: isn't the result beautiful? *Hint*: Imagine the hole is produced by the superposition of a sphere of negative density $-\rho_0$ and calculate its repulsive field from Eq. (1.15).

1.5 Point particles, rings, and disks are often encountered in problems of stellar dynamics. Prove that the three-dimensional density of a *particle* of mass m placed on the z-axis at distance a from the origin, in spherical and cylindrical coordinates, can be written respectively as

$$\rho = m\frac{\delta(r-a)\delta(\vartheta)}{2\pi r^2 \sin\vartheta} = m\frac{\delta(z-a)\delta(R)}{2\pi R} \tag{1.16}$$

Moreover, show that for a *homogeneous ring* of total mass M, radius a, linear density $\lambda = M/(2\pi a)$, placed on the $z = 0$ plane, and with the center at the origin,

$$\rho = M\frac{\delta(r-a)\delta(\vartheta - \pi/2)}{2\pi r^2 \sin\vartheta} = M\frac{\delta(R-a)\delta(z)}{2\pi R}. \tag{1.17}$$

Finally, show that for a *razor-thin disk* of surface density $\Sigma(R)$ on the $z = 0$ plane,

$$\rho = \Sigma(r)\frac{\delta(\vartheta - \pi/2)}{r\sin\vartheta} = \Sigma(R)\delta(z), \tag{1.18}$$

so that from Eqs. (1.17) and (1.18) it follows that the "surface density" of the homogeneous ring can be written as

$$\Sigma_{\text{ring}} = \frac{M\delta(r-a)}{2\pi r} = \frac{M\delta(R-a)}{2\pi R}. \tag{1.19}$$

1.6 What happens to Eq. (1.7) in the special case of the point mass placed on the boundary $\partial\Omega$? *Hint*: Suppose the tangent plane to $\partial\Omega$ exists at the particle's position. Draw a little semisphere of radius ϵ centered on the particle and use Eq. (1.7) to evaluate the total flux on the new resulting surface made by $\partial\Omega$, minus the disk of radius ϵ, plus the semisphere of radius ϵ. Conclude that, no matter whether the semisphere includes or excludes the particle, the limit of the total flux through $\partial\Omega$ for $\epsilon \to 0$ is $-2\pi Gm$, so that from Eq. (1.8) it follows that integration of the Dirac δ-function on a regular point of a boundary of a closed volume evaluates to 1/2.

1.7 By using Newton's second theorem and Eq. (1.2) twice, prove that two nonoverlapping spheres of masses M_1 and M_2 and centers at $\mathbf{x}_1$ and $\mathbf{x}_2$ attract each other as two material points of masses M_1 and M_2 placed at $\mathbf{x}_1$ and $\mathbf{x}_2$.

1.8 This problem illustrates the danger of the naive use of symmetry arguments when calculating integrals. It is commonly said that the gravitational (or electrostatic) field of a homogeneous, infinite, razor-thin density distribution of surface density σ is constant and directed normally to the plane "because the field components parallel to the plane vanish by obvious symmetry arguments." Actually, it is easy to prove that within Newtonian gravity (or electrostatics) the horizontal field is *undetermined* (see also Footnote 3). Consider in the (x, y) plane four material squares of surface density σ, with a common vertex at the origin. Let a be the length of the side of the square in the first quadrant, and by direct integration show that the gravitational field components at the point $(0, 0, z)$ above the origin are

$$g_x = g_y = G\sigma\left(\text{arcsinh}\frac{a}{z} - \text{arcsinh}\frac{a}{\sqrt{a^2+z^2}}\right), \tag{1.20}$$

where $\text{arcsinh}(x) = \ln(x + \sqrt{1+x^2})$, and

$$g_z = -G\sigma \arctan\frac{a^2}{z\sqrt{a^2+z^2}}, \tag{1.21}$$

(e.g., see Kellogg 1953; McMillan 1958). Consider now the behavior of the total field $\mathbf{g}$ produced by the four squares of side length a, b, c, and d, when the sides are independently extended to infinity. Show that g_z reduces to the value obtained from the Gauss theorem. What happens to the tangential components?

1.9 As is well known, Newton's first theorem can be seen as the result of a perfect cancellation at a generic point inside a homogeneous spherical shell of the r^{-2} force produced by the two infinitesimal mass elements determined by the intersection of the shell itself and the two sides of the conus with an infinitesimal opening angle and a vertex in the considered point, followed by a rotation of the conus's axis over the whole solid angle (e.g., see Binney and Tremaine 2008; Chandrasekhar 1995). Repeat this wonderful geometric argument for the case of a generic point inside a homogenous ring and conclude that the resulting attractive force is directly radially toward the nearest point of the ring (see also Exercise 2.33): therefore, when computing the gravitational field of a disk, its external regions cannot be ignored, a fact of great importance for the interpretation of the rotation of disk galaxies.

2
The Gravitational Potential

In this second introductory chapter, the concept of *gravitational potential* is presented and then developed up to the level usually encountered in applications of stellar dynamics, such as the computation of the gravitational fields of disks and heterogeneous triaxial ellipsoids, the construction of the far-field multipole expansion of the gravitational field of generic mass distributions, and finally the expansion in orthogonal functions of the Green function for the Laplace operator.

2.1 The Gravitational Potential

In Chapter 1, we obtained the integral expression for the gravitational field **g** produced by a generic mass distribution, and we derived important information about its properties by considering the Gauss divergence theorem jointly with the superposition principle. In a certain sense, we could content ourselves with these results that in principle would allow us to compute (e.g., by using the powerful numerical techniques and computers available today) the gravitational fields of astronomical objects such as galaxies, stars, and planets. However, this "brute force" approach would lead almost immediately to a conceptual dead end: yes, we can certainly compute the gravitational field of each specific object with the desired accuracy, yet a full grasp of the general properties of the gravitational field would be precluded. As already stressed in Chapter 1, it is worth recalling that in physics we usually learn about a problem as much from the equations as from the solution, and accordingly we will now look at the mathematical problem behind the solution given by Eq. (1.2).

Therefore, we repeat the approach in Section 1.3, where the results have been obtained by considering the divergence (*div*) of field **g** of a point mass, focusing now on the effects on **g** of the *rotor* (also known as the *curl* or *rot*) when applied to **g**, as well as on the consequences that can be derived from the Stokes theorem (Appendix A.4). First, however, it is important to answer a question that quite often arises in discussions with students: Why are div and rot so special that even the Maxwell equations for the electric and magnetic fields are written in terms of these two operators (e.g., Feynman et al. 1977; Jackson 1998)? We could certainly invent other differential operators at our leisure! The reason for this is elucidated by the Helmholtz decomposition theorem (see Appendix A.6 for more details). Here, it suffices to recall that this theorem states that in three-dimensional space every reasonably well-behaved vector field can be almost uniquely[1] represented as the sum of two vector fields, one with null divergence (or *solenoidal*; i.e., in a simply connected domain this component can be written as the rotor of a *vector potential*) and the other with null rotor (or *irrotational* or *closed*; i.e., in a simply connected domain it can be written as the gradient of a *scalar potential*, and so this second component is not only closed, but also *exact*). In practice, the Helmholtz theorem ensures that, when the divergence and the rotor of a vector field are known, then we know "everything" about the field itself.

Therefore, we now consider the rotor of the gravitational field $\mathbf{g}$ produced by a generic density distribution $\rho(\mathbf{x})$. From Eq. (1.1), it follows that for a point mass rot $\mathbf{g} = 0$ for $\mathbf{x} \neq \mathbf{y}$, because the rotor of a radial field vanishes (prove it!); moreover, $\Re^3$ minus a point is simply connected, and so the field produced by a point mass is not only closed, but also exact (see Appendix A.4). Without loss of generality, for the field of a point mass we can then write

$$\mathbf{g} = -\nabla\phi, \tag{2.1}$$

where $\phi(\mathbf{x})$ is the *gravitational potential*, and the minus sign in front of it is a standard convention in physics. The superposition principle now guarantees that Eq. (2.1) also holds for the gravitational field of a generic density distribution given in Eq. (1.2) (i.e., that the potential can always be defined once $\mathbf{g}$ exists); alternatively, the student could evaluate the rotor under the integral and show that in fact the rotor vanishes (i.e., that $\mathbf{g}$ is closed). Having established the existence of the potential, now the question is how to obtain the explicit expression of $\phi(\mathbf{x})$ for an assigned $\rho(\mathbf{x})$. Following the rules of vector analysis in Appendix A.4,

$$\phi(\mathbf{x}) - \phi(\mathbf{x}_0) = -\int_{\gamma(\mathbf{x}_0,\mathbf{x})} \langle \mathbf{g}(\mathbf{x}), d\mathbf{x} \rangle, \tag{2.2}$$

where $\mathbf{x}_0$ is an arbitrary point of space (with the only obvious condition that $\mathbf{g}$ is defined there), and $\gamma(\mathbf{x}_0,\mathbf{x})$ is a well-behaved but otherwise arbitrary path connecting the point $\mathbf{x}_0$ to the point $\mathbf{x}$. The student will surely note that the potential actually depends on *two* arbitrary choices (i.e., the position $\mathbf{x}_0$ *and* the value of the gravitational potential at $\mathbf{x}_0$). For example, in the case of the point mass placed at $\mathbf{y}$, the integration path in Figure 2.1 shows that the most general expression of the potential is

$$\phi(\mathbf{x}) = \phi(\mathbf{x}_0) + \frac{Gm}{\|\mathbf{x}_0 - \mathbf{y}\|} - \frac{Gm}{\|\mathbf{x} - \mathbf{y}\|}, \quad \mathbf{x}_0 \neq \mathbf{y}. \tag{2.3}$$

The usual formula for the potential of a point mass is then recovered by fixing $\mathbf{x}_0 = \infty$ and $\phi(\mathbf{x}_0) = 0$.

Along the same lines, we can finally show that in the case of extended mass distributions the field is not only closed, but also exact (as expected from the superposition principle),

[1] The decomposition is *almost* unique because it is possible to add to the decomposition the gradient of a generic *harmonic* function (a *gauge*) without changing the divergence and the rotor of the resulting field. However, if the vector field is *also* assigned on a boundary $\partial\Omega$ of some prescribed (simply connected) volume Ω (the Dirichelet problem in Figure 2.2), the decomposition *is* unique.

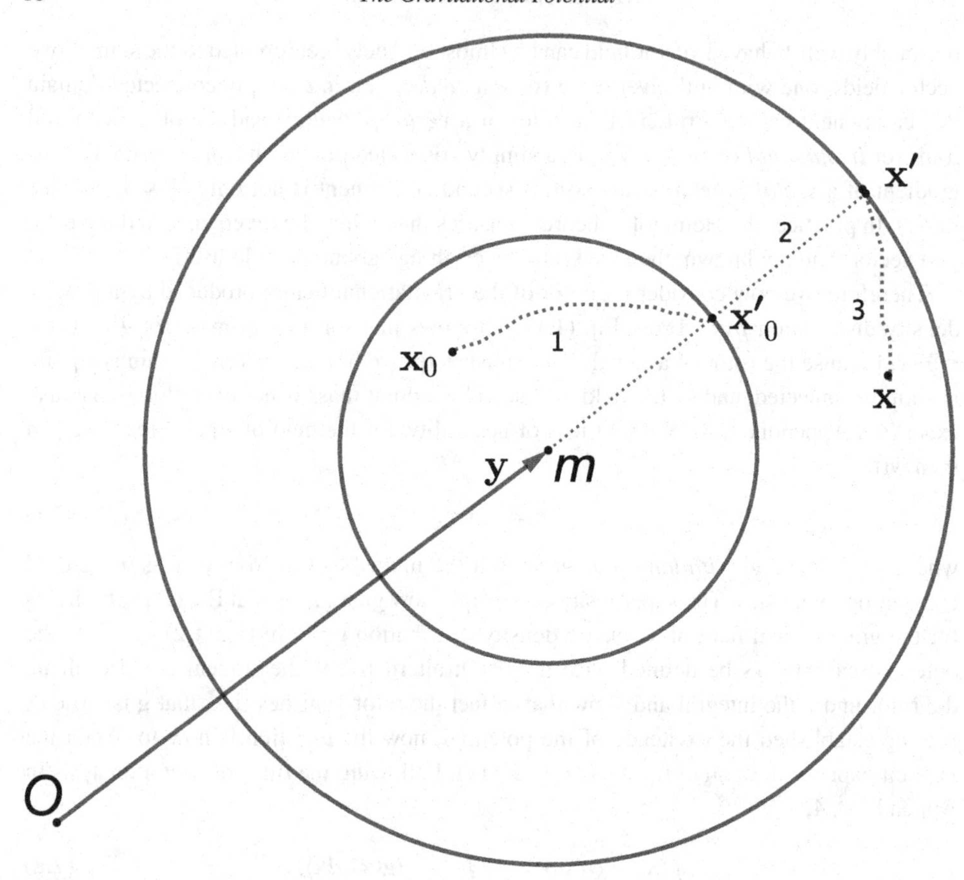

Figure 2.1 The integration procedure used to establish Eq. (2.3). The two points $\mathbf{x}_0$ and $\mathbf{x}$ are connected by Path 1 from $\mathbf{x}_0$ to $\mathbf{x}_0'$ on the surface of the sphere centered on $\mathbf{y}$ and of radius $\|\mathbf{x}_0 - \mathbf{y}\|$, followed by the radial Path 2 from $\mathbf{x}_0'$ to $\mathbf{x}'$ aligned with the origin O and the position $\mathbf{y}$ of the point of mass m, and finally by Path 3 from $\mathbf{x}'$ to $\mathbf{x}$ on the surface of the sphere centered on $\mathbf{y}$ and of radius $\|\mathbf{x} - \mathbf{y}\|$. The integrals along Paths 1 and 3 are zero because $\langle \mathbf{g}, d\mathbf{x}\rangle = 0$ by construction. For a generic spherical mass distribution centered at the origin, an identical treatment shows that $\phi(r) = \phi(r_0) - \int_{r_0}^{r} g_r(r')dr'$, where g_r is given by Eq. (1.11), so that Eq. (2.5) is finally proved.

simply by constructing the function $\phi(\mathbf{x})$. This is done by using Eqs. (1.2) and (2.2), exchanging the order between line and volume integration, and performing the path integration as in the case of the point mass:

$$\phi(\mathbf{x}) = \phi(\mathbf{x}_0) - G\int_{\Re^3}\left(\frac{1}{\|\mathbf{x}-\mathbf{y}\|} - \frac{1}{\|\mathbf{x}_0-\mathbf{y}\|}\right)\rho(\mathbf{y})d^3\mathbf{y}. \tag{2.4}$$

Notice that if the density distribution decreases faster than $\|\mathbf{y}\|^{-2}$ for $\|\mathbf{y}\| \to \infty$ (the common situation of systems of *finite* total mass, when ρ decreases faster than $\|\mathbf{y}\|^{-3}$, is contained in this case), it is possible to fix $\mathbf{x}_0 = \infty$ and $\phi(\mathbf{x}_0) = 0$. However, in stellar dynamics, sometimes idealized stellar systems of infinite total mass, not included in the previous family, are encountered (e.g., see Exercise 2.1 and Chapter 13 for relevant examples). In these special cases, $\mathbf{x}_0$ cannot be placed at infinity: however, if one expands the integrand in Eq. (2.4) for $\|\mathbf{y}\| \to \infty$ with $\mathbf{x}_0$ at some *finite* distance from the origin, a cancellation of the leading term occurs and convergence is reestablished even for density distributions decreasing faster than $\|\mathbf{y}\|^{-1}$ for $\|\mathbf{y}\| \to \infty$ (prove it!).

We conclude this section by proving a particularly important (and simple) formula for the potential of spherically symmetric mass distributions centered on the origin. By using again the approach depicted in Figure 2.1 from Newton's second theorem, and after an exchange in the order of integration, one gets

$$\phi(r) = \phi(r_0) + \frac{G\,M(r_0)}{r_0} - \frac{G\,M(r)}{r} - 4\pi G \int_r^{r_0} \rho(t) t\,dt, \tag{2.5}$$

where $M(r) = 4\pi \int_0^r \rho(t) t^2 dt$ is the total mass contained in a sphere of radius r. The formula becomes particularly simple in the case of systems for which the limit $M(r_0)/r_0$ is finite for $r_0 \to \infty$, so that the sum of the first two quantities on the right-hand side of Eq. (2.5) can be set equal to zero by considering $r_0 = \infty$ and $\phi(\infty) = \lim_{r_0 \to \infty} GM(r_0)/r_0$. Finally, it can be useful to remind the student that Newton's second theorem applies to the *force*, not to the potential (i.e., the potential at distance r from the center of a spherical mass distribution is *not* the potential of a point of mass $M(r)$ placed at the origin unless $\rho(r)$ is truncated and r is outside the mass distribution itself).

2.1.1 The Poisson Equation

We can now combine Eqs. (1.9) and (2.1) and obtain one of the most important equations in physics, the *Poisson equation*

$$\Delta\phi(\mathbf{x}) = 4\pi G\rho(\mathbf{x}), \tag{2.6}$$

where $\Delta \equiv \operatorname{div}\operatorname{grad}$ is the so-called *Laplacian* (Appendix A.4); incidentally, by considering the special case of the point mass, from Eqs. (1.10) and (2.6) we recover the correspondent of Eq. (1.8); in other words, we can prove that in $\Re^3$

$$\Delta_{\mathbf{x}} \frac{1}{\|\mathbf{x} - \mathbf{y}\|} = -4\pi\delta(\mathbf{x} - \mathbf{y}), \tag{2.7}$$

a fundamental identity in physics and mathematics. We will come back to this identity in Section 2.4.

As we will see in the next section, from Helmholtz's theorem the Poisson equation, when supplemented with the appropriate boundary conditions,[2] encapsulates all of the properties of the gravitational field, and it can be considered as the field equation of classical gravity

[2] It is elementary to realize the importance of boundary conditions: for a given pair (ρ, ϕ), the Poisson equation remains unchanged if the harmonic function $\langle A\mathbf{x}, \mathbf{x}\rangle + \langle \mathbf{b}, \mathbf{x}\rangle + c$ (with A being a 3×3 matrix with $\mathrm{Tr}A = 0$, $\mathbf{b}$ being a constant vector, and c being a constant scalar) is added to ϕ, but $\mathbf{g} = -\nabla\phi$ would be different!

(and of electrostatics). In an empty space, Eq. (2.6) is called the *Laplace equation*, and its solutions are called *harmonic* functions. As anticipated, we can learn much more about **g** from inspection of Eq. (2.6) than from its solution given by Eq. (1.2).

2.1.1.1 The Gravitational Field in Cavities

We now show how Green's first identity (Appendix A.5) can be used to prove that the solution of the Poisson equation (2.6) in a given simply connected region $\Omega \subseteq \Re^3$, with assigned dependence $\phi(\mathbf{x}) = h(\mathbf{x})$ on the boundary $\partial\Omega$ (the *Dirichelet* problem; see Figure 2.2), if it exists, is *unique*. In fact, suppose that two solutions, ϕ_1 and ϕ_2, exist for the problem. Then, due to the linearity of the Laplacian, their difference $\psi \equiv \phi_1 - \phi_2$ is a solution of the Laplace equation inside the volume, with null boundary conditions, because $\phi_1 = \phi_2 = h$ on $\partial\Omega$. From the vanishing of the surface integral in Eq. (A.127) with $f = g = \psi$, it follows that $\|\nabla\psi\| = 0$ on Ω (i.e., $\psi = const.$ inside the volume). But $\psi = 0$ on $\partial\Omega$, and so $\psi = 0$ everywhere on Ω, and finally $\phi_1 = \phi_2$. Notice that a similar (but not identical) result is obtained if, instead of ϕ, its *normal derivative* is assigned on $\partial\Omega$ (the *Neumann* problem): in fact, in this case the surface integral in the Green identity vanishes; however, we can now only conclude that $\psi = const.$ (and so $\phi_2 = \phi_1 + const.$) on Ω, as expected because only the normal derivative of ϕ on $\partial\Omega$ is assigned, but not the value of ϕ. Finally, note that even if the Laplace operator is of second order, the unicity result proves that in general it is not possible to assign both the value of ϕ *and* its normal derivative on $\partial\Omega$, illustrating how in general boundary value problems are different from initial value Cauchy problems (where for a second-order operator both the function and its derivative must be assigned).

A fundamental consequence of the unicity theorem is now apparent: suppose we have an empty region of space, bounded by a surface density coincident with an isopotential surface. Surely, $\phi = const.$ is *a* solution for the Laplace problem, and the unicity theorem proves that this is *the* solution (i.e., inside an empty cavity bounded by an isopotential surface density distribution, the gravitational field produced by the density distribution is zero). We therefore obtained a third proof of Newton's first theorem as a spherical homogeneous shell of matter that coincides with an isopotential surface. It should now be clear why the gravitational field inside a closed cavity bounded by an infinitesimally thin density distribution σ is not zero (even if the flux over all of the closed surfaces contained inside the cavity *is* zero from the Gauss theorem), except for very special geometries (e.g., the homogeneous spherical shell, or, with the precautionary notes in Chapter 1, inside a homogeneous and infinite cylindrical shell, or between two infinite and parallel homogeneous planes with the same surface density), while inside a cavity of arbitrary shape carved in a perfect conductor at equilibrium, the electrostatic field is always zero, independent of the shape of the cavity. This because in general a given surface material density *does not* coincide with an equipotential surface, while in the case of perfect conductors (when charges can freely move until equilibrium is reached in the tangential direction to the surface) the bounding surface *is necessarily* an equipotential, even in presence of external charges (the so-called Faraday's cage; see the beautiful discussion of the problem

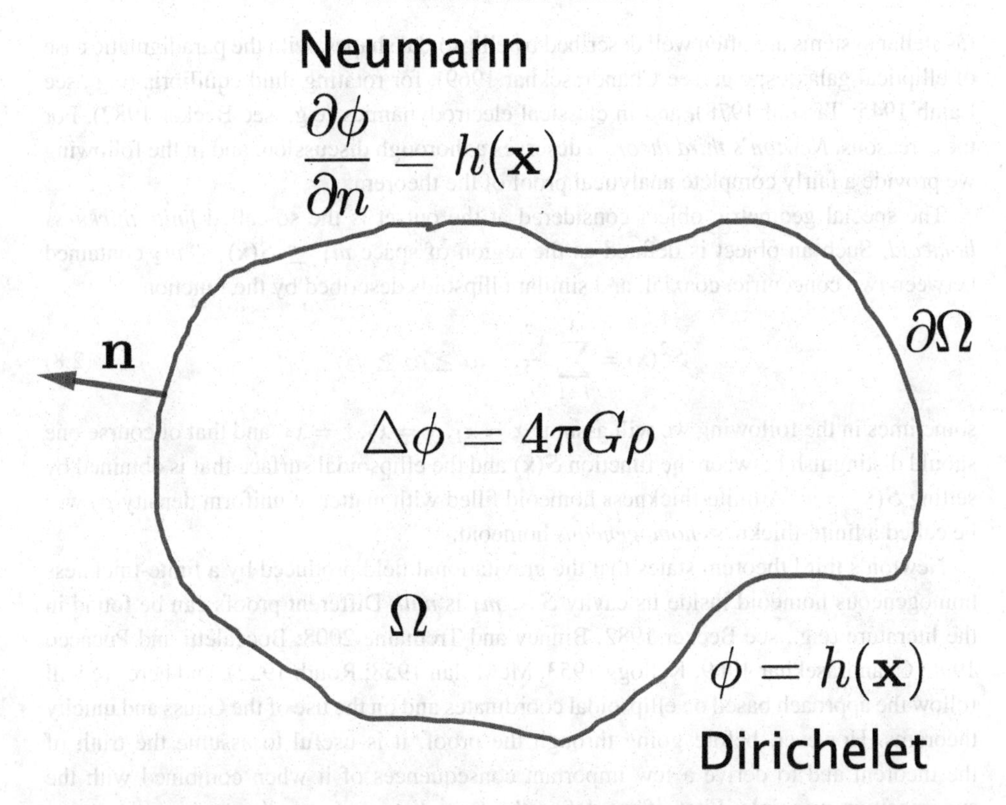

Figure 2.2 The Dirichelet and Neumann problems for the Poisson equation (2.6). As usual, we indicate the normal derivative of ϕ on $\partial\Omega$ with $\partial\phi/\partial n \equiv \langle \mathbf{n}, \nabla\phi \rangle$. As explained in the text, in the Dirichelet problem if the solution exists, it is unique, while in the Neumann problem the solution is unique up to the addition of a constant.

in volume II of Feynman et al. 1977). The interested student is invited to meditate on the following question: We reached the conclusion that a conducting surface shields the region inside from the electrostatic field of an external charge. However, if we put a charge inside a conducting surface, the field outside cannot be zero from the Gauss theorem. Why does the equipotential surface distribution of charges not shield the external region from the charges inside? After all, isn't the exterior space a region "bounded at the bottom" by an equipotential?

2.2 Newton's Third Theorem

Considering the deep nature of the unicity theorem, the fact that Newton was able to imagine and then prove by means of pure geometry the theorem that will now be illustrated is evident proof of his immense genius. This theorem concerns the gravitational fields produced by ellipsoidal figures, and it has far-reaching applications in stellar dynamics (as stellar systems are often well described by ellipsoidal shapes, with the paradigmatic case of elliptical galaxies; e.g., see Chandrasekhar 1969), for rotating fluid equilibria (e.g., see Lamb 1945; Tassoul 1978), and in classical electrodynamics (e.g., see Becker 1982). For these reasons, *Newton's third theorem* deserves a thorough discussion, and in the following we provide a fairly complete analytical proof of the theorem.

The special geometric object considered at the outset is the so-called *finite-thickness homeoid*. Such an object is defined as the region of space $m_1 \leq \mathcal{S}(\mathbf{x}) \leq m_2$ contained between two concentric, coaxial, and similar ellipsoids described by the function

$$\mathcal{S}^2(\mathbf{x}) = \sum_{i=1}^{3} \frac{x_i^2}{a_i^2}, \quad a_1 \geq a_2 \geq a_3; \tag{2.8}$$

sometimes in the following we will assume $x = x_1$, $y = x_2$, $z = x_3$, and that of course one should distinguish between the function $\mathcal{S}(\mathbf{x})$ and the ellipsoidal surface that is obtained by setting $\mathcal{S}(\mathbf{x}) = m$. A finite-thickness homeoid filled with matter of uniform density ρ_0 will be called a finite-thickness *homogeneous* homeoid.

Newton's third theorem states that the gravitational field produced by a finite-thickness homogeneous homeoid inside its cavity $\mathcal{S} < m_1$ is *null*. Different proofs can be found in the literature (e.g., see Becker 1982; Binney and Tremaine 2008; Boccaletti and Pucacco 1996; Chandrasekhar 1969; Kellogg 1953; McMillan 1958; Routh 1922), and here we will follow the approach based on ellipsoidal coordinates and on the use of the Gauss and unicity theorems. However, before going through the proof, it is useful to assume the truth of the theorem and to derive a few important consequences of it when combined with the superposition principle. First, if we define the *finite heterogeneous homeoid* as the union of homogeneous homeoids of density $\rho = \rho_i$ for $m_i \leq \mathcal{S}(\mathbf{x}) \leq m_{i+1}$, and $0 \leq m_1 \leq m_2 \leq \cdots$, we immediately deduce that the field produced in the cavity $\mathcal{S}(\mathbf{x}) < m_1$ is zero. Second, as nothing is said about the "thickness" of the regions $m_i \leq \mathcal{S}(\mathbf{x}) \leq m_{i+1}$, it is clear that the previous conclusion also holds when m is a continuous variable and $\rho = \rho(m)$: we will refer to this object, which is of great importance in stellar dynamics, as the *heterogeneous ellipsoid*. Finally, if we consider an arbitrary point $\mathbf{x}_0$ inside a heterogeneous ellipsoid $\rho(m)$, it also follows that $\mathbf{g}(\mathbf{x}_0)$ is determined only by the density distribution enclosed by the ellipsoidal surface passing thorough $\mathbf{x}_0$ (i.e., for $\mathcal{S}(\mathbf{x}) < \mathcal{S}(\mathbf{x}_0) = m_0$, a wonderful generalization of Newton's first theorem).

A very elegant way to prove Newton's third theorem, up to the level of obtaining the general expression for the gravitational potential of a heterogeneous ellipsoid, is to adopt ellipsoidal coordinates (see Appendix A.8; see also Bertin 2014; Binney and Tremaine 2008; Boccaletti and Pucacco 1996; de Zeeuw and Lynden-Bell 1985 and references therein). These coordinates, for the generic point $\mathbf{x} = (x, y, z)$, are defined by the three solutions for τ of the cubic equation in τ

$$\sum_{i=1}^{3} \frac{x_i^2}{a_i^2 + \tau} = 1, \tag{2.9}$$

whose relation with Eq. (2.8) is apparent. Notice that for $\tau = 0$ the left-hand side of Eq. (2.9) reduces to the *unitary* ellipsoidal surface $\mathcal{S}(\mathbf{x}) = 1$; therefore, for a point on the unitary surface, $\tau = 0$ is one of the three solutions. It is an interesting exercise (do it!) to show that for arbitrary $\mathbf{x}$ the three solutions (λ, μ, ν) of Eq. (2.9) are *real*, labeled with the conventional ordering $-a_1^2 \leq \nu \leq -a_2^2 \leq \mu \leq -a_3^2 \leq \lambda$. For assigned (λ, μ, ν), the substitutions $\tau = \nu$, $\tau = \mu$, and $\tau = \lambda$ in Eq. (2.9) produce three families of confocal quadrics mutually intersecting at right angles at the position $\mathbf{x}$, whose ellipsoidal coordinates are in fact (λ, μ, ν): the student is encouraged to visualize the geometry of the coordinate surfaces, and possibly to work out the axisymmetric cases, leading to *oblate* and *prolate* ellipsoidal coordinates. In particular, $\tau = \lambda \geq -a_3^2$ corresponds to the family of triaxial confocal ellipsoids, and for very large values of λ the ellipsoidal surface becomes more and more spherical, with a radius asymptotically approximated by $\sqrt{\lambda}$, so that the pair (μ, ν) can be to some extent interpreted as the pair of "angles" in spherical coordinates, determining the position of the point on the surface of the ellipsoid $\lambda = const$. In Appendix A.8 (where we also derive the explicit formulae for all of the relevant differential operators expressed in ellipsoidal coordinates to be used in the following discussion), we derive the explicit transformation formulae between (λ, μ, ν) and (x, y, z): from Eq. (A.146), note how the physical units of the ellipsoidal coordinates are *length squared*, how their values depend on the semiaxes (a_1, a_2, a_3) of the unitary ellipsoid $\mathcal{S}(\mathbf{x}) = 1$, and finally how they are *degenerate* (i.e., a given triplet (λ, μ, ν) actually corresponds to eight points in $\Re^3$ with all of the combinations of $\pm x$, $\pm y$, and $\pm z$).

Having introduced the ellipsoidal coordinates, we are now in a position to prove Newton's third theorem. We first restrict ourselves to the surface $\mathcal{S}(\mathbf{x}) = 1$, and we solve the Laplace equation $\Delta\phi = 0$ in the two distinct regions $\mathcal{S}(\mathbf{x}) < 1$ and $\mathcal{S}(\mathbf{x}) > 1$. In particular, we require that in the external region $(\lambda > 0)$ the potential can be written as $\phi = \phi(\lambda)$ (i.e., the isopotential surfaces are ellipsoids that are concentric, coaxial, and confocal with the surface $\mathcal{S}(\mathbf{x}) = 1$). Therefore, once $\phi(\lambda)$ is found, $\phi(0)$ is the value of the potential on the unitary surface, and from the unicity theorem this is also the constant value of the potential for points with $\mathcal{S}(\mathbf{x}) < 1$ (i.e., for $-a_3^2 \leq \lambda < 0$). After the determination of $\phi(\lambda)$, we will search for the infinitesimally thin surface density distribution σ_1 coincident with the unitary ellipsoid $\mathcal{S}(\mathbf{x}) = 1$ and producing such a potential: the superposition of an infinite number of such isopotential surface density distributions will finally prove Newton's third theorem.

From Eqs. (A.147) and (A.157), the Laplacian in ellipsoidal coordinates can be written as

$$\Delta\phi = \frac{4\sqrt{\Delta(\lambda)}}{(\lambda-\mu)(\lambda-\nu)}\frac{\partial}{\partial\lambda}\left[\sqrt{\Delta(\lambda)}\frac{\partial\phi}{\partial\lambda}\right] + \frac{4\sqrt{-\Delta(\mu)}}{(\lambda-\mu)(\mu-\nu)}\frac{\partial}{\partial\mu}\left[\sqrt{-\Delta(\mu)}\frac{\partial\phi}{\partial\mu}\right] + \frac{4\sqrt{\Delta(\nu)}}{(\lambda-\nu)(\mu-\nu)}\frac{\partial}{\partial\nu}\left[\sqrt{\Delta(\nu)}\frac{\partial\phi}{\partial\nu}\right], \tag{2.10}$$

where[3]

$$\Delta(\tau) \equiv (a_1^2 + \tau)(a_2^2 + \tau)(a_3^2 + \tau). \tag{2.11}$$

From the assumptions $\phi = \phi(\lambda)$ for $\lambda \geq 0$ and $\phi(\lambda) \to 0$ for $\lambda \to \infty$, the solution of the resulting Laplace equation for $\mathcal{S}(\mathbf{x}) > 1$ is immediately obtained as

$$\frac{d\phi}{d\lambda} = \frac{GC_1}{\sqrt{\Delta(\lambda)}}, \quad \Rightarrow \quad \phi(\lambda) = -GC_1 \int_{\lambda(\mathbf{x})}^{\infty} \frac{d\tau}{\sqrt{\Delta(\tau)}}, \tag{2.12}$$

where C_1 is an arbitrary constant with the dimensions of a *mass* (prove it!) and $\lambda(\mathbf{x}) > 0$ is the positive solution of Eq. (2.9) for generic $\mathbf{x}$ outside the unitary ellipsoid. The potential $\phi(0)$ in the region $\mathcal{S}(\mathbf{x}) \leq 1$ is simply given by Eq. (2.12) with $\lambda = 0$. As is shown in Eq. (2.67), in the triaxial case $\phi(\lambda)$ is expressed in terms of elliptic integrals, while in the axisymmetric case the integral involves standard trigonometric functions.

Having determined the function $\phi(\lambda)$, we now face the problem of the *existence* of a surface mass density σ_1, geometrically coincident with the surface $\mathcal{S}(\mathbf{x}) = 1$, which produces such a potential; by construction, σ_1 will then coincide with an isopotential surface. We apply the Gauss theorem to each element of the surface $\mathcal{S}(\mathbf{x}) = 1$ as follows: by construction, $\lambda = 0$ labels the isopotential surface $\mathcal{S}(\mathbf{x}) = 1$, and so the outer gradient of ϕ is perpendicular to the surface itself, and from Eq. (A.149)

$$\nabla\phi|_{\lambda=0^+} = \frac{\mathbf{f}_\lambda}{h_\lambda} \frac{d\phi}{d\lambda}\bigg|_{\lambda=0^+}. \tag{2.13}$$

Moreover, inside the surface, the potential is constant with $\nabla\phi|_{\lambda=0^-} = 0$, and from the Gauss theorem (1.7) applied to an infinitesimal cylinder with a lateral surface parallel to $\mathbf{f}_\lambda$ and crossing the surface,

$$\sigma_1(\mu, \nu) = \frac{1}{4\pi G h_\lambda} \frac{d\phi}{d\lambda}\bigg|_{\lambda=0^+} = \frac{C_1}{2\pi\sqrt{\mu\nu}}, \tag{2.14}$$

where the final beautiful identity derives from Eqs. (2.12) and (A.147). Notice that the obtained density distribution is positive everywhere, so it can be intepreted as a mass density. Of course, Eq. (2.14) can now be used the other way around: for a *given* ellipsoidal surface density of the family $\sigma_1 = A_1/\sqrt{\mu\nu}$, we must then fix $C_1 = 2\pi A_1$ in Eq. (2.12).

We now have to find the *physical* meaning $\sigma_1(\mu, \nu)$. This is obtained from the *co-area theorem* (Appendix A.9). In fact, from Eq. (A.162), we can rewrite the potential (2.4) of the heterogeneous ellipsoid (for the moment in the case of finite total mass) by breaking it into a sum over infinitesimally thin density layers of *nonconstant* surface density:

$$\phi(\mathbf{x}) = -G \int_{\Re^3} \frac{\rho(m) d^3\mathbf{y}}{\|\mathbf{x} - \mathbf{y}\|} = -G \int_0^{\infty} dm \int_{\mathcal{S}(\mathbf{y})=m} \frac{\rho(m) d^2\mathbf{y}}{\|\nabla\mathcal{S}\| \, \|\mathbf{x} - \mathbf{y}\|} \equiv \int_0^{\infty} \phi_m(\mathbf{x})\, dm, \tag{2.15}$$

[3] Note that from the ordering of λ, μ, and ν, it follows that $\Delta(\lambda) \geq 0$, $\Delta(\mu) \leq 0$, and $\Delta(\nu) \geq 0$; see also Eq. (A.147) and the associated footnote.

where $d^2\mathbf{y}$ indicates the surface area element of the ellipsoid $\mathcal{S}(\mathbf{y}) = m$ and ϕ_m indicates the potential produced by the associated co-area surface density $\sigma_m = \rho(m)/\|\nabla\mathcal{S}\|_{\mathcal{S}=m}$. The fact that σ_m is *not* constant over the surface $\mathcal{S} = m$, even though $\rho(m)$ *is* constant, is explained by the geometric interpretation of the co-area theorem (see Figure A.2). In Exercise 2.4, we prove the amazing result that for the $m = 1$ ellipsoidal surface associated with the density distribution $\rho(m)$,

$$\sigma_1 \equiv \frac{\rho(1)}{\|\nabla\mathcal{S}\|_{\mathcal{S}=1}} = \frac{a_1a_2a_3\rho(1)}{\sqrt{\mu\nu}} \Rightarrow C_1 = 2\pi a_1a_2a_3\rho(1), \tag{2.16}$$

in other words, that the co-area surface density σ_1, once expressed in ellipsoidal coordinates, has *exactly* the functional dependence of the isopotential $\sigma_1(\mu,\nu)$ in Eq. (2.14), so that Eq. (2.12) also provides the potential generated by σ_1 once C_1 is fixed as in Eq. (2.16).

In order to integrate Eq. (2.15), we now resort to the superposition principle, combined with the result just established. In practice, for given $\mathbf{x}$ we add all of the contributions $\phi_m(\mathbf{x})$ produced by σ_m of each stratum $\mathcal{S} = m$. Some care is needed, because it should be obvious that each stratum $\mathcal{S} = m$ is characterized by its own ellipsoidal coordinates, which we indicate generically with τ_m. Notice that the ellipsoid $\mathcal{S} = m$ corresponds to a *unitary* ellipsoid of semiaxes ma_i, so that the equivalent of Eq. (2.9) becomes

$$\sum_{i=1}^{3} \frac{x_i^2}{m^2a_i^2 + \tau_m} = 1; \tag{2.17}$$

the three coordinates[4] of $\mathbf{x}$ are labeled as $(\lambda_m, \mu_m, \nu_m)$, and the transformation formulae are immediately obtained from Eqs. (A.146) and (A.147) with the substitutions $a_i \mapsto ma_i$ and $\tau \mapsto \tau_m$. By repeating the treatment used for the unitary ellipsoid $m = 1$, one obtains the generalization of Eqs. (2.12), (2.14), and (2.16), where (prove it!)[5]

$$\sigma_m = \frac{C_m}{2\pi\sqrt{\mu_m\nu_m}} = \frac{a_1a_2a_3\rho(m)m^2}{\sqrt{\mu_m\nu_m}} \Rightarrow C_m = 2\pi a_1a_2a_3\rho(m)m^2, \tag{2.18}$$

so that finally

$$\phi_m(\mathbf{x}) = -2\pi G a_1a_2a_3\rho(m)m^2 \int_{\lambda_m(\mathbf{x})}^{\infty} \frac{d\tau_m}{\sqrt{\Delta_m(\tau_m)}}, \tag{2.19}$$

where $\Delta_m(\tau_m)$ is obtained from Eq. (2.11) with the substitution $a_i \mapsto ma_i$ and $\tau \mapsto \tau_m$, and finally $\lambda_m(\mathbf{x})$ is the positive solution of Eq. (2.17) if the point $\mathbf{x}$ is outside the surface $\mathcal{S} = m$, and zero otherwise. Therefore, by combining Eqs. (2.15) and (2.19), we have a quite formidable expression for the potential of a triaxial ellipsoid, and we can proceed to integrate ϕ_m over m in Eq. (2.15).

First, for the given position $\mathbf{x}$, we compute the critical value $m_c \equiv \mathcal{S}(\mathbf{x})$. The integral over m then splits into two integrals, the first over $0 \leq m \leq m_c$ and the second over

[4] In the nomenclature of Eq. (2.17), the parameter τ in Eq. (2.9) would be indicated as $\tau = \tau_1$.

[5] In Exercise 2.5, we prove by direct integration that the surface integral of σ_m over the ellipsoid $\lambda_m = 0$ evaluates to the identity $2C_m = M_m$, where M_m is the mass of the shell. We will return to this point when discussing multipole expansion.

$m_c \leq m \leq \infty$. For the latter integral, $\mathbf{x}$ is inside the ellipsoidal shells, and the value of $\lambda_m(\mathbf{x})$ in Eq. (2.19) is zero. In the first integral, we proceed as follows: we recognize that Eq. (2.17) can be rewritten as

$$\sum_{i=1}^{3} \frac{x_i^2}{a_i^2 + \tau} = m^2, \quad \tau \equiv \frac{\tau_m}{m^2}, \tag{2.20}$$

so that, if $\lambda(m, \mathbf{x})$ is the positive solution for τ in Eq. (2.20), then $\lambda(m, \mathbf{x}) = \lambda_m(\mathbf{x})/m^2$. We are almost done: we now change the variable in the inner integrals to $\tau_m = m^2\tau$, so that $\sqrt{\Delta_m(\tau_m)} = m^3\sqrt{\Delta(\tau)}$, then we invert the order of integration between m and τ from Eq. (2.20) due to the fact that in the integral over $0 \leq m \leq m_c$ we have $\lambda(m_c, \mathbf{x}) = 0$ and $\lambda(0, \mathbf{x}) = \infty$, and finally we combine all of the resulting expressions to obtain $\phi(\mathbf{x})$ (and the proof of Newton's third theorem!). It is quite astonishing that the illustrated procedure is finally summarized the following remarkable formula (e.g., see Binney and Tremaine 2008; Chandrasekhar 1969; Kellogg 1953):

$$\phi(\mathbf{x}) - \phi(\mathbf{x}_0) = -G\pi a_1 a_2 a_3 \int_0^\infty \frac{\Delta\Psi(\tau)}{\sqrt{\Delta(\tau)}} d\tau, \tag{2.21}$$

where

$$\Delta\Psi(\tau) \equiv 2\int_{m(\mathbf{x},\tau)}^{m(\mathbf{x}_0,\tau)} \rho(t)\, t\, dt, \quad m^2(\mathbf{x}, \tau) \equiv \sum_{i=1}^{3} \frac{x_i^2}{a_i^2 + \tau}, \tag{2.22}$$

and where we introduced (by subtraction) the potential at the reference point $\mathbf{x}_0$ in order to also allow for the treatment of infinite-mass ellipsoids. Due to the highly nontrivial proof of Eq. (2.21), it should be reassuring for the student to be able to show that $\phi(\mathbf{x})$ does in fact obey the Poisson equation (2.6), and Exercises 2.6 and 2.7 are devoted to this important aspect. Moreover, in some astronomical application, it is convenient to consider the potential and the gravitational field outside a *truncated* ellipsoidal distribution, and in Exercise 2.8 we derive a specialized version of Eq. (2.21) tailored to these cases.

Two comments are in order. The first is that, with $\phi(\mathbf{x})$ being given by the sum of the potentials ϕ_m produced by each isopotential surface density σ_m and with each ϕ_m being confocal with σ_m, it should not be a surprise that the isopotential surfaces produced by a generic ellipsoidal distribution *are not* in general ellipsoidal (see Exercise 2.9 and Figure 2.3). Second, a notable situation is represented by the special family of *constant-density ellipsoids* in Eq. (2.74) when the resulting potential inside the ellipsoid is a quadratic function of the Cartesian coordinates, because from Eq. (2.75)

$$\Delta\Psi(\tau) = \rho(0) \times [m_t^2 - m^2(\mathbf{x}, \tau)]. \tag{2.23}$$

As is shown in Eqs. (2.67), (3.11), (3.12), (3.62), and (3.63), the coefficients of the quadric are in general elliptic integrals in the ratios $q_z = a_3/a_1$ and $q_y = a_2/a_1$, while in the axisymmetric cases, they reduce to inverse circular functions. Physically, inside a constant-density ellipsoid, orbits are then harmonic oscillators (in general with different frequencies

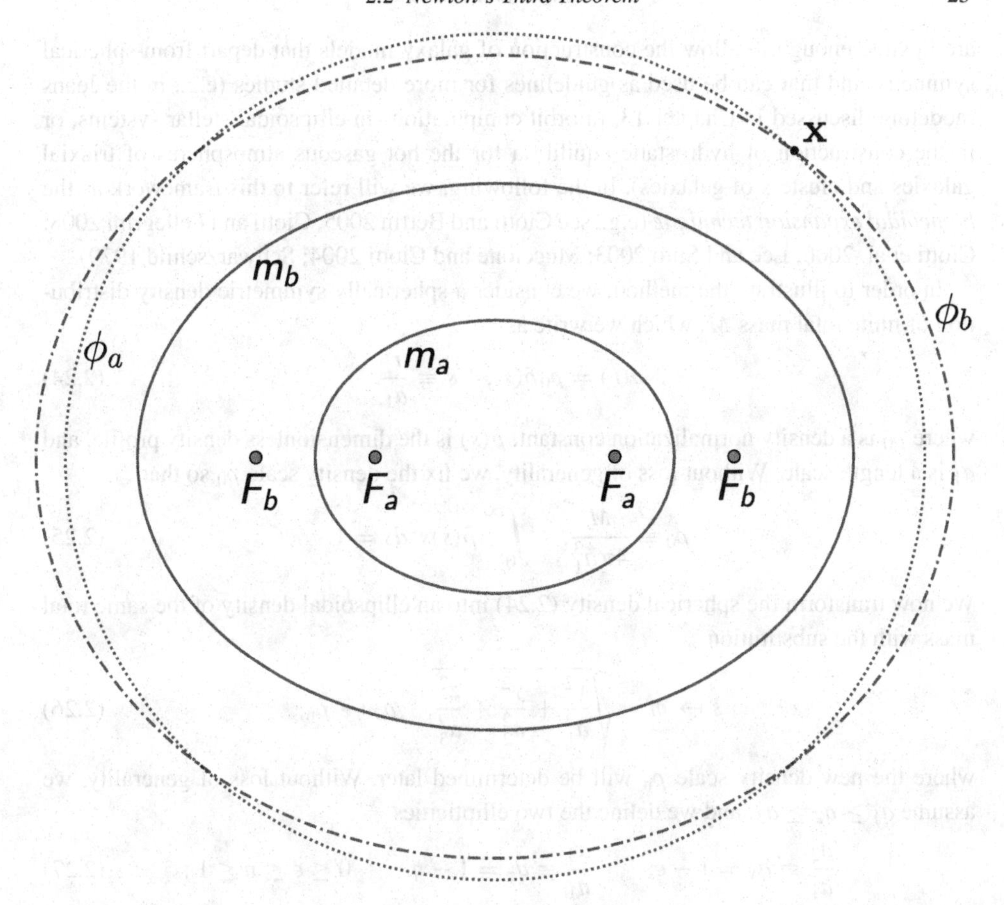

Figure 2.3 Qualitative illustration of the relative shapes of two ellipsoidal surfaces $\mathcal{S} = m_a$ and $\mathcal{S} = m_b$ (with foci at F_a and F_b) and the two associated isopotential surfaces ϕ_a (dotted line) and ϕ_b (dot-dashed line), contributing to the potential at the position **x**. Note how the isopotential $\phi_a(\mathbf{x})$ is rounder than $\phi_b(\mathbf{x})$, consistent with the confocal shapes of ϕ_m in Eq. (2.19).

along the three Cartesian axes). In Chapter 13, we will consider another special family of ellipsoidal galaxy models, the so-called *Ferrers ellipsoids*, whose potential can be written explicitely from Eq. (2.21). We will also briefly describe the important family of galaxy models obtained when starting from a gravitational potential written by using ellipsoidal coordinates, and among them the very special class of *Stäckel models*.

2.2.1 Homeoidal Expansion

It should be fairly evident that the potential generated by a generic ellipsoidal density distribution usually cannot be expressed in explicit, closed form. However, and not unexpectedly, the limit for small flattenings of Eq. (2.21) is remarkably simple, and the obtained formulae are flexible enough to allow the construction of galaxy models that depart from spherical symmetry and that can be used as guidelines for more detailed studies (e.g., in the Jeans modeling discussed in Chapter 13, in orbit computations in ellipsoidal stellar systems, or in the construction of hydrostatic equilibria for the hot gaseous atmospheres of triaxial galaxies and clusters of galaxies). In the following, we will refer to this framework as the *homeoidal expansion technique* (e.g., see Ciotti and Bertin 2005; Ciotti and Pellegrini 2008; Ciotti et al. 2006; Lee and Suto 2003; Muccione and Ciotti 2004; Schwarzschild 1979).

In order to illustrate the method, we consider a spherically symmetric density distribution of finite total mass M, which we write as

$$\rho(r) = \rho_0 \tilde{\rho}(s), \quad s \equiv \frac{r}{a_1}, \tag{2.24}$$

where ρ_0 is a density normalization constant, $\tilde{\rho}(s)$ is the dimensionless density profile, and a_1 is a length scale. Without loss of generality, we fix the density scale ρ_0 so that

$$\rho_0 \equiv \frac{M}{4\pi a_1^3}, \quad \int_0^\infty \tilde{\rho}(s) s^2 ds = 1. \tag{2.25}$$

We now transform the spherical density (2.24) into an ellipsoidal density of the same total mass with the substitution

$$s \mapsto m = \sqrt{\frac{x^2}{a_1^2} + \frac{y^2}{a_2^2} + \frac{z^2}{a_3^2}}, \quad \rho_0 \mapsto \rho_{\rm n}, \tag{2.26}$$

where the new density scale $\rho_{\rm n}$ will be determined later. Without loss of generality, we assume $a_1 \geq a_2 \geq a_3$, and we define the two ellipticities

$$\frac{a_2}{a_1} \equiv q_y = 1 - \epsilon, \qquad \frac{a_3}{a_1} \equiv q_z = 1 - \eta, \qquad 0 \leq \epsilon \leq \eta \leq 1 : \tag{2.27}$$

in the spherical limit, $q_y = q_z = 0$ and $\epsilon = \eta = 0$. It is an elementary exercise (do it!) to show that in order to conserve the total mass M of the new ellipsoidal distribution, independently of the values of ϵ and η, we must have[6]

$$\rho(\mathbf{x}) = \rho_{\rm n} \tilde{\rho}(m), \quad \rho_{\rm n} = \frac{\rho_0}{(1-\epsilon)(1-\eta)}, \quad \rho_0 = \frac{M}{4\pi a_1^3}. \tag{2.28}$$

We now perform the expansion for small values of ϵ and η of the density and of the associated potential: as the Poisson equation is linear in the pair (ρ, ϕ) and powers in ϵ and η are linearly independent, it follows that by performimg expansions at any order and selecting the functional coefficients of given orders in ϵ and η, the obtained functions

[6] Notice that M, instead of being the *total* mass, could be the mass contained in a sphere of some prescribed radius s_0 and inside the ellipsoid $m = s_0$: in this case, the normalization integral (2.25) extends up to s_0. Moreover, the assumptions of a finite total mass M for the density profile in Eq. (2.24) and the independence of M from the density flattening can be easily relaxed. The student is encouraged to work out the formulae for models of infinite total mass and/or when the total mass is not fixed. A simple way to proceed is to consider $\rho_{\rm n}$ in the first part of Eq. (2.28) as a fixed-density scale independent of flattening, so that $\Psi_{\rm n} = 4\pi G \rho_{\rm n} a_1^2 (1-\epsilon)(1-\eta)$ in Eq. (2.30).

describe *exact, nonspherical* density–potential pairs obeying the Poisson equation. For example, at the first order in the flattenings,

$$\frac{\tilde{\rho}(m)}{(1-\epsilon)(1-\eta)} \sim (1+\epsilon+\eta)\tilde{\rho}(s) + (\epsilon\tilde{y}^2+\eta\tilde{z}^2)\frac{\tilde{\rho}'(s)}{s}, \quad \tilde{y} \equiv \frac{y}{a_1}, \quad \tilde{z} \equiv \frac{z}{a_1}. \tag{2.29}$$

Of course, the density at the right-hand side of Eq. (2.29) is physically acceptable only if it is positive, and in Exercise 2.11 we derive a general inequality to be satisfied by ϵ and η in order to ensure the positivity of the expansion. Furthermore, in Exercise 2.12 we prove that the potential in Eq. (2.21), produced by the density in Eq. (2.28), can be written as

$$\phi(\mathbf{x}) = -\Psi_{\mathrm{n}}\tilde{\phi}(\mathbf{x}), \quad \Psi_{\mathrm{n}} = \frac{GM}{a_1}, \tag{2.30}$$

and that for $\epsilon \to 0$ and $\eta \to 0$

$$\tilde{\phi}(\mathbf{x}) \sim I_0(s) + (\epsilon+\eta)I_1(s) + (\epsilon\tilde{y}^2+\eta\tilde{z}^2)I_2(s), \tag{2.31}$$

where

$$\begin{cases} I_0(s) = \dfrac{1}{s}\displaystyle\int_0^s \tilde{\rho}(t)t^2 dt + \int_s^\infty \tilde{\rho}(t)t\, dt, \\[2ex] I_1(s) = \dfrac{1}{3s^3}\displaystyle\int_0^s \tilde{\rho}(t)t^4 dt + \frac{1}{3}\int_s^\infty \tilde{\rho}(t)t\, dt, \\[2ex] I_2(s) = -\dfrac{1}{s^5}\displaystyle\int_0^s \tilde{\rho}(t)t^4 dt. \end{cases} \tag{2.32}$$

A couple of explicit examples based on galaxy models often used in theoretical and observational works are discussed in Exercises 2.13 and 2.14, while further considerations are deferred to Chapter 13.

Notice that the homeoidal expansion can be used not only to calculate with good approximation the potential of slightly ellipsoidal systems, but also to generate genuine nonspherical density–potential pairs when considering the truncated expansions for finite (not necessarily small, provided positivity is guaranteed) values of ellipticities. For example, it is an instructive exercise to show that, by combining the ordering argument and the superposition principle, the following is a density–potential pair for arbitrary values of the constants α, β, and γ:

$$\begin{cases} \tilde{\rho}(\mathbf{x}) = (\alpha\tilde{x}^2+\beta\tilde{y}^2+\gamma\tilde{z}^2)\dfrac{\tilde{\rho}'(s)}{s}, \\[2ex] \tilde{\phi}(\mathbf{x}) = (\alpha+\beta+\gamma)[I_1(s)-I_0(s)] + (\alpha\tilde{x}^2+\beta\tilde{y}^2+\gamma\tilde{z}^2)I_2(s). \end{cases} \tag{2.33}$$

From this point of view, the presented technique can be seen as a particular case of a general approach based on the linearity of the Laplace operator, which can be exploited to generate families of density–potential pairs by the introduction of suitable parameters in a known density–potential pair and then performing differentiation (or integration) of both members of the Poisson equation with respect to the parameter. As a few examples, we may recall models presented in Miyamoto and Nagai (1975), Nagai and Miyamoto (1976), Satoh (1980), and Vogt and Letelier (2009a), obtained starting from the Kuzmin (1956) disks (see also Binney and Tremaine 2008; Toomre 1963), the models of Williams and Evans (2017) (see also Long and Murali 1992), and the model of Letelier (2007), or finally the complexified models discussed in Chapter 13.

2.3 Multipole Expansion

One lesson from the previous section is that we cannot hope to obtain closed-from expressions for the potential of a generic density distribution (except in very special cases), and so more powerful and flexible methods must be developed to tackle the problem when dealing with the modeling of stellar systems. A first significant step in this direction is to consider the gravitational field at large distances from an object of finite size and total mass M but otherwise of arbitrary shape: at the present stage, the only requirement (to be relaxed) is that the density $\rho(\mathbf{y})$ is zero outside some sufficiently large spherical volume enclosing the body (see Figure 2.4). In fact, in this case we can obtain the expansion of the potential for points outside the enclosing sphere as an absolutely (and so uniformly) convergent series.

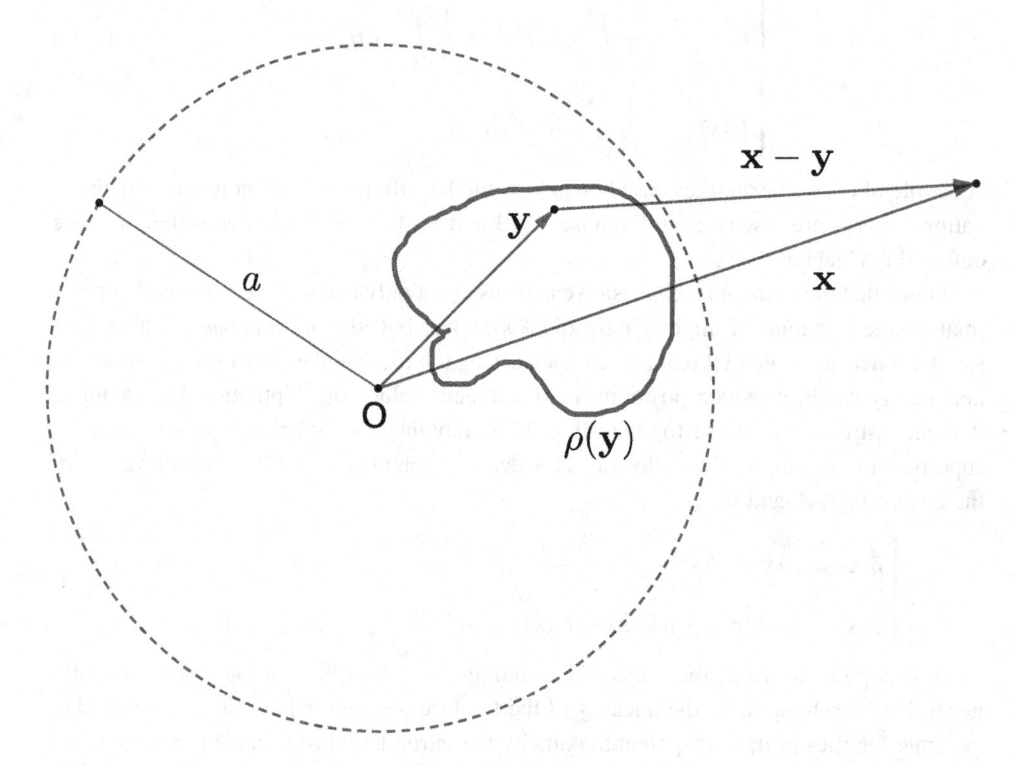

Figure 2.4 The geometry of the multipole expansion for the potential $\phi(\mathbf{x})$ produced by a density distribution $\rho(\mathbf{y})$ vanishing outside a sphere of radius a.

This is obtained as follows: we consider Eq. (2.4) with $\phi(\mathbf{x}_0) = 0$ and $\|\mathbf{x}_0\| \to \infty$, we expand the function $\|\mathbf{x} - \mathbf{y}\| = \sqrt{\|\mathbf{x}\|^2 + \|\mathbf{y}\|^2 - 2\langle \mathbf{x}, \mathbf{y}\rangle}$ for $\|\mathbf{x}\| > a$ in ascendig powers of $t \equiv \|\mathbf{y}\|/\|\mathbf{x}\| < 1$ by using Newton's binomial theorem in Eq. (A.50), and finally we perform termwise volume integration of the resulting series. The obtained expansion can be used to obtain the far-field expansion of the gravitational field of a generic body, so it is particularly useful in the investigation (for example) of the motion of distant globular clusters around a galaxy, or of a satellite around a planet, and so on. In practice, the general far-field expansion of the gravitational potential produced by $\rho(\mathbf{y})$ is given by

$$\phi = \phi_0 + \phi_1 + \phi_2 + \cdots, \tag{2.34}$$

where it is an elementary execise of algebra to show (do it!) that the terms up to quadratic order are

$$\phi_0 = -\frac{GM}{\|\mathbf{x}\|}, \quad \phi_1 = -GM\frac{\langle \mathbf{x}, \mathbf{R}_{\mathrm{CM}}\rangle}{\|\mathbf{x}\|^3}, \quad \phi_2 = -GM\frac{\langle \mathcal{Q}\mathbf{x}, \mathbf{x}\rangle}{2\|\mathbf{x}\|^5}, \tag{2.35}$$

where

$$\mathcal{Q}_{ij} \equiv \frac{1}{M}\int_{\Re^3} (3y_i y_j - \|\mathbf{y}\|^2\delta_{ij})\rho(\mathbf{y})d^3\mathbf{y}, \tag{2.36}$$

which is the *quadrupole tensor*; higher-order terms become algebraically more and more complicated. Note that ϕ_0 (the *monopole*) is independent of the position, shape, and orientation of the body, ϕ_1 (the *dipole*) depends on the position $\mathbf{R}_{\mathrm{CM}}$ of the center of mass (and so can always be set to zero by fixing the origin in $\mathbf{R}_{\mathrm{CM}}$), and ϕ_2 (the *quadrupole*) depends on the shape and orientation; more and more details about the structure of $\rho(\mathbf{y})$ are captured by higher-order multipoles. Note also that $\mathcal{Q}_{ij}$ is a second-order symmetric tensor, with $\mathrm{Tr}\mathcal{Q}_{ij} = 0$, so that in general only five integrals must be computed. However, as $\mathcal{Q}_{ij}$ can always be expressed in diagonal form with a suitable choice of the orientation of the reference system, actually only two integrals must be computed in the reference system aligned with the eigenvectors, reducing to just one integral in the case of axisymmetric systems. Finally, for a spherical system centered at the origin, the previous considerations show that $\mathcal{Q}_{ij} = 0$. Therefore, at large distances from a body with its center of mass in the origin, ϕ_0 is already a quite good approximation to the true field, as the first nonzero term is the quadrupole (or higher-order multipole) term. Of course, Newton's second theorem proves immediately that outside a spherical body with $\mathbf{R}_{\mathrm{CM}}$ in the origin, *all* of the multipole terms, except the monopole, vanish identically. A peculiar property (that we will not prove) of multipole terms, which is particularly relevant in electrostatics, is that *if* the first n terms in the expansion (2.34) are zero, *then* the value of the $n+1$-th term is *independent* of the position of the origin (e.g., see Jackson 1998). Of course, in gravity this property is not applied, as $\phi_0 \neq 0$ for mass distributions.[7] We conclude this discussion with a comment about the potential of ellipsoids discussed in the previous section. As proved in Exercise 2.5

[7] In electrostatics, when the total charge of the system is zero, the multipoles cannot be defined "per unit charge" as we do for gravity by explicitly factoring out the total mass M, as, for example, in Eq. (2.35).

with a direct and nontrivial integration, for the ellipsoidal stratum in Eq. (2.18), the constant C_m evaluates to $M_m/2$, half of the stratum's mass. Now, from multipole expansion we know that at large distance r from the stratum, $\phi_0 \sim -GM_m/r$: at the same time, we can integrate Eq. (2.19) by noting that for large r, $\lambda_m(\mathbf{x}) \sim r$ and so $\sqrt{\Delta_m(\tau_m)} \sim \tau_m^{3/2}$, and a trivial integration proves again, in a much simpler way, that $2C_m = M_m$.

2.4 The Green Function for the Poisson Equation

The idea of multipole expansion can be seen as a first (and extremely useful) attempt to elaborate a general and systematic strategy to find the solution to the Poisson equation (2.6) for arbitrary mass distributions. In fact this is possible, and it is based on the construction of the so-called *Green function* for the Laplacian: the logic behind the method is described in Appendix A.7. Of course, the literature on Green functions is immense, as the theory applies to generic linear differential operators, with applications in physics ranging from gravitation, to electromagnetism, to radiative transport, to quantum mechanics and electrodynamics, and so on. According to Eq. (A.133), it is apparent that the Green function for the Laplacian is given by the solution of the equation

$$\Delta \mathcal{G} = \delta(\mathbf{x} - \boldsymbol{\xi}). \tag{2.37}$$

The student may wonder why we should invest time in figuring out a method to solve Eq. (2.37): after all, from Eq. (2.7) we *already* know a solution:

$$\mathcal{G}(\mathbf{x}, \boldsymbol{\xi}) = -\frac{1}{4\pi \|\mathbf{x} - \boldsymbol{\xi}\|}, \tag{2.38}$$

and in fact its substitution in Eq. (A.134) reproduces (as expected) Eq. (2.4). Are we just going around in circles? Why bother with additional mathematics if we reobtain a known result? There are two main reasons for this. The first is that in the present case of the Laplacian operator, we are in the fortunate and quite exceptional situation of knowing $\mathcal{G}$ in advance, but in general the constructive approach is the *only* way to find such a solution. The second reason is that once we know how to construct $\mathcal{G}$ for a given operator, we can do this in different coordinate systems, and so obtain different *representations* for the Green function. In our case, we will obtain different representations for the function in the right-hand side of Eq. (2.38) and, even more importantly, in terms of special (orthogonal) functions, with a sort of "factorized" dependency on the coordinates $\mathbf{x}$ and $\boldsymbol{\xi}$, instead of having them "mixed" in a nonlinear way as in Eq. (2.38).

2.4.1 Cartesian Coordinates

We begin with the case of Cartesian coordinates (x, y, z), when from Eq. (A.119)

$$\Delta = \frac{\partial^2}{\partial x^2} + \frac{\partial^2}{\partial y^2} + \frac{\partial^2}{\partial z^2}. \tag{2.39}$$

Cartesian coordinates are perhaps less interesting from the point of view of stellar dynamics; however, their simplicity allows us to illustrate a few important points of the procedure without additional mathematical difficulties. Moreover, a few situations of astrophysical interest can be imagined, such as the computation of the potential of inhomogeneous mass distributions as in thin disk galaxies. As is pointed out in Appendix A.7, the first step is the search for solutions $\mathcal{G}_0$ of the homogeneous problem, which we then must combine in some clever way to obtain a solution for Eq. (2.37). A standard approach that works in several coordinate systems is to use the method of the *separation of variables*, and we then assume that $\mathcal{G}_0 = U(x)V(y)W(z)$, with the three factors to be determined (e.g., see Jackson 1998). Quite obviously, as we are searching for *a* solution in factorized form, there is ample room for different choices of the constants of separation, and in general the choices are suggested by the nature of the problem and by experience. In the present case, we separate first the x and y variables by computing $\Delta\mathcal{G}_0/\mathcal{G}_0 = 0$, and we call the two constants $-k_x^2$ and $-k_y^2$; moreover, we indicate with $\mathbf{k} = (k_x, k_y)$ a two-dimensional vector. In this way, we obtain for the functions U and V (stationary) *plane waves*,[8] while the remaining function W is given by a combination of real exponentials (do it!). As the Laplacian is a linear operator, we are now free to sum an arbitrary number of such solutions, with arbitrary weights (in general dependent on $\mathbf{k}$), and a general solution for the homogeneous problem can be written as

$$\mathcal{G}_0(\mathbf{x}) = \frac{1}{2\pi}\int_{\Re^2}\left[\hat{\mathcal{G}}_0^+(\mathbf{k})e^{i\langle\mathbf{k},\mathbf{y}\rangle+kz} + \hat{\mathcal{G}}_0^-(\mathbf{k})e^{i\langle\mathbf{k},\mathbf{y}\rangle-kz}\right]d^2\mathbf{k}, \tag{2.40}$$

where $k = \|\mathbf{k}\|$ and $\mathbf{y} = (x, y)$. We now take advantage of the special nature of plane waves, the building blocks of Fourier transforms (see Appendix A.3), to attack the Green function. In fact, we now write

$$\mathcal{G} = \frac{1}{2\pi}\int_{\Re^2}\hat{\mathcal{G}}(\mathbf{k}, z)e^{i\langle\mathbf{k},\mathbf{y}\rangle}\,d^2\mathbf{k}, \tag{2.41}$$

and we determine $\hat{\mathcal{G}}(\mathbf{k}, z)$ as follows: we insert Eq. (2.41) into Eq. (2.37) and we compute the Laplacian, we Fourier transform both members of the resulting identity, and we obtain a one-dimensional Green problem that can be treated with the method described in Appendix A.7 (see Exercise 2.21). Finally, by inserting the result into Eq. (2.41), we obtain

$$\frac{1}{\|\mathbf{x}-\boldsymbol{\xi}\|} = \frac{1}{2\pi}\int_{\Re^2}\frac{e^{i\langle\mathbf{k},\mathbf{y}-\boldsymbol{\eta}\rangle-k|z-\xi_3|}}{k}\,d^2\mathbf{k}, \quad \boldsymbol{\eta} = (\xi_1, \xi_2), \quad k = \|\mathbf{k}\|. \tag{2.42}$$

2.4.2 Spherical Coordinates

Thanks to the experience gained in the previous section, we now consider the (slightly) more complicated case of spherical coordinates (r, ϑ, φ), when from Eq. (A.157)

[8] Stationary plane waves are therefore *eigenfunctions* for the Laplacian, with *eigenvalue* $-k^2 = -\|\mathbf{k}\|^2$.

$$\Delta = \frac{1}{r^2}\frac{\partial}{\partial r}\left(r^2\frac{\partial}{\partial r}\right) + \frac{1}{r^2 \sin\vartheta}\frac{\partial}{\partial\vartheta}\left(\sin\vartheta\frac{\partial}{\partial\vartheta}\right) + \frac{1}{r^2\sin^2\vartheta}\frac{\partial^2}{\partial\varphi^2} = \Delta_r + \frac{\Delta_{\Omega}}{r^2}, \quad (2.43)$$

where the meanings of the radial (Δ_r) and the angular (Δ_{Ω}) components of the Laplacian are apparent. We search again for a solution for the empty space in the family $\mathcal{G}_0 = U(r)V(\vartheta)W(\varphi)$ (i.e., we solve $\Delta\mathcal{G}_0/\mathcal{G}_0 = 0$). We separate first the azimuthal component and we require, for obvious reasons, that the resulting W function is periodic (i.e., we require that after a rotation of 2π around the z-axis the solution is unaffected[9]), obtaining (do it!)

$$W_m(\varphi) = \frac{e^{im\varphi}}{\sqrt{2\pi}}, \quad m \in Z, \quad (2.44)$$

or in other words, rotational invariance of space *requires* that the separation constant is $-m^2$, with m being an integer number, the classical *azimuthal quantum number*! The obtained functions are an *orthonormal, complete* set (i.e., they can be used to represent functions (provided convergence is assured) over the interval $0 \leq \varphi < 2\pi$ as Fourier series; see Appendix A.3). We then move to separate the resulting equation for the function V, and the student is invited to prove that we obtain the Legendre differential equation (A.71): as discussed in Appendix A.2.3, the only acceptable solutions for our problem are given by the *Legendre associate functions* of the first kind $P_l^m(\cos\vartheta)$, with $l = 0, 1, 2, \ldots$, and $-l \leq m \leq l$, the classical *second quantum number*. This because these are the only Legendre functions that are regular at $\vartheta = 0$ and $\vartheta = \pi$; as in our problem there are no reasons for a singularity along the arbitrary choice of the z direction (again, this is a condition to be relaxed in several problems of electromagnetism, such as a conic conductor), this *forces* a quantization of the V_l functions. Remarkably, the P_l^m functions are also an *orthogonal, complete* set at fixed m for $l \geq |m|$, and again they allow us to expand in Legendre series functions of ϑ. With these two families of functions, and after normalization to the norm of P_l^m, we can construct one of the most important objects of physics and mathematics: the *spherical harmonics*

$$Y_l^m(\vartheta, \varphi) = \sqrt{\frac{2l+1}{4\pi}\frac{(l-m)!}{(l+m)!}} P_l^m(\cos\vartheta)e^{im\varphi}, \quad (2.45)$$

and it is now a simple exercise to show that integration over the whole solid angle (do it, first integrating over φ)

$$\int_{4\pi} Y_l^m(\vartheta,\varphi)Y_p^{q*}(\vartheta,\varphi)d^2\mathbf{\Omega} = \delta_{lp}\delta_{mq}, \quad Y_l^{m*} = (-1)^m Y_l^{-m}, \quad (2.46)$$

where $d^2\mathbf{\Omega} = \sin\vartheta^2 d\vartheta d\varphi$ (i.e., the spherical harmonics are a complete system of orthonormal functions on the surface of the sphere). Therefore, for a given function $f(\vartheta, \varphi)$ with $0 \leq \varphi < 2\pi$ and $0 \leq \vartheta \leq \pi$, we can write

[9] In several problems of electromagnetism, this requirement is not necessary, such as in a conductor with the shape of an orange segment.

$$f(\varphi,\vartheta)=\sum_{l=0}^{\infty}\sum_{m=-l}^{l}\hat{f}_{lm}\,\mathrm{Y}_l^m(\vartheta,\varphi),\quad \hat{f}_{lm}=\int_{4\pi}f(\vartheta,\varphi)\mathrm{Y}_l^{m*}(\vartheta,\varphi)d^2\mathbf{\Omega}, \tag{2.47}$$

the so-called Fourier–Legendre expansion[10] of f. The student should also prove that

$$\Delta_{\mathbf{\Omega}}\mathrm{Y}_l^m=-l(l+1)\mathrm{Y}_l^m, \tag{2.48}$$

a fundamental identity in the discussion of angular momentum in quantum mechanics (e.g., see Messiah 1967). In Exercise 2.24 we solve the radial equation for $U(r)$ produced by separation of variables, and from Eq. (2.99) we finally obtain the general solution for the homogeneous problem as an arbitrary linear combination of the particular solutions

$$\mathcal{G}_0(\mathbf{x})=\sum_{l=0}^{\infty}\sum_{m=-l}^{l}\left[\hat{\mathcal{G}}_0^+(l,m)r^l+\frac{\hat{\mathcal{G}}_0^-(l,m)}{r^{l+1}}\right]\mathrm{Y}_l^m(\vartheta,\varphi); \tag{2.49}$$

we will use this identity in Exercise 2.25.

We now proceed along the same lines of the previous section, and thanks to spherical harmonics we represent the still unknown Green function as a Fourier–Legendre series with generic weight functions depending on the radius:

$$\mathcal{G}=\sum_{l=0}^{\infty}\sum_{m=-l}^{l}\hat{\mathcal{G}}_{lm}(r)\mathrm{Y}_l^m(\vartheta,\varphi). \tag{2.50}$$

In order to determine $\mathcal{G}$, we now require that Eq. (2.37) is obeyed: as we are using spherical coordinates, in the following we indicate with (r,ϑ,φ) the coordinates of $\mathbf{x}$, and with $(\xi,\vartheta',\varphi')$ the coordinates of $\boldsymbol{\xi}$. Therefore, we compute (do it!) the Laplacian of $\mathcal{G}$ in Eq. (2.50) by using Eq. (2.48), we transform the three-dimensional δ-function in spherical coordinates from Eq. (A.97), and we Fourier–Legendre transform the obtained identity by multiplication for a generic Y_p^{q*} and integration over the solid angle. The procedure leads to a one-dimensional Green problem for the function $\hat{\mathcal{G}}_{lm}(r)$, already solved in Exercise 2.24, and from Eq. (2.101) we finally prove that in spherical coordinates

$$\frac{1}{\|\mathbf{x}-\boldsymbol{\xi}\|}=4\pi\sum_{l=0}^{\infty}\sum_{m=-l}^{l}\frac{\mathrm{Y}_l^m(\vartheta,\varphi)\mathrm{Y}_l^{m*}(\vartheta',\varphi')}{2l+1}\frac{r_<^l}{r_>^{l+1}}, \tag{2.51}$$

where $r_<\equiv\min(r,\xi)$ and $r_>\equiv\max(r,\xi)$, and (prove it!) from Eqs. (2.45) and (A.74)

$$\mathrm{Y}_l^{m*}(\vartheta,\varphi)=(-1)^m\mathrm{Y}_l^{-m}(\vartheta,\varphi). \tag{2.52}$$

Thus, for a generic density distribution expressed in spherical coordinates $\rho=\rho(\xi,\vartheta',\varphi')$, if we define

$$\hat{\rho}_{lm}(\xi)=\int_{4\pi}\rho(\xi,\vartheta',\varphi')\mathrm{Y}_l^{m*}(\vartheta',\varphi')d^2\mathbf{\Omega}', \tag{2.53}$$

[10] In practice, the double series in Eq. (2.47) is obtained by first Fourier expanding $f(\varphi,\vartheta)$ over φ, then Legendre expanding $f_m(\vartheta)$ over ϑ by using P_l^m at fixed m, and finally recognizing that (prove it!) $\sum_{m=-\infty}^{\infty}\sum_{l\geq|m|}=\sum_{l=0}^{\infty}\sum_{m=-l}^{l}$.

$$\phi(\mathbf{x}) = -4\pi G \sum_{l=0}^{\infty} \sum_{m=-l}^{l} \frac{Y_l^m(\vartheta,\varphi)}{2l+1} \int_0^{\infty} \frac{r_<^l}{r_>^{l+1}} \hat{\rho}_{lm}(\xi)\xi^2 d\xi. \tag{2.54}$$

Before concluding this section, it is worth pointing out two important consequences of what we have done. The first is that in Eq. (2.51) the direction of the z-axis is fully arbitrary, and of course the identity also holds when the z-axis is aligned along $\boldsymbol{\xi}$: it is a nice exercise (do it!) to show that in this special case the spherical harmonics in Eq. (2.51) reduce to Legendre polynomials, i.e.,

$$\frac{1}{\|\mathbf{x} - \xi \mathbf{e}_3\|} = \sum_{l=0}^{\infty} \mathrm{P}_l(\cos\vartheta) \frac{r_<^l}{r_>^{l+1}}. \tag{2.55}$$

We obtained a *closed-form* expression for the (messy) case of the computation of higher-order terms of multipole expansion in Cartesian coordinates (consider $r > \xi$) and the proof of the generating function of Legendre polynomials in Eq. (A.83). Notice that a comparison of Eq. (2.51) with Eqs. (2.34) and (2.35) shows that the multipole term of order n contains in general (at most) $2n+1$ different quantities, given by the spherical harmonics at fixed $l = n$ (e.g., compare with the generic monopole, dipole, and quadrupole terms); for example, the octupole potential (which we did not compute) contains a third-order tensor of 27 components, but the previous argument shows that actually at most only seven genuinely different quantities are involved! Finally, following Jackson (1998), we obtain the very important *spherical harmonics addition formula* by starting from Eq. (2.51), rotating the coordinates in the left-hand side to place $\boldsymbol{\xi}$ along the new z-axis (and this of course leaves the value of the left-hand side unchanged), and then comparing with Eq. (2.55): as powers are linearly independent, we have

$$\frac{4\pi}{2l+1} \sum_{m=-l}^{l} Y_l^m(\vartheta,\varphi) Y_l^{m*}(\vartheta',\varphi') = \mathrm{P}_l(\cos\gamma), \tag{2.56}$$

where γ is the angle between $\mathbf{x}$ and $\boldsymbol{\xi}$ (i.e., from the inner product) and $\cos\gamma = \cos\vartheta\cos\vartheta' + \sin\vartheta\sin\vartheta'\cos(\varphi - \varphi')$.

2.4.3 Cylindrical Coordinates

We can finally solve the problem of the construction of the Green function for the Laplacian in cylindrical coordinates (R, φ, z), one of the most used coordinate systems in stellar dynamics. From Eq. (A.157) we have

$$\triangle = \frac{1}{R}\frac{\partial}{\partial R}\left(R\frac{\partial}{\partial R}\right) + \frac{1}{R^2}\frac{\partial^2}{\partial\varphi^2} + \frac{\partial^2}{\partial z^2}. \tag{2.57}$$

We search again for a solution for the empty space in the family $\mathcal{G}_0 = U(R)V(z)W(\varphi)$ (i.e., we solve $\triangle\mathcal{G}_0/\mathcal{G}_0 = 0$ by using the separation of variables). As for spherical coordinates, we first separate for the azimuthal angle φ and, repeating identical arguments, we again obtain for the constant of separation $-m^2$ for the family of the complete set of orthonormal Fourier W_m functions given in Eq. (2.44). Next, we separate the resulting equation for the z-coordinate, and we use as a separation constant the quantity k^2, with k being a positive real constant; this is done in order to obtain V functions that can present an exponential decline[11] for $|z| \to \infty$. We can now move to studying the last differential equation for the radial function $U(R)$: the student is invited to prove that we obtained the Bessel equation (A.84). As discussed in Appendix A.2.4, the second-order, linear differential equation admits two independent solutions, and Fuch's theorem shows that the only regular solutions at the origin are the Bessel function of the first kind J_m with m integer. These functions are orthogonal with weight R over the range $0 \leq R < \infty$ and complete, and they allow us to represent radial functions as *Hankel transforms*, as shown in Eqs. (A.88) and (A.89). We now combine an arbitrary number of the obtained solutions, and we recover a general expression (do it!) for the homogeneous solution as

$$\mathcal{G}_0(\mathbf{x}) = \sum_{m=-\infty}^{\infty} \frac{\mathrm{e}^{im\varphi}}{\sqrt{2\pi}} \int_0^{\infty} \left[\hat{\mathcal{G}}_0^{+}(k,m)\mathrm{e}^{kz} + \hat{\mathcal{G}}_0^{-}(k,m)\mathrm{e}^{-kz}\right] \mathrm{J}_m(kR)\, dk. \tag{2.58}$$

We now proceed along the same lines as the previous section, and we represent the still unknown Green function in the Fourier–Bessel form, with generic weight functions depending on z, i.e.,

$$\mathcal{G} = \sum_{m=-\infty}^{\infty} \frac{\mathrm{e}^{im\varphi}}{\sqrt{2\pi}} \int_0^{\infty} \hat{\mathcal{G}}_m(k,z)\mathrm{J}_m(kR)k\, dk. \tag{2.59}$$

In order to determine $\mathcal{G}$, we now require that Eq. (2.37) is obeyed: as we are using spherical coordinates, in the following we indicate with (R,φ,z) the coordinates of $\mathbf{x}$, and with (ξ,z',φ') the coordinates of $\boldsymbol{\xi}$. Therefore, we compute (do it!) the Laplacian of $\mathcal{G}$ in Eq. (2.59), we transform the three-dimensional δ-function in cylindrical coordinates from Eq. (A.97), and we Fourier–Hankel transform the obtained identity by multiplication for a generic $\mathrm{e}^{-in\varphi}/\sqrt{2\pi}$ element and integrating over φ, followed by multiplication for a generic $\mathrm{J}_n(lR)$ element and performing a final integration over $R\,dR$. The procedure leads to a one-dimensional Green problem for the function $\hat{\mathcal{G}}_m(k,z)$, which is already solved in Exercise 2.21, so that in cylindrical coordinates we finally have

$$\frac{1}{\|\mathbf{x}-\boldsymbol{\xi}\|} = \sum_{m=-\infty}^{\infty} \mathrm{e}^{im(\varphi-\varphi')} \int_0^{\infty} \mathrm{e}^{-k|z-z'|}\mathrm{J}_m(kR)\mathrm{J}_m(k\xi)\, dk. \tag{2.60}$$

It follows that for a generic density distribution expressed in cylindrical coordinates $\rho = \rho(\xi,\varphi',z')$,

$$\hat{\rho}_m(k,z') = \int_0^{\infty} \mathrm{J}_m(k\xi)\xi\, d\xi \int_0^{2\pi} \frac{\mathrm{e}^{-im\varphi'}}{\sqrt{2\pi}} \rho(\xi,\varphi',z')\, d\varphi', \tag{2.61}$$

[11] Again, this choice is motivated by the fact that in astronomical systems the potential is needed over the whole space: in electromagnetism, often the value of the potential is assigned on planes at some finite distance from the equatorial plane, and other choices for the separation constant are more appropriate (e.g., see Jackson 1998).

the potential can be written as

$$\phi(\mathbf{x}) = -2\pi G \sum_{m=-\infty}^{\infty} \frac{e^{im\varphi}}{\sqrt{2\pi}} \int_0^\infty J_m(kR)dk \int_{-\infty}^{\infty} e^{-k|z-z'|}\hat{\rho}_m(k,z')dz'. \tag{2.62}$$

The identities obtained are of especially important in Stellar Dynamics for the mathematical treatment of disks and rings, and in the exercices some of the most famous cases of razor-thin disks are presented. The student is reminded that other methods exist to recover the gravitational potential of disks, and some of these additional techniques are discussed in more advanced literature (e.g., see Binney and Tremaine 2008; Evans and de Zeeuw 1992; Freeman 1966; Hunter 1963; Kalnajs 1976a,b; Mestel 1963; Toomre 1963 and references therein; see also references in the exercises in Chapter 5).

A final comment is in order: the student will notice in the exercises a quite significant use of tables of integrals, and they could ask why, with the availability of powerful computer algebra systems, one should spend time looking through deeply boring tables. Experience shows that looking through tables greatly helps us to understand the properties of our integrals, and that while computer algebra is an invaluable help for research and numerical experiments, it should not (like tables) be trusted blindly when searching for the closed forms of difficult integrals, especially when they depend on parameters. For these reasons, the student is invited to enjoy the experience of repeating at least a few of the suggested exercises.

Exercises

2.1 Consider the so-called *singular isothermal sphere* density profile

$$\rho(r) = \frac{v_c^2}{4\pi G r^2}, \quad M(r) = \frac{v_c^2 r}{G}, \tag{2.63}$$

where v_c is the constant circular velocity of the model (prove it!). By using Eq. (2.5), show that the potential can be written as

$$\phi(r) = v_c^2 \ln\frac{r}{r_0}, \tag{2.64}$$

where r_0 is an arbitrary scale length.

2.2 Consider the gravitational field produced by the razor-thin, planar mass distribution

$$\rho(\mathbf{x}) = \sigma(\mathbf{y})\delta(z), \tag{2.65}$$

where $\mathbf{y} = (x, y)$ and $\sigma(\mathbf{y})$ is the surface density of the stratum. By using Gauss theorem, show that the jump of the z-component of the field across the surface is

$$g_z^+ - g_z^- = -4\pi G\sigma(\mathbf{y}). \tag{2.66}$$

Repeat the convergence argument in Eq. (2.4) and find the asymptotic behavior of σ as a function $R = ||\mathbf{y}||$ so that $\|\mathbf{x}_0\| \to \infty$ can be assumed. Discuss the result along with that of Exercise 1.8.

2.3 Prove that the integral appearing in Eq. (2.12) when calculating the potential produced by the unitary isopotential ellipsoidal shell $\mathcal{S}(\mathbf{x}) = 1$ can be written as

$$w_0[\lambda(\mathbf{x})] \equiv \int_{\lambda(\mathbf{x})}^{\infty} \frac{d\tau}{\sqrt{\Delta(\tau)}} = \frac{2}{a_1\sqrt{1-q_z^2}} \begin{cases} \arctan\sqrt{\dfrac{1-q_z^2}{q_z^2+\tilde{\lambda}}}, \\ \operatorname{arctanh}\sqrt{\dfrac{1-q_z^2}{1+\tilde{\lambda}}}, \\ F(\varphi,k), \end{cases} \tag{2.67}$$

where $q_y \equiv a_2/a_1$, $q_z \equiv a_3/a_1$,

$$\tilde{\lambda} \equiv \frac{\lambda(\mathbf{x})}{a_1^2}, \quad \sin\varphi = \sqrt{\frac{1-q_z^2}{1+\tilde{\lambda}}}, \quad k = \sqrt{\frac{1-q_y^2}{1-q_z^2}}, \tag{2.68}$$

and $F(\varphi,k)$ is the elliptic integral of the first kind in Eq. (A.62). The three expressions in Eq. (2.67) hold respectively for the oblate case around z ($q_y = 1$), the prolate case around x ($q_y = q_z$), and the triaxial case with $\tilde{\lambda} = 0$ for points inside and on the shell (i.e., for $\mathcal{S}(\mathbf{x}) \leq 1$). Evaluate the spherical limit of the expressions above and show that Newton's first and second theorems are recovered. *Hints*: The integration of the axisymmetric cases is elementary, while for the triaxial case, use eq. (238.00) in Byrd and Friedman (1971) or eq. (3.131.8) in Gradshteyn et al. (2007). For the evaluation of the spherical limit, use Eq. (2.9).

2.4 Prove that the surface density associated via the co-area theorem with the unitary ellipsoid for the three-dimensional density $\rho(m)$, once expressed in ellipsoidal coordinates, is given by Eq. (2.16). *Hints*: First show that in Cartesian coordinates

$$\|\nabla\mathcal{S}\|_{\mathcal{S}=1} = \sqrt{\frac{x^2}{a_1^4} + \frac{y^2}{a_2^4} + \frac{z^2}{a_3^4}}, \tag{2.69}$$

and then use Eq. (A.146) with $\lambda = 0$. A more elegant proof can be obtained by differentiation of the two sides of Eq. (A.145) with respect to τ

$$\begin{aligned} &\frac{x^2}{(a_1^2+\tau)^2} + \frac{y^2}{(a_2^2+\tau)^2} + \frac{z^2}{(a_3^2+\tau)^2} \\ &= \frac{(\tau-\mu)(\tau-\nu)}{\Delta(\tau)} + (\tau-\lambda)\frac{d}{d\tau}\frac{(\tau-\mu)(\tau-\nu)}{\Delta(\tau)}, \end{aligned} \tag{2.70}$$

and then setting $\tau = \lambda = 0$.

2.5 Prove by direct integration in ellipsoidal coordinates that the surface integral of σ_m in Eq. (2.18), over the ellipsoid $\lambda_m = 0$, evaluates to $2C_m = M_m$, where M_m is the mass of the shell. *Hints:* From Appendix A.8,

$$M_m = 8 \int_{\mathcal{S}(\mathbf{x})=m} \sigma_m h_{\mu_m} h_{\nu_m} d\mu_m d\nu_m$$

$$= \frac{C_m}{\pi} \times \int_{a_2^2}^{a_1^2} d\nu \int_{a_3^2}^{a_2^2} \frac{(\nu - \mu)\, d\mu}{\sqrt{(a_1^2 - \nu)(\nu - a_2^2)(\nu - a_3^2)(a_1^2 - \mu)(a_2^2 - \mu)(\mu - a_3^2)}}, \tag{2.71}$$

where the factor 8 takes into account the degeneracy of ellipsoidal coordinates, the Lamé coefficients are evaluated at $\lambda_m = 0$, and we changed variables as $\mu_m = -m^2\mu$ and $\nu_m = -m^2\nu$. The double integral can be expressed in terms of complete elliptic integrals, and from eqs. (3.131.3-5), (3.132.2-5), and (8.122) in Gradshteyn et al. (2007), it evaluates to 2π, completing the proof.

2.6 Show by differentiation of Eq. (2.21) that the gravitational field produced at $\mathbf{x}$ by an heterogeneous ellipsoid, along the direction $\mathbf{e}_i$, is given by

$$g_i(\mathbf{x}) = -\frac{\partial \phi}{\partial x_i} = -2\pi G a_1 a_2 a_3 x_i \int_0^\infty \frac{\rho(m_\tau)\, d\tau}{(a_i^2 + \tau)\sqrt{\Delta(\tau)}}, \tag{2.72}$$

where $m_\tau \equiv m(\mathbf{x}, \tau)$ is defined in Eq. (2.22).

2.7 Show that $\phi(\mathbf{x})$ in Eq. (2.21) obeys the Poisson equation. *Hints*: Use Eq. (2.72) and the fact that for a generic function $f(m_\tau)$, where $m_\tau \equiv m(\mathbf{x}, \tau)$, we have (prove it!)

$$\frac{\partial f}{\partial x_i} = \frac{f'}{m_\tau} \frac{x_i}{a_i^2 + \tau}, \quad \frac{\partial f}{\partial \tau} = -\frac{f'}{2m_\tau} \sum_{i=1}^{3} \frac{x_i^2}{(a_i^2 + \tau)^2}, \quad f' \equiv \frac{df}{dm_\tau}. \tag{2.73}$$

2.8 With this exercise we derive the explicit integral formula for the potential of a *truncated* ellipsoid

$$\rho_t(m) = \rho(m) \times \theta(m_t - m), \tag{2.74}$$

where θ is Heaviside's function in Eq. (A.99). Show that when $\mathbf{x}$ is outside the ellipsoid (i.e., $\mathcal{S}(\mathbf{x}) > m_t$), Eq. (2.21) can be written as

$$\phi(\mathbf{x}) = -G\pi a_1 a_2 a_3 \int_{\lambda_t(\mathbf{x})}^\infty \frac{\Delta\Psi(\tau)}{\sqrt{\Delta(\tau)}} d\tau, \quad \Delta\Psi(\tau) = 2 \int_{m(\mathbf{x},\tau)}^{m_t} \rho(t)\, t\, dt, \tag{2.75}$$

where $\lambda_t(\mathbf{x}) > 0$ is the positive solution of Eq. (2.20) for $m = m_t$, while for $\mathcal{S}(\mathbf{x}) \le m_t$ the formula above holds with $\lambda_t(\mathbf{x}) = 0$. Finally, in analogy with Eq. (2.72), obtain the expression for the field $\mathbf{g}$ outside the ellipsoid considering that

$$\frac{\partial \lambda_t}{\partial x_i} = \frac{2x_i}{\lambda_t + a_i^2} \left[\sum_{i=1}^{3} \frac{x_i^2}{(\lambda_t + a_i^2)^2} \right]^{-1}. \tag{2.76}$$

Hints: As the ellipsoid is truncated, its mass is finite and we can fix $\mathbf{x}_0 = \infty$ and $\phi(\mathbf{x}_0) = 0$. Then, from the obvious identity

$$\Delta\Psi_t(\tau) = 2\int_{m(\mathbf{x},\tau)}^{\infty} \rho_t(t)\, t dt = \Delta\Psi(\tau) \times \theta[m_t - m(\mathbf{x},\tau)], \tag{2.77}$$

Eq. (2.75) is recovered. Equation (2.76) is proved by differentiation of Eq. (2.20) with $m = m_t$ and $\tau = \lambda_t(\mathbf{x})$.

2.9 Consider an *ellipsoidal potential*, i.e.,

$$\phi = \phi(m), \quad m \equiv \sqrt{\frac{x^2}{a_1^2} + \frac{y^2}{a_2^2} + \frac{z^2}{a_3^2}}, \tag{2.78}$$

and show that

$$\Delta\phi = \left(\frac{1}{a_1^2} + \frac{1}{a_2^2} + \frac{1}{a_3^2}\right) h(m) + \left(\frac{x^2}{a_1^4} + \frac{y^2}{a_2^4} + \frac{z^2}{a_3^4}\right) \frac{h'(m)}{m}, \tag{2.79}$$

where $h(m) \equiv \phi'(m)/m$ and $\phi'(m) = d\phi(m)/dm$. Can an ellipsoidal density distribution produce an ellipsoidal potential? What is the mass of a system associated with an ellipsoidal potential? *Hint*: Consider the monopole term in Eq. (2.35).

2.10 From Eqs. (2.23) and (2.75), show that the gravitational potential at interior points of a homogeneous ellipsoid is a quadratic form of the coordinates. Use the superposition principle to prove that the field in an arbitrary ellipsoidal cavity carved inside a homogeneous ellipsoid is a linear function of coordinates (see also Exercise 1.4).

2.11 Show that the positivity of the homeoidally expanded density distribution in Eq. (2.29), obtained from a spherical parent distribution with $d\rho/ds \leq 0$, is guaranteed for

$$\frac{1+\epsilon+\eta}{\max(\epsilon,\eta)} \geq A_{\rm M}, \quad A_{\rm M} \equiv \sup_{0\leq s} \left|\frac{d\ln\tilde{\rho}(s)}{d\ln s}\right|. \tag{2.80}$$

Prove that for a parent density profile $\tilde{\rho} \propto s^{-a}(1+s^b)^{-c}$ one has $A_{\rm M} = a + bc$, so that $A_{\rm M} = 4$ for the important family of ellipsoidal γ-models in Eq. (2.83), with $\eta \leq 1/3$ in the axisymmetric oblate case ($\epsilon = 0$) and $\eta \leq 1/2$ in the axisymmetric prolate case ($\epsilon = \eta$). *Hint:* Rewrite Eq. (2.29) in spherical polar coordinates and recast the positivity request as

$$1 \geq A_{\rm M} \frac{\epsilon \sin^2\vartheta \sin^2\varphi + \eta\cos^2\vartheta}{1+\epsilon+\eta} \tag{2.81}$$

over the closed region $0 \leq \vartheta \leq \pi$ and $0 \leq \varphi \leq 2\pi$. What is the analogous inequality if mass conservation is *not* required, as indicated in Footnote 6 of this chapter (see Ciotti and Bertin 2005; Ciotti et al. 2020)?

2.12 By using the general formula (2.21) for the potential of ellipsoidal distributions, shows that in Eq. (2.30)

$$\tilde{\phi}(\mathbf{x}) = \frac{1}{4}\int_0^{\infty} \frac{\widetilde{\Delta\Psi}(\tilde{\tau})}{\sqrt{\tilde{\Delta}(\tilde{\tau})}} d\tilde{\tau}, \quad \tilde{\tau} \equiv \frac{\tau}{a_1^2}, \tag{2.82}$$

where all of the lengths are normalized to a_1 and $\widetilde{\Delta\Psi} \equiv \Delta\Psi/\rho_n$ is obtained from Eq. (2.22) applied to $\tilde{\rho}$. By expansion for $\epsilon \to 0$ and $\eta \to 0$, then recover Eqs. (2.29)–(2.31) and show that the Poisson equation is satisfied.

2.13 From Eq. (2.28), show that the mass-conserving ellipsoidal generalization of the spherical γ-models (Dehnen 1993; Tremaine et al. 1994; see also Chapter 13) is given by

$$\rho(\mathbf{x}) = \frac{M}{4\pi a_1^3} \frac{3-\gamma}{(1-\epsilon)(1-\eta)m^{\gamma}(m+1)^{4-\gamma}}, \quad 0 \leq \gamma < 3, \tag{2.83}$$

where M is the total galaxy mass, so that from Eq. (2.29)

$$\rho(\mathbf{x}) \sim \frac{M}{4\pi a_1^3} \frac{3-\gamma}{s^{\gamma}(s+1)^{4-\gamma}} \left[1 + \epsilon + \eta - \frac{(\epsilon\tilde{y}^2 + \eta\tilde{z}^2)(\gamma + 4s)}{s^2(1+s)}\right]. \tag{2.84}$$

Moreover, from Eq. (2.32), show that for the ellipsoidal Hernquist (1990) model ($\gamma = 1$),

$$\begin{cases} I_0(s) = \dfrac{1}{1+s}, \\ I_1(s) = \dfrac{2+s}{s^2(1+s)} - \dfrac{2\ln(1+s)}{s^3}, \\ I_2(s) = -\dfrac{2s^2+9s+6}{s^4(1+s)^2} + \dfrac{6\ln(1+s)}{s^5}, \end{cases} \tag{2.85}$$

while for the ellipsoidal Jaffe (1983) model ($\gamma = 2$),

$$\begin{cases} I_0(s) = \ln\dfrac{1+s}{s}, \\ I_1(s) = \dfrac{2-s}{3s^2} + \dfrac{1}{3}\ln\dfrac{1+s}{s} - \dfrac{2\ln(1+s)}{3s^3}, \\ I_2(s) = -\dfrac{2+s}{s^4(1+s)} + \dfrac{2\ln(1+s)}{s^5}. \end{cases} \tag{2.86}$$

The formulae for generic values of γ are given in Ciotti and Bertin (2005). Repeat the exercise for the model in Eq. (2.93), and for the ellipsoidal generalizations of all the spherical models in Section 13.2.

2.14 By direct computation of the Poisson equation (spherical coordinates are suggested), show that the toroidal density

$$\rho = \rho_n \frac{\tilde{R}^2}{\tilde{r}^{\alpha}}, \quad \tilde{R} \equiv \frac{R}{r_*}, \quad \tilde{r} \equiv \frac{r}{r_*}, \tag{2.87}$$

where $2 < \alpha < 5$ and ρ_n and r_* are density and a length scales, respectively generates the potential (obtained from homeoidal expansion in Ciotti and Bertin 2005)

$$\phi = 4\pi G\rho_n r_*^2 \times \begin{cases} -\dfrac{\tilde{r}^{2-\alpha}}{(\alpha-2)(7-\alpha)}\left[\dfrac{4\tilde{r}^2}{(\alpha-4)(5-\alpha)} + \tilde{R}^2\right], \\ \dfrac{2}{3}\ln\tilde{r} - \dfrac{\tilde{R}^2}{6\tilde{r}^2}, \quad (\alpha = 4). \end{cases} \tag{2.88}$$

2.15 Convince yourself that if *all* of the multipole terms in Eq. (2.34) vanish up to the order n, then ϕ_{n+1} is independent of the position of the origin of the reference system. Of course, this theorem is only relevant in electromagnetism (e.g., see Jackson 1998).

2.16 Consider a spatially *untruncated* density distribution. What happens to Eqs. (2.35) and (2.36) and to higher-order terms? Discuss the possible divergence of $\mathcal{Q}_{ij}$ *Hint*: Also consider the expansion of $\|\mathbf{x}-\mathbf{y}\|$ for $\|\mathbf{y}\| > \|\mathbf{x}\|$ and compare it with the expansion in spherical harmonics in Eq. (2.51).

2.17 Show that (when they exist) the quadrupole tensor $\mathcal{Q}_{ij}$ and the *inertia tensor* $\Im_{ij}$ are related as

$$M\mathcal{Q}_{ij} = \mathrm{Tr}(\Im_{ij})\delta_{ij} - 3\Im_{ij}, \quad \Im_{ij} = \int_{\Re^3} (\|\mathbf{y}\|^2\delta_{ij} - y_i y_j)\rho(\mathbf{y})d^3\mathbf{y}. \tag{2.89}$$

Express $\Im_{ij}$ and $\mathcal{Q}_{ij}$ in terms of the second-order mass virial tensor I_{ij} in Eq. (10.45) and show that

$$\Im_{ij} = \mathrm{Tr}(I_{ij})\delta_{ij} - I_{ij}, \quad I_{ij} = \frac{\mathrm{Tr}(\Im_{ij})}{2}\delta_{ij} - \Im_{ij}, \quad M\mathcal{Q}_{ij} = 3I_{ij} - \mathrm{Tr}(I_{ij})\delta_{ij}. \tag{2.90}$$

Deduce that $\mathcal{Q}_{ij}$, $\Im_{ij}$, and I_{ij} are all diagonal in the same reference system and find the relation between their eigenvalues. Note that $\mathrm{Tr}(\Im_{ij}) = 2\mathrm{Tr}(I_{ij})$.

2.18 Show that for the heterogeneous ellipsoid in Eq. (2.28), with finite mass M and finite extension (i.e., $\rho = 0$ for $m \geq m_t$), $\Im_{ij}$ and $\mathcal{Q}_{ij}$ are in diagonal form, with

$$\Im_{ii} = M\frac{a_j^2 + a_k^2}{3}h, \quad \mathcal{Q}_{ii} = \frac{2a_i^2 - a_j^2 - a_k^2}{3}h, \tag{2.91}$$

where $i \neq j \neq k$, and

$$h \equiv \int_0^\infty \tilde{\rho}(m)m^4 dm, \quad \int_0^\infty \tilde{\rho}(m)m^2 dm = 1. \tag{2.92}$$

Under the convention $a_1 \geq a_2 \geq a_3$, deduce that $0 < \Im_{11} \leq \Im_{22} \leq \Im_{33}$ and $\mathcal{Q}_{11} \geq \mathcal{Q}_{22} \geq \mathcal{Q}_{33}$. Finally, use Eq. (2.35) to write the potential $\phi = \phi_0 + \phi_2$ in spherical coordinates and discuss the shape of the isopotentials $\phi = c$ for $c \to 0$. *Hint*: Solve a cubic equation for $r(\vartheta, \varphi, c)$, or (better) perform a standard asymptotic analysis to obtain $r \sim r_0(c) + \delta r(\vartheta, \varphi, c)$.

2.19 An important ellipsoidal density profile with peculiar orbital properties is the *perfect ellipsoid* (de Zeeuw and Lynden-Bell 1985):

$$\rho(\mathbf{x}) = \frac{M}{4\pi a_1^3}\frac{4}{\pi(1-\epsilon)(1-\eta)(m^2+1)^2}. \tag{2.93}$$

Modify this density profile, and those of the ellipsoidal γ-models in Eq. (2.83), introducing a truncation ellipsoidal surface m_t so that $\rho = 0$ for $m > m_t$. Call M_t the total mass inside m_t, and for these truncated models calculate h in Eq. (2.92).

2.20 By using Eq. (2.91), show that for a finite-mass, spatially truncated ellipsoid axisymmetric around the z-axis (i.e., $a_1 = a_2$), $\mathcal{Q}_{11} = \mathcal{Q}_{22} = -\mathcal{Q}_{33}/2$, and discuss the sign of $\mathcal{Q}_{33}$ as a function of flattening a_3/a_1. Show that in the equatorial plane the quadrupole potential at distance R from the origin is

$$\phi_2 = \frac{GM\mathcal{Q}_{33}}{4R^3} = -\frac{GM(a_1^2 - a_3^2)}{6R^3} h. \tag{2.94}$$

Therefore, at radius R in the equatorial plane, the gravitational field obtained from $\phi_0 + \phi_2$ of an oblate ellipsoid is stronger than that of a sphere of the same mass, and in turn this is stronger than that of a prolate ellipsoid of identical mass.

2.21 Show that the problem of the Green function for the Laplacian in Cartesian coordinates, with the separation of variables adopted in Section 2.4.1, leads to a one-dimensional Green problem of the family

$$\frac{d^2\hat{\mathcal{G}}}{dz^2} - k^2\hat{\mathcal{G}} = A\delta(z - \xi_3). \tag{2.95}$$

Determine the constant A. By using the method in Appendix A.7 and restricting to solutions so that $\hat{\mathcal{G}} \to 0$ for $|z| \to \infty$, show that

$$\hat{\mathcal{G}} = -A\frac{\mathrm{e}^{-k|z-\xi_3|}}{2k}. \tag{2.96}$$

Hint: The associated homogeneous problem is a constant-coefficient differential equation.

2.22 Consider a razor-thin density distribution $\rho = \sigma(\boldsymbol{\eta})\delta(\xi_3)$, where $\boldsymbol{\eta} = (\xi_1, \xi_2)$. From Eq. (2.42), show that the gravitational potential (provided convergence is assured) can be written as

$$\phi(\mathbf{x}) = -G\int_{\Re^2} \frac{\mathrm{e}^{i\langle\mathbf{k},\mathbf{y}\rangle - k|z|}}{\|\mathbf{k}\|}\hat{\sigma}(\mathbf{k})\, d^2\mathbf{k}, \quad \mathbf{x} = (x, y, z) = (\mathbf{y}, z), \tag{2.97}$$

where $\hat{\sigma}$ is the Fourier transform of the surface density. Then show that the Gauss theorem holds. Finally, reobtain the same result by using Eq. (2.40) to compute the field above and below the plane and by imposing the validity of the Gauss theorem on the jump of the field through the plane.

2.23 Show that the Green function for the n-dimensional Laplace operator in Cartesian coordinates $\Delta\mathcal{G} = \delta(\mathbf{x} - \boldsymbol{\xi})$ can be obtained by direct Fourier transform:

$$\mathcal{G} = -\frac{1}{(2\pi)^n}\int_{\Re^n} \frac{e^{i\langle\mathbf{k},\mathbf{x}-\boldsymbol{\xi}\rangle}}{\|\mathbf{k}\|^2} d^n\mathbf{k}. \tag{2.98}$$

In case of $\Re^3$, show the equivalence of the obtained expression with that in Eq. (2.42) by using the residue theorem of complex analysis by integrating over k_3.

Hints: Write $\mathcal{G}$ as a Fourier antitransform from Eq. (A.102), apply the Laplacian and obtain $\hat{\mathcal{G}}(\mathbf{k}, \mathbf{x}_0)$ by Fourier transforming the obtained identity, and finally resum the result. Notice that the integration path depends on the sign of $z - \xi_3$.

2.24 Show that the separation of variables for the Laplacian in spherical coordinates leads to the homogeneous equation for the function $U(r)$

$$r^2 \Delta_r U = l(l+1)U, \quad U = Ar^l + \frac{B}{r^{l+1}}, \tag{2.99}$$

Then consider the Green function and show that the procedure leads us to solve the one-dimensional radial Green problem

$$\Delta_r \hat{\mathcal{G}}_{lm} - \frac{l(l+1)\hat{\mathcal{G}}_{lm}}{r^2} = A \frac{\delta(r-\xi)}{r^2}. \tag{2.100}$$

Determine the constant A. By using the method in Appendix A.7, using the result in Eq. (2.99), and finally restricting to solutions that are regular at $r = 0$ and $r \to \infty$, show that

$$\hat{\mathcal{G}}_{lm}(r) = -\frac{A}{2l+1} \frac{r_<^l}{r_>^{l+1}}, \quad r_< \equiv \min(r, \xi), \quad r_> \equiv \max(r, \xi). \tag{2.101}$$

Hints: The radial equation for U is a second-order, linear equidimensional Euler equation that can be solved using standard methods (e.g., Bender and Orszag 1978; Lomen and Mark 1988).

2.25 Reobtain Eq. (2.54) from the homogeneous solution in Eq. (2.49) by using the Gauss theorem. *Hint*: Fix a radius ξ and impose that the jump in the radial component of the field is due to a infinitesimally thin material shell.

2.26 Consider the razor-thin density distribution $\rho = \sigma(\xi, \varphi')\delta(\xi_3)$ and obtain the formula for the potential from Eqs. (2.61) and (2.62). Show that the obtained expression obeys the Gauss theorem. Then show that the same formula can be obtained from the homogeneous solution in Eq. (2.58) by evaluating the gravitational field above and below the disk and imposing the jump condition from the Gauss theorem.

2.27 From Eqs. (2.53) and (2.54), show that the potential of an *axisymmetric system* in spherical coordinates $\rho(r, \vartheta)$ can be written as

$$\begin{cases} \phi(r,\vartheta) = -2\pi G \sum_{l=0}^{\infty} \mathrm{P}_l(\cos\vartheta) \int_0^{\infty} \frac{r_<^l}{r_>^{l+1}} \hat{\rho}_l(\xi)\, \xi^2 d\xi, \\ \hat{\rho}_l(\xi) = \int_0^{\pi} \rho(\xi, \vartheta') \mathrm{P}_l(\cos\vartheta') \sin\vartheta'\, d\vartheta', \end{cases} \tag{2.102}$$

where $r_< = \min(r, \xi)$, $r_> = \max(r, \xi)$, and $\hat{\rho}_l$ is the the *Legendre integral transform* of ρ; then show that in the spherical limit Eq. (2.5) is reobtained. From the same system in cylindrical coordinates $\rho(R, z)$, show that Eqs. (2.61) and (2.62) reduce to

$$\begin{cases} \phi(R,z) = -2\pi G \int_0^\infty \mathrm{J}_0(kR)dk \int_{-\infty}^\infty \mathrm{e}^{-k|z-z'|}\hat{\rho}(k,z')dz', \\ \hat{\rho}(k,z') \equiv \dfrac{\hat{\rho}_0(k,z')}{\sqrt{2\pi}} = \int_0^\infty \rho(\xi,z')\mathrm{J}_0(k\xi)\xi\, d\xi, \end{cases} \tag{2.103}$$

where $\hat{\rho}(k,z')$ is the zero-th-order *Hankel transform* of ρ according to Eq. (A.89) and $\hat{\rho}_0(k,z')$ is the zero-th-order Fourier–Hankel transform of ρ.

2.28 *Axisymmetric, razor-thin disks* are idealized but very important astrophysical configurations. By using Eqs. (1.18) and (2.102), show that in spherical coordinates, for a razor-thin disk with surface density $\Sigma(r)$ in the equatorial plane,

$$\begin{cases} \hat{\rho}_{2l}(\xi) = \dfrac{\Sigma(\xi)\mathrm{P}_{2l}(0)}{\xi}, \quad \hat{\rho}_{2l+1}(\xi) = 0, \\ \phi(r,\vartheta) = -2\pi G \displaystyle\sum_{l=0}^{\infty} \mathrm{P}_{2l}(\cos\vartheta)\mathrm{P}_{2l}(0) \int_0^\infty \frac{r_<^{2l}}{r_>^{2l+1}}\Sigma(\xi)\xi\, d\xi, \end{cases} \tag{2.104}$$

where $r_< = \min(a,r)$ and $r_> = \max(a,r)$; then specialize the formula for the potential $\phi(r,0)$. For the same disk, show that in cylindrical coordinates Eq. (2.103) reduces to

$$\begin{cases} \hat{\rho}(k,z') = \delta(z')\hat{\Sigma}(k) = \delta(z') \displaystyle\int_0^\infty \Sigma(\xi)\mathrm{J}_0(k\xi)\xi\, d\xi, \\ \phi(R,z) = -2\pi G \displaystyle\int_0^\infty \mathrm{e}^{-k|z|}\hat{\Sigma}(k)\mathrm{J}_0(kR)dk, \end{cases} \tag{2.105}$$

and that particularly in the equatorial plane

$$\phi(R,0) = -2\pi G \int_0^\infty \hat{\Sigma}(k)\mathrm{J}_0(kR)dk = -4G \int_0^\infty \Sigma(\xi)\xi\, \mathbf{K}\left(\frac{r_<}{r_>}\right)\frac{d\xi}{r_>}, \tag{2.106}$$

where $r_< = \min(R,\xi)$, $r_> = \max(R,\xi)$, and $\mathbf{K}$ is the complete elliptic integral of the first kind in Eq. (A.65). *Hints*: Use Eq. (A.78) to prove that in Eq. (2.104) only even terms appear. Equation (2.106) is established from Eq. (2.105) by setting $z = 0$, inverting the order of integration, and finally using eqs. (6.512.1) and (8.113.1) in Gradshteyn et al. (2007) to prove that

$$\int_0^\infty \mathrm{J}_0(ka)\mathrm{J}_0(kb)dk = \frac{2}{\pi a}\mathbf{K}\left(\frac{b}{a}\right), \quad a > b. \tag{2.107}$$

2.29 By using Eqs. (1.17) and (2.102), show that for a *homogeneous ring* of radius a and total mass M, in spherical coordinates

$$\begin{cases} \hat{\rho}_{2l}(\xi) = M\dfrac{\delta(\xi - a)\mathrm{P}_{2l}(0)}{2\pi\xi^2}, \quad \hat{\rho}_{2l+1}(\xi) = 0, \\ \phi(r,\vartheta) = -GM \displaystyle\sum_{l=0}^{\infty} \mathrm{P}_{2l}(\cos\vartheta)\mathrm{P}_{2l}(0)\frac{r_<^{2l}}{r_>^{2l+1}}, \end{cases} \tag{2.108}$$

where $r_< = \min(a,r)$ and $r_> = \max(a,r)$. For the same ring, show that in cylindrical coordinates Eq. (2.103) reduces to

$$\begin{cases} \hat{\rho}(k,z') = M\dfrac{\delta(z')\mathrm{J}_0(ka)}{2\pi}, \\ \phi(R,z) = -GM\displaystyle\int_0^\infty \mathrm{e}^{-k|z|}\mathrm{J}_0(ka)\mathrm{J}_0(kR)dk. \end{cases} \tag{2.109}$$

Alternatively, obtain the same expressions by using Exercise 2.28 and $\Sigma_{\rm ring}$ in Eq. (1.19).

2.30 When observed face on, the stellar (surface) density distribution of disk galaxies can be qualitatively represented as (Freeman 1970)

$$\Sigma(R) = \Sigma_0 \mathrm{e}^{-\tilde{R}}, \quad \tilde{R} = R/R_{\rm d}, \tag{2.110}$$

where $R_{\rm d}$ is the disk scale length and $R_{\rm h} \simeq 1.6784R_{\rm d}$ is the half-mass radius.[12] Calculate the total mass of the disk and then the quadrupole tensor, showing that

$$M_{\rm d} = 2\pi\Sigma_0 R_{\rm d}^2, \quad \mathcal{Q}_{11} = \mathcal{Q}_{22} = 3R_{\rm d}^2, \quad \mathcal{Q}_{33} = -6R_{\rm d}^2. \tag{2.112}$$

Evaluate the potential in cylindrical coordinates from Eq. (2.105) and show that in the equatorial plane

$$\phi(R,0) = -\pi G\Sigma_0 R_{\rm d}\tilde{R}\left[\mathrm{I}_0\left(\frac{\tilde{R}}{2}\right)\mathrm{K}_1\left(\frac{\tilde{R}}{2}\right) - \mathrm{I}_1\left(\frac{\tilde{R}}{2}\right)\mathrm{K}_0\left(\frac{\tilde{R}}{2}\right)\right], \tag{2.113}$$

where I_m and K_m are the modified Bessel functions in Appendix A.2.4. Repeat the exercise in spherical coordinates by using Eqs. (2.104) and (A.51). *Hints*: Use Gradshteyn et al. (2007). First, from eq. (6.621.1), prove that

$$\hat{\Sigma}(k) = \frac{\Sigma_0 R_{\rm d}^2}{(1+\lambda^2)^{3/2}}, \quad \lambda = kR_{\rm d}, \tag{2.114}$$

and insert $\hat{\Sigma}(k)$ in the second identity of Eq. (2.105) with $z=0$. Consider eq. (6.552.1), differentiate both sides of the identity with respect to the parameter a, and set $a = 1$; the left-hand side is the desired integral and the right-hand side is reduced to its final form with the aid of Eqs. (A.92)–(A.94). Notice that Eq. (2.113) can be obtained in a different, clever way, as is shown in Eq. (2.164a) in Binney and Tremaine (2008).

[12] The half-mass radius can be expressed (do it!) in terms of the *Lambert-Euler W* function, a multivalued complex function defined implicitly as

$$W\mathrm{e}^W = z, \tag{2.111}$$

(e.g., Corless et al. 1996).

2.31 Mestel (1963) introduced a *truncated* disk with the surface density in cylindrical cordinates

$$\Sigma(R) = \frac{\Sigma_d R_d}{R}\theta\left(1 - \frac{R}{R_t}\right), \quad M_d = 2\pi\Sigma_d R_d R_t, \tag{2.115}$$

where Σ_d and R_d are the surface density and length scales, respectively, $\theta(x)$ is the Heaviside function in Eq. (A.99), R_t is the disk truncation radius, and M_d is the total mass. When $R_t \to \infty$, we obtain the so-called *untruncated* Mestel disk. By using Eq. (2.104), obtain the potential in the equatorial plane and show that

$$\phi(r,0) = -2\pi G\Sigma_d R_d \times \begin{cases} \dfrac{R_t}{r} + \displaystyle\sum_{l=1}^{\infty} \frac{[P_{2l}(0)]^2}{2l+1}\left(\frac{R_t}{r}\right)^{2l+1}, \\ \ln\dfrac{R_t}{r} + 1 + \displaystyle\sum_{l=1}^{\infty} \frac{[P_{2l}(0)]^2}{2l}\left[\frac{4l+1}{2l+1} - \left(\frac{r}{R_t}\right)^{2l}\right], \end{cases} \tag{2.116}$$

where the first expression holds for $r \geq R_t$ and the second holds for $0 \leq r \leq R_t$ (discuss the convergence of the series). Show that for $r \to \infty$ and fixed R_t, the expected expression of the monopole potential is obtained. Finally, discuss the behavior of the potential for the untruncated case ($R_t \to \infty$). *Hint*: For the treatment of the untruncated case, reread the discussion after Eq. (2.4) and use the second part of Eq. (2.116).

2.32 Due to the conceptual importance of the Mestel disk, we now obtain the closed form of the potential in the disk by using cylindrical coordinates, and we show the equivalence with the expression in Eq. (2.116). Show that Eq. (2.106) for the Mestel disk becomes

$$\phi(R,0) = -4G\Sigma_d R_d \int_0^{\delta} \mathbf{K}\left(\frac{\tilde{r_<}}{\tilde{r_>}}\right)\frac{d\tilde{\xi}}{\tilde{r_>}}, \tag{2.117}$$

where $\delta = R_t/R_d$, $\tilde{r_<} = \min(\tilde{R},\tilde{\xi})$, $\tilde{r_>} = \max(\tilde{R},\tilde{\xi})$, $\tilde{\xi} = \xi/R_d$, and $\tilde{R} = R/R_d$. Verify the equivalence with Eq. (2.116). *Hints*: The equivalence of Eqs. (2.116) and (2.117) is established thanks to the identity

$$\mathbf{K}(k) = \frac{\pi}{2}\sum_{l=0}^{\infty}[P_{2l}(0)]^2 k^{2l}, \quad |k| < 1, \tag{2.118}$$

proved by series expansion of $\mathbf{K}(k)$ in Eq. (A.65) for $|k| < 1$, according to Eqs. (A.49) and (A.50). Term-by-term integration, the use of Eqs. (A.53), (A.54), and (A.46) in this order, and finally comparison with Eq. (A.78) prove the result. Notice that the function

$$\hat{\Sigma}(k) = \Sigma_0 R_d R_t \left\{J_0(\lambda) + \frac{\pi}{2}[J_1(\lambda)\mathbf{H}_0(\lambda) - J_0(\lambda)\mathbf{H}_1(\lambda)]\right\}, \quad \lambda = kR_t, \tag{2.119}$$

which is not used to obtain Eq. (2.117), is expressed in terms of *Struve functions* $\mathbf{H}_\nu$ by using eq. (6.511.6) in Gradshteyn et al. (2007).

2.33 Show that the potential of an axisymmetric system $\rho(R,z)$ in cylindrical coordinates can be written in terms of a complete elliptic integral of the first kind $\mathbf{K}$ in Eq. (A.65) as

$$\phi(R,z) = -4G\int_0^\infty \xi d\xi \int_{-\infty}^\infty \frac{\rho(\xi,z')dz'}{\sqrt{(R+\xi)^2+(\Delta z)^2}}\mathbf{K}\left[\sqrt{\frac{4R\xi}{(R+\xi)^2+(\Delta z)^2}}\right], \tag{2.120}$$

where $\Delta z \equiv z - z'$ (e.g., see Binney and Tremaine 2008); note that, at variance with the functions appearing in orthogonal expansions, the integrand is positive everywhere. Specialize Eq. (2.120) to the equatorial plane of the razor-thin disk in Eq. (1.18) and to a generic point of space for the ring in Eq. (1.17), and show that

$$\phi(R,z) = -\frac{2GM}{\pi\sqrt{(R+a)^2+z^2}}\mathbf{K}\left[\sqrt{\frac{4Ra}{(R+a)^2+z^2}}\right]. \tag{2.121}$$

Compute the radial force of the ring as a function of R in the $z = 0$ plane (see also Exercises 1.9 and 2.29). *Hints*: Due to rotational symmetry, compute ϕ at $\mathbf{x} = (R,0,z)$ and reduce integration over φ' to the first quadrant with the duplication formula $\cos\varphi' = 2\cos^2(\varphi'/2) - 1$. Notice that the *complete* elliptic integrals in Legendre form can be expressed indifferently by using sin or cos in the integrand.

2.34 In a nonrelativistic formulation, the field equation of modified Newtonian dynamics (the so-called MOND, a proposed alternative to dark matter; e.g., see Bekenstein and Milgrom 1984; Milgrom 1983; see also Sanders 2010) is

$$\mathrm{div}\left[\mu\left(\frac{\|\nabla\phi_\mathrm{M}\|}{a_0}\right)\nabla\phi_\mathrm{M}\right] = 4\pi G\rho, \quad \mu(x) = \frac{x}{\sqrt{1+x^2}}, \tag{2.122}$$

where a_0 is a characteristic acceleration. In the *strong* (i.e., Newtonian) regime ($x \gg 1$, $\mu \sim 1$), Eq. (2.122) reduces to the Poisson equation, while in the *deep* MOND regime ($x \ll 1$, $\mu \sim x$)

$$\mathrm{div}\,(\|\nabla\phi_\mathrm{M}\|\nabla\phi_\mathrm{M}) = 4\pi G a_0\rho, \tag{2.123}$$

where the left-hand side is a *p-Laplacian*. Show that the MOND potential ϕ_M and the Newtonian potential ϕ of a given barionic distribution, are related as

$$\mu\left(\frac{\|\nabla\phi_\mathrm{M}\|}{a_0}\right)\nabla\phi_\mathrm{M} = \nabla\phi + \mathbf{h}, \quad \mathbf{h} \equiv \nabla\wedge\mathbf{A}_\mathrm{M}, \tag{2.124}$$

where $\mathbf{A}_\mathrm{M}$ is a vector potential, and that in the spherical case the solenoidal field $\mathbf{h}$ can be fixed to zero. Deduce that in spherical systems the MOND and Newtonian fields are related as

$$\|\nabla\phi_{\rm M}\| = \sqrt{a_0\|\nabla\phi\|} \Longrightarrow \nabla\phi_{\rm M} = \frac{\sqrt{Ga_0 M(r)}}{r}\mathbf{e}_r, \tag{2.125}$$

where $M(r)$ is the mass of the distribution inside radius r. Conclude that the MOND potential for a point mass is logarithmic and that the rotation curve at large radii for finite-mass systems stays flat without need of a dark matter halo. For an application of homeoidal expansion to the construction of analytical nontrivial axisymmetric and triaxial MOND density–potential pairs, see Ciotti et al. (2006).

2.35 Show that for $\alpha < 3$ the *Riesz potential* produced by the density distribution $\rho(\mathbf{x})$ with the additive force law in Eq. (1.13) is

$$\phi(\mathbf{x}) - \phi(\mathbf{x}_0) = \begin{cases} -\dfrac{G}{\alpha - 1}\displaystyle\int_{\Re^3}\left(\frac{1}{\|\mathbf{x}-\mathbf{y}\|^{\alpha-1}} - \frac{1}{\|\mathbf{x}_0-\mathbf{y}\|^{\alpha-1}}\right)\rho(\mathbf{y})d^3\mathbf{y}, \\ G\displaystyle\int_{\Re^3}\ln\frac{\|\mathbf{x}-\mathbf{y}\|}{\|\mathbf{x}_0-\mathbf{y}\|}\rho(\mathbf{y})d^3\mathbf{y}, \qquad (\alpha = 1). \end{cases} \tag{2.126}$$

Then show that for the spherical homogeneous shell in Eq. (1.4)

$$\phi(r) = -\frac{GM}{2Rr} \times \begin{cases} \dfrac{r_+^{3-\alpha} - r_-^{3-\alpha}}{(\alpha-1)(3-\alpha)}, \\ Rr + \dfrac{r_-^2\ln(r_-/a) - r_+^2\ln(r_+/a)}{2}, \qquad (\alpha = 1), \end{cases} \tag{2.127}$$

where $r_+ \equiv r + R$ and $r_- \equiv |r - R|$ (see Exercise 1.2; see also Di Cintio and Ciotti 2011). Finally, discuss the case for $r = R$ and show by differentiation that Eq. (1.14) is recovered.

3
Tidal Fields

In astronomy in general, and in the study of stellar systems in particular, one is often led to consider the effects of an "external" gravitational field on a body of some spatial extension: examples are satellites around planets, binary stars, open and globular clusters in galaxies, and galaxies in clusters of galaxies. The general problem can be mathematically very difficult; however, when the extension of the body of interest is small compared to the characteristic length scale of the external gravitational field (i.e., when the system is in the *tidal regime*), the problem becomes more tractable. In this chapter, we provide the basic ideas and tools that can be used in stellar dynamics when dealing with tidal fields. Among other things, we will find that tidal fields are not always expansive (as in the familiar case of the Earth–Moon system), as they can also be compressive (e.g., for stellar systems inside galaxies or for galaxies in galaxy clusters).

3.1 The Tidal Potential and the Tidal Field

The basic idea behind the concept of a tidal field experienced by an astronomical body of finite extension in an external[1] gravitational field (e.g., a globular cluster in the field of the host galaxy) involves performing a Taylor expansion of the external gravitational potential at some place near or inside the object of interest and then retaining just a few terms in the expansion. As we will see, the second order is the optimal truncation of the potential: a higher-order expansion would be more precise, but the second-order expansion of the potential leads to *linear* tidal forces, with all of the mathematical advantages of a linear force field. We therefore start by considering the external potential $\phi(\mathbf{x})$ produced by some external density distribution $\rho(\mathbf{x})$ and calculated from some of the techniques presented in Chapter 2.

[1] "External" should not be necessarily interpreted in a geometric sense, it simply means that the "external" system is not the direct object of our investigation. For example, a galaxy hosting a globular cluster will be called external if we are interested in the dynamics of the globular cluster.

Let $\mathbf{x}_0$ be the position of interest in some inertial reference system S_0 and let $\mathbf{x} = \mathbf{x}_0 + \boldsymbol{\xi}$. By Taylor expansion, up to second-order inclusive, we have

$$\phi(\mathbf{x}_0 + \boldsymbol{\xi}) = \phi(\mathbf{x}_0) + < \nabla\phi(\mathbf{x})\Big|_{\mathbf{x}=\mathbf{x}_0}, \boldsymbol{\xi} > + \frac{\langle \mathcal{T}(\mathbf{x}_0)\boldsymbol{\xi}, \boldsymbol{\xi}\rangle}{2} + \mathcal{O}\left(\|\boldsymbol{\xi}\|^3\right), \tag{3.1}$$

where the dependence of the various quantities on $\mathbf{x}_0$ is made explicit for clarity. The second-order *tidal tensor* $\mathcal{T}$ is manifestly symmetric, and in Cartesian coordinates

$$\mathcal{T}_{ij}(\mathbf{x}_0) = \frac{\partial^2\phi}{\partial x_i \partial x_j}\Big|_{\mathbf{x}=\mathbf{x}_0}, \tag{3.2}$$

for $i, j = 1, 2, 3$. The quadratic form[2]

$$\phi_{\mathcal{T}}(\boldsymbol{\xi}) = \frac{\langle \mathcal{T}\boldsymbol{\xi}, \boldsymbol{\xi}\rangle}{2}, \tag{3.3}$$

appearing in Eq. (3.1), is the (second-order) *tidal potential*: $\phi_{\mathcal{T}}$ is naturally associated with quadric isopotential surfaces, and due to its symmetry $\mathcal{T}$ can be diagonalized with some appropriate rotation of the reference system.

Taking the gradient of ϕ in Eq. (3.1) with respect to $\boldsymbol{\xi}$, we obtain the linearization of the external gravitational field for points near $\mathbf{x}_0$, with

$$\mathbf{g}(\mathbf{x}) = -\nabla\phi \sim \mathbf{g}(\mathbf{x}_0) - \mathcal{T}(\mathbf{x}_0)\boldsymbol{\xi}, \quad \mathbf{g}(\mathbf{x}_0) = -\nabla\phi(\mathbf{x})\Big|_{\mathbf{x}=\mathbf{x}_0}, \tag{3.4}$$

where in particular $-\mathcal{T}\boldsymbol{\xi}$ is the *tidal field* along the displacement $\boldsymbol{\xi}$: notice that the tidal field at fixed $\mathbf{x}_0$ depends on the direction and length of $\boldsymbol{\xi}$. Also notice that from Eq. (3.2) the (invariant) *trace* of $\mathcal{T}$ is none other than the Laplacian of the external potential ϕ, so that the Poisson equation (2.6) dictates that at the generic point $\mathbf{x}$

$$\mathrm{Tr}\,\mathcal{T}(\mathbf{x}) = 4\pi G\rho(\mathbf{x}). \tag{3.5}$$

It follows that $\mathrm{Tr}\,\mathcal{T} = 0$ where $\rho = 0$, so at these points the eigenvalues of the $\mathcal{T}$ produced by ρ cannot all have the same sign, and as a consequence the tidal field is *expansive* along some direction and *compressive* along some other direction. However, for points with $\rho \neq 0$ (i.e., inside a galaxy or a galaxy cluster), the three eigenvalues *may* all be positive, the isopotential quadric is an ellipsoid, and the tidal field is compressive along all directions. It follows that the common expectation of an extensive tidal field is actually limited to quite special cases (e.g., a satellite orbiting a planet or a binary star), and also in these cases only along particular directions.

3.1.1 The Tidal Field Produced by Spherical Systems

In order to gain some confidence with the geometric properties of $\mathcal{T}$, we now study the tidal fields produced by spherical and ellipsoidal systems, two of the most common idealizations encountered in stellar dynamics. We begin by considering the case of a spherical system

[2] Note that $\mathcal{T}$ in Ciotti and Giampieri (1997) is defined as the opposite of $\mathcal{T}$ in Eq. (3.1), so that the quantities $\mathcal{T}_i$ in Eq. (3.10) are the same therein and in this book. See also Ciotti and Dutta (1994).

with density profile $\rho(r)$, cumulative mass profile $M(r)$, and potential $\phi(r)$. It is a simple exercise (do it!) to show that at distance $r = \|\mathbf{x}\|$ from the center

$$\mathcal{T}_{ij}(r) = \frac{4\pi G\bar{\rho}(r)}{3}\left[\delta_{ij} + 3x_i x_j \frac{q(r)-1}{r^2}\right], \tag{3.6}$$

where δ_{ij} is the Kronecker symbol. Moreover,

$$q(r) \equiv \frac{\rho(r)}{\bar{\rho}(r)}, \quad \bar{\rho}(r) \equiv \frac{3M(r)}{4\pi r^3} = \frac{3\Omega^2(r)}{4\pi G}, \tag{3.7}$$

where $\bar{\rho}(r)$ is the average density, within a sphere of radius r, of the body producing the field, and the second identity relates the average density at r and the angular velocity $\Omega(r)$ of circular orbits at r (see Chapter 5). In the exercises, the explicit form of the function $q(r)$ associated with some of the most widely used models of spherical stellar systems is derived; here, we simply note that outside a spherical body (and specifically for the case of the tidal field produced by a point mass), $q(r) = 0$, and that for monotonically decreasing density profiles (the situation in almost all astrophysical systems), $q(r) \leq 1$.

Without loss of generality, we can obtain a simple geometric interpretation of $\mathcal{T}$ considering an arbitrary point $\mathbf{x}_0 = (r,0,0)$, so that from Eqs. (3.6) and (3.7)

$$\mathcal{T}(r) = \frac{4\pi G\bar{\rho}(r)}{3}\begin{pmatrix} 3q-2 & 0 & 0 \\ 0 & 1 & 0 \\ 0 & 0 & 1 \end{pmatrix} = \Omega^2(r)\begin{pmatrix} 3q-2 & 0 & 0 \\ 0 & 1 & 0 \\ 0 & 0 & 1 \end{pmatrix}. \tag{3.8}$$

Notice how in the present case Eq. (3.5) is of immediately verified. From Eq. (3.3), the tidal potential at r is given by

$$\phi_{\mathcal{T}}(\boldsymbol{\xi}) = \Omega^2(r)\frac{[3q(r)-2]\xi_1^2 + \xi_2^2 + \xi_3^2}{2}, \tag{3.9}$$

so that the tidal isopotential surfaces are rotationally symmetric about the ξ_1 axis (i.e., the radial direction) and reflection symmetric with respect to the $\xi_1 = 0$ plane (i.e., the plane tangent to the spherical surface of radius r). For $2/3 < q < 1$ the surfaces of constant $\phi_{\mathcal{T}}$ are prolate ellipsoids with the major axis along the radial direction, and the resulting tidal field is compressive all around $\mathbf{x}$. The limit case $q = 1$ corresponds to spherical $\phi_{\mathcal{T}}$, while for $q = 2/3$ the equipotentials are cylinders with axes along the radial direction ξ_1, and the tidal field is (cylindrically) radially compressive toward these axes, with no tidal forces along ξ_1. Finally, for $0 \leq q < 2/3$ the coefficient of ξ_1 is negative; the isopotential surfaces are two-sheet and one-sheet similar hyperboloids of revolution around the radial direction for negative and positive $\phi_{\mathcal{T}}$ values, respectively. The two families are separated by the $\phi_{\mathcal{T}} = 0$ cone, the opening angle of which gets smaller as q increases. The associated tidal field is expansive along ξ_1 and radially compressive in the perpendicular plane (ξ_2, ξ_3), a familiar situation occuring, for example, in the Earth–Moon system.

We conclude this preliminary section by noticing an important and quite general property of $q(r)$ as a function of the density profile $\rho(r)$. In fact, the cases in Eqs. (3.58) and (3.59) show that the transition from the region where $q(r) < 2/3$ to $q(r) > 2/3$ (corresponding to a transition from a radially expansive to a radially compressive tidal field) will be experienced by a stellar system along its orbit in a host system when entering the shallow density "core" of the external density distribution $\rho(r)$. For example, a globular cluster in radial orbit inside an elliptical galaxy or crossing the galactic disk of a spiral galaxy will be subjected to repeated compression and expansion phases, with significant structural and dynamical consequences (for a detailed discussion of the dynamical phenomena involved, see, e.g., Binney and Tremaine 2008; Spitzer 1987).

3.1.2 The Tidal Field Produced by Triaxial Systems

We now extend the previous discussion to the case of the tidal field produced by an ellipsoidal density distribution (e.g., a triaxial elliptical galaxy or a triaxial dark matter halo) at a generic point of space. We assume that the density is stratified on homeoidal surfaces $\rho = \rho(m)$ as in Eq. (2.28), with the origin at the center of the distribution and axes parallel to the axes of the reference system. Moreover, we adopt, without loss of generality, the same convention as in Eq. (2.27) (i.e., $q_z = a_3/a_1 \leq q_y = a_2/a_1 \leq 1$). The associated potential is given in Eqs. (2.21) and (2.22) and the general expression of $\mathcal{T}_{ij}$ at a generic point $\mathbf{x}$ is given in Eq. (3.60): it follows that at the center of a sufficiently regular density distribution

$$\mathcal{T}_{ii}(0) = 2\pi G\rho(0)w_i \equiv -T_i, \quad \phi_{\mathcal{T}} = -\frac{T_1\xi_1^2 + T_2\xi_2^2 + T_3\xi_3^2}{2}, \tag{3.10}$$

the dimensionless factors w_i are given by

$$w_i = a_1a_2a_3 \int_0^\infty \frac{d\tau}{(a_i^2+\tau)\sqrt{\Delta(\tau)}}, \tag{3.11}$$

and finally $\mathcal{T}_{ij}(0) = 0$ for $i \neq j$. Note that with T_i we indicate the *opposite* of the element $\mathcal{T}_{ii}$ when $\mathcal{T}$ is in diagonal form, so that from Eqs. (3.4) and (3.10), the tidal field at $\boldsymbol{\xi}$ near the center is $(T_1\xi_1, T_2\xi_2, T_3\xi_3)$. Note also that the specific form of the density distribution does not enter Eq. (3.11), and that with the adopted convention of $a_1 \geq a_2 \geq a_3$, we have $w_1 \leq w_2 \leq w_3$ and $0 \geq T_1 \geq T_2 \geq T_3$, so the maximum (in absolute value) compressive force is along the short-axis direction of the density $\rho(m)$. Finally, again from Eq. (3.60), we conclude that from a constant-density ellipsoid the tidal field obtained above actually coincides for all points inside the ellipsoid with the gravitational field produced by the ellipsoid, because from Chapter 2 we know that the potential inside a constant-density ellipsoid is a quadratic function of coordinates.

In applications, it is quite useful to have the explicit expressions of the w_i functions, and with some work it can be proved that

$$\begin{cases} w_1 = 2q_yq_z\dfrac{F(\varphi,k) - E(\varphi,k)}{k^2\sin^3\varphi}, \\[2ex] w_2 = 2q_yq_z\dfrac{E(\varphi,k) - k'^2F(\varphi,k) - k^2(q_z/q_y)\sin\varphi}{k^2k'^2\sin^3\varphi}, \\[2ex] w_3 = 2q_yq_z\dfrac{(q_y/q_z)\sin\varphi - E(\varphi,k)}{k'^2\sin^3\varphi}, \end{cases} \tag{3.12}$$

where $\varphi \equiv \arcsin\sqrt{1-q_z^2}$, $k^2 \equiv (1-q_y^2)/(1-q_z^2)$, $k'^2 = 1-k^2$, and finally $F(\varphi,k)$ and $E(\varphi,k)$ are the elliptic integrals of the first and second kind in Eqs. (A.62) and (A.63), respectively (see also Byrd and Friedman 1971; Gradshteyn et al. 2007). Finally, as is shown in Eqs. (3.62) and (3.63), for axisymmetric systems the functions w_i reduce to a combination of elementary functions (see also Binney and Tremaine 2008). In Chapters 10 and 13, we will again encounter the functions w_i.

Admittedly, the formulae in Eq. (3.12) are not very illuminating; however, a qualitative understanding of the dependence of the functions w_i on the flattenings can be achieved by considering the series expansion near the spherical limit. By setting $q_y = 1-\epsilon$ and $q_z = 1-\eta$ (see Section 2.2.1), with $0 \leq \epsilon \leq \eta \leq 1$, it is a simple exercise (do it!) to show that the series expansion of Eq. (3.11) reads (see also Exercise 3.10)

$$\begin{cases} w_1 \sim 2(1-\epsilon)(1-\eta)\left(\dfrac{1}{3}+\dfrac{\epsilon+\eta}{5}\right), \\ w_2 \sim 2(1-\epsilon)(1-\eta)\left(\dfrac{1}{3}+\dfrac{3\epsilon+\eta}{5}\right), \\ w_3 \sim 2(1-\epsilon)(1-\eta)\left(\dfrac{1}{3}+\dfrac{\epsilon+3\eta}{5}\right). \end{cases} \tag{3.13}$$

Of course, the same result can also be established (with substantially more work) from the series expansion of Eq. (3.12). Note that from the adopted choice of the axis labeling, the oblate case (with the short axis along z) corresponds to $\epsilon = 0$, while the prolate case (with the long axis along x) corresponds to $\epsilon = \eta$.

3.2 Rigid Bodies in Tidal Fields

We now consider a few consequences of astrophysical interest obtained under the (more or less justified) assumption that the stellar system experiencing the external gravitational field behaves as a rigid body, with density distribution ρ_* and total mass M_*, in motion in the tidal field. This approximation may appear unreasonable for a system made of stars, but in fact in certain circumstances the predictions compare quite well with the results of N-body simulations (for a more detailed discussion, see, e.g., Ciotti and Dutta 1994; Ciotti and Giampieri 1997; Muccione and Ciotti 2004; see also D'Ercole et al. 2000 for tidal effects on gas flows in elliptical galaxies).

To set the stage, we consider an inertial reference system S_0 and the time-independent density distribution $\rho(\mathbf{x})$ producing the external field $\mathbf{g} = -\nabla\phi(\mathbf{x})$, as well as the associated tidal field. We assume that the characteristic dimensions of ρ_* are much smaller than the typical length scale over which the external potential changes, and we indicate with $\mathbf{R}_{\rm CM} = \int_{\Re^3} \mathbf{x}\rho_* d^3\mathbf{x}/M_*$ the instantaneous position of the center of mass of ρ_*. From classical mechanics, the acceleration of $\mathbf{R}_{\rm CM}$ is given by (see also Chapter 6)

$$\begin{aligned} M_*\mathbf{A}_{\rm CM} &= \int_{\Re^3} \mathbf{g}(\mathbf{x})\rho_*(\mathbf{x};t)d^3\mathbf{x} \sim M_*\mathbf{g}(\mathbf{x}_0) - \mathcal{T}(\mathbf{x}_0)\int_{\Re^3}(\mathbf{x}-\mathbf{x}_0)\rho_*(\mathbf{x};t)d^3\mathbf{x} \\ &= M_*\left[\mathbf{g}(\mathbf{x}_0) - \mathcal{T}(\mathbf{x}_0)(\mathbf{R}_{\rm CM}-\mathbf{x}_0)\right] = M_*\mathbf{g}(\mathbf{R}_{\rm CM}), \end{aligned} \tag{3.14}$$

where the possible time dependence of ρ_* in S_0 (e.g., due to a change of orientation) has been made explicit. In Eq. (3.14), the first integral gives the full and exact contribution of the external field, the second and third expressions hold according to Eq. (3.4) for the linearized external field around the generic point $\mathbf{x}_0$, and the last expression is finally obtained for $\mathbf{x}_0 = \mathbf{R}_{\rm CM}$. From the first identity in Eq. (3.14), it is clear how $\mathbf{A}_{\rm CM}$ in general depends on the system extension and orientation, so that translations and rotations of extended bodies in an external field may be coupled in a very complicated (i.e., nonlinear) way. However, from the last identity in Eq. (3.14), it also follows that, at the second order in the tidal field, the motion of $\mathbf{R}_{\rm CM}$ is not affected by the shape and orientation of ρ_* when the external potential is expanded around $\mathbf{R}_{\rm CM}$, and so under this approximation we can decouple the motion of $\mathbf{R}_{\rm CM}$ from the orientation and study the translation of the center of mass of $\mathbf{R}_{\rm CM}$ as the motion of a point mass M_* in the external field $\mathbf{g}$.

The second relevant identity of classical mechanics concerns the angular momentum of the system computed with respect to the reduction point $\mathbf{x}_0$ (see again Chapter 6)

$$\mathbf{J} = \int_{\Re^3} (\mathbf{x} - \mathbf{x}_0) \wedge \mathbf{v}(\mathbf{x}, t)\rho_*(\mathbf{x}; t) d^3\mathbf{x}, \quad \frac{d\mathbf{J}}{dt} = M_* \mathbf{V}_{\rm CM} \wedge \mathbf{v}_0 + \mathbf{N}, \tag{3.15}$$

where $\mathbf{V}_{\rm CM} = d\mathbf{R}_{\rm CM}/dt$ and $\mathbf{v}_0 = d\mathbf{x}_0/dt$ are the velocities of the center of mass of ρ_* and of $\mathbf{x}_0$, $\mathbf{v}$ is the velocity of the distribution ρ_* at each position, and finally

$$\mathbf{N} = \int_{\Re^3} (\mathbf{x} - \mathbf{x}_0) \wedge \mathbf{g}(\mathbf{x})\rho_*(\mathbf{x}; t) d^3\mathbf{x} \sim - \int_{\Re^3} (\mathbf{x} - \mathbf{x}_0) \wedge [\mathcal{T}(\mathbf{x}_0)(\mathbf{x} - \mathbf{x}_0)]\rho_*(\mathbf{x}; t) d^3\mathbf{x}, \tag{3.16}$$

which is the torque[3] due to the external field in its exact and tidal expansion, respectively.

Equipped with these important identities, we now consider two astrophysically interesting applications of the tidal field approximation: namely, the study of the motion of a triaxial galaxy at the center of a triaxial cluster and of a triaxial galaxy in circular orbit around/inside a spherically symmetric density distribution.

3.2.1 A Triaxial Galaxy at the Center of a Triaxial Cluster

In this section, we consider the simple case of a triaxial galaxy, modeled as a rigid body of density profile ρ_* and total mass M_*, with its center of mass $\mathbf{R}_{\rm CM}$ at rest at the equilibrium point $\mathbf{x}_0 = 0$ of an external potential (e.g., at the center of a triaxial galaxy cluster that is sufficiently regular near the origin; that the origin is a point of equilibrium is proved immediately by Eq. (2.72)). Therefore, Eq. (3.14) shows that $\mathbf{R}_{\rm CM}$ will remain fixed in the origin of S_0, and only rotational motions of ρ_* around $\mathbf{x}_0 = 0$ will be possible. Finally, near the equilibrium position we have $\mathbf{x} = \boldsymbol{\xi}$, and Eqs. (3.1) and (3.4) reduce to

$$\phi(\mathbf{x}) \sim \phi(0) + \frac{\langle \mathcal{T}(0)\mathbf{x}, \mathbf{x}\rangle}{2}, \quad \mathbf{g}(\mathbf{x}) \sim -\mathcal{T}(0)\mathbf{x}, \tag{3.17}$$

where without loss of generality we can assume that $\mathcal{T}$ is diagonal in S_0.

[3] Actually, Eqs. (3.14)–(3.16) hold for a generic density distribution ρ_*, not necessarily a rigid body, as can be proved from the general equations of fluid dynamics; see Eqs. (10.70) and (10.71).

A natural question of obvious astrophysical interest relates to the existence of equilibrium configurations of ρ_* in the tidal field. Mathematically, this is equivalent to searching for the vanishing of the tidal torque in Eq. (3.16), which reads

$$N_i = -\epsilon_{ijk} \int_{\Re^3} x_j \frac{\partial \phi(\mathbf{x})}{\partial x_k} \rho_*(\mathbf{x};t) d^3\mathbf{x} \sim -\epsilon_{ijk} \mathcal{T}_{kl}(0) \int_{\Re^3} x_j x_l \rho_*(\mathbf{x};t) d^3\mathbf{x}, \tag{3.18}$$

where ϵ_{ijk} is the Levi–Civita tensor and the integral in the last expression can be expressed in terms of the tensor I_{ij} appearing in Exercise 2.17. Using Eqs. (2.90) and (3.10), the antisymmetry of ϵ_{ijk}, and the symmetry of the inertia tensor $\Im$ of ρ_*, it is a nice exercise (do it!) to show that

$$N_1 = \Im_{23} \Delta T_{23}, \quad N_2 = \Im_{31} \Delta T_{31}, \quad N_3 = \Im_{12} \Delta T_{12}, \tag{3.19}$$

where $\Delta T_{ij} \equiv T_i - T_j$ is a manifestly antisymmetric tensor; if the external potential and/or ρ_* are spherically symmetric (and so $\Delta T_{ij} = 0$ and/or $\Im_{12} = \Im_{13} = \Im_{23} = 0$), then no net torque acts on the stellar system and all of the orientations of ρ_* are at equilibrium. What happens in the more realistic case of a nonspherical ρ_* in a nonspherical external potential, when $\Delta T_{ij} \neq 0$ for $i \neq j$? From Eq. (3.19), it follows that ρ_* is in an equilibrium as a rigid body if and only if $\Im_{ij} = 0$ for $i \neq j$ (i.e., if and only if the principal axes of ρ_* are aligned with the axes of the cluster distribution (the axes of S_0)). In practice, a "rigid" triaxial galaxy (modeled as a rigid body) at the center of a triaxial cluster will be in a configuration of equilibrium when its axes are aligned with the cluster axes.

We can now move to study the stability of the equilibrium positions and, for stable positions, to determine the oscillation frequencies of ρ_*. Let S' be an orthogonal reference system with its axes directed along the axes of inertia of ρ_* and with the origin coincident with the origin of S_0 (and with $\mathbf{R}_{\mathrm{CM}}$). Moreover, let $\mathcal{R}(t)$ be the orthogonal transformation matrix between S' and S_0, so that $\mathbf{x} = \mathcal{R}\mathbf{x}'$, where $\mathbf{x}'$ is a generic position vector in S' (see Appendix A.1.1). In S' the inertia tensor $\Im'$ of ρ_* is diagonal by construction, and we define $I_i \equiv \Im'_{ii}$ (no sum intended); for simplicity, from now on we consider only nondegenerate cases (i.e., $I_1 \neq I_2 \neq I_3$ and $T_1 \neq T_2 \neq T_3$). Instead of using the usual representation of $\mathcal{R}$ in terms of Euler's angles (i.e., the successive combination of the 3-1-3 rotations,[4] where the number indicates the axis of the system around which the specific rotation is performed), it is convenient to adopt the counterclockwise 1-2-3 rotations (e.g., see Ciotti and Giampieri 1997):

$$\mathcal{R} \equiv \mathcal{R}_1(\varphi)\mathcal{R}_2(\vartheta)\mathcal{R}_3(\psi), \tag{3.20}$$

with

$$\mathcal{R}_1(\varphi) = \begin{pmatrix} 1 & 0 & 0 \\ 0 & \cos\varphi & -\sin\varphi \\ 0 & \sin\varphi & \cos\varphi \end{pmatrix}, \tag{3.21}$$

[4] This choice would complicate the treatment due to the indeterminacy of the angles φ and ψ for a null inclination ($\vartheta = 0$). See also Eqs. (11.46)–(11.48) for a different choice of $\mathcal{R}$ as a 3-2-3 rotation.

$$\mathcal{R}_2(\vartheta) = \begin{pmatrix} \cos\vartheta & 0 & \sin\vartheta \\ 0 & 1 & 0 \\ -\sin\vartheta & 0 & \cos\vartheta \end{pmatrix}, \tag{3.22}$$

$$\mathcal{R}_3(\psi) = \begin{pmatrix} \cos\psi & -\sin\psi & 0 \\ \sin\psi & \cos\psi & 0 \\ 0 & 0 & 1 \end{pmatrix}. \tag{3.23}$$

In particular, from Eqs. (3.20) and (A.26), it is easy to prove (do it!) that the angular velocity of ρ_* expressed in the reference system S' is (see Footnote 1 in Appendix A.1)

$$\begin{aligned} \boldsymbol{\omega} &= \boldsymbol{\omega}_\psi + \mathcal{R}_3^{\mathrm{T}}(\psi)[\boldsymbol{\omega}_\vartheta + \mathcal{R}_2^{\mathrm{T}}(\vartheta)\boldsymbol{\omega}_\varphi] \\ &= (\dot\varphi\cos\vartheta\cos\psi + \dot\vartheta\sin\psi, -\dot\varphi\cos\vartheta\sin\psi + \dot\vartheta\cos\psi, \dot\varphi\sin\vartheta + \dot\psi), \end{aligned} \tag{3.24}$$

where $\boldsymbol{\omega}_\varphi = (\dot\varphi, 0, 0)$, $\boldsymbol{\omega}_\vartheta = (0, \dot\vartheta, 0)$, and $\boldsymbol{\omega}_\psi = (0, 0, \dot\psi)$ are the three angular velocities associated with the three rotations and the suffix "T" means "transpose."

It should be obvious that, apart from a renaming of the inertia axes, the equilibrium positions previously determined are given, without loss of generality (see also Exercise 3.7), by $(\varphi, \vartheta, \psi)_{\mathrm{eq}} = (0,0,0)$, and we now study the motion of ρ_* around this configuration. We then construct the Lagrangian $\mathcal{L} = T - U$ of the problem (e.g., Arnold 1978; Goldstein et al. 2000; Landau and Lifshitz 1969) by using the three angles as generalized coordinates. It is not diffcult to show (see Exercise 3.8) that the potential energy of ρ_* in the external potential can be written by using repeated indices as

$$U = \int_{\Re^3} \phi(\mathbf{x})\rho_*(\mathbf{x};t)d^3\mathbf{x} \sim M_*\phi(0) + \pi G\rho(0)\mathrm{Tr}(\Im) + \frac{1}{2}\mathcal{R}^2_{i\mu}(t)T_i I_\mu. \tag{3.25}$$

The first two terms on the right-hand side are constant, and so they do not affect the equations of motion of ρ_*. In the same exercise we also obtain the explicit expression for the kinetic energy T of ρ_*. The linearized potential and kinetic energies for small displacements from the equilibrium position $(\varphi, \vartheta, \psi)_{\mathrm{eq}} = (0,0,0)$ are easily obtained from Exercise 3.9 (do it!) as

$$\begin{cases} T \sim \dfrac{I_1}{2}\dot\varphi^2 + \dfrac{I_2}{2}\dot\vartheta^2 + \dfrac{I_3}{2}\dot\psi^2, \\ U \sim -\dfrac{\Delta T_{32}\Delta I_{32}}{2}\varphi^2 - \dfrac{\Delta T_{31}\Delta I_{31}}{2}\vartheta^2 - \dfrac{\Delta T_{21}\Delta I_{21}}{2}\psi^2, \end{cases} \tag{3.26}$$

where $\Delta I_{ij} \equiv I_i - I_j$ is the antisymmetric tensor built with the principal components of $\Im$ and ΔT_{ij} is defined in Eq. (3.19). Therefore, from the Lagrange–Euler equations $d(\partial\mathcal{L}/\partial\dot q_i)/dt = \partial\mathcal{L}/\partial q_i$, the linearized equations of motion for the galaxy near the equilibrium configurations are given by

$$\ddot\varphi = \frac{\Delta T_{32}\Delta I_{32}}{I_1}\varphi, \quad \ddot\vartheta = \frac{\Delta T_{31}\Delta I_{31}}{I_2}\vartheta, \quad \ddot\psi = \frac{\Delta T_{21}\Delta I_{21}}{I_3}\psi. \tag{3.27}$$

Following Eq. (3.10), let us assume, without loss of generality, that $0 \geq T_1 \geq T_2 \geq T_3$ (as obtained for an ellipsoidal cluster distribution with $a_1 \geq a_2 \geq a_3$; i.e., that ΔT_{32}, ΔT_{31}, and ΔT_{21} are all positive quantities). From Eq. (3.27), stable motions (*librations*) are possible if and only if $I_1 \leq I_2 \leq I_3$ (i.e., according to Eq. (2.91), when the longest axis of ρ_* is aligned with the long axis of the triaxial cluster distribution, the intermediate with the intermediate, and the short with the short). In summary, a stable equilibrium configuration (in the nondegenerate case) is achieved when:

(1) The principal inertia axes of ρ_* are aligned with the principal axes of the external tidal field.
(2) If $T_1 \geq T_2 \geq T_3$, then $I_1 \leq I_2 \leq I_3$.

Near the stable equilibrium positions, the solution of Eq. (3.27), corresponding to maximum angular displacement and zero velocity at $t = 0$, is given by

$$\varphi(t) = \varphi_{\rm M}\cos(2\pi\nu_\varphi t),\ \vartheta(t) = \vartheta_{\rm M}\cos(2\pi\nu_\vartheta t),\ \psi(t) = \psi_{\rm M}\cos(2\pi\nu_\psi t), \tag{3.28}$$

where the libration frequencies and associated periods are given respectively by

$$\begin{cases} \nu_\varphi = \dfrac{1}{2\pi}\sqrt{\dfrac{\Delta T_{23}\Delta I_{32}}{I_1}} = \dfrac{\sqrt{\Delta T_{23}\,(1-v)}}{2\pi\sqrt{u}}, & P_\varphi = \dfrac{1}{\nu_\varphi}, \\ \nu_\vartheta = \dfrac{1}{2\pi}\sqrt{\dfrac{\Delta T_{13}\Delta I_{31}}{I_2}} = \dfrac{\sqrt{\Delta T_{13}\,(1-u)}}{2\pi\sqrt{v}}, & P_\vartheta = \dfrac{1}{\nu_\vartheta}, \\ \nu_\psi = \dfrac{1}{2\pi}\sqrt{\dfrac{\Delta T_{12}\Delta I_{21}}{I_3}} = \dfrac{\sqrt{\Delta T_{12}\,(v-u)}}{2\pi}, & P_\psi = \dfrac{1}{\nu_\psi}, \end{cases} \tag{3.29}$$

and

$$u \equiv \frac{I_1}{I_3};\quad v \equiv \frac{I_2}{I_3}. \tag{3.30}$$

Note that only the ratios of the inertia moments of ρ_* appear in the expressions above, and that the stable equilibrium condition corresponds to the inequality $u < v < 1$; as expected, in the spherical limits, the libration frequencies vanish.

Some insight into the dependence of the libration frequencies on the cluster and galaxy flattenings can be obtained by considering the limit of small flattenings. First, we model the cluster as a triaxial ellipsoidal density distribution with semiaxes $a_1 \geq a_2 \geq a_3$, so that the tidal field is given by Eqs. (3.10) and (3.11) and

$$\Delta T_{ij} = -2\pi G\rho(0)(w_i - w_j), \tag{3.31}$$

where $\rho(0)$ is the cluster's central density. Second, for the galaxy density we assume that $\rho_* = \rho_*(m)$ in the system S', with m given in Eq. (2.28), and semiaxes $\alpha_1 \geq \alpha_2 \geq \alpha_3$, so that

$$\frac{\alpha_2}{\alpha_1} \equiv 1 - \epsilon_*,\quad \frac{\alpha_3}{\alpha_1} \equiv 1 - \eta_*,\quad \epsilon_* \leq \eta_* \leq 1, \tag{3.32}$$

are the corresponding ellipticities. In S', the galaxy inertia tensor is diagonal, with the explicit expression for $I_i \equiv \Im_{ii}$ given in Eq. (2.91) so that $I_1 \leq I_2 \leq I_3$; notice that the ratios u and v *do not depend on the specific density profile* $\rho_*(m)$*, but only on the ellipticities*, because the shape function h cancels out. In Exercise 3.10, the student is guided toward obtaining the leading term of Eq. (3.29) in the limit of small cluster and galaxy flattenings. The resulting formula will be used in Chapter 10 where, with the aid of the virial theorem, we will be able to estimate the libration frequencies, and we will discover that they can be of the same order of magnitude as the orbital periods of stars in the external regions of elliptical galaxies. In Section 3.3, we briefly discuss the possibility that associated resonances significantly affect the stellar orbits at the galaxy outskirts, leading to what can be called *collisionless evaporation.*

3.2.2 A Triaxial Galaxy in Circular Orbit in a Spherical Cluster

We now consider the more complicated case of a triaxial galaxy (again modeled as a rigid body of density distribution ρ_*) with the center of mass $\mathbf{R}_{\rm CM}$ on a circular orbit inside a spherically symmetric potential produced by a density distribution $\rho(r)$. This situation could represent the idealized case of an elliptical galaxy orbiting inside a galaxy cluster. Note that the assumption that $\mathbf{R}_{\rm CM}$ remains on the initial circular orbit independently of the orientation of ρ_* is justified by Eq. (3.14); of course, the rotation of $\mathbf{R}_{\rm CM}$ does affect the equilibrium positions and oscillation times of the galaxy near stable equilibria.

To set the stage, we assume without loss of generality that the orbit of $\mathbf{R}_{\rm CM}$ is in the (x, y) plane of the inertial system S_0 centered on $\rho(r)$, as in Figure 3.1. Therefore, in S_0, the galaxy center of mass $\mathbf{R}_{\rm CM}$ rotates with constant angular velocity

$$\boldsymbol{\Omega} = (0, 0, \Omega), \qquad \Omega^2 = \left.\frac{GM(r)}{r^3}\right|_{r=\mathbf{R}_{\rm CM}}, \tag{3.33}$$

where $M(r)$ is the cumulative mass of the cluster density distribution inside the sphere of radius r. As is shown in Figure 3.1, the natural frame of reference for our problem is the noninertial reference system $S_{\rm CM}$, with its origin placed at $\mathbf{R}_{\rm CM}$, the $\mathbf{e}_1 = \mathbf{R}_{\rm CM}/\|\mathbf{R}_{\rm CM}\|$ axis directed along the cluster radial direction, and $\mathbf{e}_3$ parallel to angular velocity (i.e., the z-axis of S_0); the direction of $\mathbf{e}_2$ is finally fixed so that $(\mathbf{e}_1, \mathbf{e}_2, \mathbf{e}_3)$ are positively oriented.

We indicate with $\mathbf{x}$ and $\mathbf{y}$ the generic position vectors in S_0 and $S_{\rm CM}$, respectively, and from Eq. (A.25)

$$\mathbf{x} = \mathbf{R}_{\rm CM} + \mathcal{C}\mathbf{y}, \quad \mathcal{C} = \begin{pmatrix} \cos\Omega t & -\sin\Omega t & 0 \\ \sin\Omega t & \cos\Omega t & 0 \\ 0 & 0 & 1 \end{pmatrix}, \tag{3.34}$$

where without loss of generality we assume that at the arbitrary time $t = 0$, $\mathbf{R}_{\rm CM}(0)$ is aligned with the x-axis of S_0, and so S_0 and $S_{\rm CM}$ are coincident with $\mathcal{C}(0) = I$. It follows that in $S_{\rm CM}$

$$\boldsymbol{\omega}_{\rm CM} = \mathcal{C}^{\rm T}\boldsymbol{\Omega} = \boldsymbol{\Omega}, \quad \mathcal{T}_{\rm CM} = \mathcal{C}^{\rm T}\mathcal{T}\mathcal{C}, \tag{3.35}$$

where $\boldsymbol{\omega}_{\rm CM}$ is the angular velocity of $S_{\rm CM}$ as observed from $S_{\rm CM}$, $\mathcal{T}_{\rm CM}$ is the cluster tidal tensor at $\mathbf{R}_{\rm CM}$ expressed in the system $S_{\rm CM}$ and so given by Eq. (3.8), and finally $\mathcal{T}$ is the tidal tensor in the system S_0.

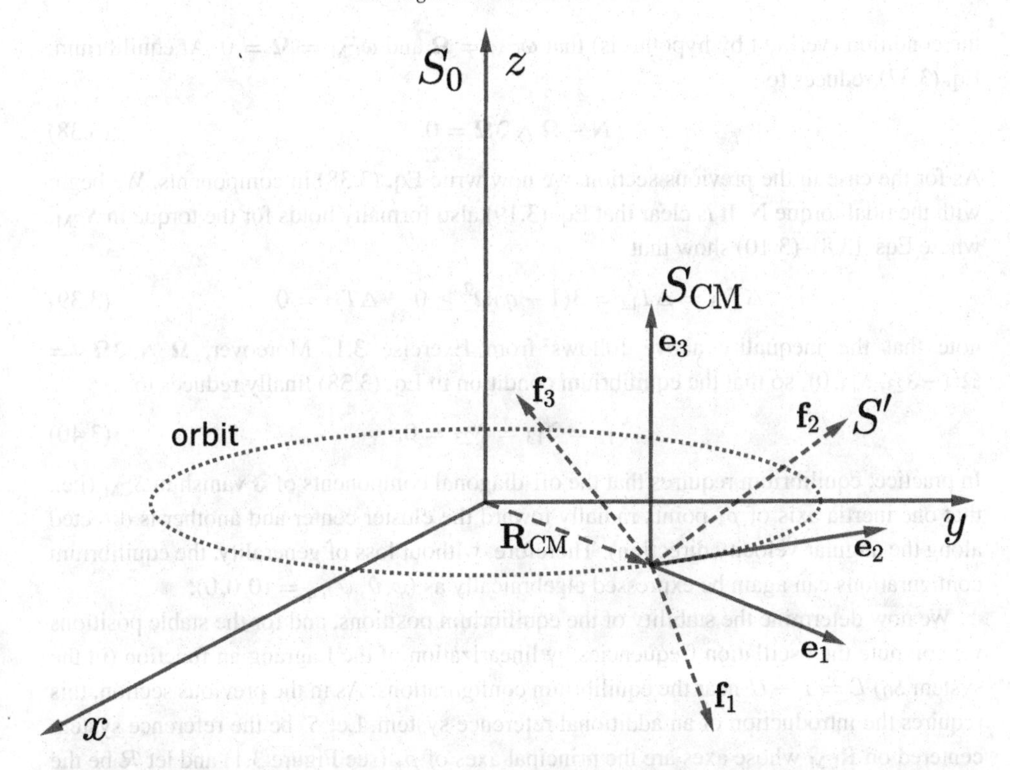

Figure 3.1 The definition of the three references systems needed to determine the equilibrium position of a triaxial body in circular orbit inside a spherically symmetric potential. S_0 is the inertial reference system centered on the spherical distribution producing the tidal field, $S_{\rm CM}$ is centered on the center of mass of the triaxial body ρ_* and rotating on a circular orbit of radius $\mathbf{R}_{\rm CM}$, and finally S' (dashed axes) is the noninertial reference system fixed in ρ_* where the inertia tensor is in diagonal form.

We can now start with a discussion of the possible equilibrium positions of ρ_* in $S_{\rm CM}$ as follows: we consider Eqs. (A.33)–(A.36) with $\mathbf{x}_0 = \mathbf{R}_{\rm CM}$, $\mathcal{R} = \mathcal{C}$, $\boldsymbol{\omega} = \boldsymbol{\omega}_{\rm CM}$, and where we define $\mathbf{J} = \mathbf{J}_0$, $\mathbf{N} = \mathbf{N}_0$, the angular momentum and the torque in S_0 with reduction point $\mathbf{R}_{\rm CM}$. Finally, we set $\Im' = \Im$, $\mathbf{J}' = \mathbf{J}$, and $\mathbf{N}' = \mathbf{N}$, the angular momentum and torque in $S_{\rm CM}$, so that

$$\mathbf{J} = \int_{\Re^3} \mathbf{y} \wedge \dot{\mathbf{y}}\, \rho_*(\mathbf{y},t) d^3\mathbf{y}, \qquad \mathbf{N} = -\int_{\Re^3} \mathbf{y} \wedge (\mathcal{T}_{\rm CM}\mathbf{y})\, \rho_*(\mathbf{y},t) d^3\mathbf{y}. \tag{3.36}$$

We can now evaluate the Newton's equation $\dot{\mathbf{J}}_0 = \mathbf{N}_0$ in S_0, and after easy algebra (do it!), we obtain that in $S_{\rm CM}$

$$\dot{\mathbf{J}} = \mathbf{N} - \dot{\Im}\boldsymbol{\omega}_{\rm CM} - \Im\dot{\boldsymbol{\omega}}_{\rm CM} - \boldsymbol{\omega}_{\rm CM} \wedge \mathbf{J} - \boldsymbol{\omega}_{\rm CM} \wedge \Im\boldsymbol{\omega}_{\rm CM}, \tag{3.37}$$

where $\Im$ is the (in general time-dependent) inertia tensor of ρ_*. In $S_{\rm CM}$, the equilibrium positions of ρ_* are then determined by the condition that $\mathbf{J} = \dot{\mathbf{J}} = \dot{\Im} = 0$, along with the condition (verified by hypothesis) that $\boldsymbol{\omega}_{\mathrm{CM}} = \boldsymbol{\Omega}$ and $\dot{\boldsymbol{\omega}}_{\mathrm{CM}} = \dot{\boldsymbol{\Omega}} = 0$. At equilibrium, Eq. (3.37) reduces to

$$\mathbf{N} - \boldsymbol{\Omega} \wedge \mathfrak{I}\boldsymbol{\Omega} = 0. \tag{3.38}$$

As for the case in the previous section, we now write Eq. (3.38) in components. We begin with the tidal torque $\mathbf{N}$. It is clear that Eq. (3.19) also formally holds for the torque in S_{CM}, where Eqs. (3.8)–(3.10) show that

$$\Delta T_{13} = \Delta T_{12} = 3(1-q)\Omega^2 \geq 0, \quad \Delta T_{23} = 0; \tag{3.39}$$

note that the inequality above follows from Exercise 3.1. Moreover, $\boldsymbol{\Omega} \wedge \mathfrak{I}\boldsymbol{\Omega} = \Omega^2(-\mathfrak{I}_{23}, \mathfrak{I}_{13}, 0)$, so that the equilibrium condition in Eq. (3.38) finally reduces to

$$\mathfrak{I}_{12} = \mathfrak{I}_{13} = \mathfrak{I}_{23} = 0. \tag{3.40}$$

In practice, equilibrium requires that the off-diagonal components of $\mathfrak{I}$ vanish in S_{CM} (i.e., that one inertia axis of ρ_* points radially toward the cluster center and another is directed along the angular velocity direction). Therefore, without loss of generality, the equilibrium configurations can again be expressed algebraically as $(\varphi, \vartheta, \psi)_{\mathrm{eq}} = (0,0,0)$.

We now determine the stability of the equilibrium positions, and for the stable positions we compute the oscillation frequencies by linearization of the Lagrangian function (in the system S_0) $\mathcal{L} = T - U$ near the equilibrium configurations. As in the previous section, this requires the introduction of an additional reference system. Let S' be the reference system centered on $\mathbf{R}_{\mathrm{CM}}$ whose axes are the principal axes of ρ_* (see Figure 3.1) and let $\mathcal{R}$ be the transformation matrix between S_{CM} and S', again defined as in Eq. (3.20) (i.e., $\mathbf{y} = \mathcal{R}\mathbf{x}'$), so that the total rotation matrix between S_0 and S' is given by $\mathcal{CR}$ and $\mathbf{x} = \mathbf{R}_{\mathrm{CM}} + \mathcal{CR}\mathbf{x}'$. The potential energy in S_0 is given again by Eqs. (3.25) and (3.26), where the expansion point is now $\mathbf{R}_{\mathrm{CM}}$, while the kinetic energy in S_0, expressed by using the quantities in S', is obtained from Eq. (A.38) that in the present case reads

$$T = M_* \frac{\|\mathbf{V}_{\mathrm{CM}}\|^2}{2} + \frac{\langle \mathfrak{I}'\boldsymbol{\omega}_{\mathrm{T}}, \boldsymbol{\omega}_{\mathrm{T}} \rangle}{2}, \qquad \boldsymbol{\omega}_{\mathrm{T}} = \mathcal{R}^{\mathrm{T}}\boldsymbol{\Omega} + \boldsymbol{\omega}, \tag{3.41}$$

where $\mathfrak{I}'$ is the diagonal intertia tensor of ρ_* in S', $\boldsymbol{\omega}_{\mathrm{T}}$ is the angular velocity of S' with respect to S_0 (prove it!), and $\boldsymbol{\omega}$ is the angular velocity of S' with respect to S_{CM} given again by Eq. (3.24), while the explicit expression of $\mathcal{R}^{\mathrm{T}}\boldsymbol{\Omega}$, the angular velocity of S_{CM} as seen from S', is finally given in Eq. (3.72).

The linearized potential and kinetic energies for small displacements from the equilibrium position are easily obtained (do it!), and the resulting differential equations are:

$$\begin{cases} \ddot{\varphi} = -\dfrac{\Omega^2 \Delta I_{32}}{I_1}\varphi + \dfrac{\Omega(I_1 + I_2 - I_3)}{I_1}\dot{\vartheta}, \\[2ex] \ddot{\vartheta} = -\dfrac{(\Omega^2 + \Delta T_{13})\Delta I_{31}}{I_2}\vartheta - \dfrac{\Omega(I_1 + I_2 - I_3)}{I_2}\dot{\varphi}, \\[2ex] \ddot{\psi} = \dfrac{\Delta T_{21}\Delta I_{21}}{I_3}\psi. \end{cases} \tag{3.42}$$

The ψ equation immediately provides the first condition for a stable equilibrium, by virtue of $\Delta T_{21} \leq 0$ from Eq. (3.39), so that $I_2 \geq I_1$. Then (see Exercise 3.12), the φ and ϑ equations can be separated into two identical linear fourth-order equations with constant coefficients, which are solved by setting

$$\varphi(t) \equiv X e^{i\omega t}, \quad \vartheta(t) \equiv Y e^{i\omega t}. \tag{3.43}$$

The resulting biquadratic characteristic equation for ω is

$$\omega^4 - \alpha\omega^2 + \beta = 0, \tag{3.44}$$

where

$$\begin{cases} \alpha = \left[\dfrac{(u+v-1)^2}{uv} + \dfrac{1-v}{u} + \dfrac{(4-3q)(1-u)}{v}\right] \times \Omega^2, \\ \beta = \dfrac{(4-3q)(1-u)(1-v)}{uv} \times \Omega^4. \end{cases} \tag{3.45}$$

u and v are given in Eq. (3.30), so that solutions of Eq. (3.44) can be written as $\omega = \tilde{\omega}(u, v, q) \times \Omega$, where $\tilde{\omega}$ is a dimensionless function, and the dependence on r is contained in Ω and q.

Stability requires that the four solutions to Eq. (3.44) be real (i.e., that the two solutions $\lambda_\pm$ of the associated quadratic equation are real and positive). For reality

$$\Delta \equiv \alpha^2 - 4\beta > 0, \tag{3.46}$$

and for positivity Descartes' rule of signs dictates that

$$\alpha > 0; \quad \beta > 0. \tag{3.47}$$

Note that the conditions (3.46) and (3.47) are formally equivalent to the single condition $\alpha \geq 2\sqrt{\beta}$; of course, the *numerical* verification of stability for assigned values of u and v is trivial, while in the following we discuss the problem from the analytical point of view.

The positivity of β in Eq. (3.45) is easily addressed, because $4 - 3q > 1$ from Exercise 3.1, so that we are left with the discussion of two possible stable configurations:

(1) $u < v < 1 \Rightarrow I_3 > I_2 > I_1$,
(2) $1 < u < v \Rightarrow I_2 > I_1 > I_3$,

where the stability of ψ motion ($I_2 > I_1$; i.e., $v > u$) has already been assumed.

Case (1) is easy to discuss: in fact, for $u < v < 1$, the coefficient α is a sum of positive quantities, and some algebra (do it!) shows that Δ can also be written as a sum of positive quantities. Thus, *the motion near the equilibrium position is stable for* $I_3 > I_2 > I_1$ (see Figure 3.2) (i.e., for the minor axis of ρ_* perpendicular to the orbital plane, for the intermediate axis directed tangentially to the orbit, and for the major axis directed toward the center of the external potential).

Case (2) is far more interesting, but its analytical (algebraic) discussion is not easy and is quite long, and so it is not reported here. A detailed study of the conditions in Eqs. (3.46)

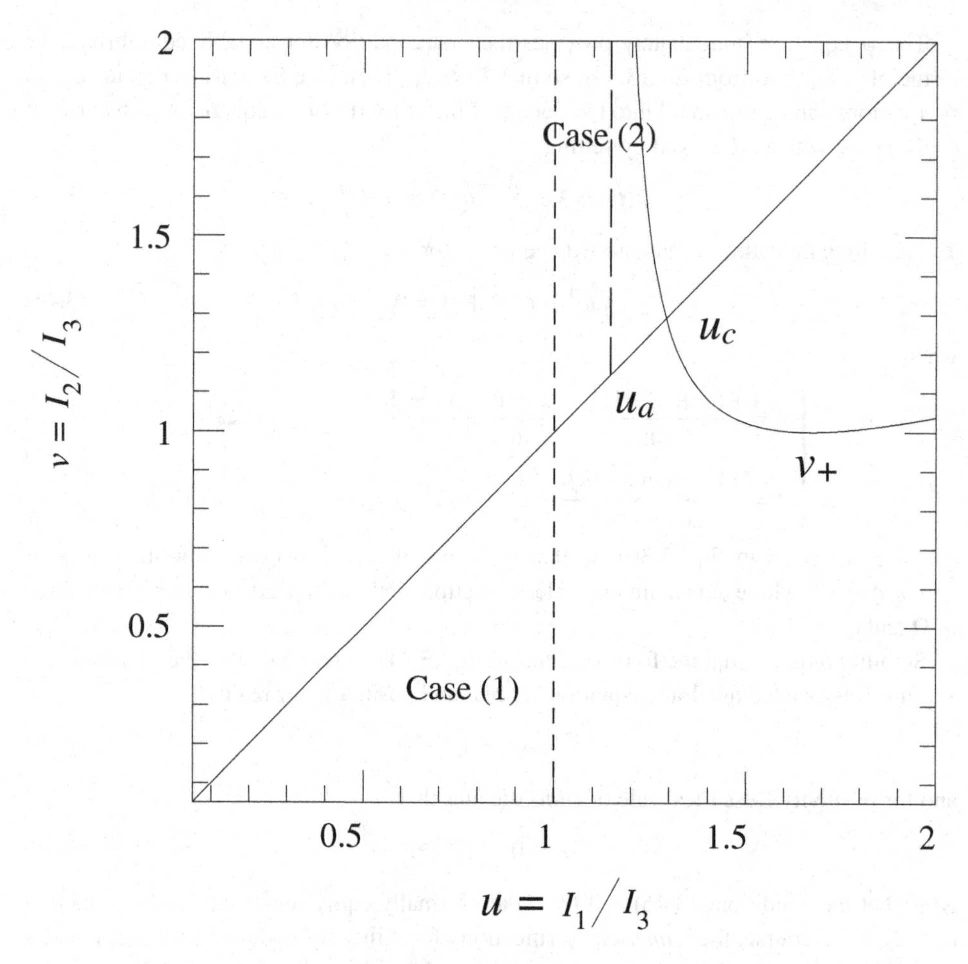

Figure 3.2 The parametric stability plane for the case of a triaxial galaxy in circular orbit about the center of a spherical cluster. The triangular region extending up to $u = 1$ and delimited by the vertical dashed line and by the line $u = v$ corresponds to the Case (1) stable configurations. The Case (2) stable region is bounded at the left by the vertical line $u = 1$, at the right by the curve $v = v_+(u)$, and at the bottom by the line $u = v$ with $1 < u < u_{\rm c}$; it is separated into two subregions by the vertical long-dashed line at $u = u_{\rm a}$. In the figure, we adopted $q = 1/2$, so that $u_{\rm a} \simeq 1.15$ and $u_{\rm c} \simeq 1.29$.

and (3.47) shows that for $1 < u < v$ (i.e., when the major axis of ρ_* is perpendicular to the orbital plane, the medium axis is directed toward the center, and the short axis is parallel to the orbital velocity), the condition $\alpha > 0$ is verified once $\Delta > 0$. Moreover, the stable region of Case (2) (see Figure 3.2) is made up of two parts separated by the vertical asymptote at

$$u_{\rm a}(q) = \frac{1}{2} + \frac{1}{2}\sqrt{\frac{4-3q}{3(1-q)}}, \tag{3.48}$$

with $u_a \to \infty$ for $q \to 1$. In the region $1 < u < u_a$ there is no upper bound on I_2/I_3. For $q < 1$ there exists another stable region spanning the range $u_a < u < u_c$, whose lower bound is again represented by $v = u$, while the upper bound on I_2/I_3 is given by

$$v_+ = (u-1)\frac{1+9(1-q)u-6(1-q)u^2+6(1-q)u\sqrt{u(u-1)(4-3q)}}{12(1-q)u^2-12(1-q)u-1}. \tag{3.49}$$

The point of intersection between $v_+(u)$ and $v = u$ is obtained by solving a fourth-degree equation, and it is given by

$$u_c(q) = \frac{1}{2} + \frac{1}{2}\sqrt{\frac{-9q^2+6q+11+8\sqrt{4-3q}}{3(1-q)(5+3q)}}. \tag{3.50}$$

A very important case is obtained for $q = 0$ (i.e., in the case of a triaxial body orbiting *outside* a spherical body): in this case, the Case (2) region is very small, with $u_a(0) = 1/2 + 1/\sqrt{3} \simeq 1.08$ and $u_c(0) = 1/2 + 3\sqrt{5}/10 \simeq 1.17$.

Having found the criteria for stable equilibrium, we now focus our attention on the associated libration frequencies. From the last differential of Eq. (3.42), the frequency for the ψ motion is given again by the last differential of Eq. (3.29), where now

$$\nu_\psi = \frac{1}{2\pi}\sqrt{\frac{\Delta T_{12}\Delta I_{21}}{I_3}} = \frac{\Omega}{2\pi}\sqrt{k(v-u)}, \quad k \equiv 3(1-q), \quad P_\psi = \frac{1}{\nu_\psi}, \tag{3.51}$$

and u and v are defined in Eq. (3.30). From Exercise 3.12, the φ and ϑ librations result from the superposition of two independent oscillations, with frequencies

$$\nu_\varphi^\pm = \nu_\vartheta^\pm = \frac{\Omega}{2\pi}\sqrt{\frac{1+(\sigma+k)\sigma \pm \sqrt{1+\sigma^2(\tau+k)^2-2\sigma(\tau+2k\tau-k)}}{2}}, \tag{3.52}$$

where

$$\sigma \equiv \frac{1-u}{v}; \qquad \tau \equiv \frac{1-v}{u}. \tag{3.53}$$

Notice that the presence of the coefficient Ω indicates that the periods of libration are of the same order as the orbital period of the galaxy around the cluster center. The student is invited to compute these times for some representative galaxy and cluster models, repeating the approach of Exercise 3.10, and using Exercises 10.20 and 10.21.

3.3 Stellar Orbits in Tidal Fields

From the analysis of the two highly idealized cases in the previous sections, we concluded that the libration times of galaxies around their equilibrium configurations in the tidal field of the parent cluster can be of the same order of magnitude as the average stellar orbital times at the galaxy outskirts, and this makes it interesting to investigate in more detail the responses of such stellar orbits to periodic variations of the external (tidal) field. It is not unreasonable to expect that some fraction of the galaxy stellar population will "resonate" with the galaxy libration periods and may be lost in the cluster intergalactic space: perhaps some fraction of the so-called intracluster stellar population (e.g., see Gerhard et al. 2005) is produced by such *collisionless evaporation* (e.g., see Muccione and Ciotti 2004 and references therein).

The problem of stellar motions inside a galaxy with density distribution ρ_* in a cluster can be described in the highly idealized framework adopted in this chapter. We call S_0 the inertial reference system with its origin in the cluster center, and we consider the noninertial reference system S' with its origin in the galaxy center of mass and axes aligned along the principal axes of the galaxy inertia tensor. Each star moves under the action of the physical force produced by the galaxy itself ($\mathbf{g}_*$) and by the cluster ($\mathbf{g}_{\rm ext}$), so that from Eq. (A.29) we obtain that in S'

$$\ddot{\mathbf{x}}' = \mathcal{R}_{\rm T}^{\rm T}(\mathbf{g}_* + \mathbf{g}_{\rm ext} - \mathbf{A}_{\rm CM}) - 2\boldsymbol{\omega}_{\rm T} \wedge \dot{\mathbf{x}}' - \dot{\boldsymbol{\omega}}_{\rm T} \wedge \mathbf{x}' - \boldsymbol{\omega}_{\rm T} \wedge (\boldsymbol{\omega}_{\rm T} \wedge \mathbf{x}'), \tag{3.54}$$

where $\mathbf{x}'$ is the position of a star in S', $\mathcal{R}_{\rm T} = \mathcal{C}\mathcal{R}$ is the total rotation matrix between S_0 and S', and $\boldsymbol{\omega}_{\rm T}$ is the associated angular velocity of S' as given in Eq. (3.41). From the previous discussions, it immediately follows that at the second-order tidal expansion, the physical force field in S' is

$$\mathcal{R}_{\rm T}^{\rm T}(\mathbf{g}_* + \mathbf{g}_{\rm ext} - \mathbf{A}_{\rm CM}) \sim -\nabla_{\mathbf{x}'}\phi'_* - \mathcal{T}'\mathbf{x}', \tag{3.55}$$

where $\nabla_{\mathbf{x}'}$ is the gradient operator in S', $\mathcal{T}' = \mathcal{R}_{\rm T}^{\rm T}\mathcal{T}\mathcal{R}_{\rm T} = \mathcal{R}^{\rm T}\mathcal{T}_{\rm CM}\mathcal{R}$ is the dependent tidal field in S', and finally $\mathcal{T}$ is the tidal tensor expressed in S_0 and $\mathcal{T}_{\rm CM}$ in $S_{\rm CM}$. Quite obviously, the two cases of galaxy libration considered in the previous sections are contained in Eq. (3.55).

Consistently with the adopted "small oscillations approximation," the linearized rotation matrix $\mathcal{R}$ and angular velocity $\boldsymbol{\omega}_{\rm T}$ are obtained from the expansion of their full expressions up to the linear terms in $\varphi_{\rm M}$, $\vartheta_{\rm M}$, and $\psi_{\rm M}$, so that (prove it!)

$$\mathcal{R} \sim \begin{pmatrix} 1 & -\psi & \vartheta \\ \psi & 1 & -\varphi \\ -\vartheta & \varphi & 1 \end{pmatrix}, \quad \boldsymbol{\omega}_{\rm T} \sim (\dot{\varphi} - \vartheta\Omega, \dot{\vartheta} + \varphi\Omega, \dot{\psi} + \Omega), \tag{3.56}$$

where the two cases of a triaxial galaxy at the center of a triaxial cluster and of a triaxial galaxy in circular orbit in a spherical cluster are obtained by inserting the explicit time dependence of the three angles in the two cases, and in the first case $\Omega = 0$. The resulting differential equation (3.54), with periodic time-dependent coefficients, can be easily integrated numerically (e.g., see Muccione and Ciotti 2004), and the simulations nicely confirm that the outskirts of galaxies can be significantly affected by the cooperative effects of librations and of the cluster tidal field, with stars easily doubling their apocenters and being finally lost in the cluster, suggesting that collisionless evaporation may play some role in the production of the intracluster stellar population.

As can be easily imagined, the literature on the effects of tidal fields on the structure and dynamics of stellar systems is enormous; for additional readings discussing problems of great astrophysical relevance, the student is invited to consult more specialized books, such as Binney and Tremaine (2008), Chandrasekhar (1942), Ogorodnikov (1965), and Spitzer (1987).

Exercises

3.1 Prove that Eq. (3.6) gives the tidal field produced by a spherically symmetric density profile $\rho(r)$ and show that $q(r) = 1 - \gamma/3$ for the power-law density profile $\rho(r) \propto r^{-\gamma}$ in Eq. (13.29) with $\gamma < 3$. Finally, prove that $q(r) \leq 1$ for monotonically decreasing $\rho(r)$. *Hint*: The inequality $\rho(r) \leq \bar{\rho}(r)$ can be proved from integration by parts of the integral expression of $\bar{\rho}(r)$ in Eq. (3.7) and then using the assumption $d\rho/dr \leq 0$.

3.2 King (1972) proposed the density distribution

$$\rho(r) = \frac{\rho(0)}{(s^2+1)^{3/2}}, \quad s = \frac{r}{r_c}, \tag{3.57}$$

in order to describe clusters of galaxies (see also Eq. (13.34)), where r is the distance from the cluster center, r_c is the "core" radius, and the distribution is truncated at some r_t. From Eq. (3.6), show that for $r \leq r_t$

$$q(r) = \frac{s^3}{3(s^2+1)^{3/2}\left(\mathrm{arcsinh} s - s/\sqrt{s^2+1}\right)}, \tag{3.58}$$

where $\mathrm{arcsinh} x = \ln(x + \sqrt{x^2+1})$, so that $q = 2/3$ for $s \simeq 1.027$.

3.3 Show that in the spherical limit ($a_1 = a_2 = a_3 = r_c$), for γ models in Eq. (2.83), and for the perfect ellipsoid in Eq. (2.93), we have

$$q(r) = \frac{1-\gamma/3}{s+1}, \quad q(r) = \frac{2s^3}{3(s^2+1)^2}\left(\arctan s - \frac{s}{s^2+1}\right)^{-1}, \tag{3.59}$$

where $s = r/r_c$, so that $q = 2/3$ for $s = (1-\gamma)/2$ $(0 \leq \gamma \leq 1)$ and $s \simeq 0.824$, respectively. Do the exercise for *all* of the spherical models from Eqs. (13.29)–(13.43) and determine for each of them the critical radius so that $q(r) = 2/3$.

3.4 By using Eqs. (2.72) and (3.2), show that the tidal tensor $\mathcal{T}$ at position $\mathbf{x}$, produced by a triaxial ellipsoid with $\rho = \rho(m)$, is given by

$$\mathcal{T}_{ij} = 2\pi G a_1 a_2 a_3 \int_0^\infty \left[\delta_{ij}\rho(m_\tau) + \frac{x_i x_j \rho'(m_\tau)}{(a_j^2+\tau)m_\tau}\right] \frac{d\tau}{(a_i^2+\tau)\sqrt{\Delta(\tau)}}, \tag{3.60}$$

where $m_\tau = m(\mathbf{x}, \tau)$ and $\rho' = d\rho/dm$. Show that at the center ($m_\tau = 0$) of a sufficiently regular density distribution or in the case of a constant-density ellipsoid, Eqs. (3.10) and (3.11) are recovered. Finally, prove that the Poisson identity (3.5) holds. *Hint*: To prove the Poisson identity, write the trace of Eq. (3.60) and express the second term of the resulting integrand by using the second identity in Eqs. (2.73). An integration by parts of the second integral finally establishes the result.

3.5 By using Eq. (3.11), prove that

$$w_1 + w_2 + w_3 = 2. \tag{3.61}$$

Hint: Write the sum above adding the three corresponding integrals and compare the resulting integrand with the quantity $d\Delta(\tau)^{-1/2}/d\tau$.

3.6 From direct integration of Eq. (3.11), prove that in the *oblate* axisymmetric case $a_1 = a_2 \geq a_3$ (i.e., for $q_y = 1$),

$$\begin{cases} w_1 = w_2 = \dfrac{\sqrt{1-e^2}}{e^2}\left(\dfrac{\arcsin e}{e} - \sqrt{1-e^2}\right), \\ w_3 = 2\dfrac{\sqrt{1-e^2}}{e^2}\left(\dfrac{1}{\sqrt{1-e^2}} - \dfrac{\arcsin e}{e}\right), \end{cases} \tag{3.62}$$

while in the *prolate* axysimmetric case $a_1 \geq a_2 = a_3$ (i.e., for $q_y = q_z$),

$$\begin{cases} w_1 = 2\dfrac{1-e^2}{e^2}\left(\dfrac{\operatorname{arctanh} e}{e} - 1\right), \\ w_2 = w_3 = \dfrac{1-e^2}{e^2}\left(\dfrac{1}{1-e^2} - \dfrac{\operatorname{arctanh} e}{e}\right), \end{cases} \tag{3.63}$$

where in both cases we adopt the definition $e = \sqrt{1-q_z^2} = \sin\varphi$ as in Eq. (3.12), and we recall that for $|x| \leq 1$, $\operatorname{arctanh} x = \ln\sqrt{(1+x)/(1-x)}$. Show by direct evaluation that identity (3.61) is verified.

3.7 Show that the inertia tensor of ρ_* in S_0 can be written in terms of $\mathcal{R}$ and I_i as

$$\Im_{jk}(t) = \sum_{\mu=1}^{3} \mathcal{R}_{j\mu}(t)\mathcal{R}_{k\mu}(t)I_\mu, \tag{3.64}$$

and discuss the vanishing of $\Im_{jk}$ for $j \neq k$ as a function of the rotation angles. Prove by direct evaluation that the trace of $\Im$ is invariant, $\mathrm{Tr}(\Im) = I_1 + I_2 + I_3$. *Hints*: Change the coordinates in Eq. (2.89) to $\mathbf{y} = \mathcal{R}\mathbf{x}'$, where $\mathcal{R}$ is the orthogonal matrix in Eq. (3.20), and recall that in S' the inertia tensor of ρ_* is diagonal, with $I_i \equiv \Im'_{ii}$ (no sum over i intended). Also recall that the rows and columns of an orthogonal matrix can be seen as vectors of unitary norm.

3.8 Prove Eq. (3.25). *Hints*: Substitute the first identity in Eq. (3.17) into the left-hand side of Eq. (3.25): the resulting integral splits into two integrals. The first is trivial. For the second integral, use the second identity in Eq. (2.90): the first integral is obtained from Eq. (3.5) and the second one is from the diagonality of $\mathcal{T}$ in S_0 and Eq. (3.64).

3.9 Show that the kinetic and potential energies of ρ_* for the problem discussed in Section 3.2.1 can be written as

$$\begin{aligned} 2T = {} & I_1(\dot\vartheta\sin\psi + \dot\varphi\cos\vartheta\cos\psi)^2 + I_2(\dot\vartheta\cos\psi - \dot\varphi\cos\vartheta\sin\psi)^2 \\ & + I_3(\dot\psi + \dot\varphi\sin\vartheta)^2, \end{aligned} \tag{3.65}$$

and

$$2U = (\cos\varphi\cos\psi - \sin\varphi\sin\vartheta\sin\psi)^2 \Delta T_{21}\Delta I_{21} + (\sin\varphi\cos\vartheta)^2 \Delta T_{21}\Delta I_{31} + (\sin\varphi\cos\psi + \cos\varphi\sin\vartheta\sin\psi)^2 \Delta T_{31}\Delta I_{21} + (\cos\varphi\cos\vartheta)^2 \Delta T_{31}\Delta I_{31}, \tag{3.66}$$

where all of the additive constants in Eq. (3.25) have been set to zero. *Hints*: The expression for the kinetic energy is obtained from Eq. (3.24). For the potential energy, focus on the last term in Eq. (3.25), and from the definitions $\Delta T_{ij} \equiv T_i - T_j$ and $\Delta I_{ij} \equiv I_i - I_j$ show that

$$\mathcal{R}^2_{i\mu}(t) T_i I_\mu = \sum_{i,\mu=2}^{3} \mathcal{R}^2_{i\mu}(t) \Delta T_{i1} \Delta I_{\mu 1} + T_1(I_1 + I_2 + I_3) + I_1(T_1 + T_2 + T_3) - 3T_1 I_1, \tag{3.67}$$

simply by using the orthogonality of $\mathcal{R}$.

3.10 With this exercise we consider the limit for small cluster and galaxy flattenings of the galaxy libration frequencies in Eq. (3.29). First, show that for a triaxial cluster with $a_1 \geq a_2 \geq a_3$ (so that $\epsilon \leq \eta \leq 1$), from Eqs. (3.13) and (3.31)

$$\Delta T_{23} \sim \frac{8\pi G\rho(0)}{5}(\eta - \epsilon), \quad \Delta T_{13} \sim \frac{8\pi G\rho(0)}{5}\eta, \quad \Delta T_{12} \sim \frac{8\pi G\rho(0)}{5}\epsilon, \tag{3.68}$$

and from Exercise 2.18, that

$$\Im_{11} = \frac{2Ma_1^2 h}{3}(1 - \epsilon - \eta), \ \Im_{22} = \frac{2Ma_1^2 h}{3}(1 - \eta), \ \Im_{33} = \frac{2Ma_1^2 h}{3}(1 - \epsilon), \tag{3.69}$$

$$\mathcal{Q}_{11} = \frac{2a_1^2 h}{3}(\epsilon + \eta), \ \mathcal{Q}_{22} = \frac{2a_1^2 h}{3}(\eta - 2\epsilon), \ \mathcal{Q}_{11} = \frac{2a_1^2 h}{3}(\epsilon - 2\eta). \tag{3.70}$$

Finally, from Eqs. (3.30) and (3.32), show that at linear order in the galaxy flattenings

$$\frac{1-v}{u} \sim \eta_* - \epsilon_*, \quad \frac{1-u}{v} \sim \eta_*, \quad v - u \sim \epsilon_*, \tag{3.71}$$

and finally obtain the leading term of the resulting Eq. (3.29). See also Exercise 10.20.

3.11 Show that in Eq. (3.41)

$$\mathcal{R}^{\mathrm{T}}\mathbf{\Omega} = \Omega(\sin\varphi\sin\psi - \cos\varphi\sin\vartheta\cos\psi, \sin\varphi\cos\psi + \cos\varphi\sin\vartheta\sin\psi, \cos\varphi\cos\vartheta). \tag{3.72}$$

3.12 With this exercise we construct the solution for assigned initial conditions $(\varphi_0, \dot\varphi_0)$ and $(\vartheta_0, \dot\vartheta_0)$ of the two first components of Eq. (3.42), which we rewrite as

$$\ddot\varphi = -A_1\varphi + B_1\dot\vartheta, \quad \ddot\vartheta = -A_2\vartheta - B_2\dot\varphi. \tag{3.73}$$

First, show that equations above can be separated into two identical fourth-order ordinary differential equations,

$$\frac{d^4\varphi}{dt^4} + \alpha\frac{d^2\varphi}{dt^2} + \beta\varphi = 0, \qquad \frac{d^4\vartheta}{dt^4} + \alpha\frac{d^2\vartheta}{dt^2} + \beta\vartheta = 0, \tag{3.74}$$

with the two constants

$$\alpha \equiv A_1 + B_1 B_2 + A_2, \quad \beta \equiv A_1 A_2, \tag{3.75}$$

given in Eq. (3.45). Then show that the four solutions of Eq. (3.44) are

$$\omega_{1,2,3,4} = \pm\omega_\pm, \quad \omega_\pm = \sqrt{\frac{\alpha \pm \sqrt{\alpha^2 - 4\beta}}{2}}, \tag{3.76}$$

so that the general solution for φ is given by

$$\varphi(t) = X_1 \mathrm{e}^{i\omega_+ t} + X_2 \mathrm{e}^{-i\omega_+ t} + X_3 \mathrm{e}^{i\omega_- t} + X_4 \mathrm{e}^{-i\omega_- t}, \tag{3.77}$$

where X_i are complex numbers to be determined from the initial conditions by evaluating up to the third derivative of $\varphi(t)$ at $t = 0$:

$$\begin{cases} X_1 + X_2 + X_3 + X_4 = \varphi_0, \\ (X_1 - X_2)\omega_+ + (X_3 - X_4)\omega_- = -i\dot{\varphi}_0, \\ (X_1 + X_2)\omega_+^2 + (X_3 + X_4)\omega_-^2 = A_1\varphi_0 - B_1\dot{\vartheta}_0, \\ (X_1 - X_2)\omega_+^3 + (X_3 - X_4)\omega_-^3 = -i(A_1\dot{\varphi}_0 - B_1\ddot{\vartheta}_0), \end{cases} \tag{3.78}$$

and a similar expression holds for $\vartheta(t)$. Finally, from the first and third identities in Eq. (3.78), show that $X_1 + X_2$ and $X_3 + X_4$ are real, and from the second and fourth identities in Eq. (3.78), show identities in Eq. (3.78), show that $X_1 - X_2$ and $X_3 - X_4$ are purely imaginary, so that $X_2 = X_1^*$ and $X_4 = X_3^*$ and $\varphi(t)$ and $\vartheta(t)$ in the stable case can be written in terms of real trigonometric functions.

3.13 With this exercise we explore (qualitatively) the important astrophysical concept of *tidal disruption*, and we estimate the size of the so-called *Roche limit*. Consider a self-gravitating spherical body (a star or a satellite) of mass m and radius R moving in the gravitational field of a larger mass M (a black hole or a planet). First, let the center of mass of the body m approaching M be on a radial orbit at distance r along the x-axis; use Eqs. (3.8) and (3.55) to determine the critical distance r_{Roche} where the tidal field along the radial direction balances the gravitational pull Gm/R^2 for a point at the equator of the falling system and show that

$$r_{\mathrm{Roche}} = \left(\frac{2M}{m}\right)^{1/3} R. \tag{3.79}$$

For $r < r_{\mathrm{Roche}}$, the object is presumably disrupted by the tidal forces due to M. Repeat the exercise for the object on a circular orbit of radius r around M, and again from Eqs. (3.8) and (3.54) show that now

$$r_{\mathrm{Roche}} = \left(\frac{3M}{m}\right)^{1/3} R. \tag{3.80}$$

Explain physically why the Roche limit in the case of a circular orbit is placed at a larger distance from M. Finally, recast the two equations obtained in terms of the averaged densities of the two bodies, using Eq. (3.7) for the body M. *Hint*: Use Eq. (3.54) with the center of mass of m on the x'-axis of the system S', with $\mathbf{x}' = (R, 0, 0)$, and in the rotating case with $\boldsymbol{\omega}_{\mathrm{T}} = (0, 0, \Omega)$.

4
The Two-Body Problem

In this chapter, we introduce and solve (by means of the Laplace–Runge–Lenz vector) the two-body problem, with an emphasis on the properties of hyperbolic orbits, and specifically on the so-called *slingshot effect*. The obtained results will be used in Chapters 7 and 8 for the derivation of two fundamental timescales characterizing the dynamical evolution of stellar systems: the *two-body relaxation time* and *dynamical friction time*.

4.1 Center of Mass and the Reduced Mass

The *two-body problem* is, togheter with the *harmonic oscillator*, one of the most celebrated cases of a "solvable" problem of classical physics. Its importance in stellar dynamics is fundamental, as will be clear from (some of) its consequences worked out in later chapters. Here, we focus our attention on the main properties of the orbits: it is assumed that the student has already been exposed to the basic results in the field (the literature on the subject is immense; suggested readings of different levels of difficulties are, e.g., Binney and Tremaine 2008; Boccaletti and Pucacco 1996; Danby 1962; Goldstein et al. 2000; Landau and Lifshitz 1969; Roy 2005; Szebehely 1967). Before addressing the problem, it may be useful to remind the reader that in stellar dynamics – at variance with celestial mechanics – attention is focused more on unbound (hyperbolic) orbits than on elliptical orbits.

As is well known, different formulations of the basic laws of mechanics allow for the use of different analytical methods for the solution of Newton's differential equations of motion. Thus, in addition to the vectorial form (the modern version of Newton's severely geometric approach; e.g., see Arnold 1990; Chandrasekhar 1995), the student certainly knows the elegant and essentially analytical (coordinate-based) Euler–Lagrange formulation, and perhaps the deeper Hamiltonian approach, in which phase-space geometry and action-angle variables are the major ingredients (e.g., see Arnold 1978; Binney and Tremaine 2008). Due to the introductory nature of this book, in our discussion we will use the usual vector approach, but we will also exploit the full power of the conserved quantitites of the problem, avoiding the need to perform explicit integrations.

We start from the equations of motion for two point masses or stars (a test mass $m_{\rm t}$ and a field mass $m_{\rm f}$) in a generic inertial reference system S_0 mutually interacting with the gravitational force

$$m_{\rm t}\ddot{\mathbf{x}}_{\rm t} = -\frac{Gm_{\rm t}m_{\rm f}}{r^3}\mathbf{r}, \quad m_{\rm f}\ddot{\mathbf{x}}_{\rm f} = \frac{Gm_{\rm t}m_{\rm f}}{r^3}\mathbf{r}, \quad r \equiv \|\mathbf{r}\|, \tag{4.1}$$

where $\mathbf{r} \equiv \mathbf{x}_{\rm t} - \mathbf{x}_{\rm f}$ is the *relative orbit* of $m_{\rm t}$ with respect to $m_{\rm f}$ and $\mathbf{v} = \dot{\mathbf{r}} = \mathbf{v}_{\rm t} - \mathbf{v}_{\rm f}$ is the *relative velocity*. Of course, the equations must be supplemented by the initial conditions $\mathbf{x}_{\rm t0}$, $\mathbf{v}_{\rm t0}$, $\mathbf{x}_{\rm f0}$, and $\mathbf{v}_{\rm f0}$.

As is well known, by adding the two equations in (4.1), we can show that the conservation of the total momentum $\mathbf{P} = m_{\rm t}\mathbf{v}_{\rm t} + m_{\rm f}\mathbf{v}_{\rm f}$ is embodied in Newton's law, and from double time integration

$$\begin{cases} m_{\rm t}\mathbf{x}_{\rm t} + m_{\rm f}\mathbf{x}_{\rm f} = M\left(\mathbf{R}_{\rm CM0} + \mathbf{V}_{\rm CM}t\right), \quad M = m_{\rm t} + m_{\rm f}, \\ \mathbf{R}_{\rm CM0} = \dfrac{m_{\rm t}\mathbf{x}_{\rm t0} + m_{\rm f}\mathbf{x}_{\rm f0}}{M}, \quad \mathbf{V}_{\rm CM} = \dfrac{m_{\rm t}\mathbf{v}_{\rm t0} + m_{\rm f}\mathbf{v}_{\rm f0}}{M}, \end{cases} \tag{4.2}$$

where $\mathbf{R}_{\rm CM0}$ is the position of the center of mass of the pair at $t = 0$ and $\mathbf{V}_{\rm CM}$ is the constant velocity of the center of mass $\mathbf{R}_{\rm CM} = \mathbf{R}_{\rm CM0} + \mathbf{V}_{\rm CM}t$. Moreover, by scalar multiplication of the first equation in (4.1) by $\dot{\mathbf{x}}_{\rm t}$ and of the second equation in (4.1) by $\dot{\mathbf{x}}_{\rm f}$ and by summing the two resulting identities, we obtain the expression of the total time derivative of a function that remains constant when evaluated along the orbits of the two masses:

$$\frac{m_{\rm t}\|\mathbf{v}_{\rm t}\|^2}{2} + \frac{m_{\rm f}\|\mathbf{v}_{\rm f}\|^2}{2} - \frac{Gm_{\rm t}m_{\rm f}}{r} = E. \tag{4.3}$$

In other words, we prove that the total energy E is conserved.[1] Finally, from summation of the cross product of the first equation in (4.1) by $\mathbf{x}_{\rm t}$ and of the second equation in (4.1) by $\mathbf{x}_{\rm f}$, it follows that the vectorial quantity

$$m_{\rm t}\mathbf{x}_{\rm t} \wedge \mathbf{v}_{\rm t} + m_{\rm f}\mathbf{x}_{\rm f} \wedge \mathbf{v}_{\rm f} = \mathbf{J} \tag{4.4}$$

is also conserved when evaluated along the orbits of the two masses: $\mathbf{J}$ is the total angular momentum referring to the origin of S_0. The constant quantities $\mathbf{R}_{\rm CM0}$, $\mathbf{V}_{\rm CM}$, E, and $\mathbf{J}$ are fixed by the initial conditions, and so their specific *values* depend on the specific reference system S_0 adopted, but their *existence* is guaranteed in *all* of the inertial reference systems; in practice, in the two-body problem we have 10 quantities conserved during the orbital evolution (also known as the *10 classical integrals of motion*; see also Chapter 6). It is a simple exercise (do it!) to show that all of the results obtained so far are not restricted to the $1/r^2$ force law, but also hold in the case of a generic force law that can be derived from an interaction energy $U(r)$, and thus automatically obeying Newton's Third Law of

[1] This approach is so relevant to the theory of ordinary differential equations that the analogous procedure used to reduce the order of differential equations is usually referred to in mathematics as the "energy method."

Dynamics, with $m_t\ddot{\mathbf{x}}_t = -\nabla_{\mathbf{x}_t}U$ and $m_f\ddot{\mathbf{x}}_f = -\nabla_{\mathbf{x}_f}U = \nabla_{\mathbf{x}_t}U$. Remarkably, as we will see in the next section, an *additional* quantity is conserved in the specific case of the central $1/r^2$ force field.

The reference system S_0 is nothing special (except for it being inertial), so that formally identical expressions will hold in any other inertial reference system S'. In particular, the coordinates of a point in the two reference systems (see Appendix A.1.1) are related by the Galilean transformation $\mathbf{x} = \mathbf{R} + \mathcal{R}\mathbf{x}'$, with $\mathbf{R}(t)$ being the origin of S' moving with constant velocity $\mathbf{V} = d\mathbf{R}/dt$ with respect to the origin of S_0, and $\mathcal{R}$ being an orthogonal, time-independent 3×3 matrix fixing the constant orientation of S' with respect to S_0. We highlight the important (but not always fully appreciated) fact that by changing inertial reference system, the values of the constant quantities $\mathbf{R}_{\mathrm{CM}0}$, $\mathbf{V}_{\mathrm{CM}}$, E, and $\mathbf{J}$ in general will change (i.e., these quantities are *constants* in different[2] inertial systems, but they are not *invariants*). Among all of the inertial reference systems, the most important are those moving with velocity $\mathbf{V} = \mathbf{V}_{\mathrm{CM}}$ with respect to S_0, and of these, the most important the subset (which we indicate with S_{CM}) with the origin coincident with the center of mass (i.e., $\mathbf{R} = \mathbf{R}_{\mathrm{CM}}$). As is shown in Exercise 4.1, in this last family (when restricting without loss of generality to $\mathcal{R} = I$; i.e., when the axes of S_0 and S_{CM} are parallel), the following identities hold at all times:

$$\begin{cases} m_t\mathbf{x}'_t + m_f\mathbf{x}'_f = 0, \quad m_t\mathbf{v}'_t + m_f\mathbf{v}'_f = 0, \\ E = M\dfrac{\|\mathbf{V}_{\mathrm{CM}}\|^2}{2} + E', \quad E' = m_t\dfrac{\|\mathbf{v}'_t\|^2}{2} + m_f\dfrac{\|\mathbf{v}'_f\|^2}{2} - \dfrac{Gm_tm_f}{r}, \\ \mathbf{J} = M\mathbf{R}_{\mathrm{CM}0} \wedge \mathbf{V}_{\mathrm{CM}} + \mathbf{J}', \quad \mathbf{J}' = m_t\mathbf{x}'_t \wedge \mathbf{v}'_t + m_f\mathbf{x}'_f \wedge \mathbf{v}'_f, \end{cases} \tag{4.5}$$

where E' and $\mathbf{J}'$ are the total energy and the total angular momentum of the pair in the barycentric reference systems S_{CM}.

We are now in a position to solve Eqs. (4.1), a process that will be completed in Section 4.3 after an interlude in Section 4.2 dedicated to the *slingshot effect*. The standard procedure to solve Eqs. (4.1) is to combine them in the differential equation for the *relative orbit* of m_t with respect to m_f: after division of the first equation by m_t and of the second by m_f, by evaluation of $\ddot{\mathbf{x}}_t - \ddot{\mathbf{x}}_f$ one gets

$$\mu\ddot{\mathbf{r}} = -\frac{Gm_tm_f}{r^3}\mathbf{r}, \quad \mu = \frac{m_tm_f}{M}, \qquad M = m_t + m_f, \tag{4.6}$$

where μ is the *reduced mass*. $\mu = m/2$ for identical masses, $m_t = m_f = m$, while in the case of a significant mass discrepancy, μ is just slightly less than the mass of the light component. Mathematically, Eq. (4.6) is identical to the equation of motion of a point of mass μ orbiting in a radial $1/r^2$ field with a fixed center and a magnitude

[2] The student should be convinced that different values of conserved quantities do not simply represent a mathematical property without physical importance. Consider a free particle moving with constant velocity in some inertial system: its (kinetic) energy is conserved, and it can be used to do some work (e.g., by heating a target in an inelastic impact). When observed in a second inertial system moving with the particle velocity, its kinetic energy is again conserved, but its value is 0, and no work can be done in this system.

equal to that between the two particles. The student should realize that the physics behind this transformation is much less trivial than the resulting mathematical expression. First, Eq. (4.6) is invariant for all of the observers, *as well as in accelerated motion, provided their reference system is not rotating* (i.e., *all* of these observers would determine the *same* relative orbit (prove it!)). Second, despite Eq. (4.6) being formally identical to that of a gravitational problem in an inertial reference system, it is actually obtained by referring the coordinates to $\mathbf{x}_\mathrm{f}(t)$ (i.e., to a noninertial reference system), and the student should ask them self (and answer) why the noninertial translational term in Eq. (A.29) seems to be missing on the right-hand side of Eq. (4.6).

The full solution of Eq. (4.6) can be obtained by following standard procedures (e.g., see Goldstein et al. 2000; Landau and Lifshitz 1969), and we will work on this in Section 4.3. For the moment we assume that the relative orbit $\mathbf{r}(t)$ has been determined, and we focus on how to recover the orbits $\mathbf{x}_\mathrm{t}$ and $\mathbf{x}_\mathrm{f}$ in S_0 from the knowledge of the relative orbit $\mathbf{r}$. We exploit the unique properties of S_CM: from the first equation in (4.5) and from the identities $\mathbf{r} = \mathbf{x}_\mathrm{t} - \mathbf{x}_\mathrm{f} = \mathbf{x}'_\mathrm{t} - \mathbf{x}'_\mathrm{f}$ and $\mathbf{v} = \dot{\mathbf{r}}$, it follows that in S_CM

$$\mathbf{x}'_\mathrm{t} = \frac{\mu\mathbf{r}}{m_\mathrm{t}}, \quad \mathbf{x}'_\mathrm{f} = -\frac{\mu\mathbf{r}}{m_\mathrm{f}}, \quad \mathbf{v}'_\mathrm{t} = \frac{\mu\mathbf{v}}{m_\mathrm{t}}, \quad \mathbf{v}'_\mathrm{f} = -\frac{\mu\mathbf{v}}{m_\mathrm{f}}, \tag{4.7}$$

and so in S_0

$$\begin{cases} \mathbf{x}_\mathrm{t}(t) = \dfrac{\mu\mathbf{r}(t)}{m_\mathrm{t}} + \mathbf{R}_\mathrm{CM}(t), & \mathbf{v}_\mathrm{t}(t) = \dfrac{\mu\mathbf{v}(t)}{m_\mathrm{t}} + \mathbf{V}_\mathrm{CM}, \\ \mathbf{x}_\mathrm{f}(t) = -\dfrac{\mu\mathbf{r}(t)}{m_\mathrm{f}} + \mathbf{R}_\mathrm{CM}(t), & \mathbf{v}_\mathrm{f}(t) = -\dfrac{\mu\mathbf{v}(t)}{m_\mathrm{f}} + \mathbf{V}_\mathrm{CM}. \end{cases} \tag{4.8}$$

It is apparent how the two barycentric orbits are simply scaled-down (homotetic) versions of the relative orbit $\mathbf{r}$ and how the reconstruction of the two "absolute" orbits in S_0 from $\mathbf{r}$ has been achieved thanks to the fact that $\mathbf{R}'_\mathrm{CM} = 0 = \mathbf{P}'$ in S_CM. Finally, the student is invited to verify that the energy E_r and the angular momentum $\mathbf{J}_\mathrm{r}$ of the relative orbit are conserved and that their values coincide, respectively, with those of E' and $\mathbf{J}'$ in S_CM, introduced in Eq. (4.5). In formulae,

$$E_\mathrm{r} = \frac{\mu\|\mathbf{v}\|^2}{2} - \frac{Gm_\mathrm{t}m_\mathrm{f}}{r} = E', \quad \mathbf{J}_\mathrm{r} = \mu\mathbf{r}\wedge\mathbf{v} = \mathbf{J}'. \tag{4.9}$$

4.2 The Slingshot Effect

Before moving on to solve the gravitational two-body problem, we consider a very general and important phenomenon associated with central forces that can be derived from an interaction energy $U(r)$ allowing for *unbound* motions (i.e., two-body interactions beginning and ending at an infinite relative separation of m_t and m_f). The $1/r^2$ force is of course the most important case, while examples of forces that *do not* share this property are the harmonic force (with $U \propto r^2$) and the modified Newtonian dymanic (MOND) force (with a qualitative $U \propto \ln r$ behavior in the deep regime; see Exercise 2.34). The phenomenon under consideration is known as the *slighshot effect*, and it is used (for example) to send probes to outer planets in the Solar System (or to slow down probes so that they can fall[3] toward the inner planets). But how does the slingshot effect work? After all, Eq. (4.9) tells us that energy is strictly conserved in the relative orbit and, perhaps making the effect even more mysterious, we will show that in $S_{\rm CM}$ the moduli of the velocities of the two objects (e.g., the planet and the probe) at the beginning and at the end of their interaction are also unchanged: How can the probe then change its energy?

The explanation is rooted in the fact that energy *is* conserved, but it *is not* invariant for a change of (inertial) reference system, and the energy balance between the probe and the planet in the reference system S_0 (e.g., the system centered on the Sun) is different from that in the system $S_{\rm CM}$ of the planet and the probe. Suppose that $\mathbf{v}_{\rm t}(-\infty)$ is the initial velocity at $t = -\infty$ of the test mass $m_{\rm t}$ and $\mathbf{v}_{\rm f}(-\infty)$ is the initial velocity of a field mass $m_{\rm f}$ in a generic inertial reference system S_0. Moreover, let us assume that the initial separation of the two masses is infinite, so that from Eq. (4.5) the total relative energy of the pair coincides with the initial total kinetic energy. From Eq. (4.9), in the relative orbit the conservation of $E_{\rm r}$ also reduces to the "conservation" for $t = \pm\infty$ of the relative kinetic energy, i.e.,

$$\|\mathbf{v}(+\infty)\| = \|\mathbf{v}(-\infty)\|, \quad \mathbf{v} = \mathbf{v}_{\rm t} - \mathbf{v}_{\rm f}. \tag{4.10}$$

Our interaction can be imagined as an *elastic collision*, even if no real collision is occurring. It should be superfluous to stress that the "conservation" of the *modulus* of $\mathbf{v}$ at the beginning and at the end of the interaction does not imply that the *direction* of the relative velocity also remains unchanged (see Figure 4.1), and so in general

$$\Delta\mathbf{v} \equiv \mathbf{v}(+\infty) - \mathbf{v}(-\infty) = \Delta\mathbf{v}_{\perp} + \Delta\mathbf{v}_{\parallel} \neq 0, \tag{4.11}$$

where $\Delta\mathbf{v}_{\perp}$ and $\Delta\mathbf{v}_{\parallel}$ are the components of the change of relative velocity perpendicular and parallel to the initial relative velocity $\mathbf{v}(-\infty)$, respectively. By definition,

$$\begin{cases} \mathbf{v}_{\perp}(-\infty) = 0, \quad \mathbf{v}_{\parallel}(-\infty) = \mathbf{v}(-\infty), \\ \Delta\mathbf{v}_{\perp} = \mathbf{v}_{\perp}(+\infty), \quad \Delta\mathbf{v}_{\parallel} = \mathbf{v}_{\parallel}(+\infty) - \mathbf{v}(-\infty), \end{cases} \tag{4.12}$$

so that from $\mathbf{v}(+\infty) = \mathbf{v}(-\infty) + \Delta\mathbf{v}$ and Eq. (4.10) we obtain the *exact* identity along the relative orbit

$$\|\Delta\mathbf{v}_{\perp}\|^2 + \|\Delta\mathbf{v}_{\parallel}\|^2 + 2\langle\Delta\mathbf{v}_{\parallel}, \mathbf{v}(-\infty)\rangle = 0, \tag{4.13}$$

with the necessary conclusion being that the inner product in (4.13) is negative (i.e., that $\Delta\mathbf{v}_{\parallel}$ is necessarily antiparallel[4] to $\mathbf{v}(-\infty)$). It is useful to consider the positive and negative

[3] It may be of some surprise to the reader to realize that sending a probe to Mercury actually requires a lot of energy. This is because the orbital velocity of Earth around the Sun is much larger than Earth's escape velocity, and the orbital velocity of Earth must be *subtracted* from that of the probe to allow its "fall" toward the Sun!

[4] By definition, $\Delta\mathbf{v}_{\parallel}$ is parallel to $\mathbf{v}(-\infty)$, so that in the limit case of $\Delta\mathbf{v}_{\perp} = 0$, from Eq. (4.13) necessarily $\Delta\mathbf{v}_{\parallel} = 0$, or $\Delta\mathbf{v}_{\parallel} = -2\mathbf{v}(-\infty)$ (prove it!).

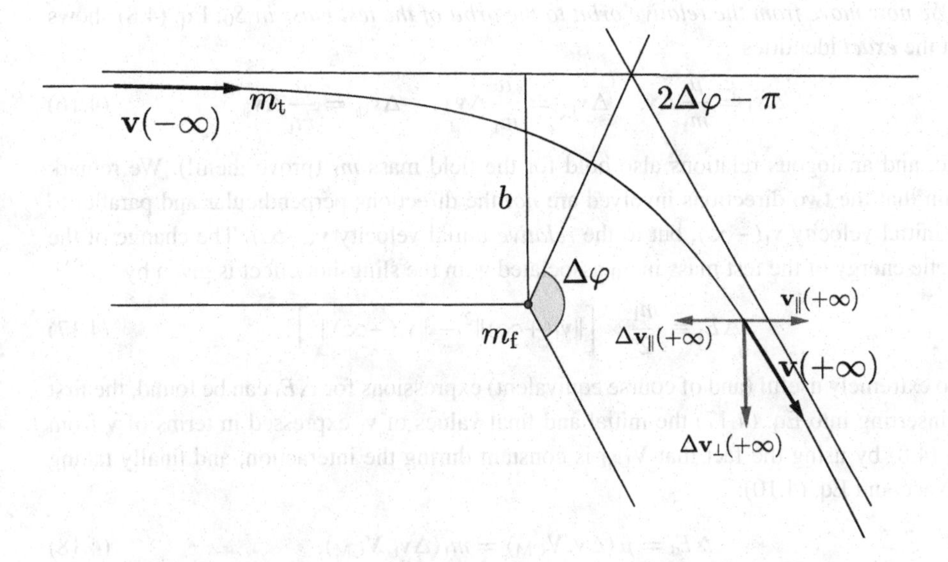

Figure 4.1 Schematic representation of an unbound two-body relative orbit in an attractive force field of potential energy $U(r)$ allowing for escape. In the case of the $1/r^2$ force with $U = -Gm_t m_f/r$ (or in Rutherford's electrostatic scattering experiment), the relative orbit is a hyperbola. For the meaning of the vectorial quantities, see Eqs. (4.10)–(4.12).

terms in Eq. (4.13), respectively, as "heating" and "cooling" of the relative orbit due to the gravitational interaction, and in Chapters 7 and 8 we will elaborate on this point.

Figure 4.1 illustrates qualitatively the geometry of the relative orbit in an unbound elastic interaction. In particular, the rotation angle $\Delta\varphi$ is usually obtained by direct integration of the equations of motion of the reduced mass in polar coordinates (see Exercise 4.7 and the next section for the $1/r^2$ case) and the relative orbit rotation angle is given by $2\Delta\varphi - \pi$ (prove it geometrically!); therefore, $\Delta\varphi = \pi/2$ corresponds to the limiting case of a straight (unperturbed) relative orbit, $\Delta\varphi = 3\pi/4$ to a rotation of the relative orbit of $\pi/2$, and $\Delta\varphi = \pi$ to a complete inversion (rebounce) of the relative velocity. The student should prove the following identities involving the rotation angle,[5]

$$\begin{cases} \|\mathbf{v}_\perp(+\infty)\| = -\|\mathbf{v}(-\infty)\| \sin(2\Delta\varphi), \\ \langle \mathbf{v}_\parallel(+\infty), \mathbf{v}(-\infty)\rangle = -\|\mathbf{v}(-\infty)\|^2 \cos(2\Delta\varphi), \end{cases} \tag{4.14}$$

and finally

$$\|\Delta\mathbf{v}_\perp\| = \|\mathbf{v}_\perp(+\infty)\|, \quad \|\Delta\mathbf{v}_\parallel\| = 2\|\mathbf{v}(-\infty)\| \cos^2(\Delta\varphi). \tag{4.15}$$

[5] Notice that $\sin(2\Delta\varphi) \leq 0$ and $\cos(2\Delta\varphi) \geq 0$; see also Eq. (4.25).

We now move from the relative orbit to the orbit of the test mass in S_0: Eq. (4.8) shows that the *exact* identities

$$\Delta \mathbf{v}_t = \frac{\mu}{m_t}\Delta \mathbf{v}, \quad \Delta \mathbf{v}_{t\perp} = \frac{\mu}{m_t}\Delta \mathbf{v}_\perp, \quad \Delta \mathbf{v}_{t\|} = \frac{\mu}{m_t}\Delta \mathbf{v}_\| \tag{4.16}$$

hold, and analogous relations also hold for the field mass m_f (prove them!). We remark again that the two directions involved are *not* the directions perpendicular and parallel to the initial velocity $\mathbf{v}_t(-\infty)$, but to the *relative* initial velocity $\mathbf{v}(-\infty)$. The change of the kinetic energy of the test mass in S_0 associated with the slingshot effect is given by

$$\Delta E_t \equiv \frac{m_t}{2} \times \left[\|\mathbf{v}_t(+\infty)\|^2 - \|\mathbf{v}_t(-\infty)\|^2\right]. \tag{4.17}$$

Two extremely useful (and of course equivalent) expressions for ΔE_t can be found, the first by inserting into Eq. (4.17) the initial and final values of $\mathbf{v}_t$ expressed in terms of $\mathbf{v}$ from Eq. (4.8) by using the fact that $\mathbf{V}_{CM}$ is constant during the interaction, and finally taking into account Eq. (4.10):

$$\Delta E_t = \mu\langle\Delta \mathbf{v}, \mathbf{V}_{CM}\rangle = m_t\langle\Delta \mathbf{v}_t, \mathbf{V}_{CM}\rangle, \tag{4.18}$$

where the last identity derives immediately from Eq. (4.16). In Eq. (4.27), we will construct a second alternative expression for ΔE_t without explicit reference to $\mathbf{V}_{CM}$, which we will use when considering the collective effects of stellar encounters in astronomical systems.

A few important comments are in order. The first is that, obviously, the total energy of the pair is conserved, with $\Delta E_f = -\Delta E_t$. The second is that, as has already been stressed, the formulae in this section hold for *generic* forces allowing unbound orbits, and for illustrative purposes in Exercises 4.8 and 4.9 we will compute the deflection angle in the limit of very large angular momentum for the case of $1/r^\alpha$ forces. The third is that, according to Eq. (4.18), ΔE_t depends on the *rotation* of the relative velocity vector with respect to $\mathbf{V}_{CM}$, so that both energy gains and losses are possible in S_0, depending on the angle made by $\Delta \mathbf{v}$ and $\mathbf{V}_{CM}$. This property is routinely used for the orbital design of interplanetary missions. This last point is further illustrated by considering a (highly idealized) application of the slingshot effect to planetary rings/protoplanetary disks. Following Figure 4.2, let us consider a body of mass M traveling from right to left in a system S_0 with velocity $\mathbf{v}_M$: it could be a shepherd moon moving counterclockwise in a circular orbit in the plane of the rings of Saturn (placed at the origin of S_0) or a planet embryo rotating in a protoplanetary disk with the protostar at the origin of S_0. Clearly, we are only looking at a little piece of the orbit, so that its curvature is negligible. The object M interacts with a pebble of mass $m \ll M$ (the focus of our attention), also rotating counterclockwise but on a slightly larger circular orbit, with velocity $\mathbf{v}_m$; in case of a Keplerian rotation curve, M will therefore reach and surpass m, and near the ecounter the two velocities will be almost parallel, with a *relative* configuration depicted in Figure 4.1, where $m_f = M$ and $m_t = m$. After the encounter, the body m is pushed down, and from Eq. (4.18) its energy is *increased* because, due to the enormous mass difference, $\mathbf{V}_{CM} \simeq \mathbf{v}_M(-\infty)$, and so the angle between $\mathbf{V}_{CM}$ and $\Delta \mathbf{v}$ is less than $\pi/2$ (see Figure 4.2). Notice that, according to Eqs. (4.30) and

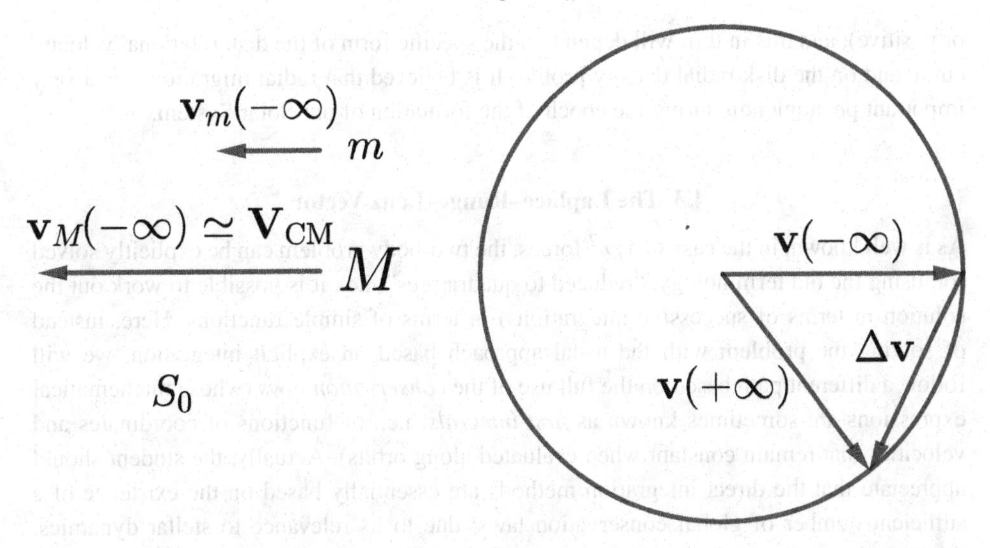

Figure 4.2 Schematic representation of a shepherd moon (M) interacting with a small rock (m) in a planetary ring. The planet center, in the inertial reference system S_0, is at the bottom of the page, and both M and m are rotating counterclockwise. In the right part of the figure, the diagram shows the orientation of the initial and final relative velocities and the total change of the relative velocity; see the text for details.

(4.31), this positive energy variation is associated with a final orbit of m around the center of S_0 with a longer semimajor axis and a longer period; in practice, the gravitational "kick" increases the kinetic energy of m, but as a net result the average kinetic energy decreases (we will come back to this fundamental property of gravity in Chapter 6). The student is encouraged to repeat this discussion for the case of the light body m initially rotating on a circular orbit with a radius smaller than that of the orbit of M, so that in this case M is surpassed by m during the interaction. The result is that now the kinetic energy of m is decreased during the interaction and m is momentarily turned away from the central planet, but the final orbit is characterized by a shorter semimajor axis and higher average kinetic energy than before the interaction.

If we now imagine millions of such encounters of M with little orbiting bodies, we realize that M will "clear" an annular region around its own orbit, pushing the pebbles further out on larger orbits, and sending the pebbles on smaller orbits toward the planet/protostar.[6] So what is the fate of M? In fact, the energy gain/loss of the population of little masses m is associated with a corresponding energy loss/gain of M: a single encounter will do nothing to M, but when we sum over an enormous number of encounters, it is quite obvious that M must *radially migrate* inward or outward as a function of its total energy balance (negative

[6] In ring dynamics, much more complicated dynamical phenomena (e.g., resonances) are at play, producing the opening of "gaps" in the rings (e.g., see Esposito 2014; Goldreich and Tremaine 1982).

or positive), and this in turn will depend on the specific form of the disk rotational velocity curve and on the disk radial density profile. It is believed that radial migration was a very important phenomenon during the epoch of the formation of our Solar System.

4.3 The Laplace–Runge–Lenz Vector

As is well known, in the case of $1/r^2$ forces, the two-body problem can be explicitly solved (or, using the old terminology, "reduced to quadratures"; i.e., it is possible to work out the solution in terms of successive integrations) in terms of simple functions. Here, instead of solving the problem with the usual approach based on explicit integration, we will follow a different path based on the full use of the *conservation laws* (whose mathematical expressions are sometimes known as *first integrals*; i.e., of functions of coordinates and velocities that remain constant when evaluated along orbits). Actually, the student should appreciate that the direct integration methods are essentially based on the existence of a sufficient number of global conservation laws: due to its relevance to stellar dynamics, this argument will be discussed again in Chapter 9. Here, we simply recall that when a dynamical system (a differential equation) admits a sufficient number of globally conserved quantitites, the system is said to be *integrable* in the sense of Lie–Jacobi. One of the most important and somewhat unexpected breakthroughs in mathematics and physics (with important consequences for classical and quantum mechanics) was the recognition, due to Poincaré (1854–1912) while working on the three-body problem, that dynamical systems may not have a sufficient number of such conservation laws (Poincaré 1892; see Barrow-Green 1997 for a very interesting and readable account). In practice, it was proved that some of the most significant problems that resisted solution did so not because the great mathematicians working on them were not sufficiently smart, but because these problems belong to the class of *nonintegrable* systems. Of course, *nonintegrable* systems can also have perfectly well-behaved solutions (in the sense of Cauchy; i.e., regular, unique, and continuously dependent on the initial conditions), but these solutions *cannot* in general be described by the intersection of global analytic varieties in phase space (obtained by fixing the values of the global conservation laws from initial conditions). The interested reader can find very clear discussions of some of these fascinating problems which are of great importance for celestial mechanics and stellar dynamics, in several excellent books (e.g., see Arnold 1978; Bertin 2014; Binney and Tremaine 2008; Buchler et al. 1988; Contopoulos et al. 1994; Evans 1990; Gutzwiller 1990; Lichtenberg and Lieberman 1992; McCauley 1997; Whittaker 1917; Wintner 1947), where other issues (e.g., how additional conservation laws, in addition to the "natural" ones from *Noether's theorem* and evident symmetries of the potential, can be discovered with specific procedures) are also discussed.

Here, due to the introductory nature of this book, we will only show that in the two-body problem with $1/r^2$ forces we have enough conservation laws (in addition to the 10 classical ones) that the solution can be determined without resorting to explicit integration (see also Chapter 9 for further discussion). In fact, from Exercise 4.3

$$\frac{d}{dt}\frac{\mathbf{r}}{r} = \frac{\mathbf{v}}{r} - \frac{\langle \mathbf{r}, \mathbf{v} \rangle \mathbf{r}}{r^3} = \frac{\mathbf{J}_{\mathrm{r}} \wedge \mathbf{r}}{\mu r^3}. \tag{4.19}$$

The next step is to multiply the previous identity by $-Gm_{\mathrm{t}}m_{\mathrm{f}}$ and then follow the surprising chain of identities:

$$\frac{d}{dt}\left(-\frac{Gm_{\mathrm{t}}m_{\mathrm{f}}\mathbf{r}}{r}\right) = \frac{\mathbf{J}_{\mathrm{r}}}{\mu} \wedge \left(-\frac{Gm_{\mathrm{t}}m_{\mathrm{f}}\mathbf{r}}{r^3}\right) = \frac{\mathbf{J}_{\mathrm{r}}}{\mu} \wedge \frac{d\mu\mathbf{v}}{dt} = -\frac{d}{dt}(\mathbf{v} \wedge \mathbf{J}_{\mathrm{r}}), \tag{4.20}$$

where we used Eq. (4.6) for the relative orbit and the time independence of $\mathbf{J}_{\mathrm{r}}$. We conclude that on the relative orbit, the vectorial quantity

$$Gm_{\mathrm{t}}m_{\mathrm{f}}\mathbf{l} \equiv \mathbf{v} \wedge \mathbf{J}_{\mathrm{r}} - \frac{Gm_{\mathrm{t}}m_{\mathrm{f}}}{r}\mathbf{r}, \quad l = \|\mathbf{l}\| \tag{4.21}$$

is conserved, where the dimensionless constant vector $\mathbf{l}$ is the so-called[7] *Laplace–Runge–Lenz vector*. First, we note that $\mathbf{l}$ is placed in the orbital plane (i.e., in the plane perpendicular to $\mathbf{J}_{\mathrm{r}}$). Second, we multiply the identity above by $\mathbf{r}$ and we call φ the angle between $\mathbf{l}$ and $\mathbf{r}$ (the *true anomaly*), so that $\langle \mathbf{l}, \mathbf{r} \rangle = r\, l \cos\varphi$, and from Exercise 4.4 we obtain the fundamental result

$$r = \frac{p}{1 + l\cos\varphi}, \tag{4.22}$$

where

$$p = \frac{\|\mathbf{J}_{\mathrm{r}}\|^2}{Gm_{\mathrm{t}}m_{\mathrm{f}}\mu}, \quad l = \sqrt{1 + \frac{2E_{\mathrm{r}}\|\mathbf{J}_{\mathrm{r}}\|^2}{G^2m_{\mathrm{t}}^2m_{\mathrm{f}}^2\mu}}. \tag{4.23}$$

Equation (4.22) is the polar form of a conic, where p is the *semilatus rectum*, so that $\mathbf{l}$ is oriented toward the pericenter of the relative orbit and its length is the *orbital eccentricity*. From Eq. (4.23), it turns out that orbital classification depends on the sign of the barycentric energy: in particular, $l > 1$ for $E_{\mathrm{r}} > 0$ and the relative orbit is hyperbolic, while $l = 1$ for $E_{\mathrm{r}} = 0$ and the corresponding orbit is parabolic. For completeness, in Exercise 4.5 the basic properties of elliptic orbits ($0 \le l < 1$, $E_{\mathrm{r}} < 0$) are finally determined.

Now, as the total energy and the total angular momentum are conserved, they can be evaluated on the relative orbit at $t = -\infty$ (i.e., on the asymptote at the beginning of the impact), and from Figure 4.1 and Eq. (4.9)

$$\|\mathbf{J}_{\mathrm{r}}\| = \mu b\|\mathbf{v}(-\infty)\|, \quad E_{\mathrm{r}} = \mu\frac{\|\mathbf{v}(-\infty)\|^2}{2}, \tag{4.24}$$

where b is the so-called *impact parameter*. The impact parameter can then be seen geometrically as the *minimum* distance that the two bodies would reach on the relative orbit *in the absence of interaction* (see Figure 4.1), or dynamically, as a measure of the barycentric angular momentum of the pair, with large b corresponding to large angular momentum at a given asymptotic relative velocity. Again from Figure 4.1, it is obvious that the direction

[7] The literature on the Laplace–Runge–Lenz vector and its possible generalizations is enormous; interesting introductory papers are Goldstein (1975) and Yoshida (1987); see also Exercise 5.21.

of the asymptote with respect to the direction of the pericenter (fixed by the direction of $\mathbf{l}$) is obtained by finding the zero of the denominator in Eq. (4.22), i.e.,

$$\cos(\Delta\varphi) = -\frac{1}{l}, \quad l = \sqrt{1 + \frac{b^2 \|\mathbf{v}(-\infty)\|^4}{G^2 M^2}}, \quad M = m_t + m_f, \tag{4.25}$$

where the expression for l is obtained for the hyperbolic orbit from Eqs. (4.23) and (4.24). As expected, when $b \to \infty$ we obtain $\Delta\varphi = \pi/2$, while $\Delta\varphi = \pi$ for $b \to 0$. Moreover, as we have seen, a rotation of the relative orbit of $\pi/2$ corresponds to $\Delta\varphi = 3\pi/4$, and from Eq. (4.25) this translates (prove it!) to the special value $b_{\pi/2}$ of the impact parameter, and finally to a particularly simple expression for l in terms of b and $b_{\pi/2}$:

$$l_{\pi/2} = \sqrt{2}, \quad b_{\pi/2} = \frac{GM}{\|\mathbf{v}(-\infty)\|^2}, \quad l = \sqrt{1 + \frac{b^2}{b_{\pi/2}^2}}. \tag{4.26}$$

Exercises

4.1 Obtain the general transformation formulae for the total energy and linear and angular momentum of the two-body problem as seen from two inertial reference systems moving with relative velocity $\mathbf{V} = d\mathbf{R}/dt$. Specialize to the case $\mathbf{V} = \mathbf{V}_{\rm CM}$, further restrict to the case $\mathbf{R} = \mathbf{R}_{\rm CM}$, and finally prove Eq. (4.5). *Hint*: See Appendix A.1.1 for the general case.

4.2 Show that the change in the kinetic energy of the test star m_t in the reference system S_0, defined in Eq. (4.17), can be written as

$$\begin{aligned}\Delta E_t &= \frac{m_t}{2}\left[\|\Delta\mathbf{v}_t\|^2 + 2\langle\Delta\mathbf{v}_t, \mathbf{v}_t(-\infty)\rangle\right] \\ &= \Delta E_{t\perp} + \Delta E_{t\parallel} + m_t\langle\Delta\mathbf{v}_t, \mathbf{v}_t(-\infty)\rangle,\end{aligned} \tag{4.27}$$

where $\Delta E_{t\perp} = m_t\|\Delta\mathbf{v}_{t_\perp}\|^2/2$ and $\Delta E_{t\parallel} = m_t\|\Delta\mathbf{v}_{t\parallel}\|^2/2$ are the energy gains associated with $\Delta\mathbf{v}_{t_\perp}$ and $\Delta\mathbf{v}_{t\parallel}$ defined in Eq. (4.16). Prove the equivalence of Eqs. (4.18) and (4.27). The first two terms in the last expression of Eq. (4.27) can be interpreted as "heating," and in Chapter 8 we will prove that the last term is a "cooling." *Hint*: Use the identity $\mathbf{v}_t(+\infty) = \mathbf{v}_t(-\infty) + \Delta\mathbf{v}_t$ in the definition of ΔE_t and then use Eq. (4.16).

4.3 Prove Eq. (4.19). *Hint*: After time differentiation, use Eq. (A.22) with $\mathbf{b} = \mathbf{v}$ and $\mathbf{a} = \mathbf{c} = \mathbf{r}$.

4.4 Prove Eqs. (4.22) and (4.23). *Hints*: For Eq. (4.22), multiply Eq. (4.21) by $\mathbf{r}$ and use the identity $\langle\mathbf{v}\wedge\mathbf{J}_r, \mathbf{r}\rangle = \langle\mathbf{r}\wedge\mathbf{v}, \mathbf{J}_r\rangle = \|\mathbf{J}_r\|^2/\mu$. This also proves the expression of p in Eq. (4.23). In order to express the modulus l, consider the squared norm of Eq. (4.21) and the fact that $\|\mathbf{v}\wedge\mathbf{J}_r\|^2 = \|\mathbf{v}\|^2\|\mathbf{J}_r\|^2$ from the orthogonality of $\mathbf{v}$ and $\mathbf{J}_r$. Obtain

$$G^2 m_t^2 m_f^2 l^2 = G^2 m_t^2 m_f^2 - \frac{2Gm_t m_f}{\mu r} \|\mathbf{J}_r\|^2 + \|\mathbf{v}\|^2 \|\mathbf{J}_r\|^2$$
$$= G^2 m_t^2 m_f^2 + \frac{2\|\mathbf{J}_r\|^2}{\mu} \left(\frac{\mu \|\mathbf{v}\|^2}{2} - \frac{Gm_t m_f}{r} \right)$$
$$= G^2 m_t^2 m_f^2 + \frac{2\|\mathbf{J}_r\|^2 E_r}{\mu}. \tag{4.28}$$

4.5 The solution of the two-body problem was one of the greatest achievements of classical physics, providing the theoretical explanation for the three Kepler (1610) Laws of planetary motions. By using Eq. (4.22), prove that in the case of negative relative energy, when the orbital eccentricity is $0 \leq l < 1$ and the orbit is elliptic, the *apocenter* and the *pericenter* are given by

$$r_{\max} = \frac{p}{1-l}, \quad r_{\min} = \frac{p}{1+l}, \tag{4.29}$$

so that the semimajor and semiminor axes are

$$a = \frac{p}{1-l^2} = \frac{Gm_t m_f}{2|E_r|}, \quad b = a\sqrt{1-l^2} = \frac{p}{\sqrt{1-l^2}} = \frac{\|\mathbf{J}_r\|}{\sqrt{2|E_r|\mu}}. \tag{4.30}$$

Finally, show that the period of the orbit is

$$P = \frac{2\mu\pi ab}{\|\mathbf{J}_r\|} = \frac{2\pi a^{3/2}}{\sqrt{GM}}, \quad M = m_t + m_f. \tag{4.31}$$

Hints: From the conservation of angular momentum, and using polar coordinates in the orbital plane, conclude that

$$\|\mathbf{J}_r\| = \mu r^2(\varphi) \frac{d\varphi}{dt} = 2\mu \frac{d\mathcal{A}}{dt}, \tag{4.32}$$

where $d\mathcal{A} = r^2(\varphi) d\varphi/2$ is the area element as a function of φ. Then integrate over the period P and recall that $\mathcal{A} = \pi ab$ is the area of the ellipse.

4.6 By using Eqs. (4.14), (4.25), and (4.26), show that

$$\|\Delta \mathbf{v}_\perp\| = 2\|\mathbf{v}(-\infty)\| \frac{b/b_{\pi/2}}{1 + b^2/b_{\pi/2}^2}, \quad \|\Delta \mathbf{v}_\parallel\| = \frac{2\|\mathbf{v}(-\infty)\|}{1 + b^2/b_{\pi/2}^2}. \tag{4.33}$$

Obtain the leading term of the asymptotic expansion in the case of very large barycentric angular momentum (i.e., for $b \to \infty$).

4.7 Consider the two-body problem with interaction energy $U(r)$ allowing for unbound orbits. Write Eq. (4.9) in polar coordinates and show that the rotation angle $\Delta\varphi$ in Figure 4.1 can be written as

$$\Delta\varphi = \frac{\|\mathbf{J}_r\|}{\sqrt{2\mu}} \int_{r_{\min}}^{\infty} \frac{dr}{r^2 \sqrt{E_r - U_{\text{eff}}(r)}}, \quad U_{\text{eff}}(r) = \frac{\|\mathbf{J}_r\|^2}{2\mu r^2} + U(r), \tag{4.34}$$

where $r_{\min}$ is the solution of the equation

$$E_{\mathrm{r}} = U_{\mathrm{eff}}(r_{\min}). \tag{4.35}$$

Then compute the integral for the $1/r^2$ force and show that Eq. (4.25) is reobtained once Eq. (4.24) is used. The student attempting this exercise will appreciate the elegance of the approach based on the Laplace–Runge–Lenz vector. *Hints*: For the derivation of Eq. (4.34), consult Landau and Lifshitz (1969) or Goldstein et al. (2000). In order to perform the integration, write E_{r} as in Eq. (4.35), change the variable to $x = 1/r$, and compute $x_{\min} = 1/r_{\min}$ by solving a quadratic equation; trigonometric identities will finally conclude the proof. See also Goldstein et al. (2000) for a list of special values of α allowing for the integration of Eq. (4.34) in terms of special functions.

4.8 Equation (4.34) cannot be solved in closed form for generic $U(r)$, even restricting to the class of power-law $1/r^{\alpha}$ forces in Eq. (1.13). However, it is possible to obtain $\Delta\varphi$ in the case of large angular momentum (i.e., for $b \to \infty$). Show that for $1 < \alpha < 3$ and $U(r) = -Gm_{\mathrm{t}}m_{\mathrm{f}}r^{1-\alpha}/(\alpha - 1)$ an asymptotic expansion[8] gives

$$\Delta\varphi \sim \frac{\pi}{2} + \frac{Gm_{\mathrm{t}}m_{\mathrm{f}}}{2\mu\|\mathbf{v}(-\infty)\|^2 b^{\alpha-1}} \mathrm{B}\left(\frac{1}{2}, \frac{\alpha}{2}\right), \tag{4.36}$$

where $\mathrm{B}(x, y)$ is the Euler beta function in Eq. (A.53). Moreover, from Eq. (4.15), prove that for distant encouters

$$\mu\|\Delta\mathbf{v}_{\perp}\| \sim \frac{Gm_{\mathrm{t}}m_{\mathrm{f}}}{b^{\alpha}} \frac{b}{\|\mathbf{v}(-\infty)\|} \mathrm{B}\left(\frac{1}{2}, \frac{\alpha}{2}\right). \tag{4.37}$$

Hints: First, show that the leading term of the solution $r_{\min}$ of Eq. (4.35) for $b \to \infty$ is $r_{\min} \sim b$, independent of the value of α. Then expand the integrand in Eq. (4.34) for $b \to \infty$. The integration of the first term is trivial, but the second requires some care, and it can be performed in parts.

4.9 With this exercise we prove the validity of a classical approach to the computation of the orbital deflection in case of a large impact parameter (e.g., see Binney and Tremaine 2008; see also chapter 4 in Landau and Lifshitz 1969). Consider again the force in Exercise 4.8, and assume that for large b the relative orbit remains straight and the relative velocity constant. Show that the time integral of the component of the force perpendicular to $\mathbf{v}(-\infty)$ along the fictitious path is

$$\mu\|\Delta\mathbf{v}_{\perp}\| \sim Gm_{\mathrm{t}}m_{\mathrm{f}}b \int_{-\infty}^{+\infty} \frac{dt}{\left[b^2 + \|\mathbf{v}(-\infty)\|^2 t^2\right]^{\frac{\alpha+1}{2}}}, \tag{4.38}$$

and that it evaluates to Eq. (4.37). *Hint*: Decompose the instantaneous force along the rectilinear orbit as $\mathbf{g} = \mathbf{g}_{\perp} + \mathbf{g}_{\parallel}$ and recast the integral as a complete beta function in Eq. (A.53).

[8] The lower limit on α is needed to allow for unbound orbits and the upper limit is needed to exclude the possibility of a collision of the two particles, even for a nonzero angular momentum.

5

Quasi-Circular Orbits

This chapter presents the basic properties of quasi-circular orbits in axisymmetric stellar systems. In fact, axisymmetric models are often sufficiently realistic descriptions – beyond the zeroth-order spherical case – of elliptical and disk galaxies. The associated potentials (with a reflection plane, the equatorial plane) admit circular orbits, and in this chapter we focus on the properties of orbits slightly departing from perfect circularity, describing in some detail the second-order epicyclic approximation. We also derive, in a geometrically rigorous way, the expression of the Oort constants, which are important kinematic quantities related to the rotation curve of disk galaxies, and in particular to the orbit of the Sun around the center of the Milky Way.

5.1 Orbits in Axisymmetric Potentials

Stellar orbits in generic gravitational potentials can be very complicated, and explicit solutions of the Newton equations of motion cannot in general be found in a closed form. This may appear to be a drawback to the student, as the availability of explicit solutions certainly seems desirable. However, even when available, a very complicated formula seldom leads to a deep understanding of the physics behind the problem, and any gains achieved from such complicated formulae may be illusory. At the same time, the work needed to understand a problem that cannot be solved in full analytical form quite often is more useful than the knowledge of the formal solution itself. As we will see, this is exactly the situation we face in this chapter (i.e., orbits in axisymmetric potentials); in fact, even if the hypothesis of perfect axisymmetry for the potential lead to great simplifications, for a generic $\phi(R,z)$ in general it is not possible to obtain a closed-form expression for the orbits, except for very special cases such as circular orbits in the equatorial plane and axial orbits along the symmetry axis. Here, thanks to a very clever idea originating from the great mathematicians of ancient Greece, we will study the properties of orbits slightly departing from circular paths, by using *epicyclic theory* (e.g., see Gallavotti 2001; Schiaparelli 1926 for presentations of remarkable clarity of how epicyclic theory translates in the language of modern physics, being a powerful mathematical tool). Of course, axisymmetric systems are of special interest in astrophysics, as many stellar systems are flattened and, at first approximation, axisymmetric. For example, spiral galaxies and S0 galaxies are perhaps the most prominent examples, but we should also remember accretion disks around black holes and planetary rings (e.g., those of Saturn).

We begin by writing the equations of motion for a particle/star (of unit mass) in a generic axisymmetric systems whose potential is $\phi(R,z)$ and (R,φ,z) are the standard cyclindrical coordinates. It is a useful exercise for the student to derive the equations from the Newton equations of dynamics in vector formulation, but here we follow the shorter Lagrangian approach. The Lagrangian function of our star is

$$\mathcal{L} = \frac{\dot{R}^2}{2} + \frac{\dot{z}^2}{2} + \frac{(R\dot{\varphi})^2}{2} - \phi(R,z), \tag{5.1}$$

so that the equations for the three generalized coordinates are obtained from the well-known Euler–Lagrange equation $d(\partial\mathcal{L}/\partial\dot{q}_i)/dt = \partial\mathcal{L}/\partial q_i$. In particular, the momentum conjugate to the coordinate φ is conserved, and this momentum is none other that the z component[1] of the angular momentum of the star, i.e.,

$$J_z = R^2\dot{\varphi}. \tag{5.3}$$

The equations for the two remaining coordinates can be written as

$$\ddot{R} = -\frac{\partial\phi_{\rm e}}{\partial R}, \quad \ddot{z} = -\frac{\partial\phi}{\partial z} = -\frac{\partial\phi_{\rm e}}{\partial z}, \tag{5.4}$$

where the *effective potential*

$$\phi_{\rm e}(R,z) \equiv \phi(R,z) + \frac{J_z^2}{2R^2} \tag{5.5}$$

has been introduced. Therefore, the motion of the star would be known if one first solves Eq. (5.4) in the *meridional plane* (R,z) and then solves Eq. (5.3) for the time evolution of the angle φ; by construction, $\dot{\varphi}$ is also the instantaneous angular velocity of the meridional plane. Notice that, independently of how complicated the time evolution of φ is, $\dot{\varphi}$ never changes sign, as J_z is constant and $R(t) \geq 0$: no "inversion" in the sense of rotation of the angular coordinate can occur in axisymmetric potentials.

An important point to be appreciated is that the reduction of the equations of motion from $\Re^3$ to the meridional plane is done by fixing the value of J_z; initial conditions associated with different values of J_z evolve into different effective potentials. We will refer to the set of orbits corresponding to a given J_z as an *orbital family*. Of course, fixing J_z by no means fully constrains the orbits, as an infinite number of different orbits (and different meridional planes!) exist for a given J_z.

From Eq. (5.4), the energy (per unit mass) in the meridional plane of each orbit

$$E_{\rm m} = \frac{\dot{R}^2}{2} + \frac{\dot{z}^2}{2} + \phi_{\rm e}(R,z) = E \tag{5.6}$$

[1] Notice that from Exercise 9.1 and Eq. (A.144) the angular momentum per unit mass in cylindrical coordinates reads

$$\mathbf{J} = \mathbf{x}\wedge\mathbf{v} = -zv_\varphi\mathbf{e}_R + (zv_R - Rv_z)\mathbf{e}_\varphi + Rv_\varphi\mathbf{e}_z = J_R\mathbf{e}_R + J_\varphi\mathbf{e}_\varphi + J_z\mathbf{e}_z. \tag{5.2}$$

is conserved, and its value coincides with the value E of the energy of the orbit in physical space, as obtained from Eq. (5.1). It immediately follows that the motion is limited to the region of the meridional plane defined by

$$E \geq \phi_e(R,z), \tag{5.7}$$

where the equality identifies the *zero-velocity curve* (i.e., the locus where $v_R = v_z = 0$); of course, when a star hits the zero-velocity curve, its velocity is in general *not* zero (think of Earth's velocity at the perihelion and at the aphelion!) because from Eq. (5.3) $v_\varphi = J_z/R$.

In order to develop our epicyclic theory, we identify the most important member of each orbital family (i.e., the *circular orbit* in the equatorial plane corresponding to the given J_z). Circular orbits for a given value of J_z are obtained by setting to zero the accelerations in the meridional plane from Eq. (5.4), and it is easy to show that for regular potentials with a reflection plane (i.e., $\phi(R,z) = \phi(R,-z)$) the radius R_0 of the circular orbit(s) of angular momentum J_z in the equatorial plane is given by the solution(s) of the equation

$$J_z^2 = R^3 \frac{\partial \phi(R,0)}{\partial R}, \tag{5.8}$$

obtained from the equilibrium points of Eq. (5.4). In particular, the rotation curve of the potential in the equatorial plane is given by

$$v_c^2(R) = \Omega^2(R)\, R^2 = R \frac{\partial \phi(R,0)}{\partial R} = \frac{J_z^2}{R^2}, \tag{5.9}$$

provided the radial derivative of the potential is positive. In the following, we will assume that this condition is satisfied (the usual condition!), but we recall that it is possible to construct nonspherical, nowhere-negative density distributions for which the positivity of v_c^2 is violated (e.g., see Exercise 5.7). Of course, in the special case of spherically symmetric mass distributions, from Newton's second theorem the rotation curve at distance r from the center depends only on the mass $M(r)$ contained in a sphere of radius r, i.e.,

$$v_c^2(r) = \frac{GM(r)}{r} \geq 0. \tag{5.10}$$

Equations (5.8) and (5.9) allow us to determine R_0 for the assigned potential and J_z. However, R_0 of an orbital family can also be determined by assigning the orbital total energy in Eq. (5.6) as

$$E_m = \phi_e(R_0,0) = \frac{v_c^2(R_0)}{2} + \phi(R_0,0) = \frac{R_0}{2} \frac{\partial \phi(R_0,0)}{\partial R_0} + \phi(R_0,0). \tag{5.11}$$

See also Exercise 5.9.

We conclude this preliminary section by reminding ourselves of the fundamental importance of the study of *rotation curves* of cold gas (in particular by using the 21-cm emission line of neutral hydrogen, HI) in spiral galaxies, one of the major diagnostics for the study of dark matter in disk galaxies[2] (e.g., van Albada et al. 1985); we do not discuss this subject here, as it is treated extensively in the literature (e.g., see Bertin 2014; Bertin and Lin 1996; Binney and Tremaine 2008 and references therein). The exercises provide several cases of the explicit construction of rotation curves in the most used (and idealized) disk models; the student is encouraged to plot the resulting curves for different choices of parameters and to verify that quite flat rotation curves can be easily produced inside a disk by the field of the disk itself without need for dark matter. Only far outside the optical (stellar) disk does the flat rotation curve require a stronger field than that provided by luminous (baryonic) matter, which in those regions would be dominated by the monopole term of the disk (see Chapter 2), with the characteristic Keplerian falloff.

5.2 Second-Order Epicyclic Approximation

The idea behind the epicyclic approximation discussed here is to expand the effective potential of an orbital family up to the second order that is inclusive of small displacements around a circular orbit of radius R_0 for fixed values of J_z. We assume $R(t) = R_0 + r(t)$ with $|r(t)|/R_0 \ll 1$ and $|z(t)|$ being sufficiently small to guarantee the expansion, so that

$$\phi_e(R,z) \sim \phi_e(R_0,0) + \frac{\kappa_R^2}{2} r^2 + \frac{\kappa_z^2}{2} z^2, \tag{5.12}$$

where the *radial* and *vertical* epicyclic frequencies are defined as

$$\kappa_R^2 \equiv \left.\frac{\partial^2 \phi_e}{\partial R^2}\right|_{(R,z)=(R_0,0)}, \qquad \kappa_z^2 \equiv \left.\frac{\partial^2 \phi_e}{\partial z^2}\right|_{(R,z)=(R_0,0)}. \tag{5.13}$$

Note that in Eq. (5.12) the two linear terms in the displacements vanish identically on the circular orbit of radius R_0, that the mixed second-order derivative vanishes for the regularity and reflection properties of the potential on the plane $z = 0$, and finally that κ_z is independent of J_z. If we now insert Eq. (5.12) into Eq. (5.4), we see that at second order the equations of motion for the displacements are those of two independent harmonic oscillators

$$\ddot{r} = -\kappa_R^2 r, \quad \ddot{z} = -\kappa_z^2 z, \tag{5.14}$$

with oscillatory solutions, provided $\kappa_R^2 > 0$ and $\kappa_z^2 > 0$. Therefore, we have stable perturbed orbits for real values of the epicyclic frequencies and (linearly) unstable motions when at least one of the two frequencies is imaginary. In particular, it is easy to prove (do it!) that

$$\kappa_R^2(R) = \left.\frac{1}{R^3}\frac{dJ_c^2(R)}{dR}\right|_{R=R_0} = \left.\frac{2\Omega(R)}{R}\frac{d\Omega(R)}{dR}R^2\right|_{R=R_0}, \tag{5.15}$$

[2] Observationally, the circular velocity of HI in disk galaxies of large radii presents a characteristic flat shape with a remarkable proportionality between the total stellar light of the disk and (approximately) the fourth power of the circular velocity: the so-called *Tully–Fisher* (Tully and Fisher 1977) relation.

where $J_c(R) = Rv_c(R)$ is the angular momentum (per unit mass) of circular orbits in the equatorial plane as a function of R; of course, for a given orbital family, $J_c(R_0) = J_z$. From Eq. (5.15) follows the very important (and very useful in practical applications) *Rayleigh criterion* (i.e., stable circular orbits can exists only if $J_c(R)$ is a *monotonically increasing* function of the radius). Such a result can be extended to fluid dynamics, leading to a stability criterion for equilibrium configurations of rotating fluids (e.g., Tassoul 1978).

From Eqs. (5.7) and (5.12), it is now elementary to obtain the expression for the zero-velocity curves in our approximation, with

$$\frac{\kappa_R^2}{2}r^2 + \frac{\kappa_z^2}{2}z^2 \leq E_m - \phi_e(R_0, 0). \tag{5.16}$$

For linearly unstable orbits the zero-velocity curves can be open, while for stable orbits the zero-velocity curves are ellipses in the meridional plane. Of course, for increasing values of E_m the shapes of the true zero-velocity curves as obtained from the full Eq. (5.7) will in general depart more and more for the ellipses of the second-order expansion (e.g., see Exercise 5.10).

We now proceed with the discussion of epicyclic orbits, limiting ourselves to the case of stability. The solution of the linearized equations of motion in Eq. (5.14) is

$$r(t) = r_0 \cos(\kappa_R t), \quad z(t) = z_0 \sin(\kappa_z t), \tag{5.17}$$

where for simplicity we assume that at time $t = 0$ the orbit is on the equatorial plane and at maximum displacement from the center (apocenter). From the geometry of the problem, the resulting orbit can be expressed in vector notation as (prove it!)

$$\mathbf{x}(t) = \mathbf{R}_0(t) + \mathcal{C}(t)\mathbf{x}' = \mathcal{C}(t)\,(R_0\mathbf{f}_1 + \mathbf{x}'), \tag{5.18}$$

where $\mathbf{x}'(t) = [X(t), Y(t), z(t)]$ is the orbit as seen from the rotating reference system of the deferent (also known as the guiding center; see Figure 5.1), $\mathbf{f}_1$ is the unitary vector aligned with the x-axis, $\mathbf{R}_0(t)$ is the position of the deferent, and finally $\mathcal{C}(t)$ is the rotation matrix in Eq. (3.34); with this equation, the student can draw the orbits in their preferred potential (e.g., see Figure 5.2).

We now move to study the properties of $\mathbf{x}'(t)$ in some detail. From Eq. (5.17), the orbit will reach the pericenter after a time of π/κ_R, return again to the apocenter after the radial period $T_R = 2\pi/\kappa_R$, and so on; in the time T_R between two passages of the orbit from the apocenter, the circular orbit of radius R_0 (the so-called *deferent*) rotates by the amount

$$\Delta\varphi = \frac{2\pi\Omega_0}{\kappa_R}. \tag{5.19}$$

As in general κ_R and κ_z depend on R_0, we can conclude that if for some special value of R_0 the equality

$$\Omega_0 = \frac{m}{n}\kappa_R \tag{5.20}$$

holds for m and n integer numbers, then the orbit will close in the inertial reference system after m revolutions of R_0 and n complete radial oscillations of the epicycle (i.e., n radial

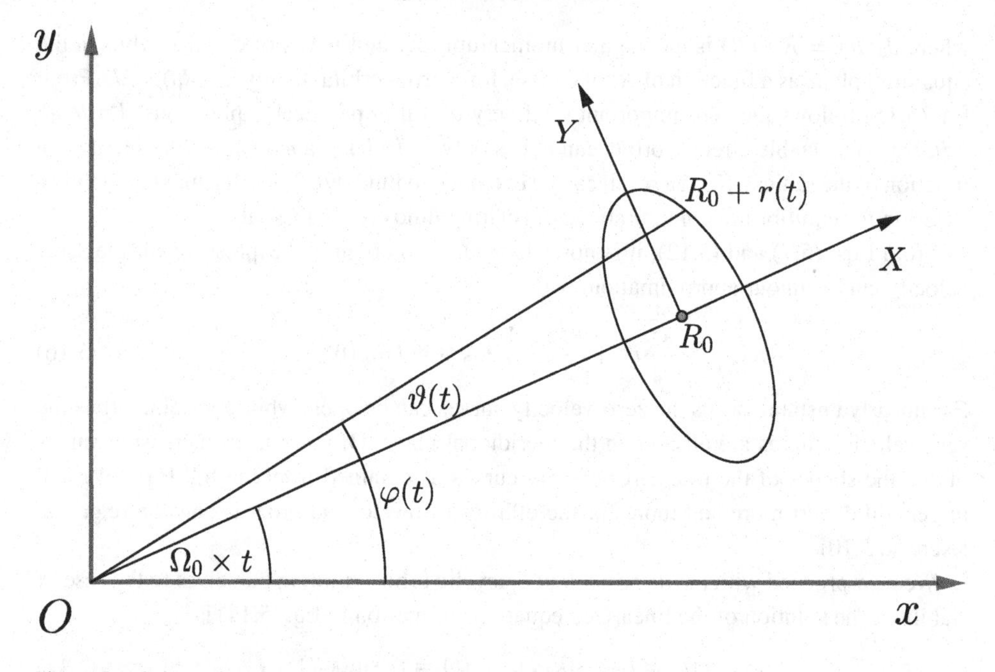

Figure 5.1 Schematic representation of a deferent orbit of radius R_0 (rotating counterclockwise) and of the associated epicyclic ellipse (rotating clockwise) in the equatorial plane of an axisymmetric potential. The instantaneous value of the angle $\vartheta(t) = \varphi(t) - \Omega_0 t$ is given by Eq. (5.22) and the distance of the orbit from the center is $R_0 + r(t)$.

"inversions" of the orbit; again see Figure 5.2). We will return to this important point at the end of this section, while in Exercise 5.20 we will use Eq. (5.19) to obtain the expression for orbital precession/recession angle, which is a very important property of quasi-circular orbits with consequences for classical mechanics and general relativity (e.g., see Landau and Lifshitz 1971).

As was already well understood in antiquity, the appearance of the epicycle is of the utmost importance when observed from the guiding center. This is particularly important in the (X, Y) plane (i.e., the equatorial plane of the potential; see Figure 5.1), and we now solve this problem using the methods of analytic geometry. The first step is to obtain the differential equation for the instantaneous relative position of the meridional plane of the star with respect to the position of the meridional plane of the deferent. Expanding at the first order Eq. (5.3) in terms of the small number r_0/R_0, one gets

$$\dot{\varphi} = \Omega_0 - \frac{2\Omega_0}{R_0} r(t), \tag{5.21}$$

so that the instantaneous angular distance between the guiding center and the star in the (X, Y) plane is given by

$$\vartheta(t) = \int_0^t (\dot{\varphi} - \Omega_0) dt = -\frac{2\Omega_0}{\kappa_R R_0} r_0 \sin(\kappa_R t). \tag{5.22}$$

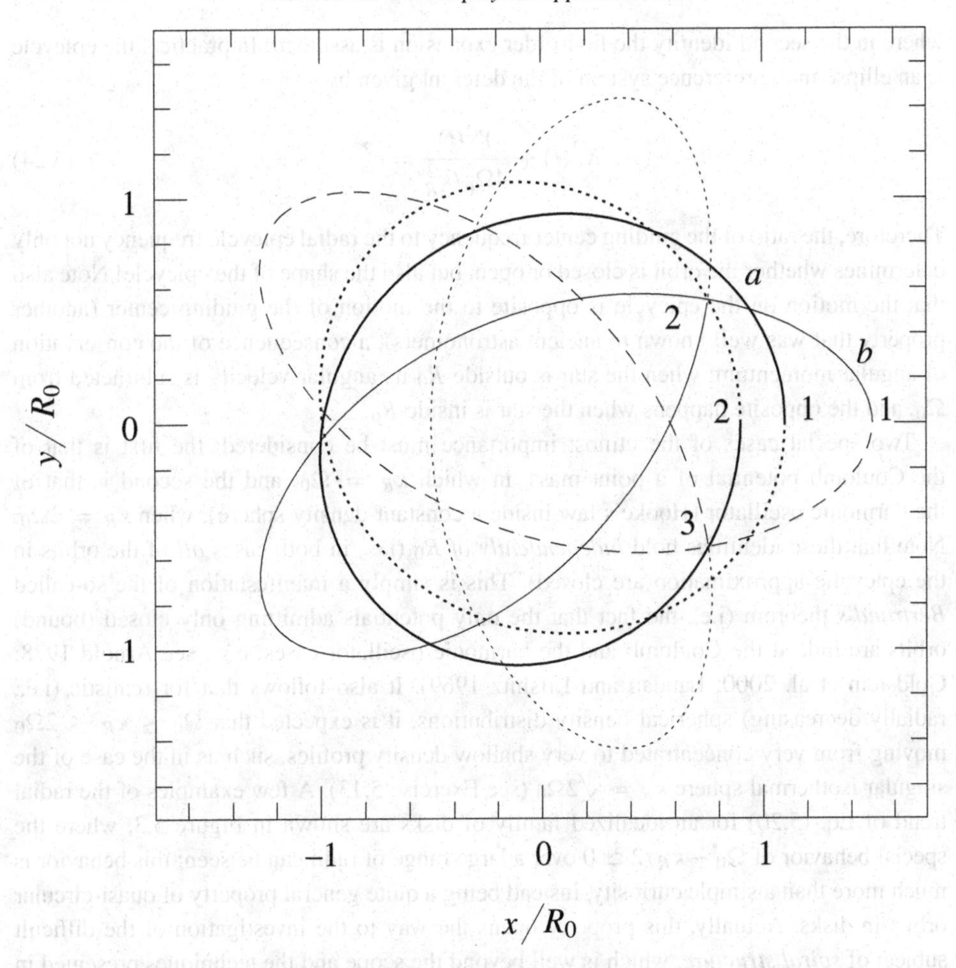

Figure 5.2 Two examples of rosette orbits $\mathbf{x}(t)$ in the equatorial plane of an axisymmetric potential built using epicyclic approximation and Eq. (5.18). The lengths are normalized to the deferent radius R_0 and $(r_0/R_0, \kappa_R/\Omega_0) = (1/10, 3/2)$ for orbit a (heavy line) and $(1/2, 5/3)$ for orbit b. Both orbits start at positions 1, and after a first rotation of the deferent they reach positions 2. After a second rotation of the deferent orbit a closes, while orbit b closes after a third rotation of the deferent. Consistently with the adopted parameters, orbit a presents three apocenters/pericenters after two rotations of the deferent, while five are presented by orbit b after three rotations of the deferent.

Now, an elementary geometric construction (see Figure 5.1) shows that the Cartesian coordinates of the epicyclic orbit as seen from the guiding center are

$$\begin{cases} X(t) = [R_0 + r(t)]\cos\vartheta(t) - R_0 \sim r(t) = r_0\cos(\kappa_R t), \\ Y(t) = [R_0 + r(t)]\sin\vartheta(t) \sim R_0\vartheta(t) = -\dfrac{2\Omega_0}{\kappa_R} r_0 \sin(\kappa_R t), \end{cases} \tag{5.23}$$

where in the second identity the first-order expansion is assumed. In practice, the epicycle is an ellipse in the reference system of the deferent given by

$$X^2(t) + \frac{Y^2(t)}{4\Omega_0^2/\kappa_R^2} = r_0^2. \tag{5.24}$$

Therefore, the ratio of the guiding center frequency to the radial epicycle frequency not only determines whether the orbit is closed or open, but also the shape of the epicycle! Note also that the motion on the epicycle is opposite to the motion of the guiding center (another property that was well known to ancient astronomers), a consequence of the conservation of angular momentum: when the star is outside R_0 its angular velocity is subtracted from Ω_0, and the opposite happens when the star is inside R_0.

Two special cases of the utmost importance must be considered: the first is that of the Coulomb potential of a point mass, in which $\kappa_R = \Omega_0$; and the second is that of the harmonic oscillator (Hooke's law inside a constant-density sphere), when $\kappa_R = 2\Omega_0$. Note that these identities hold *independently of* R_0 (i.e., in both cases *all* of the orbits in the epicyclic approximation are closed). This is simply a manifestation of the so-called *Bertrand's* theorem (i.e., the fact that the only potentials admitting only closed (bound) orbits are indeed the Coulomb and the harmonic oscillator cases; e.g., see Arnold 1978; Goldstein et al. 2000; Landau and Lifshitz 1969). It also follows that for realistic (i.e., radially decreasing) spherical density distributions, it is expected that $\Omega_0 \leq \kappa_R \leq 2\Omega_0$ moving from very concentrated to very shallow density profiles, such as in the case of the singular isothermal sphere $\kappa_R = \sqrt{2}\Omega_0$ (see Exercise 5.13). A few examples of the radial trend of Eq. (5.20) for an idealized family of disks are shown in Figure 5.3, where the special behavior of $\Omega_0 - \kappa_R/2 \simeq 0$ over a large range of radii can be seen: this behavior is much more than a simple curiosity, instead being a quite general property of quasi-circular orbits in disks. Actually, this property opens the way to the investigation of the difficult subject of *spiral structure*, which is well beyond the scope and the techniques presented in this introductory book (e.g., see Bertin 2014; Bertin and Lin 1996; Binney and Tremaine 2008 and references therein).

5.3 Oort's Constants

Another important use of circular orbits in astronomy is the deduction of the so-called *Oort's constants* (e.g., Bertin 2014; Binney and Merrifield 1998; Binney and Tremaine 2008; Chandrasekhar 1942; Ogorodnikov 1965). In this section, we will derive these constants geometrically under the idealized assumptions of perfectly circular orbits, all placed in the equatorial disk of a razor-thin galaxy, as illustrated in Figure 5.4. We indicate with $\mathbf{\Omega}(R) = \Omega(R)\mathbf{e}_z$ the angular velocity of the stars in the disk at distance $R = \|\mathbf{R}\|$ from the center, and we assume that the orbital plane is the (x, y) plane, so that $\mathbf{e}_z$ is the unit vector directed along the z-axis and the velocity of circular orbits of radius $\mathbf{R}$ is given

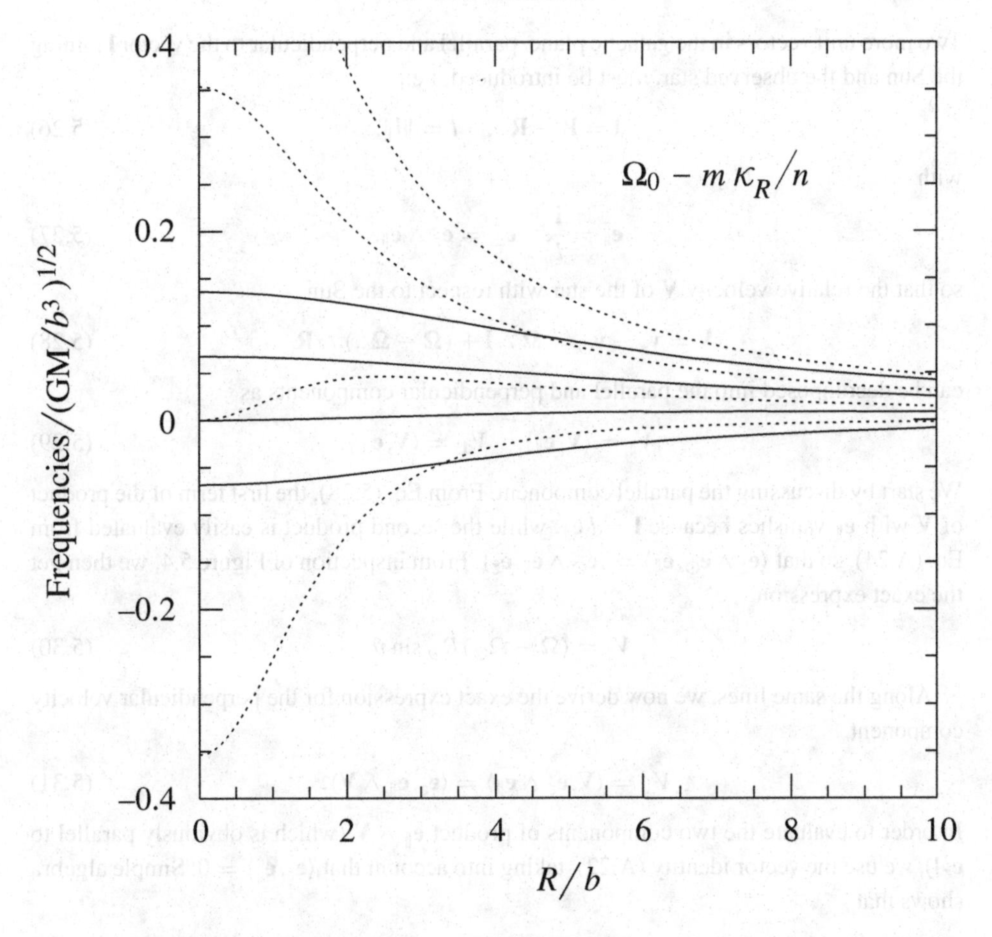

Figure 5.3 Radial trend of the quantity $\Omega_0 - m\kappa_R/n$ for the Miyamoto–Nagai disk in Eqs. (13.62) and (13.63). Dotted and solid lines refer to values of the flattening parameter $a/b = 1$ and $a/b = 5$, respectively; higher values of the flattening parameter correspond to flatter models. In each of the two families, the ratios m/n from the upper to lower curves are $-1/2$, 0, $1/2$, and 1. Note how the curve corresponding to $\Omega_0 - \kappa_R/2$ is quite flat over a large radial range. G = universal gravitational constant; M = total disk mass.

by $\mathbf{v} = \boldsymbol{\Omega} \wedge \mathbf{R}$. The Sun is located[3] at $\mathbf{R}_\odot = R_\odot \mathbf{e}_\odot$, where $\mathbf{e}_\odot$ is the unit vector in the galactic plane pointing toward the galactic *anticenter*, so that $\Omega(R_\odot) = \Omega_\odot$. It follows that the velocities of the Sun and of a generic star around the galactic center are given respectively by

$$\mathbf{v}_\odot = \Omega_\odot \mathbf{e}_z \wedge \mathbf{R}_\odot, \quad \mathbf{v}_\star = \Omega(R)\mathbf{e}_z \wedge \mathbf{R}. \tag{5.25}$$

[3] Note that in this chapter $R_\odot$ is not the radius of the Sun, but the distance (assumed constant, ≈8 kpc) of the Sun from the galactic center.

Two more unit vectors in the galactic plane, parallel and perpendicular to the vector $\mathbf{l}$ joining the Sun and the observed star, must be introduced, i.e.,

$$\mathbf{l} = \mathbf{R} - \mathbf{R}_\odot, \quad l = \|\mathbf{l}\|, \tag{5.26}$$

with

$$\mathbf{e}_\parallel = \frac{\mathbf{l}}{l}, \quad \mathbf{e}_\perp = \mathbf{e}_z \wedge \mathbf{e}_\parallel, \tag{5.27}$$

so that the relative velocity $\mathbf{V}$ of the star with respect to the Sun

$$\mathbf{V} = \mathbf{v}_\star - \mathbf{v}_\odot = \boldsymbol{\Omega} \wedge \mathbf{l} + (\boldsymbol{\Omega} - \boldsymbol{\Omega}_\odot) \wedge \mathbf{R}_\odot \tag{5.28}$$

can be decomposed into the parallel and perpendicular components as

$$V_\parallel = \langle \mathbf{V}, \mathbf{e}_\parallel \rangle, \quad V_\perp = \langle \mathbf{V}, \mathbf{e}_\perp \rangle. \tag{5.29}$$

We start by discussing the parallel component. From Eq. (5.28), the first term of the product of $\mathbf{V}$ with $\mathbf{e}_\parallel$ vanishes because $\mathbf{l} = l\,\mathbf{e}_\parallel$, while the second product is easily evaluated from Eq. (A.24), so that $\langle \mathbf{e}_z \wedge \mathbf{e}_\odot, \mathbf{e}_\parallel \rangle = \langle \mathbf{e}_\odot \wedge \mathbf{e}_\parallel, \mathbf{e}_z \rangle$. From inspection of Figure 5.4, we then get the exact expression

$$V_\parallel = (\Omega - \Omega_\odot) R_\odot \sin\vartheta. \tag{5.30}$$

Along the same lines, we now derive the exact expression for the perpendicular velocity component

$$V_\perp = \langle \mathbf{V}, \mathbf{e}_z \wedge \mathbf{e}_\parallel \rangle = \langle \mathbf{e}_z, \mathbf{e}_\parallel \wedge \mathbf{V} \rangle. \tag{5.31}$$

In order to evaluate the two components of product $\mathbf{e}_\parallel \wedge \mathbf{V}$ (which is obviously parallel to $\mathbf{e}_z$!), we use the vector identity (A.22), taking into account that $\langle \mathbf{e}_z, \mathbf{e}_\parallel \rangle = 0$. Simple algebra shows that

$$V_\perp = \Omega\, l + (\Omega - \Omega_\odot) R_\odot \cos\vartheta. \tag{5.32}$$

Equations (5.30)–(5.32) are exact under the hypotheses of the problem (i.e., they hold independently of the distance between the observing point (the Sun) and each point in the disk); moreover, they can be immediately expressed in terms of the *galactic longitude*, which is the angle $\lambda = \vartheta + \pi$ measured counterclockwise from the galactic center (see Figure 5.4).

We now restrict ourselves to places near the Sun (i.e., we consider stars with $l \ll R_\odot$), and we expand the angular velocity in Eqs. (5.30)–(5.32), retaining terms up to first order inclusive in l. From a Taylor expansion of the function $\Omega(R)$, one gets

$$\Omega\left(\|\mathbf{R}_\odot + \mathbf{l}\|\right) = \Omega_\odot + \Omega'(R_\odot)\langle \mathbf{e}_\odot, \mathbf{l} \rangle + O(l^2), \tag{5.33}$$

where $\Omega' = d\Omega/dR$ and $\mathbf{e}_\odot = \mathbf{R}_\odot / R_\odot$, so that $\langle \mathbf{e}_\odot, \mathbf{l} \rangle = l \cos\vartheta$. Inserting Eq. (5.33) into Eqs. (5.30)–(5.32), we see that in the Solar neighborhood

$$V_\parallel = \frac{\Omega'(R_\odot) R_\odot}{2}\, l \sin 2\vartheta = A\, l \sin 2\vartheta, \tag{5.34}$$

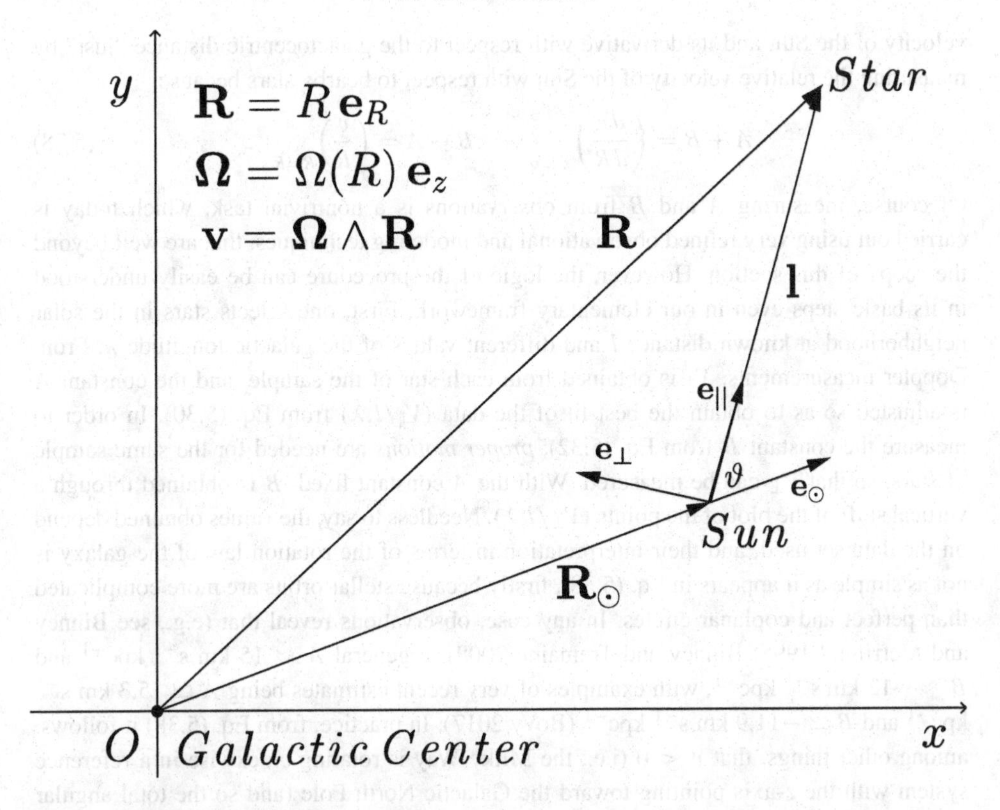

Figure 5.4 Vector decomposition of relative velocities as observed from a point on a circular orbit in a infinitesimally thin disk made by perfectly circular orbits.

$$V_{\perp} = \left[\Omega_{\odot} + \frac{\Omega'(R_{\odot})R_{\odot}}{2}\right] l + \frac{\Omega'(R_{\odot})R_{\odot}}{2} \, l \cos 2\vartheta = B\,l + A\,l \cos 2\vartheta, \tag{5.35}$$

where A and B are two functions depending on $R_{\odot}$, called the *first and second Oort's constants*. Notice that Eqs. (5.34) and (5.35), at variance with Eqs. (5.30)–(5.32), remain unchanged when the galactic longitude λ is used instead of the angle ϑ.

The Oort's constants can be rewritten in terms of circular velocity instead of angular velocity because from $\Omega(R) = v(R)/R$

$$\frac{d\Omega(R)}{dR} R = \frac{dv(R)}{dR} - \frac{v(R)}{R}, \tag{5.36}$$

and so

$$A = \frac{1}{2}\left(\frac{dv}{dR} - \frac{v}{R}\right)_{R=R_{\odot}}, \qquad B = \frac{1}{2}\left(\frac{dv}{dR} + \frac{v}{R}\right)_{R=R_{\odot}}. \tag{5.37}$$

Therefore, if from observations we can know the values of the two constants and the distance of the Sun from the galactic center, we can also know the value of the orbital velocity of the Sun and its derivative with respect to the galactocentric distance "just" by measuring the relative velocity of the Sun with respect to bearby stars because

$$A + B = \left(\frac{dv}{dR}\right)_{R=R_\odot}, \quad B - A = \left(\frac{v}{R}\right)_{R=R_\odot}. \tag{5.38}$$

Of course, measuring A and B from observations is a nontrivial task, which today is carried out using very refined observational and modeling techniques, that are well beyond the scope of this section. However, the logic of the procedure can be easily understood in its basic steps even in our elementary framework. First, one selects stars in the solar neighborhood at known distance l and different values of the galactic longitude λ. From Doppler measurements, $V_\parallel$ is obtained from each star of the sample, and the constant A is adjusted so as to obtain the best fit of the data $(V_\parallel/l, \lambda)$ from Eq. (5.30). In order to measure the constant B from Eq. (5.32), *proper motions* are needed for the same sample of stars, so that $V_\perp$ can be measured. With the A constant fixed, B is obtained through a vertical shift of the plot of the points $(V_\perp/l, \lambda)$. Needless to say, the values obtained depend on the data set used, and their interpretation in terms of the rotation law of the galaxy is not as simple as it appears in Eq. (5.38), firstly because stellar orbits are more complicated than perfect and coplanar circles. In any case, observations reveal that (e.g., see Binney and Merrifield 1998; Binney and Tremaine 2008) in general $A \approx 15$ km s^{-1} kpc^{-1} and $B \approx -12$ km s^{-1} kpc^{-1}, with examples of very recent estimates being $A \simeq 15.3$ km s^{-1} kpc^{-1} and $B \simeq -11.9$ km s^{-1} kpc^{-1} (Bovy 2017). In practice, from Eq. (5.38) it follows, among other things, that $v < 0$ (i.e., the Milky Way is rotating clockwise in a reference system with the z-axis pointing toward the Galactic North Pole (and so the total angular momentum vector of our galaxy points toward the Galactic South Pole)).

This fact is at the origin of a long and established (and sometimes confusing) tradition of *rewriting* A and B in a way that explicitly takes into account the clockwise rotation of the Milky Way. Call $\Omega_{\rm MW}$ and $v_{\rm MW} = \Omega_{\rm MW} R$ the modulus of the angular and circular velocity of the Milky Way, so that by fixing $\Omega = -\Omega_{\rm MW}$ and $v = -v_{\rm MW}$ in Eq. (5.37) we obtain the expressions of A and B that are almost universally encountered in the literature

$$A = \frac{1}{2}\left(\frac{v_{\rm MW}}{R} - \frac{dv_{\rm MW}}{dR}\right)_{R=R_\odot}, \quad B = -\frac{1}{2}\left(\frac{v_{\rm MW}}{R} + \frac{dv_{\rm MW}}{dR}\right)_{R=R_\odot}. \tag{5.39}$$

Exercises

5.1 The *rotation curve* $v_c(R)$ in a given axisymmetric potential $\phi(R,z) = \phi(R,-z)$ is the circular velocity as a function of radius R of orbits in the equatorial plane of the system given by Eq. (5.9). Show that for a stellar system made of different mass components (e.g., a bulge, a central supermassive black hole, a stellar disk, a gaseous disk, and a dark matter halo), the total circular velocity squared is the sum of the squared circular velocities of the components. By using the results of the following exercises and Eq. (5.10), the student is invited to experiment and construct the circular velocities for multicomponent galaxy models.

5.2 From Eqs. (2.104) and (5.9), show that in spherical coordinates the circular velocity in the equatorial plane of an axisymmetric, razor-thin disk of surface density $\Sigma(r)$ can be written as

$$v_c^2(r) = 2\pi G \sum_{l=0}^{\infty} [P_{2l}(0)]^2 \left[\frac{2l+1}{r^{2l+1}} \int_0^r \Sigma(\xi)\xi^{2l+1}\, d\xi - 2l r^{2l} \int_r^{\infty} \frac{\Sigma(\xi)}{\xi^{2l}}\, d\xi \right]. \tag{5.40}$$

Isolate the $l = 0$ term and show that it can be written as $GM(r)/r$, where $M(r)$ is the mass of the disk contained inside a circle of radius r; comment on the origin of the additional terms with respect to $v_c^2(r)$ of spherical systems in Eq. (5.10). Repeat the exercise with $\Sigma(R)$ in cylindrical coordinates, and from Eq. (2.105) prove that

$$v_c^2(R) = 2\pi G R \int_0^{\infty} \hat{\Sigma}(k) J_1(kR) k\, dk = 4G \int_0^{\infty} \Sigma(\xi) \mathcal{V}(\xi, R)\, \xi\, d\xi, \tag{5.41}$$

where

$$\mathcal{V}(\xi, R) = \begin{cases} \dfrac{\mathbf{K}(R/\xi)}{\xi} - \dfrac{\xi\, \mathbf{E}(R/\xi)}{\xi^2 - R^2}, & R < \xi, \\[2ex] \dfrac{R\, \mathbf{E}(\xi/R)}{R^2 - \xi^2}, & R > \xi. \end{cases} \tag{5.42}$$

Hints: Read again Exercise 1.9. By using Eq. (A.78), compare Eq. (5.40) with eq. (2.263) in Binney and Tremaine (2008). The first identity in Eq. (5.41) is established from the first identity in Eq. (2.106) following Eq. (5.9) and using Eq. (A.87) and the second identity is established by evaluating the radial derivative of the last integral in Eq. (2.106) by using Eq. (A.66) or by the radial derivative of Eq. (2.107).

5.3 Show by direct differentiation of Eq. (2.113) that the circular velocity of the razor-thin exponential disk in Eq. (2.110) with $\tilde{R} = R/R_d$ can be written as

$$v_c^2(R) = \pi G \Sigma_0 R_d \tilde{R}^2 \left[I_0\left(\frac{\tilde{R}}{2}\right) K_0\left(\frac{\tilde{R}}{2}\right) - I_1\left(\frac{\tilde{R}}{2}\right) K_1\left(\frac{\tilde{R}}{2}\right) \right]. \tag{5.43}$$

Repeat the exercise with spherical coordinates by using Eqs. (5.40) and (A.51). We recall here the relevant reference of Casertano (1983), where the rotation curve of the *finite-thickness and truncated* exponential disk (with a soft cutoff; see also the following exercises) has been presented.

5.4 By using Eq. (5.40), show that the circular velocity in the equatorial plane of the (truncated) Mestel (1963) disk in Eq. (2.115) can be written as

$$v_c^2(r) = 2\pi G \Sigma_d R_d \times \begin{cases} \dfrac{R_t}{r} + \displaystyle\sum_{l=1}^{\infty} [P_{2l}(0)]^2 \left(\frac{R_t}{r}\right)^{2l+1}, & r \geq R_t, \\[2ex] 1 + \displaystyle\sum_{l=1}^{\infty} [P_{2l}(0)]^2 \left(\frac{r}{R_t}\right)^{2l}, & 0 \leq r \leq R_t, \end{cases} \tag{5.44}$$

in agreement with the expression obtained by differentiation of Eq. (2.116). First, from the divergence of the series for $r = R_t$ (prove it!), conclude that the potential there is continuous with a vertical inflection point and $v_c(R_t) = \infty$. Then, from the general result in Exercise 5.2, show that the constant term of the series, for points with $r \leq Rt$, is $v_c^2(0) = 2\pi G\Sigma_d R_d = GM(r)/r$, and that $v_c^2(r) = v_c^2(0) = GM(r)/r$ at all radii in the *untruncated* case ($R_t \to \infty$), in formal[4] analogy with the case of the singular isothermal sphere in Eq. (2.63).

5.5 This exercise again focuses on Mestel (truncated) disk. From differentiation of Eq. (2.117), show that with cylindrical coordinates

$$v_c^2(R) = 4G\Sigma_d R_d \times \begin{cases} \dfrac{\delta}{\tilde{R}}\mathbf{K}\left(\dfrac{\delta}{\tilde{R}}\right), & \tilde{R} \geq \delta, \\[2ex] \mathbf{K}\left(\dfrac{\tilde{R}}{\delta}\right), & 0 \leq \tilde{R} \leq \delta, \end{cases} \tag{5.45}$$

where $\delta = R_t/R_d$ and $\tilde{R} = R/R_d$, and by using Eq. (2.118), verify the equivalence of Eqs. (5.44). Consider the asymptotic limits for $\tilde{R} \to \infty$ at fixed δ and for $\delta \to \infty$ at fixed $\tilde{R}$, and prove again that for the *untruncated* disk the circular velocity is constant. Finally (as a more difficult exercise), show that Eq. (5.45) can also be obtained by performing the integration in Eq. (5.41) using the second identity in Eq. (5.42). *Hints*: In order to prove Eq. (5.45) with direct differentiation, rewrite the last integral in Eq. (2.117) for points inside and outside R_t and change the integration variable to $t = \tilde{\xi}/\tilde{R}$ before differentiating. For the evaluation of the asymptotic limits, recall that $\mathbf{K}(0) = \pi/2$; notice that $\mathbf{K}(k)$ diverges for $k \to 1$, confirming again that v_c diverges on the rim of the truncated Mestel disk. Notice that in the *untruncated* case from eq. (6.511.1) of Gradshteyn et al. (2007)

$$\hat{\Sigma}(k) = \frac{\Sigma_d R_d}{k}, \tag{5.46}$$

and so the first integral in Eq. (5.41) and eq. (6.511.1) of Gradshteyn et al. (2007) prove that $v_c^2(R) = 2\pi G\Sigma_d R_d$. For the last question, the integration for points at $R > R_t$ presents no difficulties by using eq. (5.112.11) of Gradshteyn et al. (2007) or evaluating $d[k\mathbf{K}(k)]/dk$ with the aid of the first identity of Eq. (A.66). For points at R inside the disk, split the integration region as $(0, R - \epsilon)$ and $(R + \epsilon, R_t)$, evaluate the inner integral as before and the outer integral from the first identity of Eq. (A.66), and show that the logarithmic singularities cancel for $\epsilon \to 0$.

5.6 We finally consider three truncated razor-thin disks in order to elucidate the effect of truncation on the rotation curve at the disk edge. For each of these models, we only consider the equatorial plane and we use cylindrical coordinates; the student is

[4] The student should be careful to avoid the conclusion that the mass distribution of the untruncated Mestel disk at $r \geq r_0$ does not affect v_c at $r \leq r_0$ just because $v_c(r) = \sqrt{GM(r)/r}$. If this were true (as *it is* for spherical systems!), then $v_c(r)$ of the *truncated* Mestel disk, as obtained by removing the outer parts of the untruncated disk, would be constant (as happens inside the truncated isothermal sphere), which would be at variance with the second identity of Eq. (5.44).

invited to integrate Eqs. (2.104) and (2.105) and to prove the following results. First, for a *truncated constant-density disk* (Mestel 1963) of total mass $M_{\rm d} = \pi\Sigma_0 R_{\rm t}^2$

$$\Sigma(R) = \Sigma_0\,\theta(1-\tilde{R}), \quad \tilde{R} = R/R_{\rm t}, \tag{5.47}$$

$$v_{\rm c}^2(R) = 4G\Sigma_0 R_{\rm t} \times \begin{cases} \mathbf{K}(\tilde{R}) - \mathbf{E}(\tilde{R}), & 0 \le \tilde{R} < 1, \\ \tilde{R}\,\mathbf{K}(1/\tilde{R}) - \tilde{R}\,\mathbf{E}(1/\tilde{R}), & \tilde{R} > 1, \end{cases} \tag{5.48}$$

where $\mathbf{K}$ and $\mathbf{E}$ are the complete elliptic integrals of the first and second kind in Eq. (A.65). Second, for the *Maclaurin disk* (Binney and Tremaine 2008; Kalnajs 1972; Mestel 1963; Schulz 2009) of total mass $M_{\rm d} = 2\pi\Sigma_0 R_{\rm t}^2/3$

$$\Sigma(R) = \Sigma_0\sqrt{1-\tilde{R}^2}\,\theta(1-\tilde{R}), \quad \tilde{R} = R/R_{\rm t}, \tag{5.49}$$

$$v_{\rm c}^2(R) = \pi G\Sigma_0 R_{\rm t} \times \begin{cases} \dfrac{\pi\tilde{R}^2}{2}, & 0 \le \tilde{R} < 1, \\ \tilde{R}^2\arcsin(1/\tilde{R}) - \sqrt{\tilde{R}^2-1}, & \tilde{R} > 1. \end{cases} \tag{5.50}$$

Third, for the *finite Mestel disk* (not to be confused with the truncated Mestel disk; Lynden-Bell and Pineault 1978; Mestel 1963; Schulz 2012) of total mass $M_{\rm d} = 2\pi\Sigma_0 R_{\rm t}^2$

$$\Sigma(R) = \Sigma_0\frac{\arccos\tilde{R}}{\tilde{R}}\,\theta(1-\tilde{R}), \quad \tilde{R} = R/R_{\rm t}, \tag{5.51}$$

$$v_{\rm c}^2(R) = 2\pi G\Sigma_0 R_{\rm t} \times \begin{cases} \dfrac{\pi}{2}, & 0 \le \tilde{R} < 1, \\ \arcsin(1/\tilde{R}), & \tilde{R} > 1, \end{cases} \tag{5.52}$$

with a constant circular velocity inside the disk. Notice that $v_{\rm c}(R)$ in Eq. (5.48) jumps to infinity at the disk edge, as in the truncated Mestel disk in Eq. (5.45). This is a general feature of abruptly truncated razor-thin disks, as is discussed in Mestel (1963), and it can be easily proved from the last integral in Eq. (5.41) expanding the second integral of Eq. (5.42) near the disk truncation radius $R_{\rm t}$. The other two disks present soft truncations. *Hints*: We use Gradshteyn et al. (2007). For the potential of the *truncated constant-density disk*, from eq. (6.561.5) recover

$$\hat{\Sigma}(k) = \Sigma_0 R_{\rm t}^2\frac{\mathrm{J}_1(\lambda)}{\lambda}, \quad \lambda = kR_{\rm t}, \tag{5.53}$$

then use eq. (6.574.1); transform the hypergeometric functions into complete elliptic integrals with the aid of Gauss' recursion formulae (9.137) and eqs. (8.113.1) and (8.114.1). For $v_{\rm c}(R)$, use eq. (5.512.1) and convert the expressions again in terms of complete elliptic integrals. For the potential of the *Maclaurin disk*, from eqs. (6.567.1) and (8.464.3) recover

$$\hat{\Sigma}(k) = \Sigma_0 R_t^2 \sqrt{\frac{\pi}{2}} \frac{J_{3/2}(\lambda)}{\lambda^{3/2}} = \Sigma_0 R_t^2 \frac{\sin\lambda - \lambda\cos\lambda}{\lambda^3}, \qquad \lambda = kR_t, \tag{5.54}$$

then use eq. (6.574.1), and finally reduce the hypergeometric functions to elementary functions. For $v_c(R)$, use again eq. (6.574.1) and convert the expressions in terms of elementary functions. For the potential of the *finite Mestel disk*, first prove that[5]

$$\hat{\Sigma}(k) = \Sigma_0 R_t^2 \frac{\mathrm{Si}(\lambda)}{\lambda}, \qquad \lambda = kR_t, \tag{5.55}$$

where Si is the sine integral function in Eq. (A.61). The circular velocity can be obtained from eq. (2.12.48.2) of volume 2 of Prudnikov et al. (1990), or with integration by parts using Eq. (A.87) with $\mathrm{Si}(\infty) = \pi/2$, $\mathrm{Si}(0) = 0$, $J_0(\infty) = 0$, and $J_0(0) = 1$, and then eq. (6.693.6) from Gradshteyn et al. (2007).

5.7 Calculate the circular velocity for the family of power-law tori in Eqs. (2.87) and (2.88) and show that

$$v_c^2(R) = -4\pi G\rho_n r_*^2 \frac{\alpha^2 - 9\alpha + 16}{(7-\alpha)(5-\alpha)(\alpha-2)} \tilde{R}^{4-\alpha}. \tag{5.56}$$

Discuss the result for $2 < \alpha < 5$ and obtain the constant value of the circular velocity for $\alpha = 4$. What happens for $2 < \alpha \leq 9/2 - \sqrt{17}/2 \simeq 2.44$? *Hint*: Consider the sign of the radial component of the force in the equatorial plane; see also Ciotti and Bertin (2005).

5.8 From Eq. (5.4), it follows immediately that circular orbits in a given orbital family are critical points for the effective potential. From Eq. (5.6), show that circular orbits are also critical points for the energy $E_m(\dot{R}, \dot{z}, R, z)$ in the meridional plane. Finally, prove that E_m is a minimum in cases of stable circular orbits. *Hint*: Use Eq. (5.16).

5.9 Show that the circular orbits in the equatorial ($z = 0$) plane of an axisymmetric potential $\phi(R, z)$ are fully determined by assigning the value of one of the quantities R_0, $v_c(R_0)$, J_z, or E_m. By using Eqs. (5.8), (5.9), and (5.11), and for each possible choice among the four quantities above, show how to determine the values of the remaining three.

5.10 Draw the zero-velocity curves for the potential of a point mass as a function of orbital energy and for a given value of J_z, and discuss their shape compared to that obtained from epicyclic expansion in Eq. (5.16). What happens for $E = 0$ and for $E > 0$?

[5] Equation (5.55) can be proved by considering the integral $I(\lambda) = \int_0^1 J_0(\lambda t) \arccos t \, dt$, with $I(0) = 1$. Differentiate with respect to λ, use $dJ_0(\lambda t)/d\lambda = -tJ_1(\lambda t) = (t/\lambda) dJ_0(\lambda t)/dt$, and integrate by parts. From eq. (6.554.2) from Gradshteyn et al. (2007) obtain $I' = -I/\lambda + (\sin\lambda)/\lambda^2$ and solve the resulting ordinary differential equation with the given initial condition, obtaining $I = \mathrm{Si}(\lambda)/\lambda$.

5.11 Show that in a spherically symmetric stellar system the vertical epicyclic frequency is related to the potential $\phi(r)$ by the identity

$$\kappa_z^2(r) = \frac{1}{r}\frac{d\phi(r)}{dr} = \Omega^2(r). \tag{5.57}$$

Explain physically the identity above. *Hint*: In spherical systems the orbital angular momentum $\mathbf{J}$ of each star is conserved, and the orbits are forced to remain in the invariable plane perpendicular to $\mathbf{J}$ and so are necessarily planar (see Chapter 6).

5.12 Calculate the epicyclic radial frequency around a point mass with the Yukawa potential

$$\phi = -\frac{Ae^{-\mu r}}{r}, \quad \mu \geq 0, \tag{5.58}$$

and discuss the stability of circular orbits as a function of μ and r.

5.13 Calculate the epicyclic radial frequency in the potentials

$$\phi = -\frac{A}{r^\alpha}, \quad \phi = A\ln\frac{r}{r_0}, \tag{5.59}$$

and discuss the stability of circular orbits as a function of α. For the stable cases, calculate the axial ratio of the epicycles and the precession/recession angle $\Delta\varphi$. What happens for $\alpha = 2$ (i.e., for the $1/r^3$ force) and in the logarithmic case (corresponding to the singular isothermal sphere in Exercise 2.1)? Finally, plot the two Oort's constants $A(r)$ and $B(r)$ as a function of r.

5.14 A phenomenological pseudo-Newtonian potential, designed to describe accurately the gravitational field near a (Schwarzschild) black hole of mass M_{BH}, has been proposed in an influential paper by Paczyńsky and Wiita (1980):

$$\phi = -\frac{GM_{\mathrm{BH}}}{r - r_{\mathrm{S}}}, \quad r_{\mathrm{S}} = \frac{2GM_{\mathrm{BH}}}{c^2}, \quad r > r_{\mathrm{S}}, \tag{5.60}$$

where r_{S} is the Schwarzschild radius and c is the speed of light. Calculate the epicyclic radial frequency and show that circular orbits around a nonrotating black hole are unstable inside the so-called *innermost stable circular orbit* (i.e., for $r \leq r_{\mathrm{ISCO}} \equiv 3r_{\mathrm{S}}$).

5.15 By using the Poisson equation with cyclindrical coordinates, show that in any regular axisymmetric system with $\phi(R,z) = \phi(R, -z)$ the following identity holds at all radii in the equatorial plane:

$$\kappa_R^2 + \kappa_z^2 = 4\pi G\rho(R,0) + 2\Omega^2(R). \tag{5.61}$$

Conclude that (1) κ_R^2 and κ_z^2 cannot both be negative; and (2) from Exercise 5.11, that in spherical systems κ_R^2 cannot be negative. What happens outside the system (i.e., where $\rho(R,0) = 0$) and outside a spherical system? Explain the results physically.

5.16 With this exercise we determine the circular velocity and the epicyclic frequencies in the equatorial plane for the axisymmetric case of Eq. (2.28), with $a_1 = a_2$ and $a_3 = q_z a_1$, so that $\rho = \rho_0 \tilde{\rho}(m)/q_z$. First, by using Eqs. (2.21) and (5.9), show that

$$v_c^2(R) = 2\pi G R^2 a_1^3 q_z \int_0^\infty \rho \left(\frac{R}{\sqrt{a_1^2 + \tau}} \right) \frac{d\tau}{(a_1^2 + \tau)^2 \sqrt{a_3^2 + \tau}}. \tag{5.62}$$

Next, consider for simplicity the spherical case and prove that $v_c^2(r) \sim GM/r$ for $r \to \infty$. Then, with the help of Eq. (2.73), compute the vertical epicyclic frequency

$$\kappa_z^2(R) = 2\pi G a_1^3 q_z \int_0^\infty \rho \left(\frac{R}{\sqrt{a_1^2 + \tau}} \right) \frac{d\tau}{(a_1^2 + \tau)(a_3^2 + \tau)^{3/2}}. \tag{5.63}$$

Finally, obtain the radial epicyclic frequency from the sum rule in Eq. (5.61). As a more challenging exercise, compute κ_R^2 from the Rayleigh formula in Eq. (5.15) by using Eq. (5.62), and verify the sum rule.

5.17 By using the results in Exercise 5.16, show that the circular velocity in the equatorial plane of the axisymmetric (oblate) ellipsoidal power-law model

$$\rho = \frac{\rho_0}{q_z m^\gamma}, \quad m^2 = \frac{R^2}{a_1^2} + \frac{z^2}{q_z^2 a_1^2} \tag{5.64}$$

is given by

$$v_c^2(R) = 2\pi G \rho_0 a_1^2 (1 - q_z^2)^{\frac{\gamma-3}{2}} \mathrm{B}\left(\frac{3-\gamma}{2}, \frac{1}{2}; 1 - q_z^2 \right) \left(\frac{R}{a_1} \right)^{2-\gamma}, \tag{5.65}$$

where $\mathrm{B}(a, b; x)$ is the incomplete Euler beta function in Eq. (A.55) and for convergence $\gamma < 3$. Compute the frequency of the deferent Ω_0 and the radial epicyclic frequency as a function of q_z and γ, and repeat the exercise for the prolate case. Observe that for $\gamma = 2$, corresponding to the ellipsoidal generalization of the singular isothermal sphere in Eq. (2.63), v_c is independent[6] of R and the beta function reduces to $2 \arcsin \sqrt{1 - q_z^2}$; evaluate the limit for $q_z \to 1$ and show that $v_c^2 = 4\pi G \rho_0 a_1^2$, in accordance with Eq. (2.63).

5.18 For an axisymmetric, homeoidally expanded system with the potential given in Eq. (2.31), obtain the general formulae for $v_c(R)$, $\kappa_R(R)$, and $\kappa_z(R)$. Repeat the exercise by using the potential of an axisymmetric system obtained from elliptic integrals as in Eq. (2.120), and finally for the axisymmetric case of the ellipsoidal potential in Eq. (2.78).

[6] Notice that at this stage we know of four different density distributions with perfectly flat rotation curves over the whole space: the singular isothermal sphere, the $\gamma = 2$ power-law ellipsoid in this exercise, the $\alpha = 4$ torus in Exercise 5.7, and the untruncated Mestel disk. It is easy to show that the first three systems, when projected face on (i.e., along the symmetry z-axis; see Chapters 11 and 13 for the concept of projection), all reduce to the untruncated Mestel disk.

5.19 From Eq. (5.15), show that the epicyclic radial frequency at large distances from an axisymmetric ellipsoidal body, up to the quadrupole order, is given by

$$\kappa_R^2 = \frac{GM}{R_0^3} - \frac{GM(a_1^2 - a_3^2)}{2R_0^5}h \tag{5.66}$$

(i.e., the radial frequency decreases (increases) for oblate (prolate) distributions). *Hint*: From Eq. (2.94), first show that the angular velocity of the deferent is given by

$$\Omega_0^2 = \frac{GM}{R_0^3} + \frac{GM(a_1^2 - a_3^2)}{2R_0^5}h, \tag{5.67}$$

so that circular obits in the equatorial plane of oblate (prolate) distributions rotate faster (slower) than in spherical systems of the same total mass.

5.20 With this exercise we determine the formula of *orbital precession* angle $(\Delta\varphi)_{\mathrm{pert}}$ in the case of an axisymmetric perturbation ϕ_1 of the axisymmetric potential ϕ. Let $\phi_{\mathrm{pert}} = \phi + \epsilon\phi_1$, where ϵ is some small ordering parameter. First prove the obvious identities $\Omega^2_{\mathrm{pert}}(R) = \Omega^2(R) + \epsilon\Omega_1^2(R)$ and $k^2_{R,\mathrm{pert}}(R) = \kappa_R^2(R) + \epsilon k^2_{R,1}(R)$, where $\Omega(R)$ and $\kappa_R(R)$ are the circular and radial epicyclic frequencies of the unperturbed potential at radius R. Second, from Eq. (5.19), show that

$$\delta\varphi \equiv (\Delta\varphi)_{\mathrm{pert}} - (\Delta\varphi)_{\epsilon=0} \sim \epsilon\pi\left[\frac{\Omega_1^2(R)}{\Omega^2(R)} - \frac{k^2_{R,1}(R)}{\kappa_R^2(R)}\right], \tag{5.68}$$

so that $(\Delta\varphi)_{\mathrm{pert}}$ is known once $(\Delta\varphi)_{\epsilon=0}$ and $\delta\varphi$ are known. Finally, apply Eq. (5.68) to the case of the ellipsoidal quadrupole perturbation in Exercise 5.19 and show that

$$\frac{\delta\varphi}{2\pi} = \frac{a_1^2 - a_3^2}{2R^2}h, \tag{5.69}$$

(see also Landau and Lifshitz 1969, chapter 3). Notice that from Eq. (5.68) the *effects* of the *opposite* sign of the perturbation in Eqs. (5.66) and (5.67) add up, so that in the oblate case the orbit *precesses*, while in the prolate case it *recedes*.

5.21 With this exercise we prove an obvious (but nontrivial) property of epicyclic expansion. We begin by recalling that angular momentum J_z (per unit mass) is kept fixed in the second-order expansion (5.12) in terms of the small quantities $(r_0/R_0, z_0/R_0)$. Suppose one is able to solve the *true* equations of motion (5.4) for assigned r_0 and z_0: then, when computing the z component of $\mathbf{J} = \mathbf{x} \wedge \mathbf{v}$ from the obtained solution, we would recover J_z, and this identity must hold at *all* orders in the expansion of the solution. In practice, from order balance, it follows that *all* terms of the series, except the zeroth order (i.e., the circular orbit of the deferent), must be separately identically zero. Verify by direct computation of the second-order epicyclic orbit from Eq. (5.18) that in fact

$$(\mathbf{x} \wedge \mathbf{v})_z = J_z = R_0^2\Omega_0 \tag{5.70}$$

of the deferent orbit. *Hint*: Use the vector expression in Eq. (5.18) and its time derivative from Eq. (A.28).

5.22 In this exercise we consider a plausible generalization of the Laplace–Runge–Lenz vector in Eq. (4.21). First, for a mass m in a generic potential energy $U(\mathbf{x})$, define

$$\mathbf{L} = U(\mathbf{x})\mathbf{x} + \mathbf{v} \wedge \mathbf{J}, \quad m\ddot{\mathbf{x}} = -\nabla U, \tag{5.71}$$

and show that

$$\|\mathbf{L}\|^2 = U^2(\mathbf{x})\|\mathbf{x}\|^2 + \frac{2E\|\mathbf{J}\|^2}{m}. \tag{5.72}$$

Then show that in the two-body problem (when in the previous identities $\mathbf{x}$ and $\mathbf{v}$ are the relative separation and velocity of the pair, $m \to \mu$, $\mathbf{J} \to \mathbf{J}_{\rm r}$ and $E \to E_{\rm r}$; see Chapter 4) we recover $\mathbf{L} = Gm_{\rm t}m_{\rm f}\mathbf{l}$. Second, show that in a spherical system with $\|\mathbf{x}\| = r$ and $U(r)$,

$$\frac{d\mathbf{L}}{dt} = \left(U + r\frac{dU}{dr}\right)\mathbf{v}. \tag{5.73}$$

Third, show that for orbits in the equatorial plane (x, y) of axisymmetric systems with $U(R, z)$, $R = \sqrt{x^2 + y^2}$, and $\mathbf{J} = J_z\mathbf{f}_z$,

$$\mathbf{L} = (J_z v_\varphi + UR)\mathbf{f}_R - J_z v_R \mathbf{f}_\varphi = R\left[\frac{\partial R\, U(R, 0)}{\partial R}\right]\mathbf{f}_R, \tag{5.74}$$

where the last identity holds for *circular orbits*; deduce that for these latter cases $\mathbf{L}(t)$ is aligned with the star position, with the exception of the $1/r$ potential, when $\mathbf{L} = 0$. Compute $\|\mathbf{L}\|$, verify the equivalence with Eq. (5.72) evaluated for equatorial orbits, and conclude that for these orbits $\|\mathbf{L}\|$ can depend on time only through $R(t)$. Finally, discuss the time dependence of $\mathbf{L}$ for equatorial orbits in an epicyclic approximation (see also Lynden-Bell 2006).

5.23 The radial epicyclic frequency is related to the Oort's constants by a simple relation. Prove that

$$\kappa_R^2(R) = 4\Omega(R)\, B(R) = 4\Omega^2(R) + 4\Omega(R)\, A(R). \tag{5.75}$$

Hint: Use Eqs. (5.15) and (5.37) defined at a generic radius R in the galactic plane and the identity $B(R) - A(R) = \Omega(R)$.

Part II

Systems of Particles

6

The N-Body Problem and the Virial Theorem

The N-body problem, the study of the motion of N point masses (e.g., stars) under the mutual influence of their gravitational fields, is one of the central problems of classical physics, and the literature on the subject is immense, starting with Newton's *Principia* (e.g., see Chandrasekhar 1995). Conceptually, it is more the natural subject of celestial mechanics than of stellar dynamics; however, experience suggests that at least some space should be devoted to an overview of the exact results of the N-body problem in a book like this. In fact, due to the very large number of stars in stellar systems, stellar dynamics must rely on specific techniques and assumptions, and one may legitimately ask which of the obtained results hold true in the generic N-body problem; for example, these exact results represent invaluable tests for validating numerical simulations of stellar systems. The virial theorem is one such result, and in this chapter we present a first derivation of it, while an alternative derivation in the framework of stellar dynamics will be discussed in Chapter 10.

6.1 The N-Body Problem and the Lagrange–Jacobi Identity

As is well known, in a generic inertial reference frame S_0, the differential equations[1] of the N-body ($N \geq 2$) problem read

$$\ddot{\mathbf{x}}_i = -G \sum_{j=1;\, j\neq i}^{N} m_j \frac{\mathbf{x}_i - \mathbf{x}_j}{r_{ij}^3}, \quad r_{ij} = \|\mathbf{x}_i - \mathbf{x}_j\|, \quad i = 1, \ldots, N. \tag{6.1}$$

These equations are generally not *solvable* in the informal meaning of the word (i.e., they cannot be reduced to $6N - 1$ independent integrations, being the right-hand side member explicitly independent of time; e.g., see Whittaker 1917; Wintner 1947), and so we must resort to numerical methods to obtain their solutions (e.g., see Binney and Tremaine 2008; Heggie and Hut 2003). Nevertheless, many important properties of N-body systems can be explored analytically by direct use of Eq. (6.1): obviously, the two-body problem discussed in Chapter 4 is the most important special case.

[1] With $\sum_{j=1;\, j\neq i}^{N}$ we mean "sum over j from 1 to N, excluding $j = i$." With $\sum_{i\neq j=1}^{N}$ we mean "sum over i and j from 1 to N, excluding $i = j$." And finally with $\sum_{i,j=1}^{N}$ we mean "sum over i and j from 1 to N."

Let us start by introducing some standard notation. We indicate with $r_{\rm min} = \min_{i \neq j} r_{ij}$ and $r_{\rm max} = \max r_{ij}$ the minimum and the maximum interparticle distances, and with $M = \sum_{i=1}^{N} m_i$ and $\mu = \min m_i$ the total mass of the system and the minimum particle mass, respectively. The kinetic and potential energy of the system and its *polar* moment of inertia[2] are given, respectively, by

$$T = \frac{1}{2} \sum_{i=1}^{N} m_i \|\mathbf{v}_i\|^2, \quad U = -\frac{G}{2} \sum_{i \neq j=1}^{N} \frac{m_i m_j}{r_{ij}}, \quad I = \sum_{i=1}^{N} m_i \|\mathbf{x}_i\|^2. \tag{6.2}$$

Note that in the expression for U the factor $1/2$ takes care of the fact that with the sum we are counting each pair twice.

As is well known from classical mechanics (see Exercise 6.1), in any inertial frame S_0 the following quantities are conserved for an isolated N-body system (i.e., they are time independent) when evaluated along with the solution of Eq. (6.1):

$$E = T + U, \quad \mathbf{P} = \sum_{i=1}^{N} m_i \mathbf{v}_i, \quad \mathbf{J} = \sum_{i=1}^{N} m_i \mathbf{x}_i \wedge \mathbf{v}_i, \tag{6.3}$$

where E, $\mathbf{P}$, and $\mathbf{J}$ are the total energy, momentum, and angular momentum, respectively. Therefore, the general N-body problem also admits the same 10 classical integrals of motion already encountered in Chapter 4 in the study of the two-body problem. In S_0,

$$\mathbf{R}_{\rm CM} = \frac{1}{M} \sum_{i=1}^{N} m_i \mathbf{x}_i, \quad \mathbf{V}_{\rm CM} = \frac{1}{M} \sum_{i=1}^{N} m_i \mathbf{v}_i = \frac{\mathbf{P}}{M} \tag{6.4}$$

are the position and velocity of the *center of mass*.

A particularly important reference system is the inertial barycentric reference frame $S_{\rm CM}$, with its origin located at $\mathbf{R}_{\rm CM}$, moving with constant velocity $\mathbf{V}_{\rm CM}$, and with axes parallel to the axes of S_0, where from a Galilean coordinate transformation

$$\mathbf{x}_i = \mathbf{R}_{\rm CM} + \mathbf{x}'_i, \quad \mathbf{v}_i = \mathbf{V}_{\rm CM} + \mathbf{v}'_i, \tag{6.5}$$

and so, while $U = U'$, it follows that

$$T = E_{\rm CM} + T', \quad E = E_{\rm CM} + E', \quad \mathbf{J} = \mathbf{J}_{\rm CM} + \mathbf{J}', \quad \mathbf{P} = \mathbf{P}_{\rm CM}, \tag{6.6}$$

where the quantities $\mathbf{P}_{\rm CM} = M\mathbf{V}_{\rm CM}$, $\mathbf{J}_{\rm CM} = M\mathbf{R}_{\rm CM} \wedge \mathbf{V}_{\rm CM}$, and $E_{\rm CM} = M\|\mathbf{V}_{\rm CM}\|^2/2$ are obviously time independent, and the meanings of E', T', and $\mathbf{J}'$ are apparent; in Appendix A.1.1, the general transformation formulae of the quantities above for an arbitrary change of reference system are given.

Due to the time independence of $\mathbf{J}$, when $\mathbf{J}' \neq 0$, the plane defined in $S_{\rm CM}$ as

$$\Pi_{\rm inv} = \{\mathbf{x}' \in \Re^3 : \langle \mathbf{x}', \mathbf{J}' \rangle = 0\} \tag{6.7}$$

[2] Note that the polar moment of inertia $I = {\rm Tr}(I_{ij}) = {\rm Tr}(\Im_{ij})/2$ is half of the trace of the inertia tensor $\Im_{ij}$ in Eq. (2.89).

is called the *invariable plane*, and obviously $\mathbf{R}_{CM} \in \Pi_{inv}$. Moreover, the polar moment of inertia in S_0 is related to the correspondent quantity I' as

$$I = M\|\mathbf{R}_{CM}\|^2 + I', \quad I' = \sum_{i=1}^{N} m_i \|\mathbf{x}'_i\|^2 = \frac{1}{2M} \sum_{i,j=1}^{N} m_i m_j r_{ij}^2. \tag{6.8}$$

The last expression for I', proposed by Lagrange (1736–1813), is of great importance in application, and in Exercise 6.2 the student is guided toward proving it. In Exercise 6.3, we will also prove the remarkable *exact* bounds that hold $\forall t$

$$\frac{G\mu^2}{|U(t)|} \leq r_{min}(t) \leq \frac{GM^2}{2|U(t)|}, \quad \sqrt{\frac{2\,I'(t)}{M}} \leq r_{max}(t) \leq \sqrt{\frac{M\,I'(t)}{\mu^2}}. \tag{6.9}$$

Physically, the (inverse of the) gravitational energy is a "measure" of the minimum separation of particles, while the square root of the barycentric moment of inertia is a "measure" of the maximum interparticle distance.

We conclude this section with two important identities that hold in the *generic* inertial reference system S_0 (and so in S_{CM}), and these will be used in the following discussion. The first is the *Lagrange–Jacobi identity*

$$\frac{\ddot{I}}{2} = 2T + U = E + T = 2E - U. \tag{6.10}$$

The second and third identities in Eq. (6.10) derive immediately from energy conservation (6.3), once the first identity is established. Due to its relevance to stellar dynamics, in Exercises 6.4–6.6 we will prove the Lagrange–Jacobi identity in three cases of increasing generality; in the last one we will prove a general tensorial identity for systems of N bodies interacting with central power-law forces in the presence of an external force field, and we show that its trace in the self-gravitating Newtonian case reduces to the first identity in Eq. (6.10). We will encounter Eq. (6.10) again in tensorial form in Chapter 10. The second result (see Exercise 6.7) is *Sundman's (weak) inequality* (see Boccaletti and Pucacco 1996 for the strong case)

$$\|\mathbf{J}\|^2 \leq 2I\,T = I\,(\ddot{I} - 2E). \tag{6.11}$$

The next three sections will focus on *singularities*, *special solutions*, and the *long-term behavior* of the N-body problem. These arguments are only of tangential interest for stellar dynamics; nevertheless, a qualitative comprehension thereof is important for a better understanding of the (idealized) dynamical fate of stellar systems produced by stellar dynamics on (very long) timescales (see Chapters 7 and 8). As the celestial mechanics literature on the N-body problem is immense and highly technical, here we only provide the interested reader with a list of some of the most relevant references (with different degrees of difficulty): Abraham and Marsden (1978), Arnold (1997), Boccaletti and Pucacco (1996), Hagihara (1970), Meyer and Hall (1992), Roy (2005), Szebehely (1967), Valtonen and Karttunen (2006), and Wintner (1947).

6.1.1 Singularities

The investigation of the singularities of the ordinary differential equations (ODEs) describing the N-body problem is an active field of research in celestial mechanics, with seminal contributions coming from to Poincaré (1854–1912) and Painlevé (1863–1933); fascinating accounts are given in the books by Barrow-Green (1997) and Diacu and Holmes (1996). In fact, one of the main problems posed by the nonlinear equations in (6.1) is that their solutions can develop *movable singularities* (i.e., singularities that depend on the specific choice of initial conditions $\mathbf{x}_i(0)$ and $\mathbf{v}_i(0)$ and that make the analysis of the problem extremely difficult). In fact, in contrast with the nonlinear ODEs, according to a theorem from Fuchs (1833–1902), the homogeneous *linear* ODE of order n

$$\frac{d^n f(x)}{dx^n} + \sum_{i=0}^{n-1} a_i(x) \frac{d^i f(x)}{dx^i} = 0, \qquad \frac{d^0 f(x)}{dx^0} \equiv y(x), \tag{6.12}$$

where $a_i(x)$ are functions of the independent variable (see also Eq. (A.135)), can present only *fixed singularities* placed at the singularities of the coefficients a_i that are independent of the initial conditions and that therefore can be determined independently of the ability/possibility to find a solution of the equation itself. Moreover, the nature of the singular points (regular or essential) and the behavior of the solutions in their neighborhood can be determined by using Frobenius' (1849–1917) theorem (e.g., see Jackson 1998; for detailed discussions, see also Arfken and Weber 2005; Bender and Orszag 1978; Ince 1927; Lomen and Mark 1988). An illustrative example of fixed and movable singularities[3] is provided in Exercise 6.8.

Therefore, at variance with linear problems, in the highly nonlinear N-body problem, it is natural to expect movable singularities in the particle orbits that take place at times that are dependent on the assigned initial conditions; a simple example is provided by the two-body problem (see Chapter 4) that for $\mathbf{J}' \neq 0$ never develops a singularity, while for $\mathbf{J}' = 0$ the two bodies collide at a finite time. Remarkably, there are theorems providing bounds of collision times in the N-body problem as a function of the initial conditions. Here, we only report a classification of the possible singularities:

(1) *Collisions*: $\lim_{t \to t_{\rm sing}} r_{\rm min} = 0$.
(2) *Pseudo-collisions*: $\liminf_{t \to t_{\rm sing}} r_{\rm min} = 0$ and $\limsup_{t \to t_{\rm sing}} r_{\rm min} > 0$.
(3) *Blow–up*: $\lim_{t \to t_{\rm sing}} r_{\rm max} = \infty$, with $t_{\rm sing} < \infty$.
(4) For $N = 3$, all singularities are collisions.
(5) If $I' = O(1)$ for $t \to t_{\rm sing}$, then the singularity is a collision.
(6) If the N particles are on a straight line, then all singularities are collisions.
(7) A singularity at $t_{\rm sing}$ is a collision at and only if $U = O[(t - t_{\rm sing})^{-2/3}]$.

[3] We will encounter the concept of fixed singularities again in Appendix A.2 when discussing the properties of some of the most important special functions arising in potential theory.

(8) The set of initial conditions in phase space leading to a collision at a finite time has zero Lebesgue measure.
(9) *Weierstrass–Sundman theorem*: the global collapse[4] of an N-body system may occur only in a finite amount of time (i.e., $t_{\rm sing} < +\infty$) and only if $\mathbf{J}' = 0$ (see Exercise 6.9).

In particular, Point (8) shows that collisions in the N-body problem are exceedingly rare.

6.1.2 Special Solutions

Under particular circumstances (i.e., symmetries), the N-body problem can be solved explicitly. From what we have said before, these solutions are of great interest because they are the only ones where the celebrated problem can be investigated in full detail. The most famous cases concern the three-body problem and come from Euler (1707–1783) and Lagrange (1736–1813): these are not just mathematical curiosities, but have important astronomical counterparts (e.g., see Roy 2005). Here, only an illustrative overview is given, and we begin with some definitions. In the barycentric inertial frame $S_{\rm CM}$, a solution of the N-body problem is said to be:

(1) *planar* if there exists a plane Π with a *time-independent* orientation containing all of the bodies at any time;
(2) *flat* if there exists a plane $\Pi(t)$ containing all of the bodies at any time;
(3) *syzygial* at time t_0 if at $t = t_0$ the N particles all lie on the same straight line;
(4) *rectilinear* if there exists a straight line Λ with a *time-independent* orientation containing all of the bodies at any time;
(5) *collinear* if there exists a straight line $\Lambda(t)$ containing all of the bodies at any time;
(6) *homographic* if there exists a scalar function $\lambda(t) \geq 0$ and an orthogonal matrix $\mathcal{R} \in$ SO(3) so that $\forall t$ the position of each particle is related to its initial position by

$$\mathbf{x}_i(t) = \lambda(t)\mathcal{R}(t)\mathbf{x}_i^0; \tag{6.13}$$

(7) *homothetic*, if in Eq. (6.13) $\mathcal{R}$ is the identity for all times;
(8) *relative equilibrium* if in Eq. (6.13) $\lambda(t) = 1$ for all times;
(9) *central configuration* if there exists a scalar function $\sigma(t) \geq 0$ so that for $i = 1, \ldots, N$ and $\forall t$

$$\nabla_{\mathbf{x}_i} U = \sigma(t) m_i \mathbf{x}_i. \tag{6.14}$$

In addition, the following results (no proof is given) hold:

(1) if the solution is planar and $\mathbf{J}' \neq 0$ and the plane Π coincides with $\Pi_{\rm inv}$ in Eq. (6.7);
(2) a planar solution may exist even for $\mathbf{J}' = 0$;

[4] "Global collapse" means that all interparticle distances go simultaneously to zero (i.e., in $S_{\rm CM}$ $\lim_{t\to t_{\rm sing}} I'(t) = 0$ and $\lim_{t\to t_{\rm sing}} U = -\infty$).

(3) if $\mathbf{J}' = 0$, then any flat solution is planar;
(4) every planar solution is flat;
(5) not every flat solution is planar (e.g., the solution of the general three-body problem is always flat);
(6) from Results (3) and (5), any solution of the three-body problem with $\mathbf{J}' = 0$ is planar;
(7) if Π_{inv} exists, then any syzygial configuration lies on Π_{inv};
(8) every rectilinear solution is collinear;
(9) every collinear solution is flat;
(10) every collinear solution is planar: in fact, if $\mathbf{J}' \neq 0$, from Result (7) the straight line $\Lambda(t)$ lies on Π_{inv} $\forall t$, and so the problem is planar. If $\mathbf{J}' = 0$, from Results (3) and (9) the solution must be planar;
(11) *Laplace theorem*: solutions of the N-body problem are central configurations if and only if they are homographic.

As already mentioned, the most interesting solutions from a physical point of view are the homographic/central configurations, such as the Euler collinear and the Lagrange planar (equilateral) solutions (see also Appendix A.1.1).

6.1.3 Long-Term Behavior

Many exact and asymptotic results are known for the N-body problem when $t \to \infty$. Such theorems are highly technical in nature: the interested reader is again invited to consult the given list of references. In order to illustrate the nature of such results, we simply recall (see also Exercise 6.10) that in the barycentric system S_{CM}:

(1) For $E' < 0$, a positive constant A exists so that $r_{\min}(t) \leq A \; \forall t$.
(2) For $E' = 0$, $r_{\min}(t) = O(t^{2/3})$ and a positive constant A exists so that $r_{\max}(t) \geq At^{2/3}$ for $t \to \infty$.
(3) For $E' > 0$, $r_{\min}(t) = O(t)$ and a positive constant A exists so that $r_{\max}(t) \geq At$ for $t \to \infty$.

Result (1) means that when $E' < 0$ the minimum interparticle distance $r_{\min}(t)$ remains bounded, while nothing can be said about $r_{\max}(t)$ (i.e., at variance with the two-body problem, particles may escape from an N-body system with barycentric total negative energy; e.g., as happens in the phenomenon of gravitational evaporation of stars from open and globular clusters; see also Section 6.2.1 and in particular Binney and Tremaine 2008; Chandrasekhar 1942; Heggie and Hut 2003; Spitzer 1987 and references therein). Result (2) means that $r_{\min}(t)$ cannot grow faster than $t^{2/3}$ when $E' = 0$ (and of course it may well remain bounded); in contrast, $r_{\max}(t)$ cannot grow slower than $t^{2/3}$ (i.e., at least one particle must escape in case of vanishing barycentric total energy). Finally, from Result (3) it follows that, when $E' > 0$, $r_{\min}(t)$ cannot grow faster than t (and it may well remain bounded), while $r_{\max}(t)$ cannot grow slower than t; at least one particle must escape from an N-body system with barycentric total positive energy. From the point of view of stellar dynamics, the most important results can be summarized as follows: $E' < 0$ is a *necessary* condition for all interparticle distances $r_{ij}(t)$ to be bounded at all times and $E' \geq 0$ is a *sufficient* condition to have at least one escaper. Therefore, the evolution of the system is somewhat determined by the sign of E': a system with $E' \geq 0$ will necessarily eject particles (in principle up to a complete dissolution) if after each escape the energy of the remaining system remains positive or until $E' < 0$. From this time, the system can remain bounded or eject particles, with the limit case finally reducing to a pair of particles on a very tight orbit accounting for all of the initial negative energy and diminished by all of the kinetic energy carried away by the escapers.

6.2 The Scalar Virial Theorem

We are now in a position to introduce the concept of the *virial*. Due to its relevance to stellar dynamics, the virial identities will be derived again in Chapter 10 along a different path. At present, we restrict ourselves to the (second-order) scalar virial theorem for self-gravitating systems, and we begin our discussion by recalling that an N-body system is called *self-gravitating* if the total force under which the N particles move is determined by the masses themselves through Eq. (6.1). By definition, such a system is *virialized* if in the barycentric inertial reference system S_{CM}

$$0 \leq 2T' = -U, \quad \forall t. \tag{6.15}$$

In Eq. (6.15), the limit case of $T' = 0$ is of little practical interest, formally corresponding to a degenerate configuration with all particles (stars) at infinite mutual separation and motionless. Notice that if a self-gravitating system is virialized, then its barycentric total energy cannot be positive; in other words, necessarily

$$E' \leq 0, \tag{6.16}$$

because for a self-gravitating virialized N-body system the last two identities in Eq. (6.10) dictate that $T' = -E'$ and $U = 2E'$.

Therefore, from the Lagrange–Jacobi identity, it follows that if a system is at equilibrium ($\ddot{I}' = 0$), then it is necessarily virialized,[5] and that if a system is virialized, then $\ddot{I}' = 0$. The last consideration seems to leave open the possibility that, after all, we could have a virialized system with $\dot{I}' = const. \neq 0$; however, this is impossible because it can be proved (see Exercise 6.11) that a self-gravitating N-body system is virialized not only if and only if $\ddot{I}' = 0$, but actually if and only if the stricter condition

$$\frac{dI'}{dt} = 0, \quad \forall t \tag{6.17}$$

[5] Sometimes in the literature the redundant concept of "virial equilibrium" is encountered, suggesting the impossible case of a system at equilibrium but not virialized!

holds. Moreover (see Exercise 6.12), a self-gravitating N-body system is virialized if and only if the kinetic and potential energies are constant

$$T'(t) = T'(0), \qquad U(t) = U(0), \tag{6.18}$$

and two positive constants A and B exist such that

$$A \leq r_{ij}(t) \leq B, \quad \forall i \neq j = 1, \ldots, N. \tag{6.19}$$

A more general definition of virialization, extending the concept to time-dependent systems and so relaxing the strict coincidence of the concepts of equilibrium and virialization, can also be introduced (e.g., Landau and Lifshitz 1969). Let $f(t)$ be a given function of time and define

$$\langle f \rangle_t \equiv \frac{1}{t} \int_0^t f(\tau) d\tau. \tag{6.20}$$

If $\lim_{t \to \infty} \langle f \rangle_t$ exists, such a limit $\langle f \rangle$ is called the *time-average* of f. Now, for a self-gravitating N-body system

$$2\langle T' \rangle = -\langle U \rangle \quad \Leftrightarrow \quad \dot{I}'(t) = o(t) \quad \text{for } t \to \infty. \tag{6.21}$$

In fact, the time-averaged Lagrange–Jacobi identity reads

$$\frac{\dot{I}'(t) - \dot{I}'(0)}{t} = 2\langle T' \rangle_t + \langle U \rangle_t, \tag{6.22}$$

thus proving Eq. (6.21). Notice that if $\dot{I}'(t) = o(t)$, then $I'(t) = o(t^2)$ (asymptotic relations can always be integrated), and according to Eq. (6.9), $r_{\max}(t) = o(t)$ for $t \to \infty$. Point (3) in Section 6.1.3 then excludes the possibility of time-averaged virialization for systems with $E' > 0$, a conclusion that can also be reached more simply from energy conservation, because $\langle E' \rangle_t = E'$, and so in a system virialized in a time-averaged sense

$$\langle T' \rangle = -E', \qquad \langle U \rangle = 2E' \tag{6.23}$$

(i.e., $E' \leq 0$).

Of course, a virialized system is also virialized in a time-averaged sense, but the converse is not true; the classical examples are those of systems performing bounded and/or periodic motions, such as the negative energy case of the Lynden-Bell and Lynden-Bell (2004) N-body system (see Exercise 6.5) and the elliptical two-body problem, where from Eqs. (4.7) and (4.30)

$$\langle T' \rangle = G\frac{m_{\rm t} m_{\rm f}}{a}, \quad \langle T_{\rm t}' \rangle = \frac{\mu}{m_{\rm t}}\langle T' \rangle, \quad \langle T_{\rm f}' \rangle = \frac{\mu}{m_{\rm f}}\langle T' \rangle. \tag{6.24}$$

For a virialized system, the *virial velocity dispersion* $\sigma_{\rm V}$ and the *virial radius* $r_{\rm V}$ are defined as

$$\frac{M\sigma_{\rm V}^2}{2} \equiv T', \qquad \frac{GM^2}{r_{\rm V}} \equiv |U|, \tag{6.25}$$

so that in the new variables

$$\sigma_V^2 = \frac{2T'}{M} = \frac{|U|}{M} = \frac{GM}{r_V}, \tag{6.26}$$

and perfectly analogous expressions hold in a time-averaged sense. The student should appreciate that Eq. (6.26) does not add anything new to the identities (6.15) and (6.21), being simply a rewriting in terms of new variables.[6] In Chapter 13, we will compute σ_V and r_V for some of the most common models used to represent stellar systems.

Before moving to the next sections, where we will consider a few important applications of the virial theorem to stellar dynamics, it is necessary to mention two fundamental problems that arise immediately from the previous discussion. We anticipate that an enormous body of literature is available on these two subjects, and here, for reasons of space and technical difficulty (being well above the level of this book), we simply mention them and give some relevant references for further study.

The first natural question is *how* a stellar system, initially out of equilibrium (e.g., a proto-galaxy during the epoch of galaxy formation or a pair of galaxies undergoing a merging in the local universe), reaches its equilibrium configuration, and how the structural and dynamical properties of the final state depend on the specific initial conditions. In stellar dynamics, the dynamical process leading to virialization is known as *violent relaxation* (Lynden-Bell 1967; Tremaine et al. 1986; see also Bertin 2014; Binney and Tremaine 2008; Saslaw 1987). The proper way to discuss violent relaxation would be by using concepts from statistical mechanics: here, in *very* elementary terms, we simply notice that violent relaxation is *not* due to the effects of the star–star gravitational interactions that will be discussed in the next two chapters, but to global fluctuations of the system's time-dependent potential that redistribute the energies of the stars independently of their mass (see Footnote 8 in Appendix A). Moreover, violent relaxation acts on "short" timescales compared to other relaxation phenomena, of the order of a few dynamical times $t_{VR} \approx t_{dyn} \approx 1/\sqrt{G\rho}$, where ρ is the average density of the system. The important message here is that theoretical and numerical works nicely show that the virialized end states of violent relaxation describe quite well the global structural and kinematic properties of real systems such as early-type galaxies (e.g., see Bertin and Stiavelli 1984; Bertin and Trenti 2003; Stiavelli and Bertin 1985, 1987; see also the end of Section 12.2.2; from the numerical point of view, see the seminal paper of van Albada 1982; see also Nipoti et al. 2006 and references therein) and dark matter halos (e.g., Aguilar and Merritt 1990; Dubinski and Carlberg 1991; Navarro et al. 1997). Finally, and from a more general point of view, notice that violent relaxation somewhat depends on the specific nature of the gravitational force: there are systems such as the N-body system in which the particles interact with

[6] A comment is in order here. r_V and σ_V are perfectly defined for the vast majority of the models encountered in stellar dynamics; however, there are a few important peculiar cases. For example, in the untruncated King (1972) model in Eq. (13.34) and in the untruncated Navarro et al. (1997) model in Eq. (13.40), the peculiar r^{-3} density profile at large radii leads to a divergent total mass M but also to a finite U (see Exercise 13.7); for these models, $r_V = \infty$ and $\sigma_V = 0$. Even more peculiar cases of r_V and σ_V arise in systems with divergent M and U, the most famous being the singular isothermal sphere in Eq. (2.63). See Exercise 13.21 for a discussion.

the harmonic oscillator force and no *phase mixing* (e.g., see Binney and Tremaine 2008) can take place (see also Exercise 6.5), and there are other forces (e.g., modified Newtonian dynamics) where the virialization times are significantly longer than for the $1/r^2$ force (e.g., see Di Cintio and Ciotti 2011; Di Cintio et al. 2013; Nipoti et al. 2007).

Once we have a virialized stellar system, the second natural problem concerns its *stability*, as not all equilibrium configurations are stable. Very sophisticated analytical and numerical techniques can be used to address this problem, which is of the utmost importance in stellar dynamics, because not only can we expect that the vast majority of stellar systems we observe are (more or less) stable, but also instabilities (driven by internal or external phenomena) can be important for the *evolution* of these systems. An impressive number of results are known (e.g., see Bertin 2014; Binney and Tremaine 2008; Fridman and Poliachenko 1984; Palmer 1994). Among the very general and sufficiently simple stability criteria routinely used in stellar dynamics (see also Exercise 13.33), here we only mention the so-called *radial orbit instability* criterion (e.g., Bertin et al. 1994; see also Di Cintio et al. 2017; Nipoti et al. 2002, 2011; Polyachenko and Shukhman 1981; Saha 1991 and references therein), stating that stellar systems with "too many" radial orbits are unstable, and in the case of stellar disks we mention the *Toomre stability criterion* (Toomre 1964; an extension to stellar dynamics of an analogous result from fluid dynamics for rotating fluid disks). We finally recall the global *Ostriker–Peebles stability criterion* (Ostriker and Peebles 1973) for disk galaxies that, combined with the observational flat rotation curves of HI at large radii in disk galaxies (e.g., see van Albada et al. 1985), provides very strong support to the existence of extended and massive dark matter halos around galaxies.

6.2.1 The Virial Plane: Negative Specific Heat of Self-Gravitating Systems

In this section, we illustrate how the virial theorem in its simplest form allows to derive fundamental physical properties of self-gravitating stellar systems, such as their negative specific heat and their slow collapse as a consequence "gravitational evaporation" of stars (see Chapter 7), finally leading to the so-called *gravothermal catastrophe*, one of the most important concepts for our understanding of the dynamical evolution of globular clusters (e.g., see Lynden-Bell and Eggleton 1980; Lynden-Bell and Wood 1968; see also Bertin 2014; Binney and Tremaine 2008; Spitzer 1987).

All of these phenomena can be qualitatively illustrated by using the *virial plane* in Figure 6.1. As we are dealing with self-gravitating gravitational systems, we plot on the horizontal axis the quantity $|U|$; therefore, points near the origin represent very "diluted" systems, while moving to the right we encouter more and more concentrated systems. On the vertical axis we represent the system's kinetic energy: points at the bottom of the plane represent "cold" systems, while at increasing ordinate values the systems are more and more "hot." We first ask what the position is of a virialized system in Figure 6.1.

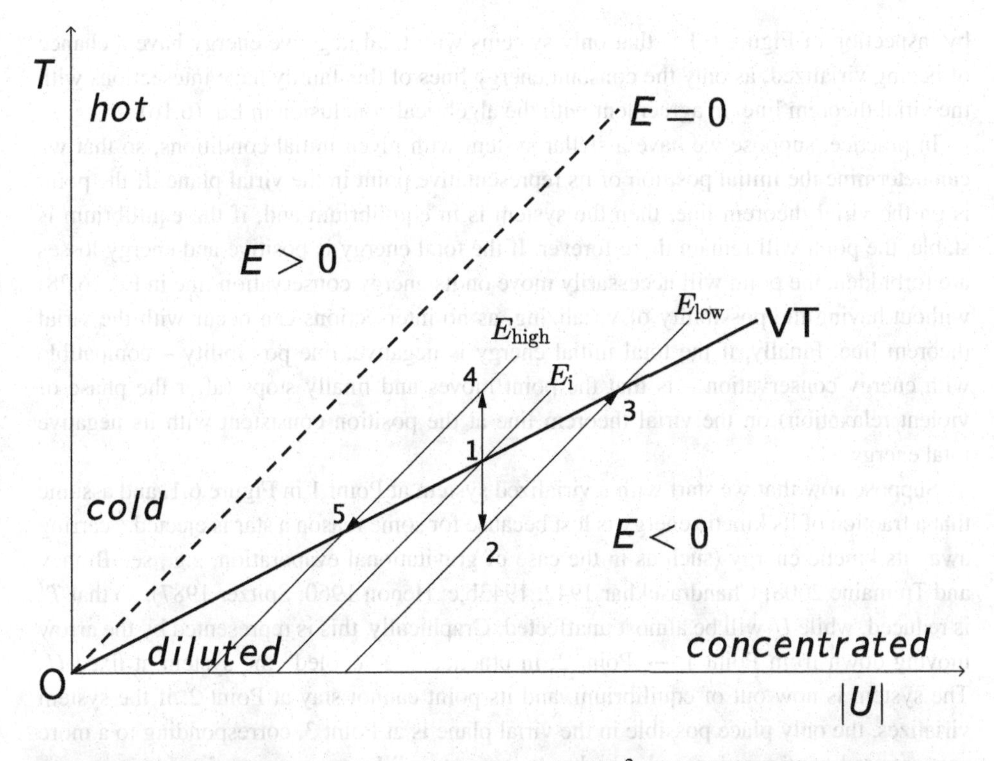

Figure 6.1 The virial plane for self-gravitating systems with $1/r^2$ forces, where the locus of virialized systems (virial theorem (VT), heavy line) and a few energy conservation lines are represented for different values of the total energy. No virialized states exist for positive values of the (barycentric) total energy E. Systems are qualitatively indicated as hot, cold, diluted, and concentrated depending on their position in the plane.

From Eq. (6.15), it follows that a self-gravitating system is virialized if and only if its representative point is placed on the *virial theorem line*

$$T' = \frac{|U|}{2}. \tag{6.27}$$

On the same plane we can now draw the family of straight lines representing energy conservation

$$T' = |U| + E'. \tag{6.28}$$

If a system evolves at constant E', then its representative point can move only on the line fixed by the initial conditions. Obviously, the line $T' = |U|$ represents systems of zero total energy, parallel lines above represent unbound systems, while parallel lines below represent systems with total negative energy. The first important consequence that can be obtained by inspection of Figure 6.1 is that only systems with total negative energy have a chance of beging virialized, as only the constant energy lines of this family have intesections with the virial theorem line, in agreement with the algebrical conclusion in Eq. (6.16).

In practice, suppose we have a stellar system with given initial conditions, so that we can determine the initial position of its representative point in the virial plane. If the point is on the virial theorem line, then the system is in equilibrium and, if the equilibrium is stable, the point will remain there forever. If the total energy is positive and energy losses are forbidden, the point will necessarily move on its energy conservation line in Eq. (6.28) without having the possibility of virializing, as no intersections can occur with the virial theorem line. Finally, if the total initial energy is negative, one possibility – compatible with energy conservation – is that the point moves and finally stops (after the phase of violent relaxation) on the virial theorem line at the position consistent with its negative total energy.

Suppose now that we start with a virialized system at Point 1 in Figure 6.1, and assume that a fraction of its kinetic energy is lost because for some reason a star is ejected,[7] carring away its kinetic energy (such as in the case of gravitational evaporation, e.g., see Binney and Tremaine 2008; Chandrasekhar 1942, 1943b,c; Hénon 1960; Spitzer 1987), so that T' is reduced, while U will be almost unaffected. Graphically, this is represented by the arrow moving down from Point 1 $\rightarrow$ Point 2; in practice, we "cooled" the system at fixed U. The system is now out of equilibrium, and its point cannot stay at Point 2: if the system virializes, the only place possible in the virial plane is at Point 3, corresponding to a more concentrated configuration and a higher "temperature." In more suggestive language, we can say that the system "cooled," but after virialization the self-gravity finally heated it to a greater temperature than the initial one. Of course, the opposite would happen along Path 1 $\rightarrow$ 4 $\rightarrow$ 5 in Figure 6.1. In a useful thermodynamic analogy, we can say that self-gravitating systems are characterized[8] by a *negative specific heat* (e.g., see Lynden-Bell and Wood 1968). What we have just described is the physical basis of the phenomenon of gravothermal catastrophe occurring in globular clusters – the progressive contraction of their inner regions associated with the continuous ejection of low-mass stars produced by the tendency toward equipartition (see Chapter 7). Unsurprisingly, similar considerations also apply to gas spheres. If we have an equilibrium gas sphere (e.g., a protostar) at a temperature higher than that of the environment, then the Second Law of Thermodynamics rules, and nothing can prevent the star from losing energy through radiation (i.e., having $dE'/dt < 0$). *If* the process is slow (i.e., if the *Kelvin time* $t_K \equiv E'/\dot{E}'$ is much longer than the gravitational time $1/\sqrt{G\rho}$), then the star configuration will readjust continuously, and macroscopically its representative point will climb along the virial theorem line, following the time derivative of Eq. (6.23)

[7] Of course, in a system with $N \geq 3$, such ejection is compatible with the system's total barycentric negative energy $E' < 0$ (see Section 6.1.3).

[8] Negative specific heats are not peculiarities of self-gravitating systems. For example, see Footnote 4 in Chapter 10 about polytropic transformations of a perfect gas.

$$\frac{d\langle T'\rangle}{dt} = -\frac{dE'}{dt}, \quad \frac{d\langle U\rangle}{dt} = 2\frac{dE'}{dt}. \tag{6.29}$$

The protostar contracts and heats up. Of course, even if it grows hotter and hotter (until contraction stops when finally the central temperature is high enough to produce nuclear reactions[9] in the core), actually the protostar becomes poorer and poorer in total energy. Unfortunately, the two relations above are sometimes "read" in a quite inappropriate way, producing some difficulty for students. In fact, if we say that in a quasi-static contraction "half of the change in gravitational energy must be radiated and half goes toward increasing the temperature" (a perfectly correct statement from the mathematical point of view!), then the physical weight of the argument is put on the gravitational energy.[10] Instead, the contraction is driven by the energy loss dE'/dt, which with virialization is *redistributed* among kinetic and gravitational energies as in Eq. (6.29). In practice, the combined effects of gravity and the virial theorem behave like a quite strange person who own two bank accounts (T' and U) and who for each expense ($\dot{E}' < 0$) not only charges it to the account U, but also takes from U the equivalent of the expense and transfers it to the account T'. Vice versa, for each gain ($\dot{E}' > 0$) the money is put into the account U and an equivalent amount is also transferred from T' to U.

6.2.2 Dissipationless Merging, Collapse, and the Virial Theorem

We conclude this chapter with the qualitative (but far-reaching) illustration of the consequences of the virial theorem for the phenomenon of *galaxy merging* (i.e., the fusion of two initially distinct galaxies or dark matter halos; see Toomre 1977) and of *structure collapse* (i.e., the evolution toward equilibrium of an energetically bound gravitational system with no or little initial kinetic energy). The relevance of these phenomena to astrophysics is so obvious that no further emphasis is needed.

For simplicity we restrict the discussion to the *dissipationless* case where only gravity is at play (i.e., we exclude the effects of energy gains and lossess due to baryonic physics) and we briefly comment on the interesting conclusions that can be obtained when simply considering the virial theorem and three of the most important empirical *scaling laws* obeyed by elliptical galaxies (e.g., see Bertin 2014; Binney and Tremaine 2008; Cimatti et al. 2019; Sparke and Gallagher 2007; see also Ciotti 2009 and references therein). The first of these relations is the Faber–Jackson law (Faber and Jackson 1976), which we write as

$$L \propto \sigma_{\circ}^{\alpha}, \quad \alpha \approx 4, \tag{6.30}$$

[9] In a beautiful analogy, we can also say that the core contraction of globular clusters due to gravothermal catastrophe stops when the clusters start to "burn" the gravitational energy of binary stars in the core (see Binney and Tremaine 2008; Spitzer 1987).

[10] There *are* cases when the physics is actually driven by U instead of E', such as when a star contracts due to the increase in the mean molecular weight in the core produced by nuclear burning – and due to the correspondent decrease in central pressure.

where $\sigma_\circ$ indicates the observed value of the velocity dispersion of a given elliptical galaxy (measured within some prescribed aperture; see Chapter 13) and L is its total (absolute) luminosity in some prescribed photometric band. Basically, this extremely important relation says that in elliptical galaxies the stellar velocity dispersion increases with their total luminosity, something the student should recognize as not obvious.[11] The second important empirical scaling law is due to Kormendy (1977),

$$L \propto R_e^\beta, \quad \beta \lesssim 1 \tag{6.31}$$

(in Chapter 13 we will encounter another very important scaling law, the fundamental plane, which contains and encompasses the two previous laws). The third and final family of scaling laws we consider relates the mass and/or velocity dispersion of the host stellar spheroids and the mass $M_{\rm BH}$ of their central supermassive black holes

$$M_{\rm BH} \propto M_*, \quad M_{\rm BH} \propto \sigma_\circ^\gamma, \quad \gamma \approx 4 \div 5 \tag{6.32}$$

(e.g., see Ferrarese and Merritt 2000; Graham 2016; Gebhardt et al. 2000; Magorrian et al. 1998; Merritt and Ferrarese 2001; Tremaine et al. 2002).

For our idealized experiment, we consider two self-gravitating, virialized stellar systems with initial separation much larger than their sizes, so that

$$E_{\rm i1} = -T_{\rm i1} = 2U_{\rm i1}, \quad E_{\rm i2} = -T_{\rm i2} = 2U_{\rm i2}. \tag{6.33}$$

For sake of generality, we also consider some extra energy ΔE (negative or positive) in the initial conditions, mimicking, for instance, the merging of a bound pair or the fusion of two stellar systems in a slightly hyperbolic orbit. During the merging, we assume that no mass is ejected, and consistent with the assumptions we enforce energy conservation $E_{\rm f} = E_{\rm i1} + E_{\rm i2} + \Delta E$; finally, we impose the virialization of the end product, with $E_{\rm f} = -T_{\rm f} = 2U_{\rm f}$ and, in terms of kinetic energies,

$$T_{\rm f} = -E_{\rm f} = T_{\rm i1} + T_{\rm i2} - \Delta E \Rightarrow \sigma_{\rm Vf}^2 = \frac{M_1\sigma_{\rm V1}^2 + M_2\sigma_{\rm V2}^2}{M_1 + M_2} - \frac{2\Delta E}{M_1 + M_2}. \tag{6.34}$$

Therefore, in a case of parabolic merging, $\sigma_{\rm Vf}$ cannot be larger than the maximum velocity dispersion of the progenitors; for bound pairs $\sigma_{\rm Vf}$ can increase, while for hyperbolic mergings $\sigma_{\rm Vf}$ decreases. If we focus on potential energy instead of kinetic energy, we obtain the analogous relation for the final virial radius

$$U_{\rm f} = \frac{E_{\rm f}}{2} = U_{\rm i1} + U_{\rm i2} + \frac{\Delta E}{2} \Rightarrow \frac{1}{r_{\rm Vf}} = \frac{M_1^2/r_{\rm V1} + M_2^2/r_{\rm V2}}{(M_1 + M_2)^2} - \frac{\Delta E}{2G(M_1 + M_2)^2}, \tag{6.35}$$

so that $r_{\rm Vf}$ will shrink for $\Delta E < 0$ and significantly expand for $\Delta E > 0$. In particular, note that in the idealized case of parabolic merging of indentical systems $\sigma_{\rm Vf}$ remains identical

[11] For example, if the size of bright galaxies had been much larger than it is, their velocity dispersion would have been much lower than observed; see Eq. (6.26).

to that of the progenitors, while r_{Vf} doubles. It is elementary to conclude that even though σ_V and r_V are related to the observed velocity dispersion $\sigma_\circ$ and to the effective radius R_e by nontrivial nonhomology effects (see Chapter 13), repeated merging poses important constraints on galaxy formation scenarios, in particular due to the difficulty of increasing the velocity dispersion and the excessive growth of the virial radius with the galaxy mass as required by Eqs. (6.30)–(6.32). These considerations add to the list of points given in the classical reference by Ostriker (1980). Of course, gas dissipation helps to increase the velocity dispersion and radius contraction (e.g., see Ciotti and van Albada 2001; Ciotti et al. 2007; González-García and van Albada 2003; Nipoti et al. 2003).

In Exercise 6.13, we will discuss the conclusions that can be reached on the problem of the *collapse* of a dissipationless, gravitationally bound structure with "cold" initial conditions; this simple computation is relevant for explaining why the virial velocity dispersion of dark matter halos formed in cosmological simulations increases with the halo mass, something that does not happen in the too-simplistic scenario where large halos form simply by the "merging" of similar small halos (e.g., see Lanzoni et al. 2004; Nipoti et al. 2006). The formation of dark matter halos should be more appropriately seen as the collapse (and homogenization) of initially "cold" and inhomogeneous structures: the primordial density fluctuations.

Exercises

6.1 By using Eq. (6.1), prove that in a generic inertial reference system S_0, $dE/dt = 0$, $d\mathbf{P}/dt = 0$ and $d\mathbf{J}/dt = 0$. Deduce that $\mathbf{V}_{CM}$, E_{CM}, and $\mathbf{J}_{CM}$ are time-independent quantities and so that the general N-body problem admits the 10 classical integrals of motion already encountered for the two-body problem in Chapter 4.

6.2 Prove Eq. (6.8). *Hints*: Note that the scalar r_{ij} is invariant for the translation of the origin from S_0 to S_{CM}. First, by using Eq. (6.5), prove that $\sum_{i=1}^N m_i r_{ij}^2 = I' + M\|\mathbf{x}'_j\|^2$, then multiply both sides by m_j and sum over j.

6.3 Show that the inequalities in Eq. (6.9) hold. *Hints*: For the first inequality, consider

$$|U| = \frac{G}{2}\sum_{i\neq j=1}^{N}\frac{m_i m_j}{r_{ij}} \geq \frac{G m_{i'} m_{j'}}{r_{\min}} \geq \frac{G\mu^2}{r_{\min}}, \tag{6.36}$$

where $m_{i'}$ and $m_{j'}$ represent (one of) the pair of particles at a minimum relative distance; then

$$|U| = \frac{G}{2}\sum_{i\neq j=1}^{N}\frac{m_i m_j}{r_{ij}} \leq \frac{G}{2r_{\min}}\sum_{i\neq j=1}^{N} m_i m_j \leq \frac{G}{2r_{\min}}\sum_{i,j=1}^{N} m_i m_j = \frac{GM^2}{2r_{\min}}. \tag{6.37}$$

For the second inequality

$$I' = \frac{1}{2M}\sum_{i,j=1}^{N} m_i m_j r_{ij}^2 \leq \frac{r_{\max}^2}{2M}\sum_{i,j=1}^{N} m_i m_j = \frac{M r_{\max}^2}{2}, \tag{6.38}$$

and

$$I' = \frac{1}{2M} \sum_{i,j=1}^{N} m_i m_j r_{ij}^2 \geq \frac{\mu^2}{2M} \sum_{i,j=1}^{N} r_{ij}^2 \geq \frac{\mu^2 r_{\max}^2}{M}. \tag{6.39}$$

6.4 Prove the scalar Lagrange–Jacobi identity (6.10) directly from Eq. (6.2). *Hint*: First obtain $\ddot{I} = 2\sum_{i=1}^{N} m_i(\langle \dot{\mathbf{x}}_i, \dot{\mathbf{x}}_i \rangle + \langle \mathbf{x}_i, \ddot{\mathbf{x}}_i \rangle)$, and then from Eq. (6.1) show that

$$W \equiv \sum_{i=1}^{N} m_i \langle \mathbf{x}_i, \ddot{\mathbf{x}}_i \rangle = -\frac{G}{2} \sum_{i \neq j=1}^{N} \frac{m_i m_j}{r_{ij}} = U. \tag{6.40}$$

6.5 With this exercise, we extend our elaboration on the Lagrange–Jacobi indentity (6.10). The particles of a system interact with the force law

$$\mathbf{F}_{ij} = m_i m_j \left(A \frac{\mathbf{x}_i - \mathbf{x}_j}{r_{ij}^{\alpha+1}} + B \frac{\mathbf{x}_i - \mathbf{x}_j}{r_{ij}^{\beta+1}} \right), \quad r_{ij} = \|\mathbf{x}_i - \mathbf{x}_j\|, \quad i \neq j, \tag{6.41}$$

so that for $\alpha \neq \beta$ and $A \cdot B \neq 0$, $\|\mathbf{F}_{ij}\|$ is *not* a power-law. From Exercise 6.4, first show that

$$W = \frac{1}{2} \sum_{i \neq j=1}^{N} m_i m_j \left(\frac{A}{r_{ij}^{\alpha-1}} + \frac{B}{r_{ij}^{\beta-1}} \right), \tag{6.42}$$

and compute W for the Newtonian force $(B,\alpha) = (0,2)$, for the harmonic oscillator $(B,\alpha) = (0,-1)$, and for the case $(B,\alpha) = (0,1)$. Then, show that the "gravitational" energy of the system is given by $U = U_\alpha + U_\beta$, where

$$U_\alpha = \frac{A}{2(\alpha - 1)} \sum_{i \neq j=1}^{N} \frac{m_i m_j}{r_{ij}^{\alpha-1}}, \quad U_1 = \frac{A}{2} \sum_{i \neq j=1}^{N} m_i m_j \ln \frac{r_{ij}}{r_0}, \tag{6.43}$$

for $\alpha \neq 1$ and $\alpha = 1$, respectively. Show that for $\alpha \neq 1$ and $\beta \neq 1$,

$$W = (\alpha - 1)U_\alpha + (\beta - 1)U_\beta. \tag{6.44}$$

For a very remarkable application of the Lagrange–Jacobi indentity obtained when $(\alpha,\beta) = (-1,3)$, see Lynden-Bell and Lynden-Bell (2004); for other applications, see Ferronsky et al. (2011).

6.6 With this exercise, we conclude our discussion of the Lagrange–Jacobi identity (6.10) and we obtain its generalization in tensorial form for an N-body system of particles interacting with central forces also in presence of an external force field $\mathbf{F}^{\text{ext}}$. Let $\mathbf{F}(kl) = -\mathbf{F}(lk)$, the force exerted by the mass $m(l)$ on the mass $m(k)$, so that for the particle k

$$m(k)\ddot{\mathbf{x}}(k) = \mathbf{F}^{\text{ext}}(k) + \sum_{l=1; l \neq k}^{N} \mathbf{F}(kl). \tag{6.45}$$

First, introduce the two tensors

$$I_{ij} = \sum_{k=1}^{N} m(k)x_i(k)x_j(k), \quad T_{ij} = \frac{1}{2}\sum_{k=1}^{N} m(k)\dot{x}_i(k)\dot{x}_j(k), \tag{6.46}$$

and consider the second-order time derivative of the tensor I_{ij}. From Eq. (6.45), show that

$$\frac{\ddot{I}_{ij}}{2} = 2T_{ij} + \frac{1}{4}\sum_{l\neq k=1}^{N} \{[x_i(k) - x_i(l)]F_j(kl) + [x_j(k) - x_j(l)]F_i(kl)\}$$

$$+ \frac{1}{2}\sum_{k=1}^{N}[x_i(k)F_j^{\text{ext}}(k) + x_j(k)F_i^{\text{ext}}(k)], \tag{6.47}$$

where the first double sum on the right-hand side is the symmetric tensor W_{ij}. Then, show that the trace of the previous identity is

$$\frac{\ddot{I}}{2} = 2T + W + \sum_{k=1}^{N}\langle \mathbf{x}(k), \mathbf{F}^{\text{ext}}(k)\rangle, \quad W = \frac{1}{2}\sum_{l\neq k=1}^{N}\langle \mathbf{r}(kl), \mathbf{F}(kl)\rangle, \tag{6.48}$$

where $W = \text{Tr}(W_{ij})$, $I = \text{Tr}(I_{ij})$ is the polar moment of inertia in Eq. (6.2), and $\mathbf{r}(kl) = \mathbf{x}(k) - \mathbf{x}(l)$. Finally, by using the results from Exercise 6.5, compute W for the forces

$$\mathbf{F}(kl) = Am(k)m(l)\frac{\mathbf{r}(kl)}{r(kl)^{\alpha+1}}, \quad r(kl) = \|\mathbf{r}(kl)\|, \quad k \neq l, \tag{6.49}$$

and prove that Eq. (6.10) is recovered for $\mathbf{F}^{\text{ext}} = 0$, $A = -G$, and $\alpha = 2$.

6.7 Prove the (weak) Sundman's inequality $\|\mathbf{J}|^2 \leq 2IT$ in Eq. (6.11). *Hint*: In a generic inertial reference system S_0,

$$\|\mathbf{J}\| = \|\sum_{i=1}^{N} m_i \mathbf{x}_i \wedge \mathbf{v}_i\| \leq \sum_{i=1}^{N} m_i \|\mathbf{x}_i \wedge \mathbf{v}_i\| \leq \sum_{i=1}^{N} m_i \|\mathbf{x}_i\| \cdot \|\mathbf{v}_i\|$$

$$= \sum_{i=1}^{N} \sqrt{m_i}\|\mathbf{x}_i\| \cdot \sqrt{m_i}\|\mathbf{v}_i\| \leq \sqrt{\sum_{i=1}^{N} m_i\|\mathbf{x}_i\|^2}\sqrt{\sum_{i=1}^{N} m_i\|\mathbf{v}_i\|^2}, \tag{6.50}$$

where the chain of inequalities follows from the Cauchy–Schwarz and triangular inequalities in Eqs. (A.6) and (A.8).

6.8 Here, with the aid of elementary differential equations amenable to explicit solutions, we illustrate the concepts of fixed and movable singularities. Let us consider a linear ODE as in Eq. (6.12) with

$$\frac{df}{dx} + \frac{f}{(1-x)^2} = 0, \quad f(0) = f_0. \tag{6.51}$$

Show that the solution is

$$f(x) = f_0 \exp\left(\frac{x}{x-1}\right). \tag{6.52}$$

The (essential) singularity occurs at $x = 1$, the pole of the coefficient is a_0, and its location is *independent* of f_0, as expected. Solve again Eq. (6.52), where now $a_0 = 1/(1-x)$ and the singularity is regular. Finally, consider the nonlinear ODE

$$\frac{df}{dx} - f^2 = 0, \quad f(0) = f_0, \tag{6.53}$$

and show that the solution is

$$f(x) = \frac{f_0}{1 - x f_0}. \tag{6.54}$$

The movable singularity occurs at $x = 1/f_0$ (i.e., its location *depends* on f_0).

6.9 Prove the Weierstrass–Sundman theorem in Section 6.1.1. *Hints*: By definition of global collapse (see Footnote 3), $\lim_{t\to t_{\rm sing}} U = -\infty$, and so $\lim_{t\to t_{\rm sing}} \ddot{I}' = +\infty$ from the Lagrange–Jacobi identity (6.10). We first prove that a global collapse must occur at a finite time. In fact, the divergence implies that a time t_+ exists so that $\ddot{I}' > 2$ for $t > t_+$ (i.e., $I'(t) > t^2 + \alpha t + \beta$) if $t_{\rm sing} = +\infty$, then $\lim_{t\to t_{\rm sing}} I' = +\infty$, which goes against the hypothesis of global collapse. The second part of the theorem can be proved as follows: first, for $t \in [t_+, t_{\rm sing}]$ from the definition of collapse we have $\dot{I}' < 0$, while again from $\lim_{t\to t_{\rm sing}} \ddot{I}' = +\infty$ one concludes that $\dot{I}'$ is a monotonically increasing function of time (i.e., $0 > \dot{I}'(t) > \dot{I}'(t_+)$ and so $\dot{I}'(t_+)^2 \geq \dot{I}'(t_+)^2 - \dot{I}'(t)^2 \geq 0$). After multiplication of the Sundman inequality (6.11) in S' by the (positive) quantity $-\dot{I}'/I'$, and after integration over $t_+ \leq t \leq t_{\rm sing}$, we have $\|\mathbf{J}'\|^2 \ln[I'(t_+)/I'(t)] \leq [\dot{I}'(t_+)^2 - \dot{I}'(t)^2]/2 + 2E'[I'(t) - I'(t_+)]$ (i.e., $\lim_{t\to t_{\rm sing}} \|\mathbf{J}'\| = 0$). But the total angular momentum is conserved, and so $\mathbf{J}' = 0$.

6.10 Prove Result 1 and the second statement of Result 3 in Section 6.1.3. *Hints*: Regarding Result 1, if $E' < 0$, then $|U| > |E'|$ from Eq. (6.3), and from Eq. (6.9) we have $r_{\rm min}(t) \leq 0.5GM^2/|E'|$. Regarding Result 3, from the Lagrange–Jacobi identity (6.10) with $E' > 0$, it follows that $\ddot{I}' = 2(T' + E') \geq E'\ \forall t$, and so $I' \geq at^2$ for $t \to \infty$ and some $a > 0$. Equation (6.9) again proves the statement.

6.11 Prove Eq. (6.17). *Hints*: The sufficient condition is a trivial application of the Lagrange–Jacobi identity (6.10). The necessary condition is again proved using the Lagrange–Jacobi identity: from $\ddot{I}' = 0$, it follows that $I'(t) = \dot{I}'(0)t + I'(0)$. If $\dot{I}'(0) = 0$, then the result is proved; if $\dot{I}'(0) \neq 0$, then for $t \to \infty$ or $t \to -\infty$ we would reach the impossible conclusion that $I'(t) \to -\infty$.

6.12 Prove that Eqs. (6.18) and (6.19) give necessary and sufficient conditions for virialization. *Hints*: First, consider necessity. If the system is virialized, then $T' = -E'$ and $U = 2E'$, so that Eq. (6.18) holds. Moreover, for $i \neq j$, $2|E'| = |U| > Gm_i m_j/r_{ij} \geq G\mu^2/r_{ij}$, and so in Eq. (6.19) $A = 0.5G\mu^2/|E'|$; in addition, from Eq. (6.17), $I'(0) = I'(t) \geq m_i m_j r_{ij}^2/M \geq \mu^2 r_{ij}^2/M$, and so $B = \sqrt{I'(0)M}/\mu$.

For the sufficient part, by integrating the Lagrange–Jacobi identity twice under the assumptions in Eqs. (6.18) and (6.19), we conclude that $I'(t) = [2T'(0) + U(0)]t^2 + \dot{I}'(0)t + I'(0)$ is a bounded function, so that the coefficients of the time-dependent part of $I'(t)$ are zero (i.e., $I'(t) = I'(0)$), and Eq. (6.17) completes the proof.

6.13 With this exercise, we evaluate the increase of σ_V after virialization of a dissipation-less, self-gravitating, collapsing bound structure: this process illustrates qualitatively why the velocity dispersion of dark matter halos obtained in numerical simulations increases with the halo mass. Let U_i be the initial gravitational energy of the system and let $2T_i = -\alpha U_i$, with $0 \leq \alpha \leq 1$, be its initial kinetic energy, so that α measures the "coldness" of the initial conditions. From energy conservation and the virial theorem, show that the final kinetic energy, in the absence of significant mass ejection, is

$$T_f = -\left(1 - \frac{\alpha}{2}\right) U_i = \left(\frac{2}{\alpha} - 1\right) T_i. \tag{6.55}$$

Compute the variation of σ_V from the initial conditions to the final state and deduce that an increase of σ_V in the mass of the system requires that more massive structures are initally more bound (per unit mass) than low-mass systems.

7

Relaxation 1

Two-Body Relaxation

In this chapter, we introduce one of the fundamental and most far-reaching concepts in stellar dynamics (and of plasma physics for the case of electric forces): that of "gravitational collisions." As an application of the developed framework, the two-body relaxation time is derived (using the Chandrasekhar approach) by using the so-called *impulsive approximation*. The concepts of the Coulomb logarithm and of infrared and ultraviolet divergence are elucidated, with an emphasis on the importance of the correct treatment of angular momentum for collisions with small impact parameters, an aspect that is sometimes puzzling for the students due to presentations in which the minimum impact parameter appears as something to be put into the theory "by hand." On the basis of the quantitative tools devised in this chapter, we will show that large stellar systems, such as elliptical galaxies, should be considered primarily as collisionless, while smaller systems, such as small globular clusters and open clusters, exhibit collisional behavior. These different regimes are rich with astrophysical consequences, both from the observational and the theoretical points of view.

7.1 The Granular Nature of Stellar Systems

How much of a star's orbit in a stellar system depends on the forces acting upon it due to more or less nearby stars? When we model a stellar system, can we ignore the fact that the true gravitational field experienced by each star is described by the discrete system of Eq. (6.1)? Or are we doomed to deal with the almost insurmountable difficulties of the N-body problem?

In order to answer these fundamental questions, we ideally replace the *discrete* (discontinuous) distribution of the N masses with a *smooth* (continuous) density distribution $\rho(\mathbf{x};t)$; at this stage, nothing is specified about how such smoothing is performed. In practice, the density distribution ρ, taken to be a satisfactory description of the true "granular" N-body system, will serve as a continuous model for the real system, and the associated smooth potential ϕ (in general time dependent) is given by Eqs. (2.4)–(2.6). As long as this continuous approximation holds, the motion of each star is determined by the smooth potential $\phi(\mathbf{x};t)$, and so the dimensionality of the phase space[1] is reduced from $6N$ to 6 (but in the last case an explicit time dependence can appear), and the mathematical problem is enormously simplified. Of course, the stellar orbits in the smoothed system are only approximations of the real orbits, because in reality each star (e.g., the i-th star) actually moves in the true potential

$$\phi_{\text{true}}(\mathbf{x}_i,t) = -G \sum_{j=1;\, j\neq i}^{N} \frac{m_j}{||\mathbf{x}_i - \mathbf{x}_j||}, \tag{7.1}$$

associated with the gravitational field in Eq. (6.1).

The problem we face can thus be summarized by the following question: How large is the difference between the fictitious orbit described by a given star under the influence of ϕ and the true orbit described under the action of ϕ_{true}? Or, stated differently, given that $\phi_{\text{true}} = \phi + (\phi_{\text{true}} - \phi)$, how long will it take for the "granularity" term $\phi_{\text{true}} - \phi$ to change significantly the orbit that would be performed by each star in the smooth potential ϕ? Of course, the exact answer to such a question would require knowledge of both orbits, and this is precisely the difficulty we would like to avoid! Therefore, we must find some clever way to obtain a quantitative estimate of this time without resorting to the solution of the N-body problem. This problem is of such great importance to physics that different approaches have been developed to answer it (e.g., see Binney and Tremaine 2008). In the following, we will use what is perhaps the simplest, approach at least in terms of visualization.

We first introduce a few basic definitions that are needed to proceed with our discussion, and these will also be useful for the rest of this book. For an N-body system made for simplicity of identical stars of the same mass m, the nowhere negative function $n = n(\mathbf{x},\mathbf{v};t)$ is defined by the property that at each time t

$$\Delta N(\Omega;t) = \int_{\Omega} n(\mathbf{x},\mathbf{v};t)\, d^3\mathbf{x}\, d^3\mathbf{v}, \tag{7.2}$$

which is the number of stars contained in a given arbitrary six-dimensional region $\Omega \subseteq \gamma$ of phase-space γ; obviously, over the whole phase space $\Delta N(\gamma;t) = N$. A function strictly related to n is the *phase-space distribution function* (DF) $f = f(\mathbf{x},\mathbf{v};t) \equiv m\, n(\mathbf{x},\mathbf{v};t)$, so that from the previous definition

$$\Delta M(\Omega;t) = \int_{\Omega} f(\mathbf{x},\mathbf{v};t)\, d^3\mathbf{x}\, d^3\mathbf{v}, \tag{7.3}$$

where ΔM is the total mass of the stars that at time t are found in the region Ω. Of course, $\Delta M(\gamma;t) = M = mN$ is the total mass of the system. Note that in an N-body system made of $k = 1,\ldots,K$ different species of stars of masses m_k, with $N = \sum_{k=1}^{K} N_k$ and $M = \sum_{k=1}^{K} m_k N_k$, the functions n_k and f_k are naturally introduced as

[1] We indicate with Γ the $6N$ dimensional phase space of the true system, and with γ the six-dimensional (one-particle) phase space of a single star (see also Chapter 9).

$$n = \sum_{k=1}^{K} n_k, \qquad f = \sum_{k=1}^{K} f_k = \sum_{k=1}^{K} m_k n_k. \tag{7.4}$$

Of course, it is also possible to extend the relations above to the case of a continuous distribution of masses (obtaining a so-called *extended* DF formalism), and the sums would be replaced by integrals over the mass spectrum (e.g., see Chapter 8). A completely analogous set of definitions also holds for the smooth density distribution ρ. In this case, the relation between the material density ρ at $\mathbf{x}$ and the DF is obtained by considering in Eq. (7.3) the phase-space region $\Omega = \Delta^3\mathbf{x} \times \Re^3$, so that for a sufficiently regular f

$$\rho(\mathbf{x};t) = \lim_{\Delta^3\mathbf{x}\to 0} \frac{1}{\text{Volume}(\Delta^3\mathbf{x})} \int_{\Delta^3\mathbf{x}} d^3\mathbf{x} \int_{\Re^3} f(\mathbf{x},\mathbf{v};t) d^3\mathbf{v} = \int_{\Re^3} f(\mathbf{x},\mathbf{v};t) d^3\mathbf{v}, \tag{7.5}$$

as the integration over all of the velocity space ensures that we are certainly "counting" all of the particles in the "volume" $d^3\mathbf{x}$. Finally, if the system consists of K different density components ρ_k,

$$f = \sum_{k=1}^{K} f_k, \qquad \rho_k(\mathbf{x};t) = \int_{\Re^3} f_k(\mathbf{x},\mathbf{v};t) d^3\mathbf{v}, \qquad \rho = \sum_{k=1}^{K} \rho_k. \tag{7.6}$$

For the sake of the present discussion, we simply assume that the function f exists as a useful mathematical object that contains all of the needed dynamical information about each star of the system under consideration. We will discuss in more depth the physical meaning[2] and the mathematical properties of the DF in Chapter 9.

7.2 The Impulsive Approximation

The standard approach to the calculation of the so-called *relaxation time* (i.e., the characteristic timescale beyond which the continuous approximation introduced in Section 7.1 is certainly no longer valid) is based on two assumptions:

(1) All encounters between particles are treated as *independent* (i.e., the effect of each encounter is simply added to that of the others).
(2) All encounters between particles are treated as *hyperbolic two-body encounters*.

For these reasons, the derived characteristic time is called the *two-body relaxation time*, and it is indicated as t_{2b}. A gravitational system for which the cumulative effects of the encounters are negligible will be called *collisionless*, or, otherwise, *collisional*. Obviously, real N-body systems are *never*, strictly speaking, collisionless. Nevertheless, we will show that models based on the equations derived under the assumptions of perfect noncollisionality play an important role in stellar dynamics. The general treatment of the evaluation of t_{2b} can be found in well-known monographs (e.g., see Bertin 2014; Binney and Tremaine 2008;

[2] Notice that the physical dimensions of the DF are those of a density divided by a velocity cubed.

Chandrasekhar 1942; Saslaw 1987; Spitzer 1987; and for a different approach based on the statistical properties of the gravitational field, see Chandrasekhar and von Neumann 1942, 1943). Here, we follow the simplest and more intuitive approach based on Assumptions (1) and (2) above, and for simplicity we further restrict ourselves to the case where all of the field particles have the same mass.

We describe our N-body system of total mass M in an inertial reference system S_0 with the aid of its phase-space DF $f = f(\mathbf{x}_f, \mathbf{v}_f)$; the particles (stars) that make up the system have identical masses m_f (where "f" stands for "field"). Moreover, we consider a "test" mass m_t that at $t = -\infty$ has initial position $\mathbf{x}_t(-\infty)$ and initial velocity $\mathbf{v}_t(-\infty)$, so that the total barycentric energy of each pair is positive and Assumption (2) is therefore satisfied. Following Chapter 4, for each pair (m_t, m_f), let μ be the *reduced mass* and $\mathbf{r}(t) = \mathbf{x}_t(t) - \mathbf{x}_f(t)$ and $\mathbf{v}(t) = \mathbf{v}_t(t) - \mathbf{v}_f(t)$, respectively, the *relative position* and the *relative velocity* of m_t at time t with respect to a given m_f. Moreover, let $\mathbf{v}(\pm\infty)$ be the *asymptotic* relative velocity of the two particles in the resulting two-body hyperbolic motion (i.e., their relative velocities for $t \to \pm\infty$, respectively), let $\Delta\mathbf{v} = \mathbf{v}(\infty) - \mathbf{v}(-\infty)$ be the total variation of the relative velocity (see Figure 4.1), and finally let $\mathbf{J}' = \mu\mathbf{r}(t) \wedge \mathbf{v}(t)$ and $E' = \mu\|\mathbf{v}(t)\|^2/2 - Gm_t m_f/\|\mathbf{r}(t)\|$ be the total (constant) angular momentum and total energy of the pair, respectively, in the inertial barycentric reference system S_{CM} of the pair. As is shown in Chapter 4, the *impact parameter b* of the resulting orbit is given by Eq. (4.24), $\|\mathbf{v}(\infty)\| = \|\mathbf{v}(-\infty)\|$, and the final relative velocity $\mathbf{v}(\infty)$ rotates with respect to the direction of the initial relative velocity $\mathbf{v}(-\infty)$ by an angle $2\Delta\varphi - \pi$, obtained from Eq. (4.25).

We are now in a position to quantify the concept of the two-body relaxation time t_{2b}. The basic idea behind Assumption (1) is to compute the sum of the effects of all of the two-body encounters evaluated in the fictitious case of the absence of the remaining $N - 2$ stars and to define some threshold so that when the total effect is larger than the threshold, a relaxation time has passed. The student should appreciate that actually nothing special abruptly happens when the system's age crosses the relaxation time, since from the beginning the granularity of the gravitational system affects the orbit of each star, such that as time increases these orbits can be described less and less accurately according to the corresponding orbits in the idealized smoothed system. Different indicators can be used to define the cumulative effects of the two-body interactions and to derive an estimate for t_{2b}: for example, Chandrasekhar (1942) adopted a suitably defined average value of the cumulative angular orbital deflections or alternatively the variation of the kinetic energy of the test particle. Other possibilities can be devised, such as by computing the so-called *diffusion coefficients* in the velocity space and their associated diffusion (relaxation) times (e.g., see Binney and Tremaine 2008, appendix L; Rosenbluth et al. 1957; Spitzer 1987). Here, we will follow the simplest approach (i.e., we will focus on the kinetic energy variations associated with the perpendicular component of the velocity of the test particle with respect to the *initial relative velocity* $\mathbf{v}(-\infty)$). This last point deserves special attention because its meaning is not as obvious as it may appear from a superficial exposition. In fact, as we are considering the phenomenon in the inertial reference system S_0, one could think that the natural decomposition of the final velocity $\mathbf{v}_t(+\infty)$ of the test star after each collision would be along the parallel and perpendicular directions with respect to the initial velocity $\mathbf{v}_t(-\infty)$. However, this is not the case: as we will see, the natural decomposition is instead along and perpendicular to the *initial relative velocity* $\mathbf{v}(-\infty)$. This may appear quite strange, as in each encounter $\mathbf{v}(-\infty)$ is obviously oriented along different directions in S_0!

So let us start by focusing our attention on a single encounter event; then we will add some (suitably choosen) properties of each of the (hypothetically independent) two-body encounters of the test star with the field stars in S_0 by using the DF of the system. Examples of such properties are represented by (but not limited to) the components of the total velocity change $\Delta\mathbf{v}_t$ of the test star in an unbound orbit in S_0 as given rigorously by Eq. (4.16) and/or by the associated change of energy ΔE_t in Eqs. (4.18) and (4.27) that in the particular case of the r^{-2} force can be obtained in closed form by using the solution of the hyperbolic two-body problem in Eqs. (4.14), (4.15), and (4.33).

In order to sum a preferred orbital property P of the test star over all of the encounters, we must first specify the phase-space DF of the field stars, as it contains the "initial conditions" of the field stars to be used in each encounter. This DF is usually written under the simplifying assumptions of a *spatially homogeneous* distribution of the field masses (i.e., the number density n_f is independent of position) and in the case of *isotropy* of their velocity distribution. In practice, the phase-space density distribution of the field stars is given by

$$f(\mathbf{x}_f, \mathbf{v}_f) = n_f\, g(v_f), \quad v_f \equiv \|\mathbf{v}_f\|, \quad 4\pi \int_0^\infty g(v_f) v_f^2\, dv_f = 1, \tag{7.7}$$

and g is a positive function dependent (as a consequence of velocity isotropy) on the modulus of $\mathbf{v}_f$; the normalization of the last integral is an obvious consequence of Eq. (7.5). Simple geometry shows that, under the previous assumptions, in the time interval Δt the number of encounters with field particles in the differential velocity volume $d^3\mathbf{v}_f$ around $\mathbf{v}_f$, and with impact parameter between b and $b + db$, is

$$\Delta n_{\mathrm{enc}} = 2\pi b db \times \|\mathbf{v}_t - \mathbf{v}_f\| \Delta t \times n_f g(v_f) d^3\mathbf{v}_f. \tag{7.8}$$

The content of the expression in Eq. (7.8) is deeper than its deceptively simple algebraic appearance. In fact, even if the direction of $\mathbf{v}_t(-\infty)$ in S_0 is fixed for all of the encounters, the *direction* of the axis of the cylinder of radius b is oriented in different directions for different field stars, and also its length is different, as these quantities are dependent on the direction of the initial relative velocity of the test star with respect to the field stars in the velocity volume $d^3\mathbf{v}_f$; in particular, note that Eq. (7.8) holds because n_f is independent of $\mathbf{x}_f$. Therefore, while the test star is moving in a given direction in S_0, the student should imagine that it is actually traveling inside a sea of cylinders with all possible orientations and lengths in the relative velocity space. Now, for a given orbital property P of the test star, in full generality we introduce the *diffusion coefficient*

$$D(P) \equiv 2\pi n_{\mathrm{f}} \int_{\Re^3} g(v_{\mathrm{f}}) \|\mathbf{v}_{\mathrm{t}} - \mathbf{v}_{\mathrm{f}}\| d^3\mathbf{v}_{\mathrm{f}} \int_0^\infty \langle P \rangle \, b db, \tag{7.9}$$

where $\langle P \rangle$ is the angle average of P around the direction of the initial relative velocity. In practice, $D(P)$ measures the time rate of change of the orbital property P as a result of the cumulative effects of the encounters with the field stars (e.g., see Binney and Tremaine 2008, appendix L, and references therein).

Therefore, suppose we solved the two-body problem of a given encounter, and so we have the expressions of $\Delta\mathbf{v}_\perp$ and $\Delta\mathbf{v}_\parallel$ for the relative orbit, and then from Eq. (4.16) we have the exact expressions of

$$\Delta\mathbf{v}_{\mathrm{t}\perp} = \frac{\mu}{m_{\mathrm{t}}} \Delta\mathbf{v}_\perp, \quad \Delta\mathbf{v}_{\mathrm{t}\parallel} = \frac{\mu}{m_{\mathrm{t}}} \Delta\mathbf{v}_\parallel \tag{7.10}$$

in S_0. At the risk of repeating ourselves, we recall one more time that the two last quantities are the components of the total variation of the test star velocity $\mathbf{v}_{\mathrm{t}}$ perpendicular and parallel to the initial relative velocity, not to $\mathbf{v}_{\mathrm{t}}(-\infty)$.

We begin with a discussion of the diffusion coefficient $D(\Delta\mathbf{v}_{\mathrm{t}\perp})$ of the perpendicular velocity component. It should be quite obvious that under the assumptions of Eq. (7.7), the angle $\langle \Delta\mathbf{v}_\perp \rangle$ vanishes, and so we deduce that in an homogeneous and isotropic distribution of field stars, $D(\Delta\mathbf{v}_\perp) = D(\Delta\mathbf{v}_{\mathrm{t}\perp}) = 0$; it can be shown that this result is also true when considering the case of the perpendicular component to the initial velocity $\mathbf{v}_{\mathrm{t}}(-\infty)$ in S_0. In the next chapter, we study $\Delta\mathbf{v}_{\mathrm{t}\parallel}$, and we will show that this quantity is *not* zero, even in our homogeneous and isotropic field distribution.

Having dealt with the case of the diffusion coefficients of the first-order velocity components, we now can move to the most basic second-order quantities (i.e., $D(\|\Delta\mathbf{v}_{\mathrm{t}\perp}\|^2)$ and $D(\|\Delta\mathbf{v}_{\mathrm{t}\parallel}\|^2)$). So we now consider the sum of the squared norm of each velocity deflection. Of course, these sums by construction are not zero. Before proceeding with the computation, it is worth understanding the physical meaning of this. Basically, the student should recognize that for a quantity such as $\Delta\mathbf{v}_{\mathrm{t}\perp}$, having a zero mean value is not inconsistent with having a finite dispersion. The visualization of this is that, as the test particle travels in the field distribution, its mean direction remains unchanged, but a larger and larger conus of possible positions is associated with the average trajectory. In some sense this is equivalent to Brownian motion, such that the statistical average position of the particle remains the initial position, but the probability of finding the particle at a given distance from the intial position increases as the rate of the square root of the number of the scattering events over the fluid molecules (e.g., see Feynman et al. 1977, vol. 1, chapter 41).

We first consider the diffusion coefficient $D(\|\Delta\mathbf{v}_\perp\|^2)$: once obtained, we also have $D(\|\Delta\mathbf{v}_{\mathrm{t}\perp}\|^2) = \mu^2 D(\|\Delta\mathbf{v}_\perp\|^2)/m_{\mathrm{t}}^2$ from the exact Eq. (7.10). The integration of Eq. (7.9) presents some important conceptual problems. In Exercise 7.1, we proceed with the "exact" integration of $D(\|\Delta\mathbf{v}_\perp\|^2)$, while here we follow the approach of *impulsive approximation*, which is often adopted in first expositions of the subject to students due to its use for presenting the basic physical arguments.

In each encounter of our test star with a field star, in the limit of a large impact parameter (i.e., high angular momentum), the change in the relative velocity *perpendicular* to the initial relative velocity $\mathbf{v}(-\infty)$ is

$$\mu \|\Delta \mathbf{v}_\perp\| \sim \frac{G m_t m_f}{b^2} \frac{2b}{v}, \qquad v = \|\mathbf{v}(-\infty)\|. \tag{7.11}$$

As can be seen by expansion of Eq. (4.33), the expression in Eq. (7.11) is asymptotically correct in the limit of a large impact parameter b or a large initial relative velocity v (see also Exercises 4.8 and 4.9). For practical purposes, the content of Eq. (7.11) can be described as the total deviation of the relative orbit obtained under the action of a force $G m_t m_f / b^2$ acting in the perpendicular direction to the initial relative velocity for a time $2b/v$, and from this interpretation it is called impulsive approximation. Later on we will see why restricting ourselves to distant encounters allows us to obtain the correct (asymptotic) result after all. From Eqs. (4.16) and (7.11), it follows that

$$\|\Delta \mathbf{v}_{t_\perp}\|^2 = \frac{\mu^2}{m_t^2} \|\Delta \mathbf{v}_\perp\|^2 \sim \frac{4G^2 m_f^2}{b^2 v^2}, \qquad b \to \infty. \tag{7.12}$$

Note how for large impact parameters there is no dependence on m_t. Equation (7.9) then would reduce to

$$D(\|\Delta \mathbf{v}_{t_\perp}\|^2) \sim 8\pi G^2 n_f m_f^2 \int_{\Re^3} \frac{g(v_f)}{v} d^3 \mathbf{v}_f \int_0^\infty \frac{db}{b}, \tag{7.13}$$

and it is apparent how integration over the impact parameter would lead to a doubly absurd result, with a logarithmic divergence both at $b = 0$ and at $b = \infty$. In particular, in the impulsive approximation an artificial divergence (the so-called *ultraviolet divergence*, similar to the analogous problem faced when computing the high-frequency component of the bremsstrahlung radiation emitted by an accelerated charge in plasma physics; e.g., see chapter 9 in Landau and Lifshitz 1971) appears for $b = 0$; however, when using the full solution of the two-body problem in Chapter 4, it is easy to show that such divergence disappears (see Exercise 7.1), and that the correct behavior of the integral near $b = 0$ is obtained by replacing the true limit in Eq. (7.13) with $b_{\pi/2}$. Instead, the logarithmic *infrared divergence* for $b \to \infty$ cannot be eliminated in an infinite homogeneous gravitational system (while in the electromagnetic analog this divergence is not present, being the upper limit of integration that is naturally truncated at the *Debye length*). In practice, we are forced to truncate[3] the integral at $b_{\max}$, a fiducial maximum impact parameter (e.g., see Binney and Tremaine 2008; Chandrasekhar 1942; Kandrup 1980; Spitzer 1987; Woltjer 1967. Also notice that even though it should be clear from the previous discussion that the question of the "exact" value of $b_{\max}$ in a infinite and homogeneous system is strictly speaking ill-posed, some discussion on the optimal truncation of the Coulomb logarithm can be

[3] The infrared divergence shows that the cumulative effect of distant encounters dominates the two-body relaxation of three-dimensional systems, and this provides an "a posteriori" justification of the use of impulsive approximation. As an interesting exercise, the student should consider two-body relaxation in an idealized infinitesimally thin disk galaxy. For the case of modified Newtonian dynamics, where the Coulomb logarithm does not exist and the two-body relaxation is even faster than in Newtonian gravity, see Ciotti and Binney (2004).

found in the literature, such as the different points of view of Chandrasekhar and Spitzer on the subject, as summarized for example in Binney 1996 and van Albada and Szomoru 2020). The final result can be expressed by introducing the *Coulomb logarithm* $\ln\Lambda$, where the quantity Λ is given in Eq. (7.25). After integration over the impact parameter, Eq. (7.13) becomes

$$D(\|\Delta\mathbf{v}_{t\perp}\|^2) \sim 8\pi G^2 n_f m_f^2 \int_{\Re^3} \frac{g(v_f)\ln\Lambda}{v} d^3\mathbf{v}_f, \quad \Lambda = \frac{b_{max}}{b_{\pi/2}}. \tag{7.14}$$

We finally integrate over the velocity space. Following Chandrasekhar (1942), we introduce the *velocity-weighted Coulomb logarithm* $\ln\bar{\Lambda}$,

$$\int_{\Re^3} \frac{g(v_f)\ln\Lambda}{\|\mathbf{v}_t - \mathbf{v}_f\|} d^3\mathbf{v}_f \equiv \Psi(v_t)\ln\bar{\Lambda}, \tag{7.15}$$

where $\ln\bar{\Lambda}$ in practical applications is estimated by using some average (typical) velocity $\overline{v_f}$ of the field stars, and

$$\Psi(v_t) \equiv \int_{\Re^3} \frac{g(v_f)}{\|\mathbf{v}_t - \mathbf{v}_f\|} d^3\mathbf{v}_f \equiv \frac{\Xi(v_t)}{v_t} + 4\pi \int_{v_t}^{\infty} g(v_f) v_f \, dv_f, \tag{7.16}$$

in analogy with the result regarding the expression of the potential of a spherical density distribution.[4] In particular,

$$\Xi(v_t) = 4\pi \int_0^{v_t} g(v_f) v_f^2 \, dv_f, \quad \Xi(\infty) = 1, \tag{7.17}$$

is the fractional velocity v_t, analogous to the "mass" of a spherical system inside a sphere of "radius" v_t. Therefore, the integral in Eq. (7.17) can be formally interpreted as the "potential" $\Psi(v_t)$ in velocity space of the "density" distribution $g(v_f)$. $\Psi(v_t)$ is known as the first *Rosenbluth potential*, and it clearly depends on the specific velocity distribution of the field particles. The normalization condition on g in Eqs. (7.7) and (7.17) shows that for large values of v_t with respect to the characteristic velocities of the field stars, the potential is dominated by the "monopole" term $1/v_t$ (e.g., see Exercise 7.3). In summary, we finally obtain the important formula

$$D(\|\Delta\mathbf{v}_{t\perp}\|^2) \sim 8\pi G^2 n_f m_f^2 \ln\bar{\Lambda}\,\Psi(v_t), \tag{7.18}$$

in perfect agreement with Eq. (7.26).

The cumulative effect of the encounters is to heat up the test particle in the direction perpendicular to the initial relative velocity. The characteristic time associated with this "heating" is defined in a natural way as

$$t_{2b} \equiv \frac{E_t(-\infty)}{D(\Delta E_{t\perp})} \sim \frac{v_t^2}{8\pi G^2 n_f m_f^2 \ln\bar{\Lambda}\,\Psi(v_t)} \sim \frac{v_t^3}{8\pi G^2 n_f m_f^2 \ln\bar{\Lambda}}, \tag{7.19}$$

the so-called *two-body relaxation time*; the last expression holds in the limit of large v_t.

[4] Remember Richard Feynman's wise advice: "The same equations have same solutions!"

A natural question arises: Why have we defined the relaxation time by using the perpendicular component and not the parallel component, as suggested by Eq. (4.27)? A simple answer is provided by Exercise 7.2, where we evaluate the "exact" formula for $D(\|\Delta \mathbf{v}_{\|}\|^2)$, and we show that for large Λ

$$D(\|\Delta \mathbf{v}_{t\|}\|^2) \sim 4\pi G^2 n_f m_f^2 \Psi(v_t), \tag{7.20}$$

in other words, for large Λ, the parallel heating term is negligible compared to the perpendicular one, so that Eq. (7.19) is asymptotically correct *also* when one would add $D(\Delta E_{t\|})$ to the denominator. Why parallel heating is less important than perpendicular heating is simply a manifestation of the geometric fact that in a right triangle, when a cathetus is vanishing, the difference between the hypothenuse and the long cathetus is an infinitesimal quantity of order 2 (think of the expansion of the functions sin and cos for a vanishing angle), combined with the fact that relaxation is dominated (for large Λ) by distant encounters; we will return to this point in the next chapter. Finally, notice that t_{2b} is (qualitatively!) inversely proportional to the DF of the field stars (see also Footnote 2).

7.3 Relaxation Time for Self-Gravitating Systems

In order to familiarize ourselves with the order of magnitude of t_{2b} in real astrophysical systems, we now proceed with a "back-of-the-envelope" estimate of Eq. (7.19) in the case of self-gravitating systems. Here, the student should focus on the physics of the problem, without looking for rigorous algebra; the obtained results fully justify the qualitative approach followed, and more careful/sophisticated/realistic (and heavy) calculations would produce only correction factors of the order of unity, not changing the logical conclusions while obscuring the physical arguments.

For the estimate of the various quantities entering Eq. (7.19), we consider a finite system, as in Exercise 1.1, of spherical shape and radius R, homogeneously filled with N identical stars of mass $m = m_f = m_t$, so that $n_f = 3N/(4\pi R^3)$ (the famous "spherical cow!"). Arbitrarily setting $r_V = R$ in the virial theorem as given in Eq. (6.26), we then have

$$\sigma_V^2 = \frac{GNm}{R}, \tag{7.21}$$

and we adopt σ_V as the fiducial value for the characteristic stellar velocities in the system (i.e., $\sigma_V = v_f = v_t$); for simplicity (but quite inconsistently!) we consider the high-velocity limit $\Psi(v_t) \sim 1/v_t$. We then express t_{2b} in the natural time unit of the system, its fiducial *crossing time*

$$t_{\rm cross} \equiv \frac{2R}{\sigma_V}, \tag{7.22}$$

so that simple algebra on Eq. (7.19) finally shows that at the leading order for large N

$$\frac{t_{2b}}{t_{cross}} \sim \frac{N}{12 \ln \bar{\Lambda}}, \quad \ln \bar{\Lambda} \equiv \ln \frac{2R}{b_{\pi/2}} = \ln N, \tag{7.23}$$

having set $b_{max} = 2R$ and used Eqs. (4.26) and (7.21) to estimate the Coulomb logarithm.

As an astrophysical application, let us consider a typical elliptical galaxy where $N \simeq 10^{11}$ and $t_{cross} \simeq 2 \times 10^8$ yrs: the characteristic two-body relaxation time is then estimated to be in the order of $10^{6\div7}$ Gyrs. Therefore, *galaxies* are essentially *collisionless systems* over cosmological timescales. In contrast, for *globular clusters*, where $N \simeq 10^{5\div6}$ and $t_{cross} \simeq 10^6$ yrs, we obtain $t_{2b} \approx 5 \times 10^9$ yrs; for such systems, the cumulative effects of two-body encounters are then expected to be important on timescales shorter than their age, so these systems are marginally collisional, and this conclusion is confirmed by observations and numerical simulations (e.g., see Spitzer 1987). Finally, in *open clusters*, with $N \simeq 10^{2\div3}$ and $t_{cross} \simeq 10^6$ yrs, $t_{2b} \approx 10^7$ yrs, an age shorter than the permanence time of the main sequence of massive stars, and this provides the explanation for why in general observed open clusters are made of young and blue stars (e.g., Binney and Merrifield 1998).

Before concluding, we comment on a seamingly paradoxical conclusion that can be obtained from Eq. (7.23). In fact, we deduce that increasing the number of particles at a *fixed* total mass of the system (i.e., at fixed t_{cross}), the effects of *granularity* of ϕ_{true} appear to become important on longer timescales (i.e., t_{2b} increases!), and this, at a superficial reading, seems to be at odds with the fact that the test star experiences *more* collisions. However, the student should consider that two-body relaxation is not only due to the number of encounters, but also to the energy variations due to the encounters:

$$\Delta E_t \propto N \times \text{energy exchange in one encounter.} \tag{7.24}$$

However, the fixed total mass from Eq. (7.12) shows that the energy exchange in one encounter *decreases* as $m_f^2 \propto N^{-2}$, so that $\Delta E_t \propto 1/N$, leading to the increase of the numerator in Eq. (7.23).

Why is the knowledge of t_{2b} important in the astrophysical context? First, this time provides a nice indicator of what techniques should be used to study the system itself. Second, and more importantly, we know from statistical mechanics (e.g., Khinchin 1949) that every time a system of particles is subjected to the prolonged phenomenon of weak and statistically uncorrelated energy exchanges (fluctuations), the system is driven toward the establishement of a thermodynamical state, characterized by the universal Maxwell–Boltzmann distribution in Eq. (7.30) and by equipartition. This fact is of extreme importance in astronomical systems that have a finite escape velocity (see Exercise 10.22 for a useful estimate). The final effect is to have *gravitational evaporation* of the low-mass stars and *core collapse* of the system (e.g., see Binney and Tremaine 2008; Spitzer 1987;

see also Chapter 6 for energetic considerations on the escape of particles from a system with total negative energy).

The last comment concerns the observational fact that astrophysical systems such as galaxies and globular clusters are *very inhomogeneous* (i.e, the densities of their central regions are many orders of magnitude larger than their *mean density*); therefore, in detailed calculations it should be taken into account that t_{2b} depends strongly on its position inside such systems.

Exercises

7.1 Show that the result of the integration over the impact parameter of Eq. (7.9) for the quantity $D(\|\Delta\mathbf{v}_{\perp}\|^2)$ obtained from the first identify in Eq. (4.33) can be written in terms of the quantities

$$\Lambda \equiv \frac{b_{\max}}{b_{\pi/2}} = \frac{b_{\max} v^2}{GM}, \quad \ln\sqrt{1+\Lambda^2} - \frac{\Lambda^2}{2(1+\Lambda^2)} \sim \begin{cases} \dfrac{\Lambda^4}{4}, & \Lambda \to 0, \\ \ln\Lambda & \Lambda \to \infty, \end{cases} \tag{7.25}$$

where $\ln\Lambda$ is the *Coulomb logarithm*, $M = m_t + m_f$, $b_{\max}$ is the fiducial maximum impact parameter (usually taken from the order of the size of the system), and $v = \|\mathbf{v}(-\infty)\|$ is the modulus of the initial relative velocity. Notice that the limit of large Λ (the standard case) refers to large values of $b_{\max}$ and nonzero values of v, while the (perculiar) case of small Λ refers to vanishing relative velocities. Then show that at the leading order for large Λ,

$$D(\|\Delta\mathbf{v}_{\perp}\|^2) \sim 8\pi G^2 n_f M^2 \Psi(v_t) \ln\bar{\Lambda}, \tag{7.26}$$

where $\ln\bar{\Lambda}$ is the velocity-averaged Coulomb logarithm[5] defined in Eq. (7.15). Finally, verify that the same result is obtained by integration over b of the impulsive approximated formula in Eq. (7.13), artificially truncating the lower limit of integration at $b_{\pi/2}$.

7.2 Show that the result of the integration over the impact parameter of Eq. (7.9) for the quantity $D(\|\Delta\mathbf{v}_{\|}\|^2)$ obtained from the second identity in Eq. (4.33) can be written in terms of the quantity

$$\frac{\Lambda^2}{2(1+\Lambda^2)} \sim \begin{cases} \dfrac{\Lambda^2}{2}, & \Lambda \to 0, \\ \dfrac{1}{2} - \dfrac{1}{2\Lambda^2}, & \Lambda \to \infty. \end{cases} \tag{7.27}$$

[5] Of course, nothing would prevent us from defining the velocity average of the full function in Eq. (7.25)!

Then show that at the leading order for large Λ,

$$D(\|\Delta\mathbf{v}_{\parallel}\|^2) \sim 4\pi G^2 n_{\rm f} M^2 \Psi(v_{\rm t}). \tag{7.28}$$

7.3 Compute the velocity potential Ψ in Eq. (7.16) for a spatially homogeneous, Gaussian[6] velocity distribution

$$g(v_{\rm f}) = \frac{e^{-v_{\rm f}^2/(2\sigma_\circ^2)}}{(2\pi)^{3/2}\sigma_\circ^3}, \qquad v_{\rm f} = \|\mathbf{v}_{\rm f}\| \tag{7.30}$$

in Eq. (7.7), and show that by setting $v_{\rm t} = \|\mathbf{v}_{\rm t}\|$ and $\tilde{v}_{\rm t} = v_{\rm t}/(\sqrt{2}\sigma_\circ)$,

$$\Psi(v_{\rm t}) = \frac{\mathrm{Erf}(\tilde{v}_{\rm t})}{\sqrt{2}\sigma_\circ \tilde{v}_{\rm t}} \sim \begin{cases} \sqrt{2/\pi}/\sigma_\circ, & \tilde{v}_{\rm t} \to 0, \\ 1/\tilde{v}_{\rm t}, & \tilde{v}_{\rm t} \to \infty. \end{cases} \tag{7.31}$$

Hints: By using Eq. (7.17), first prove that

$$\Xi(v_{\rm t}) = \mathrm{Erf}(\tilde{v}_{\rm t}) - \frac{2\tilde{v}_{\rm t} e^{-\tilde{v}_{\rm t}^2}}{\sqrt{\pi}}, \qquad 4\pi \int_{v_{\rm t}}^{\infty} g(v_{\rm f}) v_{\rm f}\, dv_{\rm f} = \frac{\sqrt{2} e^{-\tilde{v}_{\rm t}^2}}{\sqrt{\pi}\sigma_\circ}, \tag{7.32}$$

where $\mathrm{Erf}(x)$ is the standard error function in Eq. (A.57); verify that $\Xi(\infty) = 1$ and obtain the leading asymptotic term when $v_{\rm t} \to 0$ and $v_{\rm t} \to \infty$.

7.4 With this exercise, we consider the quite artificial but conceptually important case of a system with the velocities of the field particles spanning a very small range of values. Let (see e.g. Ostriker and Davidsen 1968)

$$g(v_{\rm f}) = \frac{\delta(v_{\rm f} - \sigma_\circ)}{4\pi v_{\rm f}^2}, \qquad v_{\rm f} = \|\mathbf{v}_{\rm f}\|, \tag{7.33}$$

which is the velocity distribution, where $\sigma_\circ$ is a given velocity (i.e., the distribution is an infinitesimally thin shell of "radius" $\sigma_\circ$ in velocity space). First, compute $\Psi(v_{\rm t})$ as in Eq. (7.16) and discuss the resulting expressions corresponding to Eqs. (7.26) and (7.28): What happens for $v_{\rm t} \gg \sigma_\circ$ and $v_{\rm t} \ll \sigma_\circ$? Then consider a fixed value of $b_{\rm max}$ and evaluate Eqs. (7.14) and (7.20) in closed form by using the full expressions in Eqs. (7.25) and (7.27). What happens for $v_{\rm t} \to \sigma_\circ$ (i.e., for a test particle velocity modulus equal to the modulus of the velocity of field particles)? Do we have relaxation? What about the relative importance of $D(\|\Delta\mathbf{v}_{\rm t\perp}\|^2)$ and $D(\|\Delta\mathbf{v}_{\rm t\parallel}\|^2)$? *Hints*: In order to compute the integrals over $\mathbf{v}_{\rm f}$, use spherical coordinates in velocity space with the polar axis directed along $\mathbf{v}_{\rm t}$.

[6] By considering $\sigma_\circ^2 = k_{\rm B}T/m$, where $k_{\rm B} = 1.38 \times 10^{-16}$ erg/K is the Boltzmann's constant, T is the temperature, and m is the mass of the particles, Eq. (7.30) is the *Maxwell–Boltzmann velocity distribution*, one of the most important objects of physics (e.g., see Born 1969). The associated (normalized) differential distribution in terms of the modulus of the velocity is $dn = 4\pi g(v_{\rm f}) v_{\rm f}^2 dv_{\rm f}$, so that its maximum, average, and root-mean-squared velocities, respectively, are given by

$$v_{\rm max} = \sqrt{2}\sigma_\circ, \quad \overline{v} = \sqrt{\frac{8}{\pi}}\sigma_\circ, \quad \sqrt{\overline{v^2}} = \sqrt{3}\sigma_\circ. \tag{7.29}$$

7.5 Rosenbluth et al. (1957) introduced two potential functions in velocity space. The first is given in Eq. (7.16) and the second is

$$\chi(\mathbf{v}_t) \equiv \int_{\Re^3} g(v_f)\|\mathbf{v}_t - \mathbf{v}_f\| d^3\mathbf{v}_f. \tag{7.34}$$

This occurs when constructing the diffusion coefficients in S_0 with respect to $\mathbf{v}_t$. Show that for an isotropic velocity distribution $\chi(v_t)$ is a function[7] of the velocity norm of the test stars v_t, and that from Eqs. (2.126) and (2.127) with $\alpha = 0$ (or by integration in spherical coordinates placing $\mathbf{v}_t$ along the z-axis of the velocity space), this can be written as

$$\chi(v_t) = \frac{4\pi}{3v_t}\left[\int_0^{v_t} g(v_f)v_f^2(3v_t^2 + v_f^2)dv_f + \int_{v_t}^{\infty} g(v_f)v_f v_t(3v_f^2 + v_t^2)dv_f\right]. \tag{7.35}$$

Finally, calculate $\chi(v_t)$ for the velocity distribution in Eq. (7.33) and for the Maxwell–Boltzmann distribution in Eq. (7.30) and show that

$$\chi(v_t) = \sqrt{2}\sigma_\circ\left[\frac{e^{-\tilde{v}_t^2}}{\sqrt{\pi}} + \left(\tilde{v}_t + \frac{1}{2\tilde{v}_t}\right)\mathrm{Erf}(\tilde{v}_t)\right], \quad \tilde{v}_t = \frac{v_t}{\sqrt{2}\sigma_\circ}. \tag{7.36}$$

[7] Mathematically, χ is a Riesz potential, similarly to those in Exercise 2.35. Physically, it is the analog in velocity space of the Roberts (1962) "superpotential" appearing in the evaluation of the self-gravitational energy of ellipsoidal systems, which will be encountered in Chapter 10.

8

Relaxation 2

Dynamical Friction

Dynamical friction is a very interesting physical phenomenon with important applications in astrophysics. At the simplest level, it can be described as the slowing down ("cooling") of a test particle moving in a sea of field particles due to the cumulative effect of long-range interactions (no geometric collisions are considered). Several approaches have been devised to understand the underlying physics (which is intriguing, as the final result is an irreversible process produced by time-reversible dynamics; e.g., see Bertin 2014; Binney and Tremaine 2008; Chandrasekhar 1942; Ogorodnikov 1965; Shu 1999; Spitzer 1987). In this chapter, the dynamical friction time is derived in the Chandrasekhar (1943a) approach by using the impulsive approximation discussed in Chapter 7.

8.1 The Chandrasekhar Formula

We begin with a short review of the most important logical steps used in the derivation of dynamical friction in the classical case. From Eqs. (4.13)–(4.16), we have the exact identity

$$\Delta \mathbf{v}_{\mathrm{t}\|} = \frac{\mu}{m_{\mathrm{t}}} \Delta \mathbf{v}_{\|}, \quad \Delta \mathbf{v}_{\|} = -\|\Delta \mathbf{v}_{\|}\| \frac{\mathbf{v}(-\infty)}{v}, \tag{8.1}$$

where, as usual, $v = \|\mathbf{v}(-\infty)\|$ is the modulus of the initial relative velocity and $\|\Delta \mathbf{v}_{\|}\|$ is given in Eq. (4.33). In Exercise 8.2, we will compute $D(\Delta \mathbf{v}_{\|})$ from Eq. (8.1) by using the exact expression of $\|\Delta \mathbf{v}_{\|}\|$. However, in the spirit of Chapter 7, here we discuss the problem by using impulsive approximation, and from Eq. (8.31)

$$\Delta \mathbf{v}_{\mathrm{t}\|} \sim -\frac{2G^2 m_{\mathrm{f}} M}{b^2 v^4} \mathbf{v}, \quad M = m_{\mathrm{t}} + m_{\mathrm{f}}. \tag{8.2}$$

Note an important difference between Eq. (8.2) and Eq. (7.12): while in the case of the perpendicular velocity component the mass of the test star does not appear, here we have the product $m_{\mathrm{f}}(m_{\mathrm{t}} + m_{\mathrm{f}})$, and this fact is rich in astrophysical consequences. Also note that the dependence of $\|\Delta \mathbf{v}_{\mathrm{t}\|}\|$ as the inverse of the cube of the initial relative velocity is asymptotically correct only in the impulsive approximation: for slow or grazing orbits, the functional dependence of $\|\Delta \mathbf{v}_{\mathrm{t}\|}\|$ on v is different. However, as in gravitational plasmas there is no screening effect, it can be proved that the main contribution to dynamical friction comes from distant interactions (e.g., Spitzer 1987), so that the expression in Eq. (8.2) is the leading term.

The sum over the number of encounters is performed following Chapter 7. Therefore, the diffusion coefficient of velocity of the test star, parallel to the initial relative velocity, is given by an expression that is analogous to Eq. (7.13) as

$$D(\Delta \mathbf{v}_{\mathrm{t}\parallel}) = -4\pi G^2 n_{\mathrm{f}} m_{\mathrm{f}} M \int_{\Re^3} \frac{g(v_{\mathrm{f}})}{v^3} \mathbf{v} d^3\mathbf{v}_{\mathrm{f}} \int_0^{\infty} \frac{db}{b}, \tag{8.3}$$

where again the appearance of the ultraviolet and infrared divergence of the integral over the impact parameter is apparent. We repeat the same discussion as in Chapter 7, and we integrate between $b_{\pi/2}$ and $b_{\max}$, and again in analogy with Eq. (7.14)

$$D(\Delta \mathbf{v}_{\mathrm{t}\parallel}) = -4\pi G^2 n_{\mathrm{f}} m_{\mathrm{f}} M \int_{\Re^3} g(v_{\mathrm{f}}) \ln \Lambda \frac{\mathbf{v}_{\mathrm{t}} - \mathbf{v}_{\mathrm{f}}}{\|\mathbf{v}_{\mathrm{t}} - \mathbf{v}_{\mathrm{f}}\|^3} d^3\mathbf{v}_{\mathrm{f}}. \tag{8.4}$$

We now integrate over the velocity space. As in Chapter 7, we introduce again a *velocity-weighted Coulomb logarithm*[1] $\ln \bar{\Lambda}$, and therefore Newton's theorem on spherical shells (here applied to velocity space given the assumed isotropy of the velocity distribution of field particles) leads to the wonderful identity

$$\int_{\Re^3} \ln \Lambda \, g(v_{\mathrm{f}}) \frac{\mathbf{v}_{\mathrm{t}} - \mathbf{v}_{\mathrm{f}}}{\|\mathbf{v}_{\mathrm{t}} - \mathbf{v}_{\mathrm{f}}\|^3} d^3\mathbf{v}_{\mathrm{f}} = \ln \bar{\Lambda} \frac{\Xi(v_{\mathrm{t}})}{v_{\mathrm{t}}^3} \mathbf{v}_{\mathrm{t}}, \tag{8.5}$$

where $\Xi(v_{\mathrm{t}})$ is defined in Eq. (7.17). Therefore, given the hypothesis of a constant value of $\ln \bar{\Lambda}$, the velocity-dependent factor in Eq. (8.5) would simply be the gradient (with respect to the velocity $\mathbf{v}_{\mathrm{t}}$) of the first Rosenbluth potential Ψ. *The cumulative effect of the encounters is to slow down the test particle in the direction of the test particle velocity itself.* This is not trivial, as according to Eq. (8.1) the deceleration in each single encounter is parallel to the *relative* velocity, and not to $\mathbf{v}_{\mathrm{t}}$! The dynamical friction deceleration on a test mass m_{t} moving with velocity $\mathbf{v}_{\mathrm{t}}$ in a homogeneous and isotropic distribution[2] of identical field particles of mass m_{f} and number density n_{f} is

$$D(\Delta \mathbf{v}_{\mathrm{t}\parallel}) = -4\pi G^2 n_{\mathrm{f}} m_{\mathrm{f}} (m_{\mathrm{t}} + m_{\mathrm{f}}) \ln \bar{\Lambda} \frac{\Xi(v_{\mathrm{t}})}{v_{\mathrm{t}}^3} \mathbf{v}_{\mathrm{t}} = \frac{d\mathbf{v}_{\mathrm{t}\parallel}}{dt}, \quad v_{\mathrm{t}} \equiv \|\mathbf{v}_{\mathrm{t}}\|. \tag{8.6}$$

An important comment is in order here. According to Eqs. (8.5) and (8.6), only field particles *slower* than the test particle contribute to its deceleration. This sharp "cut" in velocity space derives from two different assumptions, namely: (1) that the velocity distribution of field particles is isotropic; and that (2) that we can take the Coulomb logarithm outside the integral in Eq. (8.4). The latter hypothesis would be rigorously true only if $\ln \Lambda$ is independent of v, but this – strictly speaking – is false because $b_{\pi/2}$ depends on v. In fact, the

[1] Note that this average value is not – strictly speaking – that which appears in Eq. (7.15).

[2] The following simple but important conceptual question sometimes arises in discussions with students: *Why* in a homogeneous and isotropic distribution of field mass $D(\Delta \mathbf{v}_{\mathrm{t}\perp}) = 0$ from Chapter 7 is $D(\Delta \mathbf{v}_{\mathrm{t}\parallel})$ *not* zero? The fact is that in the reference system moving with m_{t}, the velocity distribution of field particles is *not* isotropic in the direction of v_{t}!

(small) correcting terms due to the dependence of $\ln \Lambda$ on v can be explicitly evaluated (e.g., see Chandrasekhar 1941, 1942, 1943a; Merritt 2013; White 1949, see also Exercise 8.4).

In analogy with Eq. (7.19), we are led to define the *dynamical friction time* in a natural way as

$$t_{\rm df} \equiv \frac{\|\mathbf{v}_{\rm t}(-\infty)\|}{\|D(\Delta\mathbf{v}_{\rm t\|})\|} \sim \frac{v_{\rm t}^3}{4\pi G^2 n_{\rm f} m_{\rm f}(m_{\rm f}+m_{\rm t})\ln\bar{\Lambda}\,\Xi(v_{\rm t})}. \tag{8.7}$$

Notice that in the numerator in Eq. (8.6) and in the denominator of Eq. (8.7), we can recognize the quantity $n_{\rm f}m_{\rm f} = \rho_{\rm f}$ (i.e., the mass of the field particles actually enters into the dynamical friction as the mass density, and so for $m_{\rm t} \gg m_{\rm f}$ the actual mass of the field particles does not appears). From this point of view, Eq. (8.7) can be used (for example) to study the effects of dynamical friction produced by a dark matter halo in a galaxy.

We now use all of the results obtained so far in Chapter 7 and the present chapter and we evaluate the diffusion coefficient for $\Delta E_{\rm t}$ given in Eq. (4.27)

$$D(\Delta E_{\rm t}) = D(\Delta E_{\rm t\perp}) + D(\Delta E_{\rm t\|}) + m_{\rm t}\langle D(\Delta\mathbf{v}_{\rm t\|}), \mathbf{v}_{\rm t}(-\infty)\rangle, \tag{8.8}$$

where the last expression derives from the fact that in our framework of a homegeneous and isotropic field distribution, $D(\Delta\mathbf{v}_{\rm t_\perp}) = 0$. Moreover, in the usual circumstance of a large Λ and ignoring the subdominant contribution of Eq. (7.20), from Eqs. (7.18) and (8.6), we obtain the important estimate

$$\frac{1}{t_{\Delta E_{\rm t}}} \equiv \frac{D(\Delta E_{\rm t})}{E_{\rm t}} \sim \frac{1}{t_{\rm 2b}}\left(1 - 2\frac{t_{\rm 2b}}{t_{\rm df}}\right), \tag{8.9}$$

clearly showing how the relative importance of the two-body relaxation time and of the dynamical friction time determines whether the considered test star is heated or cooled.

8.2 Dynamical Friction in the Presence of a Mass Spectrum

As we have seen in the previous section, in the classical approach to dynamical friction all of the field particles have the same mass and their distribution is uniform in configuration space and isotropic in the velocity space. However, there are astrophysical situations in which a mass spectrum can have relevant effects, namely when the test particle (even though very massive) travels with a velocity comparable to the velocity dispersion of field particles, or when its mass is of the same order of magnitude as the average mass of the field masses. A specific example is represented by the population of blue straggler stars (BSSs) in globular clusters (e.g., see Ferraro et al. 2012, 2018 and references therein). In fact, BSSs are believed to originate by the merging or mass accretion of otherwise normal stars, so that their mass is at most a factor or so larger than the average mass of the stars in the parent cluster, and their mean velocities are similar to those of normal field stars. In addition, the stars of the globular clusters are characterized by a mass spectrum, and finally, globular clusters are collisional systems with relaxation and dynamical friction times comparable to their age. Other cases of test particles (with much larger masses) moving with a velocity similar to that of field particles is represented by binary black holes in galactic nuclei (e.g., Merritt 2013 and references therein). These examples indicate that studying dynamical friction in a field particle distribution with a mass spectrum is important (e.g., see Ciotti 2010).

Given the previous preparatory work, it is now easy to generalize the classical dynamical friction formula in Eq. (8.6), derived under the assumption that all of the field particles have the same mass, to the case of a mass spectrum of field particles. A generic mass spectrum with an isotropic velocity distribution is described, by extension of the classical treatment, with an *extended phase-space distribution function*

$$\delta f \equiv m_{\rm f}\mathcal{N}(m_{\rm f})\, g(v_{\rm f}, m_{\rm f}), \quad f = \int_0^\infty \delta f \, dm_{\rm f}, \tag{8.10}$$

where $\mathcal{N}(m_{\rm f})$ is the number density per unit volume and unit mass of field stars with mass in the range between $m_{\rm f}$ and $m_{\rm f} + dm_{\rm f}$. The total number density of field stars, the average field star mass, and the mass ratio of the test star with respect to the average field star mass are given by

$$n_{\rm f} = \int_0^\infty \mathcal{N}(m_{\rm f}) dm_{\rm f}, \quad \langle m_{\rm f}\rangle = \frac{1}{n_{\rm f}} \int_0^\infty \mathcal{N}(m_{\rm f}) m_{\rm f}\, dm_{\rm f}, \quad \mathcal{R} \equiv \frac{m_{\rm t}}{\langle m_{\rm f}\rangle}. \tag{8.11}$$

Of course, the normalization of the velocity distribution for each mass component is expressed by the fact that

$$\Xi(v_{\rm t}, m_{\rm f}) = 4\pi \int_0^{v_{\rm t}} g(v_{\rm f}, m_{\rm f}) v_{\rm f}^2 dv_{\rm f}, \quad \Xi(\infty, m_{\rm f}) = 1, \quad \forall m_{\rm f}. \tag{8.12}$$

In the presence of a mass spectrum of field stars, the differential number of encounters suffered by the test star is

$$\Delta n_{\rm enc} = 2\pi b db \times ||\mathbf{v}_{\rm t} - \mathbf{v}_{\rm f}||\Delta t \times \mathcal{N}(m_{\rm f}) g(v_{\rm f}, m_{\rm f}) dm_{\rm f} d^3\mathbf{v}_{\rm f}. \tag{8.13}$$

Therefore, by summing the formula obtained in the classical treatment over all of the species, the deceleration in the mass spectrum case is given by

$$\begin{aligned} D(\Delta \mathbf{v}_{\rm t\parallel}) &= -4\pi G^2 \langle \ln \bar{\Lambda}\rangle_\Xi \frac{\mathbf{v}_{\rm t}}{v_{\rm t}^3} \int_0^\infty \mathcal{N}(m_{\rm f}) m_{\rm f} M\, \Xi(v_{\rm t}, m_{\rm f})\, dm_{\rm f} \\ &= -4\pi G^2 n_{\rm f} \langle m_{\rm f}\rangle (m_{\rm t} + \langle m_{\rm f}\rangle) \langle \ln \bar{\Lambda}\rangle_\Xi\, \Xi^*(v_{\rm t}) \frac{\mathbf{v}_{\rm t}}{v_{\rm t}^3}, \end{aligned} \tag{8.14}$$

where $\langle \ln \bar{\Lambda}\rangle_\Xi$ is the natural average of $\ln \bar{\Lambda}$ implicitly defined by the first line in Eq. (8.14).[3] In the last identity in Eq. (8.14), in contrast with the classical formula (8.6), the integration over the mass spectrum leads to the introduction of the function

[3] In all practical applications, it can be safely assumed that $\langle \ln \bar{\Lambda}\rangle_\Xi = \ln \bar{\Lambda}$.

$$\Xi^*(v_t) = \frac{\int_0^\infty \mathcal{N}(m_f) m_f M\, \Xi(v_t, m_f)\, dm_f}{n_f \langle m_f \rangle (m_t + \langle m_f \rangle)} = \frac{\mathcal{R} H_1(v_t) + H_2(v_t)}{\mathcal{R} + 1}, \tag{8.15}$$

where we recall that $\mathcal{R} = m_t / \langle m_f \rangle$ and (prove it!)

$$\begin{cases} H_1(v_t) \equiv \dfrac{1}{n_f \langle m_f \rangle} \displaystyle\int_0^\infty \mathcal{N}(m_f) m_f\, \Xi(v_t, m_f)\, dm_f, \\ H_2(v_t) \equiv \dfrac{1}{n_f \langle m_f \rangle^2} \displaystyle\int_0^\infty \mathcal{N}(m_f) m_f^2\, \Xi(v_t, m_f)\, dm_f. \end{cases} \tag{8.16}$$

In practice, from Eq. (8.16) one can derive the dynamical friction deceleration for the case of a mass spectrum by using the same formalism as the classical case, where m_f is replaced by $\langle m_f \rangle$. It is important to note that for *large* velocities of the test star $\Xi(v_t, m_f) \to 1$, and formally

$$H_1(\infty) = 1, \quad H_2(\infty) = \int_0^\infty \frac{\mathcal{N}(m_f) m_f^2}{n_f \langle m_f \rangle^2}\, dm_f, \quad \Xi^*(\infty) = \frac{\mathcal{R} + H_2(\infty)}{\mathcal{R} + 1}. \tag{8.17}$$

In order to illustrate the effects of a mass spectrum, in the following sections we present three explicit cases of mass spectra amenable to analytic solutions so that the differences from the equivalent classical cases can be quantified. In order to compare dynamical friction in the presence of a mass spectrum with the classical case (i.e., the case with identical field particles), we must carefully define the concept of the *equivalent* classical system. We will say that a classical case is equivalent to a mass spectrum case when: (1) the number density of field particles in the classical case is the same as the *total* number density in the mass spectrum case; (2) the field mass m_f in the classical case is the same as the *average* field mass $\langle m_f \rangle$; and (3) the velocity dispersion of the (Maxwellian) velocity distribution in the classical case is the same as the *equipartition* velocity dispersion of $\langle m_f \rangle$ in the mass spectrum case, i.e.,

$$r \equiv \frac{m_f}{\langle m_f \rangle}, \quad \sigma_m^2 = \frac{\sigma_\circ^2}{r}, \quad g(v_f, m_f) = \frac{e^{-v_f^2/(2\sigma_m^2)}}{(2\pi)^{3/2} \sigma_m^3} = \frac{e^{-v_f^2 r/(2\sigma_\circ^2)} r^{3/2}}{(2\pi)^{3/2} \sigma_\circ^3}, \tag{8.18}$$

so that from Eq. (7.32)

$$\Xi(v_t, m_f) = \mathrm{Erf}(\tilde{v}_t \sqrt{r}) - \frac{2 \tilde{v}_t \sqrt{r} e^{-\tilde{v}_t^2 r}}{\sqrt{\pi}}, \quad \tilde{v}_t = \frac{v_t}{\sqrt{2} \sigma_\circ}. \tag{8.19}$$

We can summarize these conditions by saying that the comparison is between two systems with the same number, mass, and kinetic energy density of field particles. Of course, alternative configurations of astrophysical relevance can be easily imagined (e.g., see Ciotti 2010); in particular, one might be interested in the study of a mass spectrum when all of the field stars have the same velocity distribution independently of their mass (e.g., a collisionless system made of stars and dark matter).

8.2.1 Exponential Mass Spectrum

In this first example, we assume that the number distribution of field stars is described by a pure exponential law: in this case, from Eq. (8.11) the mass spectrum can be written in full generality (prove it!) as

$$\mathcal{N}(m_f) = n_f \frac{e^{-m_f/\langle m_f \rangle}}{\langle m_f \rangle}. \tag{8.20}$$

In the equipartition case, the integral over masses in Eq. (8.15) can be performed analytically by inserting Eq. (8.12) into Eq. (8.16), inverting the order of integration between v_f and m_f, and finally obtaining the surprisingly simple result

$$H_1 = \frac{\tilde{v}_t^3(5 + 2\tilde{v}_t^2)}{2(1 + \tilde{v}_t^2)^{5/2}}, \quad H_2 = \frac{\tilde{v}_t^3(35 + 28\tilde{v}_t^2 + 8\tilde{v}_t^4)}{4(1 + \tilde{v}_t^2)^{7/2}}. \tag{8.21}$$

The behavior of the two functions in the limit of low- and high-velocity v_t can be easily determined, and in particular Eqs. (8.15) and (8.21) prove that the result coincides asymptotically with the classical case for fast ($H_1 \sim 1$ and $H_2 \sim 2$) and massive ($\mathcal{R} \gg 1$) test particles. In general, as can be seen from Figure 8.1, the velocity factor in the case of an exponential mass spectrum with equipartition is larger than in the corresponding classical case (heavy line in Figure 8.1); for massive test particles, the maximum drag (corresponding to $v_t \simeq 0.81\sigma_\circ$) is a factor $\approx$2 higher than in the equivalent classical case and t_{df} is correspondingly shorter.

8.2.2 Discrete Mass Spectrum

In this second experiment, we consider a system made of two species of field stars (the generalization to an arbitrary number of different field components presents no difficulties) with a mass spectrum given by

$$\mathcal{N}(m_f) = n_1\delta(m_f - m_1) + n_2\delta(m_f - m_2), \tag{8.22}$$

where δ is the Dirac δ-function. With the convenient introduction of the dimensionless parameters $x \equiv n_2/n_1$ and $y \equiv m_2/m_1$, it follows that

$$n_f = (1 + x)n_1, \quad \langle m_f \rangle = \frac{n_1 m_1 + n_2 m_2}{n_1 + n_2} = \frac{1 + xy}{1 + x} m_1, \tag{8.23}$$

and the analogous identities expressed in terms of n_2 and m_2 are immediately obtained from the definition of x and y; of course, the value $y = 1$ corresponds to the classical case. From Eq. (8.16), we now have

$$H_1 = \frac{\Xi(v_t, m_1) + xy\Xi(v_t, m_2)}{1 + xy}, \quad H_2 = \frac{(1 + x)[\Xi(v_t, m_1) + xy^2\Xi(v_t, m_2)]}{(1 + xy)^2}. \tag{8.24}$$

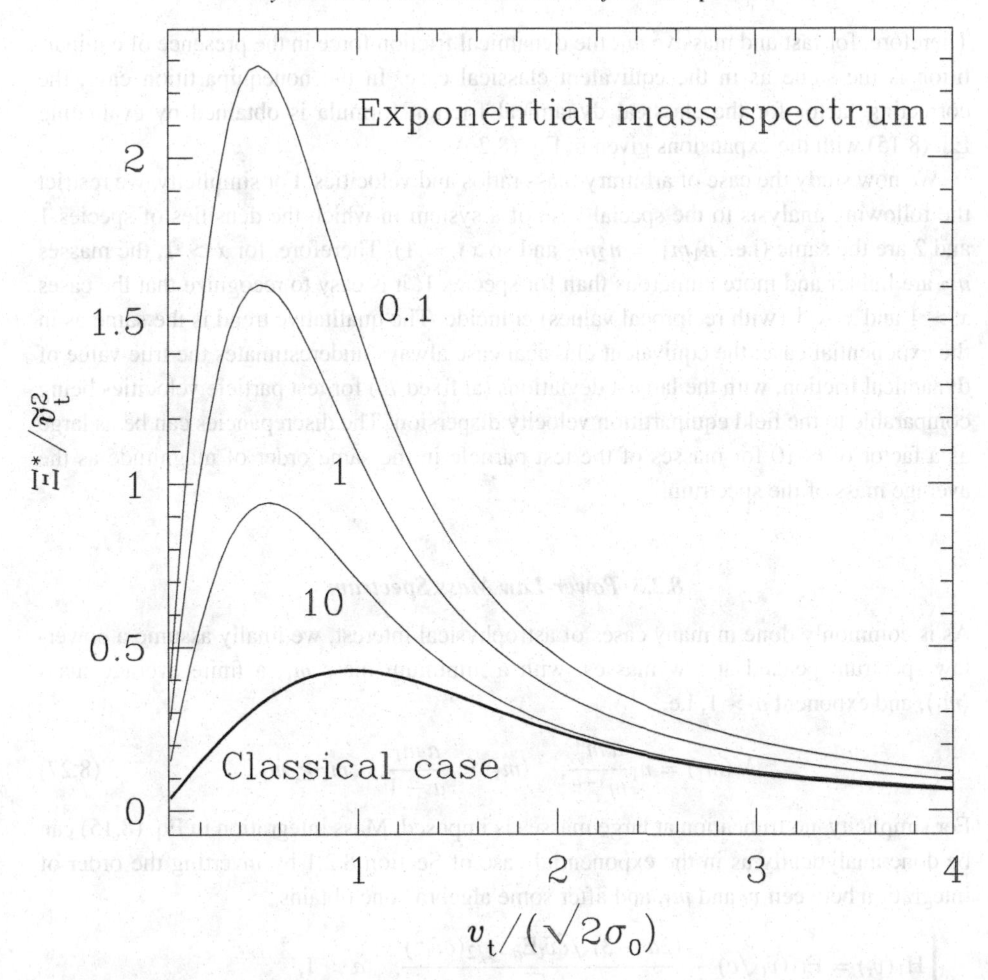

Figure 8.1 The velocity coefficient $\Xi^*/\tilde{v}_t^2$ in Eq. (8.15) for the exponential mass spectrum case of field particles with equipartition: $\tilde{v}_t = v_t/(\sqrt{2}\sigma_o)$. The curves (from top to bottom) correspond to a test particle with mass 0.1, 1, and 10 times the average mass of the spectrum. The heavy solid line represents the velocity coefficient of the equivalent classical case (i.e., when the field masses are all identical and their number density, average mass, and kinetic energy density are the same as in the mass spectrum case).

For low velocities of the test particle, one finds (do it!) the asymptotic trends

$$\mathrm{H}_1 \sim \frac{4\tilde{v}_t{}^3(1+x)^{3/2}(1+xy^{5/2})}{3\sqrt{\pi}(1+xy)^{5/2}}, \quad \mathrm{H}_2 \sim \frac{4\tilde{v}_t{}^3(1+x)^{5/2}(1+xy^{7/2})}{3\sqrt{\pi}(1+xy)^{7/2}}. \tag{8.25}$$

In turn, for large velocities of the test particle, the leading terms are

$$\mathrm{H}_1(\infty) = 1, \quad \mathrm{H}_2(\infty) = \frac{(1+x)(1+xy^2)}{(1+xy)^2}. \tag{8.26}$$

Therefore, for fast and massive m_t, the dynamical friction force in the presence of equipartition is the same as in the equivalent classical case. In the nonequipartition case, the correcting factor for the classical dynamical friction formula is obtained by evaluating Eq. (8.15) with the expansions given in Eq. (8.26).

We now study the case of arbitrary mass ratios and velocities. For simplicity, we restrict the following analysis to the special case of a system in which the densities of species 1 and 2 are the same (i.e., $n_1 m_1 = n_2 m_2$ and so $xy = 1$). Therefore, for $x > 1$, the masses m_2 are lighter and more numerous than for species 1; it is easy to recognize that the cases $x > 1$ and $x < 1$ (with reciprocal values) coincide. The qualitative trend is the same as in the exponential case: the equivalent classical case always underestimates the true value of dynamical friction, with the largest deviations (at fixed $\mathcal{R}$) for test particle velocities being comparable to the field equipartition velocity dispersion. The discrepancies can be as large as a factor of 6–10 for masses of the test particle in the same order of magnitude as the average mass of the spectrum.

8.2.3 Power-Law Mass Spectrum

As is commonly done in many cases of astrophysical interest, we finally assume a power-law spectrum peaked at low masses, with a minimum mass m_i, a finite average mass $\langle m_f \rangle$, and exponent $a > 1$, i.e.,

$$\mathcal{N}(m_f) = n_f \frac{a\, m_i^a}{m^{1+a}}, \quad \langle m_f \rangle = \frac{a\, m_i}{a-1}, \quad m_f \geq m_i. \tag{8.27}$$

For simplicity, no truncation at large masses is imposed. Mass integration in Eq. (8.15) can be done analytically as in the exponential case of Section 8.2.1 by inverting the order of integration between v_f and m_f, and after some algebra[4] one obtains

$$\begin{cases} \mathrm{H}_1(\tilde{v}_t) = \mathrm{Erf}(\tilde{v}_t\sqrt{c}) - \dfrac{(2a-3)\sqrt{c}\tilde{v}_t \mathrm{E}_{a-1/2}(c\tilde{v}_t^2)}{\sqrt{\pi}}, & a > 1, \\ \mathrm{H}_2(\tilde{v}_t) = \dfrac{(a-1)^2}{a(a-2)}\left[\mathrm{Erf}(\tilde{v}_t\sqrt{c}) - \dfrac{(2a-5)\sqrt{c}\tilde{v}_t \mathrm{E}_{a-3/2}(c\tilde{v}_t^2)}{\sqrt{\pi}}\right], & a > 2, \end{cases} \tag{8.29}$$

[4] The two equations are obtained as special cases of the general identity

$$\int_0^z x^\lambda \mathrm{E}_\nu(x^\mu)\, dx = \frac{\gamma\left(\frac{1+\lambda}{\mu}, z^\mu\right) + z^{1+\lambda}\mathrm{E}_\nu(z^\mu)}{1+\lambda+\mu(\nu-1)}, \tag{8.28}$$

where $\lambda > \max\{-1, -1-\mu(\nu-1)\}$, fixing $\lambda = \mu = 2$, $\nu = a - 3/2$, $\nu = a - 5/2$, and using the recursion relations in Appendix A (see Ciotti 2020).

where $c = 1 - 1/a$, the exponential integral function is given in Eq. (A.59), and finally $a > 2$ is required for the existence of solutions. From Eq. (8.17), it is easy to show that

$$\mathrm{H}_1(\infty) = 1, \quad \mathrm{H}_2(\infty) = \frac{(a-1)^2}{a(a-2)} \geq 1, \tag{8.30}$$

and this demonstrates again that for high-velocity v_t and large mass ratio $\mathcal{R} = m_t/\langle m_f \rangle$ of the test particle the classical result is recovered. As a nice exercise, the student may wish to obtain the behavior of the functions in Eq. (8.29) for $v_t \to 0$ and study the corresponding behavior of $\Xi^*/\tilde{v}_t^{\,2}$ as a function of $a > 2$.

Summarizing the results of the simple experiments described in the last three sections, we learned that for *fast* and *massive* test particles the results in the classical and mass spectrum cases become identical, because under these assumptions $\Xi^* \to 1$ from Eq. (8.17) because for $\mathcal{R} \gg 1$ the specific form of the mass spectrum becomes irrelevant, as all of the field particles can be considered to be vanishingly small, and so Eqs. (8.6)–(8.14) coincide for $m_f = \langle m_f \rangle$. Moreover, in the cases considered, the dynamical friction force in the mass spectrum case is larger than in the corresponding classical case and the dynamical friction times are correspondingly reduced. Notice that these results can also be extended to the case in which the mass spectrum particles are *not* at equipartition but the species are characterized by the same velocity dispersion (e.g., stars and dark matter particles in a common potential well).

8.3 Astrophysical Applications and Final Comments

As the student can easily realize, the astrophysical applications of dynamical friction in the study of the dynamical evolution of astronomical systems are counteless (e.g., see Bertin 2014; Binney and Tremaine 2008; Spitzer 1987), ranging from the dynamical evolution of binary black holes in galactic nuclei (e.g., see Merritt 2013 and references therein), to galaxy merging and the formation of type-cD galaxies (e.g., see Tremaine et al. 1975; White 1976; see also Barnes and Hernquist 1992; El-Zant et al. 2004; Nipoti et al. 2004 and references therein), to the shaping of the radial distribution of BSSs in globular clusters (e.g., Alessandrini et al. 2014 and references therein), to the sinking of globular clusters or satellites within their host galaxy (e.g., see Arena and Bertin 2007; Arena et al. 2006; Bertin et al. 2003; Bontekoe and van Albada 1987; Gnedin et al. 2014 and references therein), to modified Newtonian dynamics (Nipoti et al. 2008). We finally recall that, on the theoretical side, more powerful (and physically deeper) tools have been developed, allowing for the treatment of dynamical friction in more realistic cases than that of the infinite and homegeneous sea of field particles described in this chapter (which wever has the great merit of the physical simplicity, however!), such as the extension to anisotropic velocity distributions of field particles (e.g., Binney 1977) or interpretation in terms of "density wake" (e.g., see Binney and Tremaine 2008; Tremaine and Weinberg 1984 and references therein).

Exercises

8.1 With this exercise, we estimate the asymptotic value of $\|\Delta\mathbf{v}_{\|}\|$ for $b \to \infty$ by using impulsive approximation. Consider the exact identity in Eq. (4.13), and evaluate it for vanishingly small values of $\|\Delta\mathbf{v}_{\perp}\|$ in Eq. (7.11), imposing the second identity in Eq. (8.1). Show that for $b \to \infty$ energy conservation requires, at the lowest order,

$$\|\Delta\mathbf{v}_{\|}\| \sim \frac{\|\Delta\mathbf{v}_{\perp}\|^2}{2v^2} \sim \frac{2G^2M}{b^2v^3}, \tag{8.31}$$

where $M = m_{\mathrm{t}} + m_{\mathrm{f}}$. Reobtain the same expression by direct expansion of Eq. (4.33) in the limit of large b.

8.2 Show that the result of the integration over the impact parameter of Eq. (7.9) for the quantity $D(\Delta\mathbf{v}_{\|})$ obtained from Eq. (8.1) and the second identity in Eq. (4.33) can be written in terms of the function

$$\Lambda \equiv \frac{b_{\max}}{b_{\pi/2}} = \frac{b_{\max}v^2}{GM}, \qquad \ln\sqrt{1+\Lambda^2} \sim \begin{cases} \dfrac{\Lambda^2}{2}, & \Lambda \to 0, \\ \ln\Lambda, & \Lambda \to \infty. \end{cases} \tag{8.32}$$

Then, show that at the leading order for large $\Lambda = b_{\max}/b_{\pi/2}$,

$$D(\Delta\mathbf{v}_{\|}) \sim -4\pi G^2 n_{\mathrm{f}} M^2 \ln\bar{\Lambda}\frac{\Xi(v_{\mathrm{t}})}{v_{\mathrm{t}}^3}\mathbf{v}_{\mathrm{t}}, \tag{8.33}$$

so that Eq. (8.6) holds (see also Exercises 7.1 and 7.2).

8.3 Compare the two-body relaxation time and the dynamical friction time and show that in the limit of large Λ

$$\frac{t_{\mathrm{df}}}{t_{2\mathrm{b}}} \sim \frac{v_{\mathrm{t}}\Psi(v_{\mathrm{t}})}{\Xi(v_{\mathrm{t}})}\frac{2m_{\mathrm{f}}}{m_{\mathrm{t}}+m_{\mathrm{f}}} \sim \frac{2m_{\mathrm{f}}}{m_{\mathrm{t}}+m_{\mathrm{f}}}, \tag{8.34}$$

where the last identity holds for very large v_{t} due to the normalization condition on $\Xi(v_{\mathrm{t}})$ and to the "monopole" behavior of $\Psi(v_{\mathrm{t}})$. Then, use Eq. (7.23) and obtain the analogous expression for $t_{\mathrm{df}}/t_{\mathrm{cross}}$ for the mass m_{t} traveling in the virialized, self-gravitating system adopted in Section 7.3.

8.4 Consider again the artificial system in Eq. (7.33) and evaluate again the dynamical friction formula Eq. (8.33): What happens for $v_{\mathrm{t}} \gg \sigma_{\circ}$ and $v_{\mathrm{t}} \ll \sigma_{\circ}$? Then, following Exercise 7.4, obtain the closed-form expression of Eq. (8.4) by using the full expression in Eq. (8.32). Discuss the case $v_{\mathrm{t}} \to \sigma_{\circ}$.

8.5 Generalize the 2-body relaxation time in Eq. (7.19) to the case of field stars with a mass spectrum. First derive the expression analogous to Eq. (8.14),

$$\begin{aligned} D(\|\Delta\mathbf{v}_{\mathrm{t}\perp}\|^2) &= 8\pi G^2\langle\ln\bar{\Lambda}\rangle_{\Psi}\int_0^{\infty}\mathcal{N}(m_{\mathrm{f}})m_{\mathrm{f}}^2\Psi(v_{\mathrm{t}},m_{\mathrm{f}})\,dm_{\mathrm{f}} \\ &= 8\pi G^2 n_{\mathrm{f}}\langle m_{\mathrm{f}}\rangle^2\langle\ln\bar{\Lambda}\rangle_{\Psi}\Psi^*(v_{\mathrm{t}}), \end{aligned} \tag{8.35}$$

where $\langle \ln \bar{\Lambda} \rangle_\Psi$ is the natural average of $\ln \bar{\Lambda}$ over Ψ implicitly defined by the first line in Eq. (8.35) (see also Footnote 2), and

$$\Psi^*(v_t) = \frac{1}{n_f \langle m_f \rangle^2} \int_0^\infty \mathcal{N}(m_f) m_f^2 \Psi(v_t, m_f)\, dm_f \sim \frac{\mathrm{H}_2(\infty)}{v_t}, \tag{8.36}$$

where the last expression is obtained from Eqs. (7.16) and (8.17) in the limit of very high velocities of the test star. Then show that

$$\frac{t_{df}}{t_{2b}} \sim \frac{v_t \Psi^*(v_t)}{\Xi^*(v_t)} \frac{2\langle m_f \rangle}{m_t + \langle m_f \rangle} \sim \frac{2\mathrm{H}_2(\infty)}{\mathcal{R} + \mathrm{H}_2(\infty)}, \tag{8.37}$$

where $\mathcal{R} = m_t / \langle m_f \rangle$, the first expression is obtained for large values of Λ, and the last expression is obtained under the additional hypothesis of very large v_t.

8.6 With this exercise, we estimate how the Coulomb logarithm depends on m_t in a virialized stellar system made of stars of mass $m_f = m_*$ and of dark matter particles of mass $m_f = m_{DM}$. Consider a two-component system as in Section 7.3, with $M_* = N_* m_*$, $M_{DM} = \mathcal{R} M_*$, and radius R. By fixing all of the shape factors to unity, show that the Coulomb logarithms of m_t against stars and DM can be written, respectively, as

$$\ln \Lambda_* = \ln \frac{2N_*(1+\mathcal{R})}{r+1}, \qquad \ln \Lambda_{DM} = \ln \frac{2N_*(1+\mathcal{R})}{r + r_{DM}}, \tag{8.38}$$

where $r \equiv m_t / m_*$ and $r_{DM} \equiv m_{DM}/m_*$. *Hints*: Assume $b_{max} = 2R$ and use the estimate

$$v_t^2 = \sigma_V^2 = \frac{GM_*(1+\mathcal{R})}{R} \tag{8.39}$$

to obtain $b_{\pi/2}$ as in Eq. (7.23). What is the expression for $\ln \Lambda_{DM}$ in a system made of dark matter only?

8.7 Following Exercise 8.6, estimate how t_{2b} and t_{df} depend on dark matter in a two-component stellar system by using the same approach as in Section 7.3 (see also Exercise 8.3). By using Eqs. (7.18), (7.19), (8.6), and (8.7), considering the limit of large v_t and then setting (quite inconsistently) $v_t = \sigma_V$, show that

$$\frac{t_{2b}}{t_{cross}} = \frac{N_*(1+\mathcal{R})^2}{12(\ln \Lambda_* + \mathcal{R} r_{DM} \ln \Lambda_{DM})}, \tag{8.40}$$

$$\frac{t_{df}}{t_{cross}} = \frac{N_*(1+\mathcal{R})}{6[(r+1)\ln \Lambda_* + \mathcal{R}(r + r_{DM}) \ln \Lambda_{DM}]}, \tag{8.41}$$

respectively, where $t_{cross} \equiv 2R/v_t \propto 1/\sqrt{1+\mathcal{R}}$ from Eq. (8.39).What happens in a system made only of dark matter for $r_{DM} \to 0$? Evaluate t_{df}/t_{2b} and compare the result with Eqs. (8.34) and (8.37).

Part III

Collisionless Systems

9

The Collisionless Boltzmann Equation and the Jeans Theorem

With this chapter, the final part of the book, dedicated to collisionless stellar systems, begins. As should be clear, in order to extract information from the N-body problem, we need to move to a different approach from direct integration of the differential equations of motion, and a first (unfruitful) attempt here will be based on the Liouville equation. In fact, the basic reason for the "failure" of the Liouville approach is that, despite its apparent statistical nature, the dimensionality of the phase space[1] Γ where the function $f^{(6N)}$ is defined is the same as that of the original N-body problem. Suppose instead we find a way to replace the $6N$-dimensional $\Re^{6N}$ phase space Γ with the six-dimensional one-particle phase space γ: we can reasonably expect that the problem would be simplified enormously, and in fact along these lines we will finally obtain the collisionless boltzmann equation (CBE), one of the conceptual pillars of stellar dynamics.

9.1 The Liouville Equation

As previously mentioned, the direct approach to the N-body problem (i.e., the search for exact solutions of Eq. (6.1)) is in general hopeless, and so alternative approaches must be found. A first sensible idea could be to restrict our focus to some specific property of the N-body system, with the hope of being able to deduce and solve the associated differential equations. A particularly important family of mathematical objects that is well suited for this approach is that of *macroscopic functions*. A clear definition of a macroscopic function is founded on two basic geometric properties of the phase space Γ and on the concepts of *microstate* and *macrostate*, borrowed from statistical mechanics (e.g., see Boltzmann 1896; Khinchin 1949; Saslaw 1987). As is well known, at any time the dynamical state of an N-body system is completely determined by a point in the extended phase space, so that the dynamical evolution of the system can be imagined as a path in the extended phase space originating from its initial conditions: the curve is the solution of Eq. (6.1), represented in Figure 9.1 by the line connecting P_0 to P_t.

[1] In the following, the phase space of an N-body system $\Re^{6N}$ will be indicated with the standard name Γ. The extended phase space is then $\Gamma \times \Re$, where the time coordinate appears.

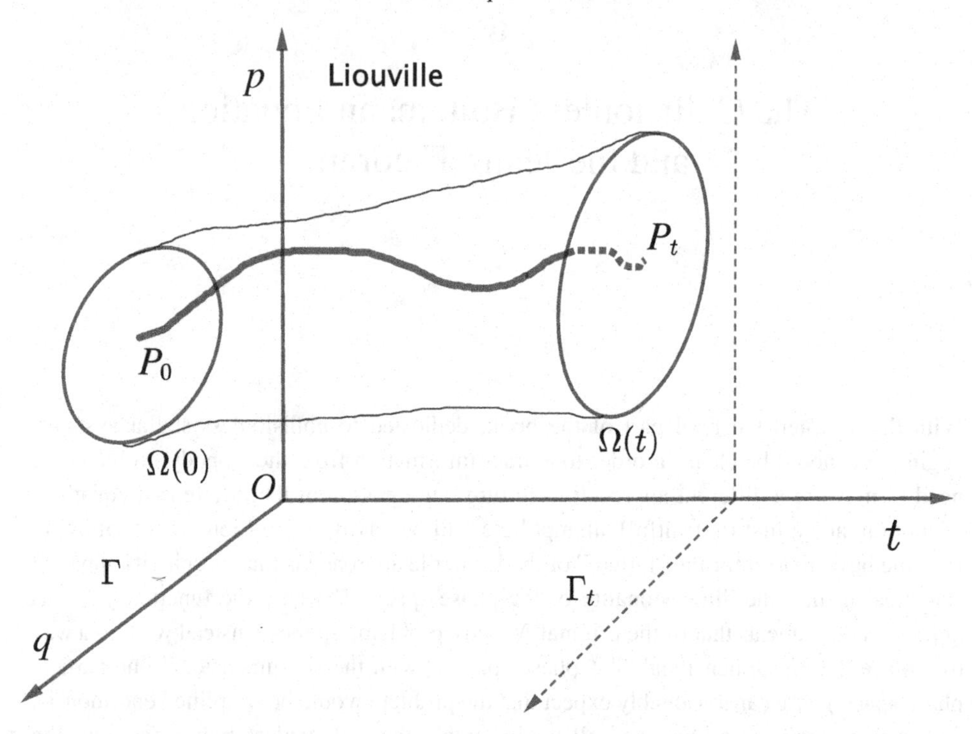

Figure 9.1 Motion of a set $\Omega(0)$ in the extended phase space of an N-body system: due to the solenoidal nature of the Hamiltonian flow, $\Omega(0)$ transforms into $\Omega(t)$, conserving its phase-space volume. Moreover, points of the frontier $\partial\Omega(0)$ remain on the frontier $\partial\Omega(t)$ (e.g., see Khinchin 1949; Meyer 1982); due to the unicity of the solution (Appendix A.10.1; see also Wintner 1947), no points can exit from $\Omega(t)$.

Notice that from the theorem on the uniqueness of solutions (see Appendix A.10.1), two different curves in the extended phase space cannot intersect each other at any time. In the following, each point of $\Gamma \times \Re$ will be called a microstate, and a function Ξ defined on $\Gamma \times \Re$ will be called a macrostate: very simple examples are the kinetic energy of the system, its angular momentum, the number of stars with velocities in some prescribed range of values, and so on. These examples should convince the student that, in general, a microstate defines unambiguously a macrostate, but the converse is not true: a given value for a prescribed macrostate determines in the extended phase space many microstates. Therefore, given an N-body system and its initial conditions, its trajectory in the extended phase space can also be interpreted as the set of all microstates of the system as a function of time; mathematically, the time evolution of an assigned macrostate Ξ is given, according to Appendix A.10.2, by its associated Lagrangian function $\Xi_{\mathcal{L}}$.

From this point of view, the question we tried to formulate earlier now becomes: Suppose that at $t = 0$ an N-body system, its initial conditions, and a desired macrostate Ξ are assigned so that $\Xi(0)$ is completely determined. *Can we predict the time evolution of $\Xi(t)$? Is the related differential equation simpler than the equations of the full N-body problem?* Note that due to the invariance of Ξ with respect to some (or many) microstates, simply assigning $\Xi(0)$ is equivalent to assigning a *set* of microstates at $t = 0$. In the following, the set of microstates that at $t = 0$ leave the macrostate $\Xi(0)$ will be called an *ensemble* $\Omega(0)$ for the macrostate Ξ. From what we have said, each microstate $\in \Omega(0)$ evolves following Eq. (6.1) and the ensemble $\Omega(0)$ transforms into $\Omega(t)$ (see Figure 9.1). Note that Ξ is by definition invariant over $\Omega(0)$, but this in general cannot be expected to be true for $\Omega(t)$! This can be easily understood by considering that the evolution of each "point" in Γ depends in general on *all* of the phase-space coordinates, and even if Ξ is independent on some coordinate, these "invisible" coordinates can also drive the evolution of the "visible" ones.

One possibility for proceeding further with this approach is to define the *mean* value for the macrostate Ξ over $\Omega(t)$; this mean value is what we call a *macroscopic function*. In order to define formally the concept of a macroscopic function and to determine its evolution equation, we introduce the *Liouville equation* (e.g., see Arnold 1978; Binney and Tremaine 2008; Chandrasekhar 1942; Ogorodnikov 1965; Saslaw 1987). Let $f^{(6N)}$ be the characteristic function for the ensemble $\Omega(t)$ (i.e., $f^{(6N)}$ equals unity if the microstate $\in \Omega(t)$, zero otherwise[2]). We now show that in phase space, from the Reynolds transport theorem in Appendix A.10.3 (and restricting for simplicity to Cartesian coordinates), the time evolution of $f^{(6N)}$ is given by the solution of the Liouville equation

$$\frac{Df^{(6N)}}{Dt} = \frac{\partial f^{(6N)}}{\partial t} + \sum_{i=1}^{N} \left(\langle \mathbf{v}_i, \nabla_{\mathbf{x}_i} f^{(6N)} \rangle - \frac{1}{m_i} \langle \nabla_{\mathbf{x}_i} U, \nabla_{\mathbf{v}_i} f^{(6N)} \rangle \right) = 0, \tag{9.1}$$

where $f_0^{(6N)}$ is the characteristic function of $\Omega(0)$, U is given in Eq. (6.2), and the meaning of the gradient operators is apparent. Equation (9.1) is obtained as follows: let $\Omega(0) \in \Gamma$ be an arbitrary ensemble. The number of microstates inside $\Omega(t)$ cannot change over the course of the evolution because, for the uniqueness of solutions, microstates cannot cross the boundary $\partial\Omega(t)$, so that the identity

$$\frac{d}{dt} \int_{\Omega(t)} f^{(6N)} d\Gamma = 0, \quad \forall \Omega(0), \tag{9.2}$$

holds at all times, where with $d\Gamma$ we indicate the phase-space volume element. From the transport theorem and from the fact that the vector field associated with U is solenoidal in Γ, being Hamiltonian, Eq. (9.1) follows. Notice that if we set $f^{(6N)} = 1$ in Eq. (9.2) and we again use the transport theorem, we prove that the phase-space volume of $\Omega(t)$ is conserved (i.e., we prove the *Liouville theorem* in Γ).

[2] Note that from its definition $f^{(6N)}$ is a nowhere negative function over the extended phase space.

We are now in a position to formulate rigorously the concept of a macroscopic function. In fact, let Ξ be a macrostate and $\Omega(0)$ its ensemble at $t = 0$. At time t, the macroscopic function associated with Ξ is given by

$$\langle \Xi \rangle_t \equiv \frac{\int_{\Omega(t)} f^{(6N)}\, \Xi \, d\Gamma}{\int_{\Omega(t)} f^{(6N)}\, d\Gamma}. \tag{9.3}$$

Therefore, $\langle \Xi \rangle_t$ is known (at least in principle) provided that the time evolution of $f^{(6N)}$ is known; from what we have said before, at $t = 0$ we have $\langle \Xi \rangle_0 = \Xi(0)$ by construction, but this is not generally true for $t > 0$, so we need $f^{(6N)}$.

The Liouville equation can be formally solved. In fact, Eq. (9.1) belongs to the class of quasi-linear partial differential equations (PDEs) that are defined as follows. Let $\mathbf{x} \in A \subseteq \Re^n$: a first-order PDE

$$a_0 \frac{\partial y}{\partial t} + \sum_{i=1}^{n} a_i \frac{\partial y}{\partial x_i} = a_0 \times \left(\frac{\partial y}{\partial t} + \sum_{i=1}^{n} \frac{a_i}{a_0} \frac{\partial y}{\partial x_i} \right) = 0, \quad y(\mathbf{x}; 0) = h(\mathbf{x}), \tag{9.4}$$

is called a *quasi-linear* homogeneous PDE with initial condition h, and in general the coefficients are functions $a_i(\mathbf{x}, y; t)$; moreover, we assume that $a_0 \neq 0$. When the coefficients a_i for $i = 0, 1, \ldots, n$ do not depend on y, the PDE is said to be *linear*. Assume for a moment that the solution y is known. The A solution $y = y(\mathbf{x}; t)$ is a function $y : \Re^n \times \Re \mapsto \Re$, so that Eq. (9.4) is identically verified, and $y(\mathbf{x}; 0) = h(\mathbf{x})$. The *characteristic* $\mathbf{c} = \mathbf{c}(\mathbf{x}_0; t)$ of Eq. (9.4) associated with $\mathbf{x}_0$ is the solution of the following system of ODEs:

$$\begin{cases} \dfrac{dx_i}{dt} = \dfrac{a_i[\mathbf{x}, y(\mathbf{x}_0; t); t]}{a_0[\mathbf{x}, y(\mathbf{x}_0; t); t]}, & i = 1, \ldots, n \\ x_i(0) = x_{0,i}, & \end{cases} \tag{9.5}$$

where $\mathbf{c}(\mathbf{x}_0; 0) = \mathbf{x}_0$. The concept of charateristics is of fundamental importance in the theory of PDEs, because we now show that a function y is a solution of Eq. (9.4) *if and only if it is constant in its characteristics*, i.e.,

$$y[\mathbf{c}(\mathbf{x}_0; t); t] = h(\mathbf{x}_0), \quad \forall \mathbf{x}_0. \tag{9.6}$$

Let us begin by showing that if y obeys Eq. (9.6), then it is a solution of Eq. (9.4). In fact, let $\mathbf{x}_0$ be fixed and y be given by Eq. (9.6): then, from Eq. (9.5), along each characteristic $\mathbf{c}$,

$$0 = \frac{dy[\mathbf{c}(\mathbf{x}_0; t); t]}{dt} = \left(\frac{\partial y}{\partial t} + \sum_{i=1}^{n} \frac{dx_i}{dt} \frac{\partial y}{\partial x_i} \right)_{\mathbf{x}=\mathbf{c}}, \tag{9.7}$$

and so Eq. (9.4) is verified. The proof that if y is a solution then it is constant along the characteristics is slightly less elementary, and it is based on a geometric interpretation of Eq. (9.4) as the inner product of the vector field of the coefficients a_i and the gradient of the solution y, whose existence is assumed. Note that for a quasi-linear PDE the characteristics depend on the solution itself, and this fact is at the basis of many remarkable properties of fluids such as shock waves (and of the difficulty of the subject!). However, for a linear PDE, the characteristics are independent of the solution y and can thus be computed (at least in principle!) without knowledge of the solution. We are finally in a position to discuss the characteristics of the Liouville equation. From Eqs. (9.1)–(9.5), it appears that the characteristics of the Liouville equation are the solutions of the ordinary differential equations (ODEs) describing the motion of each particle in the N-body system; therefore, we are forced to conclude that despite its statistical flavor, the Liouville approach is equivalent, in terms of mathematical difficulty, to the direct solution of the original problem. For our problem, we arrived at a sort of "dead end!"

9.2 The Collisionless Boltzman Equation

The lessons learned in the previous section force us to move from the quest for an exact *theory* to *modeling*, founded on additional (physically rigorous!) assumptions. This is done in the framework of the CBE that we now "deduce" from phenomenological arguments. The CBE, one of the most important mathematical tools of stellar dynamics, rigorously applies in the limit of perfectly collisionless stellar systems, and it represents the starting point for many of the topics that will be discussed in the following chapters of this book. As anticipated in Chapter 7, for a finite number N of particles (stars), the distinction between collisional and collisionless systems is not sharp, in the sense that the effects of gravitational interactions (the "collisions") build up as time increases. For this reason, one might also wish to look for a simple modification of the CBE that is applicable to stellar systems in weakly collisional regimes. Such an equation, which we will briefly present later on, is the important *Fokker–Planck equation*. It aims to describe stellar systems where the cumulative effects of (weak) collisions cannot be neglected while still using the smoothed potential ϕ.

The first goal is to determine, in the perfect collisionless regime, the differential equation for the evolution of the smooth phase-space distribution function (DF) f introduced in Eq. (7.5). In Chapter 7, we obtained indications that, by increasing the number of particles in a system dominated by gravitational forces, the collisionless approximation (i.e., the substitution of the true discrete system with the continuum approximation) is better and better realized over longer and longer timescales. In the ideal limit of $N \to \infty$, we expect the collisionless approximation to be valid for any time. Our first goal is achieved with the CBE (also known as the *Vlasov equation* in plasma physics). We begin by considering Cartesian coordinates. Priority is given here to the illustration of the physical and mathematical ideas, without the additional difficulty of more advanced formulations; however, in the next section, we will consider the CBE in curvilinear (orthogonal) coordinates, the usual framework adopted in several applications of stellar dynamics. For the sake of generality and for future applications, let us assume that in addition to the potential ϕ associated with the smooth density ρ as in eq. (2.4), an external smooth potential $\phi_{\text{ext}} = \phi_{\text{ext}}(\mathbf{x}; t)$ is considered; in other words, each element of γ moves under the action of the total potential

$\phi_{\rm T} = \phi + \phi_{\rm ext}$ (when $\phi_{\rm ext} = 0$ the system is called *self-gravitating*) with the phase-space velocity field

$$(\dot{\mathbf{x}}, \dot{\mathbf{v}}) = (\mathbf{v}, -\nabla_{\mathbf{x}}\phi_{\rm T}). \tag{9.8}$$

The CBE dictates that in the collisionless regime f evolves according to

$$\frac{Df}{Dt} = \frac{\partial f}{\partial t} + \langle \mathbf{v}, \nabla_{\mathbf{x}} f\rangle - \langle \nabla_{\mathbf{x}}\phi_{\rm T}, \nabla_{\mathbf{v}} f\rangle = 0, \tag{9.9}$$

where the meaning of the Cartesian operators $\nabla_{\mathbf{x}}$ and $\nabla_{\mathbf{v}}$ is obvious, and

$$\Delta\phi(\mathbf{x};t) = 4\pi G \int_{\Re^3} f d^3\mathbf{v}, \quad f(\mathbf{x},\mathbf{v};0) = f_0(\mathbf{x},\mathbf{v}). \tag{9.10}$$

In fact, in the perfect collisionless regime, we can use the same mathematical arguments used in the derivation of the Liouville equation: for any arbitrary phase-space region $\Omega(t)$ in γ, the mass contained in $\Omega(t)$ at any time t is given by Eq. (7.3) with $\Omega = \Omega(t)$. Each point of $\Omega(t)$ – and each point on its boundary $\partial\Omega(t)$ – moves according to the vector field determined by the gradient of the potential $\phi_{\rm T}$: from the uniqueness of the solutions of the ODEs, no orbits can leave this volume. In other words, the mass contained in $\Omega(t)$ is conserved at all times and the vector field induced by $\phi_{\rm T}$ is solenoidal, so that from the Reynolds transport theorem we finally obtain Eq. (9.9) (for more thorough discussions, see, e.g., Binney and Tremaine 2008; Chandrasekhar 1942; Ogorodnikov 1965; Saslaw 1987; Spitzer 1987; Woltjer 1967).

Of course, as with the Liouville equation, the CBE solution can be formally obtained in terms of its characteristics, now given by the solutions of the system in Eq. (9.8) with initial conditions $\mathbf{x}(0) = \mathbf{x}_0$ and $\mathbf{v}(0) = \mathbf{v}_0$; therefore, the characteristics associated with the CBE are curves in the extended one-particle phase space, and the problem is now enormously simplified with respect to that posed by the solution of the Liouville equation. Unfortunately, we do not yet have a complete mathematical understanding of the properties of orbits in three-dimensional time-dependent (or even time-independent!) potentials. Therefore, the problem posed by the general solution of Eq. (9.9) – despite its apparent simplicity – is still too difficult to be solved in general cases.

9.2.1 The CBE in Curvilinear (Orthogonal) Coordinates

Now, it is obvious that in "practical" applications of the CBE Cartesian coordinates are not the preferred choice, and so the student may legitimately ask how to rewrite the CBE in curvilinear coordinates. As would become clear to anyone bravely attempting to change coordinates directly in the CBE (or in general in PDEs), this is *not* an easy task. A first possibility (e.g., see Bertin 2014; Binney and Tremaine 2008; Woltjer 1967) is to use *canonical coordinates* and the transport theorem. In fact, let us consider Eq. (A.174), and suppose we change coordinates from Cartesian to another set of canonical coordinates $(\mathbf{q},\mathbf{p})$: of course, from Eq. (7.3) the function

$$\Delta M[\Omega(t)] = \int_{\Omega(t)} f(\mathbf{x}, \mathbf{v}; t) d^3\mathbf{x} d^3\mathbf{v} = \int_{\Omega'(t)} f'(\mathbf{q}, \mathbf{p}; t) J(\mathbf{q}, \mathbf{p}) d^3\mathbf{q} d^3\mathbf{p}, \tag{9.11}$$

where J is the determinant of the Jacobian[3] of the transformation, is *independent* of the coordinates (canonical or not!) adopted. Finally, in the last integral

$$\mathbf{W}' = (\nabla_{\mathbf{p}} H', -\nabla_{\mathbf{q}} H') \tag{9.12}$$

is the velocity field in phase space produced by the new Hamiltonian function, under which the region $\Omega'(t)$ moves.[4] Now, the use of the canonical coordinates leads to two important consequences:

$$J(\mathbf{q}, \mathbf{p}) = 1, \quad \sum_{i=1}^{3} \left(\frac{\partial W'_i}{\partial q_i} + \frac{\partial W'_{n+i}}{\partial p_i} \right) = 0. \tag{9.13}$$

For a general proof of the first identity see Binney and Tremaine (2008), and for the special case of curvilinear orthogonal coordinates see Exercise 9.2. The latter is simply the manifestation of the solenoidal nature of Hamilton equations in phase space. Repeating the same line of reasoning adopted for the derivation of the CBE in Cartesian coordinates, we conclude that under the assumption of perfect noncollisionality ΔM is conserved for arbitrary $\Omega(t)$, and again the transport theorem in phase space γ for a DF f expressed in canonical coordinates leads to the formal rewriting of Eq. (9.9) as

$$\frac{Df'}{Dt} = \frac{\partial f'}{\partial t} + \sum_{i=1}^{3} \left(\frac{\partial H'}{\partial p_i} \frac{\partial f'}{\partial q_i} - \frac{\partial H'}{\partial q_i} \frac{\partial f'}{\partial p_i} \right) = \frac{\partial f'}{\partial t} + [f', H'] = 0, \tag{9.14}$$

where $[f', H']$ is the *Poisson bracket* between[5] f' and H'. This result is really beautiful, and it allows us to write the CBE in arbitrary canonical coordinates. As we did in the case of the Liouville equation, in the present case we can also fix $f = f' = 1$ in Eq. (9.11) so that $\Delta M[\Omega(t)] = \text{Volume}[\Omega(t)]$, and using the transport theorem again, we obtain the proof of the *Liouville theorem* in γ.

[3] In a canonical coordinate transformation, the Jacobian matrix is *symplectic* (see Footnote 2 in Appendix A); therefore, its determinant evaluates to 1 and no absolute values is required in the integral.

[4] We recall that in general the quantities $\mathbf{q} = (q_1, \ldots, q_n)$, $\dot{\mathbf{q}} = (\dot{q}_1, \ldots, \dot{q}_n)$, and $\mathbf{p} = (p_1, \ldots, p_n)$, which, for simplicity, we indicate with bold face letters as true vectors, are *not* vectors (e.g., like $\mathbf{x}$, $\mathbf{v}$, and $\boldsymbol{\nu}$), but simply lists. Similarly, symbols such as $\nabla_{\mathbf{q}}$ are used to indicate in a compact and suggestive way the list of partial derivatives.

[5] For a given Hamiltonian system with $H = H(\mathbf{q}, \mathbf{p})$, the Poisson bracket of two regular functions U, V is defined as

$$[U, V] \equiv \sum_{i=1}^{n} \left(\frac{\partial U}{\partial q_i} \frac{\partial V}{\partial p_i} - \frac{\partial U}{\partial p_i} \frac{\partial V}{\partial q_i} \right). \tag{9.15}$$

Notice that the Poisson brackets can be interpreted as the advective term of the material derivative in phase space, where the "velocity" is given by the Hamiltonian function. Finally, U and V are said to be in *involution* if their Poisson bracket vanishes, i.e.,

$$[U, V] = 0. \tag{9.16}$$

However, two comments are in order here. The first is that the student should avoid the erroneous idea that the CBE can be transformed in curvilinear coordinates *only* through the procedure just described. Second, in many applications the CBE is written by using curvilinear coordinates ($\mathbf{q}$) and velocities $\boldsymbol{\nu}$ (see Exercise 9.1), not conjugate momenta $\mathbf{p} = \nabla_{\dot{\mathbf{q}}}\mathcal{L}$; therefore, it is natural to ask what happens to Eq. (9.14) because generalized velocities are not (in general) canonical variables, and so Eq. (9.13) in general does not hold! In order to answer this important question, we can proceed as follows. First, notice that

$$\Delta M[\Omega(t)] = \int_{\Omega(t)} f(\mathbf{x}, \mathbf{v}; t) d^3\mathbf{x} = \int_{\Omega'(t)} f'(\mathbf{q}, \boldsymbol{\nu}; t)|J(\mathbf{q}, \boldsymbol{\nu})| d^3\mathbf{q} d^3\boldsymbol{\nu} \tag{9.17}$$

is obviously true, but now the Jacobian J is in general different from unity and the transformed six-dimensional "velocity" field $\mathbf{W}' = (\dot{\mathbf{q}}, \dot{\boldsymbol{\nu}})$ is in general not solenoidal. However, $\Delta M[\Omega(t)]$ is conserved, being the same as in Eq. (9.11), and so from transport theorem

$$\frac{Df'|J|}{Dt} + f'|J| \operatorname{div} \mathbf{W}' = 0, \quad \operatorname{div} \mathbf{W}' \equiv \sum_{i=1}^{3} \left(\frac{\partial \dot{q}_i}{\partial q_i} + \frac{\partial \dot{\nu}_i}{\partial \nu_i} \right). \tag{9.18}$$

We can simplify the first expression in Eq. (9.18). In fact, consider again Eq. (9.17) with $f = f' = 1$: the first integral gives only the phase-space volume of $\Omega(t)$, and as we have seen, this volume is conserved in Cartesian (or canonical) coordinates according to the Liouville theorem (see also Appendix A.10.3), so that the last integral is also independent of time. The corresponding equation for $|J|$ is then obtained immediately from Eq. (9.18) with $f' = 1$. If we now expand the material derivative[6] in Eq. (9.18) and we collect the vanishing term associated with volume conservation, we see that in generic curvilinear coordinates the CBE reads

$$\frac{Df'}{Dt} = \frac{\partial f'}{\partial t} + \sum_{i=1}^{3} \left(\dot{q}_i \frac{\partial f'}{\partial q_i} + \dot{\nu}_i \frac{\partial f'}{\partial \nu_i} \right) = 0 \tag{9.19}$$

(e.g., see Ogorodnikov 1965). Restricting ourselves to curvilinear orthogonal coordinates, from Eq. (9.41) we finally prove that the CBE can be written as

$$\frac{\partial f'}{\partial t} + \sum_{i=1}^{3} \frac{\nu_i}{h_i} \frac{\partial f'}{\partial q_i} - \sum_{i=1}^{3} \frac{1}{h_i} \frac{\partial \phi_{\mathrm{T}}}{\partial q_i} \frac{\partial f'}{\partial \nu_i} + \sum_{i,j,k=1}^{3} \alpha_{ijk} \frac{\nu_j \nu_k}{h_j} \frac{\partial f'}{\partial \nu_i} = 0, \tag{9.20}$$

where $\alpha_{ijk} = \langle \partial \mathbf{f}_i / \partial q_j, \mathbf{f}_k \rangle$. As a useful exercise, the student should repeat the analysis and obtain the CBE for $f' = f'(\mathbf{q}, \dot{\mathbf{q}}; t)$ (i.e., for the DF expressed in terms of generalized coordinates and velocities).

[6] Notice that $D(a\,b)/Dt = b\,Da/Dt + a\,Db/Dt$ from the definition of the material derivative in Eq. (A.173).

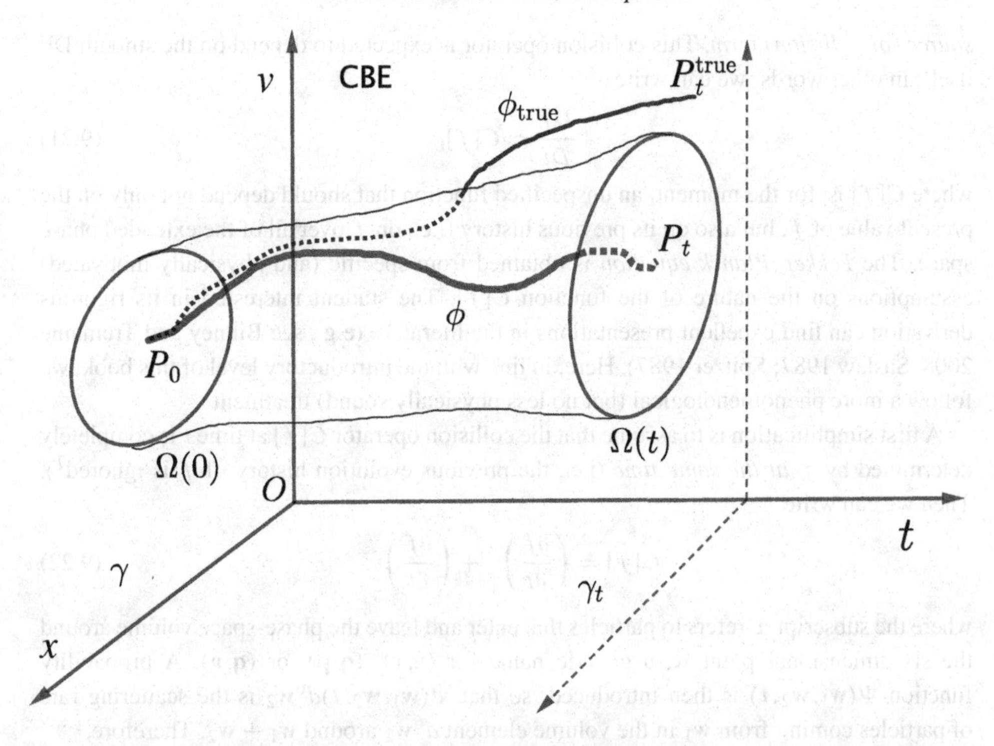

Figure 9.2 Motion of a set $\Omega(0)$ in the one-particle phase space γ, where $\Omega(0)$ is transformed into $\Omega(t)$ under the action of the smooth potential ϕ: no points can leave the set, the volume is conserved, and the CBE in Eq. (9.9) follows. However, the true evolution of the system is not determined by the smooth potential ϕ but by the true (granular) potential ϕ_{true} in Eq. (7.1), and after some time point (P_t^{true}) it can leave/enter $\Omega(t)$ without violating the unicity of the solution of the differential equation: the associated loss and gain of particles is described by the sink/source terms in the collision operator in Eqs. (9.21) and (9.22).

9.2.2 The Fokker–Planck Equation

We recall again the main differences between the CBE and the Liouville equation. Both equations hold rigorously in their domain of application (i.e., the Liouville equation is the *exact* equation describing the motion of a general N-body system in phase space Γ and the CBE is the *exact* equation describing the motion of a perfectly collisionless system in phase space γ). As anticipated, we now discuss how to modify the CBE for the description of systems with finite N over time scales longer than t_{2b} (i.e., gravitational systems in the *collisional* regime), while retaining the advantages of working with ϕ instead of ϕ_{true}.

From what we said at the beginning of Section 9.2, the fact that the number of particles of the true system can change inside the control volume $\Omega(t)$ associated with the smooth potential ϕ (see Figure 9.2) can be modeled by adding to the right-hand side of the CBE a *source (or collision) term*. This collision operator is expected to depend on the smooth DF itself; in other words, we can write

$$\frac{Df}{Dt} = C[f], \tag{9.21}$$

where $C[f]$ is, for the moment, an unspecified function that should depend not only on the present value of f, but also on its previous history (i.e., on f over all of the extended phase space. The *Fokker–Planck equation* is obtained from specific (and physically motivated) assumptions on the nature of the function $C[f]$. The student interested in its rigorous derivation can find excellent presentations in the literature (e.g., see Binney and Tremaine 2008; Saslaw 1987; Spitzer 1987). Here, in line with the introductory level of this book, we follow a more phenomenological (but no less physically sound) treatment.

A first simplification is to assume that the collision operator $C[f]$ at time t is completely determined by f *at the same time* (i.e., the previous evolution history of f is ignored[7]). Then we can write

$$C[f] = \left(\frac{\partial f}{\partial t}\right)_+ + \left(\frac{\partial f}{\partial t}\right)_-, \tag{9.22}$$

where the subscript $\pm$ refers to particles that enter and leave the phase-space volume around the six-dimensional point $\mathbf{w}$, a generic name for $(\mathbf{x},\mathbf{v})$, $(\mathbf{q},\mathbf{p})$, or $(\mathbf{q},\boldsymbol{\nu})$. A probability function $\Psi(\mathbf{w}_1,\mathbf{w}_2;t)$ is then introduced, so that $\Psi(\mathbf{w}_1,\mathbf{w}_2;t)d^6\mathbf{w}_2$ is the scattering rate of particles coming from $\mathbf{w}_1$ in the volume element $d^6\mathbf{w}_2$ around $\mathbf{w}_1+\mathbf{w}_2$. Therefore,

$$\begin{cases} \left(\dfrac{\partial f}{\partial t}\right)_+ = \displaystyle\int_\gamma \Psi(\mathbf{w}-\mathbf{w}',\mathbf{w}';t)f(\mathbf{w}-\mathbf{w}';t)d^6\mathbf{w}', \\ \left(\dfrac{\partial f}{\partial t}\right)_- = -\displaystyle\int_\gamma \Psi(\mathbf{w},\mathbf{w}';t)f(\mathbf{w};t)d^6\mathbf{w}'. \end{cases} \tag{9.23}$$

Equation (9.21) with the resulting collision term is called the *Master equation*.

A second simplification is motivated by the fact that the main effects of the gravitational scattering are due to *weak* encounters (see Chapters 7 and 8; i.e., events that change by small amounts the coordinates $\mathbf{w}$). Mathematically, this means that the main contributors to the collision terms are from phase-space regions "near" to $\mathbf{w}$ (i.e., with "small" $\mathbf{w}'$). By expanding up to the second order the first identity of Eq. (9.23) and by adding the second identity, we obtain (do it!) the collision term in the Fokker–Planck approximation

$$C[f] = -\sum_{i=1}^{6}\frac{\partial f(\mathbf{w};t)\,D_i(\mathbf{w})}{\partial w_i} + \frac{1}{2}\sum_{i,j=1}^{6}\frac{\partial^2 f(\mathbf{w},t)\,D_{ij}(\mathbf{w})}{\partial w_i \partial w_j}, \tag{9.24}$$

where the *diffusion coefficients* are given by

$$D_i(\mathbf{w}) \equiv \int_\gamma w_i'\Psi(\mathbf{w},\mathbf{w}';t)d^6\mathbf{w}', \quad D_{ij}(\mathbf{w}) \equiv \int_\gamma w_i' w_j'\Psi(\mathbf{w},\mathbf{w}';t)d^6\mathbf{w}', \tag{9.25}$$

for $i,j = 1,\ldots,6$. A third and final simplification is discussed in Exercise 9.6.

[7] In statistics, this is called a *Markov property*.

We conclude this brief excursus about the Fokker–Planck equation by recalling that it is an invaluable tool for numerical studies of the evolution of stellar systems in a weakly collisional regime, but we will not discuss these issues any further. The remaining part of this book, starting with the next section, is dedicated to the idealized class of stellar systems that are (1) stationary and (2) perfectly collisionless. Even though these two assumptions can be considered to be very strong, a significant part of the results in stellar dynamics are obtained within this framework.

9.3 Integrability and the Jeans Theorem

From a physical point of view, the assumption of stationarity is interesting because the majority of stellar systems such as galaxies and globular clusters appear to be described by regular morphologies, smooth surface brightness distributions, and relatively simple stellar kinematics. All of these properties suggest that the objects are in a stationary (or quasi-stationary) state. From a mathematical point of view, stationarity is also interesting because a powerful theorem (the *Jeans theorem*) restricts the possible dependence on the phase-space coordinates of the equilibrium phase-space DFs. In particular, we will show that for stationary collisionless stellar systems the DF must be expressed in terms of regular and global integrals of motion; unfortunately, with the exception of the *classical* integrals related to the obvious symmetries of the potential, their mathematical description can be very complicated, and in many cases these additional integrals may not exist at all.

In order to prepare for the following discussion, we begin by writing the CBE $Df/Dt = 0$ for a stationary stellar system; for example, from Eq. (9.9) we see that the mathematical formulation of our two assumptions, once expressed in Cartesian coordinates, is

$$\frac{\partial f}{\partial t} = 0 \quad \forall t \quad \text{i.e.,} \quad [f, H] = \langle \mathbf{v}, \nabla_{\mathbf{x}} f \rangle - \langle \nabla_{\mathbf{x}} \phi_{\mathrm{T}}, \nabla_{\mathbf{v}} f \rangle = 0. \tag{9.26}$$

Of course, we can (and we will!) also consider the equivalent formulations obtained by imposing time independence in Eqs. (9.14) and (9.19). In addition, it should be easy to see that once f is independent of time, all of the macroscopic quantities derived from f (see Chapter 10), in particular density and potential from Eq. (9.10), are independent of time (i.e., $\partial\phi/\partial t = 0\ \forall t$); clearly, in a stationary system with an external potential, $\partial\phi_{\mathrm{ext}}/\partial t = 0$.

9.3.1 Integrability

As we have seen in Sections 9.1 and 9.2, there are geometric objects naturally associated with an equation such as Eq. (9.26) (i.e., the characteristics) that in the present stationary case are the solutions of the general system in Eq. (9.5) specialized to the *autonomous* (i.e., time-independent) case and with the "velocity" field given by Eq. (9.8), as we are using Cartesian coordinates for simplicity. However, the following discussion is independent of the coordinates adopted; moreover, the student is warned that the remaining part of this chapter is necessarily more technical than other parts of this book, and the interested student is encouraged to consult more advanced treatises (e.g., Arnold 1978; McCauley 1997; Whittaker 1917; Wintner 1947; see also Ciotti 2000).

We begin with a concise review of some basic facts about the *integrals of motion* (sometimes also called *first integrals*) for a generic ODE – fundamental geometric objects associated with the characteristics. An *integral of motion* for a generic autonomus ODE such as that in Eq. (A.166), with solution $\Psi = \Psi(\mathbf{x}_0; t)$ as in Eq. (A.167), is introduced as follows:

(1) A function $I(\mathbf{x}) : A \subseteq \Re^n \mapsto \Re$, so that $\forall\, \mathbf{x}_0 \in A$

$$\frac{dI_{\mathcal{L}}(\mathbf{x}_0;t)}{dt} = 0, \qquad I_{\mathcal{L}}(\mathbf{x}_0;t) \equiv I[\Psi(\mathbf{x}_0;t)], \tag{9.27}$$

is called a *integral of motion* for the generic field $\mathbf{W}(\mathbf{x})$ over A; $I_{\mathcal{L}}$ is the Lagrangian function associated with I, and for the moment $\mathbf{x}$ indicates generic coordinates in $\Re^n$.

(2) If I is an integral of motion and $I \in C^{(r)}(A)$, with $r \geq 1$, then I is called a *regular integral of motion* for $\mathbf{W}$ over A.

(3) If I_i for $i = 1, \ldots, k$ are regular integrals of motion over A and the vectors $\nabla_{\mathbf{x}} I_i$ are *linearly independent* over A, then the k integrals I_i are said to be *functionally independent* over A.

(4) If I is a regular integral of motion for $\mathbf{W}$ and satisfies the implicit function theorem over A with respect to one variable x_i (e.g., Edwards 1994; i.e., if $I^0 = I(\mathbf{x}_0)$), then a function h_i exists so that

$$x_i = h_i(I^0, x_1, \ldots, x_{i-1}, x_{i+1}, \ldots, x_n),\ I(x_1, \ldots, x_{i-1}, h_i, x_{i+1}, \ldots, x_n) = I^0. \tag{9.28}$$

Then I is an *isolating integral of motion* for $\mathbf{W}$ over A with respect to x_i.

(5) In the above definitions, the integrals of motion are *global* (i.e., they exist over all A); when they exist only in a local sense (i.e., when A is a "small" region around $\mathbf{x}_0$), they are called *local integrals of motion.*

The student may ask why we are so concerned with the properties of the integrals of motion. The answer is provided by the concept of Jacobi–Lie integrability (e.g., see Eq. (1.6)). An autonomous ODE such as that in Eq. (A.166) is said to be *integrable in the sense of Jacobi–Lie* if and only if $n-1$ functionally independent, global, isolating integrals exist. Otherwise, it is called *nonintegrable*. It should be clear that the property of "integrability" does not refer to our ability to "solve" some difficult integral and/or to define some "new" transcendental function in terms of which one can express the solution. On the contrary, integrability is an *intrinsic* property of a system related to properties of $\mathbf{W}$. Let us discuss the geometric meaning of Jacobi–Lie integrability. When the assumptions given in Eq. (9.28) are verified for the first $n-1$ coordinates, one can write for a generic initial condition $\mathbf{x}_0 \in A$, and globally over A:

$$\begin{cases} x_i = h_i(I_1^0, I_2^0, \ldots, I_{n-1}^0, x_n), \quad i = 1, \ldots, n-1, \\ \dfrac{dx_n}{dt} = W_n(h_1, h_2, \ldots, h_{n-1}, x_n) = w_n(I_1^0, I_2^0, \ldots, I_{n-1}^0, x_n), \end{cases} \tag{9.29}$$

where we used the integrals of motion to express[8] the coordinates $i = 1, \ldots, n-1$ in terms of x_n, and the values I_i^0 are fixed by the initial condition $\mathbf{x}_0$. The remaining differential equation for x_n can be integrated (at least in principle!) by separation of variables. Geometrically, the solution of the ODE in the case of Jacobi–Lie integrability can therefore be seen as the intersection of global functions; a very important example is given in Chapter 4, where the solution of the two-body problem was in fact obtained *without* performing any integration! In other words, the Jacobi–Lie integrability condition is the technical translation of the old-fashioned name of *integration by quadratures*. Note that a generic autonomous ODE satisfying the conditions for the existence, uniqueness, and regularity of the solutions (see Appendix A.10.1) possesses global, unique, and regular solutions, even *without* being Jacobi–Lie integrable. In other words, *a nonintegrable ODE (in the Jacobi–Lie sense) is characterized by a vector field that does not possess a sufficient number* $(n-1)$ *of functionally independent global isolating integrals of motion.*

The requirement of globality is of fundamental importance for integrability, as we now illustrate. In fact, let us consider a system (not necessarily integrable) for which a solution exists. Then the following *flow-box* (or *rectification*) theorem holds (e.g., see McCauley 1997): let $\mathbf{x}_0 \in A \subseteq \Re^n$ be an *ordinary point* for the vector field $\mathbf{W} \in C^{(r)}(A)$ with $r \geq 1$ (i.e., $\mathbf{W}(\mathbf{x}_0) \neq 0$). A neighborhood A_0 of $\mathbf{x}_0$ and a function $\boldsymbol{\phi} \in C^{(r)}(A_0)$ exist so that, with the coordinate transformation

$$\mathbf{y} = \boldsymbol{\phi}(\mathbf{x}), \quad U_{ik}(\mathbf{x}) \equiv \frac{\partial \phi_i}{\partial x_k}, \tag{9.30}$$

Eq. (A.166) transforms into

$$\dot{y}_i = U_{ik}(\mathbf{x}_0) W_k(\mathbf{x}) = 0, \quad \dot{y}_n = U_{nk}(\mathbf{x}_0) W_k(\mathbf{x}) = 1, \tag{9.31}$$

for $i = 1, \ldots, n-1$. The solution of Eq. (9.31) is straightforward, i.e.,

$$y_i = y_i^0 = \phi_i(\mathbf{x}_0), \quad y_n = t. \tag{9.32}$$

In practice, with the change of coordinates in Eq. (9.30), the solution near the regular point $\mathbf{x}_0$ is transformed into a linear "parallel" flow. Obviously, the specific form of $\boldsymbol{\phi}$ is determined by $\mathbf{W}$. Note also that the first $n-1$ components of $\boldsymbol{\phi}(\mathbf{x}_0)$ are regular, *local* integrals of motion for $\mathbf{W}$; in other words, the regular *local* integrals of motion are simply the ϕ_i functions for $i = 1, \ldots, n-1$. A natural question arises: Could we use the initial conditions as global integrals of motion for any ODE? After all, the n coordinates of the initial conditions are conserved for any time along the orbit, not only locally! Unfortunately, as we will now prove by using two different lines of reasoning, the answer is negative.

The basic idea in the geometric argument is that of the *time elimination* between the n components of the general solution by using initial conditions as integrals of motion. In fact, when the solution Ψ exists (and is unique), from Eq. (A.168) it follows that $\mathbf{x}_0 = \Psi(\mathbf{x}; -t)$. The possibility of using initial conditions as global integrals of motion is equivalent to the

[8] Obviously, the functions h_i in Eqs. (9.28) and (9.29) are not the same.

time elimination *for all times* between $n-1$ components of Ψ. In fact, let us assume that for some k we can obtain globally $t=\Psi_k^{-1}(\mathbf{x},x_k^0)$; then, after substitution, we would obtain

$$\Psi_i[\mathbf{x},-\Psi_k^{-1}(\mathbf{x},x_k^0)]=x_i^0 \tag{9.33}$$

for $i=1,\ldots,n$, and so the functions above would be a family of global integrals of motion for $\mathbf{W}$. The problem here is that there is no reason to expect that the function Ψ_k^{-1} will be defined any better than in a local sense. Notice that this remark is true even for systems that are actually Jacobi–Lie integrable (see Exercises 9.7 and 9.8)! The general impossibility of using the initial conditions as regular, global integrals of motion, even though they are conserved along the solution of a generic (well-behaved) ODE, can also be shown in a more formal way by means of the flow-box theorem. In fact, using initial conditions as global integrals of motion is equivalent to requiring the validity of the rectification in Eq. (9.31) globally; in other words, that the transformation $U_{ik}(\mathbf{x})$ satisfies *globally* (i.e., for all $\mathbf{x}$) the identities

$$U_{ik}(\mathbf{x})W_k(\mathbf{x})=0,\quad i=1,\ldots,n-1;\quad U_{nk}(\mathbf{x})W_k(\mathbf{x})=1. \tag{9.34}$$

Now, let us consider what the consequences are of the combination of Eqs. (9.30) and (9.34). From Eq. (9.30), the following differential 1-forms are associated with the change of coordinates

$$dy_i=U_{ik}(\mathbf{x})dx_k=0. \tag{9.35}$$

In order to define a *function* (and so to be global, regular integrals of motion, because for each solution $dy_i=0$), the functions y_i also need to be *path independent* and the vectors fields U_{ik} need to be globally exact, in turn requiring the closure condition

$$\frac{\partial U_{il}(\mathbf{x})}{\partial x_m}=\frac{\partial U_{im}(\mathbf{x})}{\partial x_l} \tag{9.36}$$

for $i=1,\ldots,n$ and $l\neq m$, a condition that surely cannot be expected to be satisfied for general vector fields $\mathbf{W}$. This is the reason why initial conditions cannot be used more than locally (i.e., for small t) as regular integrals of motion. In other words, for nonintegrable ODEs, the initial conditions do not define global holonomic coordinates. In contrast, in Jacobi–Lie integrable systems, the global vector field generated by the flow-box theorem is exact, and so

$$\phi_i(\mathbf{x})=I_i(\mathbf{x}),\quad i=1,\ldots,n-1. \tag{9.37}$$

The interested student is strongly recommended to consult the remarkably clear book of McCauley (1997), from which the following passage is quoted verbatim:

... every non-integrable flow in phase-space has $n-1$ *time-independent global conservation laws,... but they are singular: during a finite time* $0\leq t<t_1$ *the flow can be parallelized via a Lie transformation, but at time* t_1 *one of the conservation laws has a singularity and so there is no parallelization by that particular transformation for* $t\geq t_1$*. According to the Flow-Box theorem, however, there is nothing special about the time* t_1*: the flow can again be parallelized locally for* $t\geq t_1$*, and the same argument applies again. This means that the singularities of the conservation laws of a non-integrable flow must be like branch cuts or phase singularities: moveable and arbitrary, like the international date line in an attempt to impose linear time on circular time.*

9.3.1.1 Integrability in Hamiltonian Systems

How do the previous general considerations apply to the special class of fields **W** in Eq. (9.12) derived from Hamiltonian functions? We can legitimately expect that the special (symplectic) nature of the differential equations adds structure (and consequences) to the integrability of Hamiltonian systems. This is in fact the case, because from Eq. (9.26) it is apparent how a time-independent function U is a regular integral of the motion for the time-independent Hamiltonian H if and only if $[U,H]=0$ (i.e., if and only if U is in *involution* with the Hamiltonian; see Footnote 5). This makes clear the importance of Poisson brackets when dealing with integrals of motion. In particular, let U,V,Z be regular functions defined over the $2n$-dimensional phase space Γ and let a,b be constants: it is not difficult to prove (do it!) that the Poisson brackets satisfy the following relations (a *Lie algebra*; e.g., see Arnold 1978; Landau and Lifshitz 1969; McCauley 1997):

$$\begin{cases} [U,V] = -[V,U], \\ [aU+bV,Z] = a[U,Z]+b[V,Z], \\ [UV,Z] = U[V,Z]+V[U,Z], \\ [U,[V,Z]]+[Z,[U,V]]+[V,[Z,U]] = 0 \end{cases} \quad (9.38)$$

The last identity is known as the *Jacobi identity* (see also Exercise 13.25). Notice that from the Jacobi identity with $U=H$, one obtains that if V and Z are two regular integrals of motion for H, so it is[9] $[V,Z]$. The relevance of the concept of involution for Hamiltonian integrability is made clear by the following fundamental result: let $H = H(\mathbf{q},\mathbf{p})$ be a Hamiltonian function over Γ. It can be proved that (e.g., see again Arnold 1978; Landau and Lifshitz 1969) the field **W** generated by H is Jacobi–Lie integrable (or *completely canonically integrable*) if and only if

(1) There exist n global isolating integrals of motion I_i.
(2) I_i are functionally independent over Γ.
(3) They are in involution over Γ (i.e., $[I_i,I_j]=0$ for $i,j=1,\ldots,n$).

Thus, the symplectic nature of the $2n$-dimensional Hamiltonian fields **W** reduces the number of independent integrals of motion required for Jacobi–Lie integrability (provided they are in involution!) from $2n-1$ (as would be required for a generic field of $2n$ dimensions) to n. Note that when the Hamilton–Jacobi equation is *separable* in some coordinate system, the constants of separation are global isolating integrals of motion (e.g., see Arnold 1978; Goldstein et al. 2000; Landau and Lifshitz 1969), even though not all integrable systems are separable (e.g., see Gutzwiller 1990). We also recall that special Hamiltonian systems exist with a number of integrals greater than n: these systems are called *superintegrable* (e.g., see Evans 1990).

[9] Unfortunately, the integrals generated with this procedure are in general not "new." Also notice that the first identity in Eq. (9.38) confirms that, for an autonomous system, H itself is an integral of motion for $[H,H]=0$.

We conclude this section by recalling that in integrable Hamiltonian systems there exist a special class of integrals of motion: the *actions* $\mathbf{J}$ (not to be confused with angular momentum!). Actions are the natural momenta in integrable Hamiltonian systems, and the associated coordinates are the so-called *angle variables* $\boldsymbol{\theta}$. Among their other remarkable properties, the beauty of actions is that in integrable systems the Hamiltonian function can be rewritten so as to depend only on actions, with

$$\dot{\boldsymbol{\theta}} = \nabla_{\mathbf{J}} H(\mathbf{J}), \quad \dot{\mathbf{J}} = -\nabla_{\boldsymbol{\theta}} H(\mathbf{J}) = 0, \tag{9.39}$$

so that the natural motion in integrable Hamiltonian systems is characterized by constant actions and by the angle coordinates evolving at constant angular velocity (e.g., see Arnold 1978; Binney and Tremaine 2008; Goldstein et al. 2000; Landau and Lifshitz 1969; Lichtenberg and Lieberman 1992). In a certain sense, this result proves the quite amazing fact that, after all, we are only able (when using appropriate coordinates) to solve motion completely in cases of uniform "velocity!"

9.3.2 The Jeans Theorem

How do the above results apply to stationary collisionless stellar systems? This question is answered by the so-called *Jeans theorem* (e.g., Binney and Tremaine 2008; see also Lynden-Bell 1962c). Let f be the DF over γ of a stationary collisionless stellar system. Then f depends on the phase-space coordinates *only* through the regular integrals of motion of ϕ_{T}, i.e.,

$$f = f(I_1, I_2, \ldots) \tag{9.40}$$

(see Exercise 9.10). With the aid of this powerful result, in the following chapters we will be able to explore some exact models for stationary collisionless stellar systems.

Unfortunately, even though an enormous body of literature on this topic is available and many deep results are known, no *general* methods able to determine whether a given system is completely canonically integrable are presently known. In practical examples, the simplest way to find integrals of motion makes use of known (more or less evident) symmetries of the system (*Noether theorem*; e.g., see Landau and Lifshitz 1969). For example, let us assume ϕ_{T} to be the potential of a gravitating system. Then:

(1) If the potential is time independent, $\partial \phi_{\mathrm{T}}/\partial t = 0$, then $H = E$ is a global isolating integral (i.e., energy E is conserved along the orbit of each star).
(2) If the potential is time independent and spherically symmetric, $\phi_{\mathrm{T}} = \phi_{\mathrm{T}}(r)$, then $H = E$ and $\mathbf{J}$ are global isolating integrals (i.e., energy E and angular momentum $\mathbf{J}$ are conserved along the orbit of each star).
(3) If the potential is time independent and axisymmetric, $\phi_{\mathrm{T}} = \phi_{\mathrm{T}}(R, z)$, then $H = E$ and J_z are global isolating integrals (i.e., energy E and the axial component J_z of the angular momentum are conserved along the orbit of each star).
(4) If the potential is time independent and separable in ellipsoidal coordinates (i.e., of the Stäckel family; see Section 13.2.3), then $H = E$ and the other two constants of

separation of the Hamilton–Jacobi equation are conserved along the orbit of each star (e.g., see de Zeeuw 1985a; de Zeeuw and Lynden-Bell 1985; Dejonghe and de Zeeuw 1988 and references therein).

(5) More generally, if the potential is completely canonically integrable, then according to Eq. (9.39) the three integrals can always be identified with the three actions (J_1, J_2, J_3) (see chapter 4 in Binney and Tremaine 2008 and references therein).

Exercises

9.1 Consider a system of curvilinear orthogonal coordinates $\mathbf{q} = (q_1, q_2, q_3)$ (see Appendix A.8) and consider the orbit $\mathbf{x}[\mathbf{q}(t)]$ of a point under the action of the potential ϕ. First, show that the components v_i of the velocity $\mathbf{v} = \dot{\mathbf{x}}$ along the unitary curvilinear versors $\mathbf{f}_i$ and their time derivatives $\dot{v}_i$ can be written using the Lamé coefficients h_i in Eq. (A.140) as

$$v_i = h_i \dot{q}_i, \quad \dot{v}_i = -\frac{1}{h_i}\frac{\partial \phi}{\partial q_i} + \sum_{j,k=1}^{3} \alpha_{ijk}\frac{v_j v_k}{h_j}, \quad \alpha_{ijk} \equiv \left\langle \frac{\partial \mathbf{f}_i}{\partial q_j}, \mathbf{f}_k \right\rangle, \tag{9.41}$$

where we suspended the convention of summing over repeated indices.[10] In particular, compute the α_{ijk} symbols in spherical and cylindrical coordinates and show that in the two coordinate systems (in the usual notation for generalized velocities)

$$\begin{cases} v_r = \dot{r}, \quad v_\vartheta = r\dot{\vartheta}, \quad v_\varphi = r\sin\vartheta\,\dot{\varphi}, \\ \dot{v}_r = -\dfrac{\partial \phi}{\partial r} + \dfrac{v_\vartheta^2 + v_\varphi^2}{r}, \\ \dot{v}_\vartheta = -\dfrac{1}{r}\dfrac{\partial \phi}{\partial \vartheta} - \dfrac{v_r\, v_\vartheta}{r} + \cot\vartheta\,\dfrac{v_\varphi^2}{r}, \\ \dot{v}_\varphi = -\dfrac{1}{r\sin\vartheta}\dfrac{\partial \phi}{\partial \varphi} - \dfrac{v_r\, v_\varphi}{r} - \cot\vartheta\,\dfrac{v_\vartheta\, v_\varphi}{r}, \end{cases} \tag{9.42}$$

while in cylindrical coordinates

$$\begin{cases} v_R = \dot{R}, \quad v_\varphi = R\dot{\varphi}, \quad v_z = \dot{z}, \\ \dot{v}_R = -\dfrac{\partial \phi}{\partial R} + \dfrac{v_\varphi^2}{R}, \\ \dot{v}_\varphi = -\dfrac{1}{R}\dfrac{\partial \phi}{\partial \varphi} - \dfrac{v_R\, v_\varphi}{R}, \\ \dot{v}_z = -\dfrac{\partial \phi}{\partial z}. \end{cases} \tag{9.43}$$

[10] The coefficients α_{ijk} could be expressed in terms of Christoffel symbols (e.g., Narashiman 1993). From the orthonormality of the basis $\mathbf{f}_i$, it follows immediately that $\alpha_{ijk} = -\alpha_{kji}$, so that $\alpha_{ijk} = 0$ for $i = k$.

Hints: By definition $v_i = \langle \mathbf{v}, \mathbf{f}_i \rangle$, so that from Eqs. (A.139)–(A.141)

$$\mathbf{v} = \frac{d\mathbf{x}}{dt} = \sum_{i=1}^{3} \mathbf{x}_i \dot{q}_i = \sum_{i=1}^{3} h_i \dot{q}_i \mathbf{f}_i, \tag{9.44}$$

and this proves the first identity in Eq. (9.41). The generalized accelerations are then obtained starting from the obvious identity $\dot{v}_i = \langle \dot{\mathbf{v}}, \mathbf{f}_i \rangle + \langle \mathbf{v}, \dot{\mathbf{f}}_i \rangle$ and using

$$\frac{d\mathbf{v}}{dt} = -\nabla \phi, \quad \frac{d\mathbf{f}_i}{dt} = \sum_{j,k=1}^{3} \langle \frac{\partial \mathbf{f}_i}{\partial q_j} \dot{q}_j, \mathbf{f}_k \rangle \mathbf{f}_k, \tag{9.45}$$

where in the last espression the derivative of the unitary versor is decomposed in the $\mathbf{f}_i$ basis itself, and finally $\dot{q}_j = v_j / h_j$ from the first identity of Eq. (9.41).

9.2 The determinant of the Jacobian J of canonical coordinates transformations evaluates to 1, being J symplectic (e.g., see Arnold 1978; Binney and Tremaine 2008; Lichtenberg and Lieberman 1992). With this exercise, we prove by direct evaluation that this in fact is true when $(\mathbf{q}, \mathbf{p})$ are associated with curvilinear orthogonal coordinates. First, from the coordinate transformation $\mathbf{x}(\mathbf{q})$, show that for a natural Lagrangian

$$\mathbf{v} = \sum_{i=1}^{3} v_i \mathbf{f}_i = \sum_{i=1}^{3} \frac{p_i}{h_i} \mathbf{f}_i, \tag{9.46}$$

where $h_i(\mathbf{q})$ are the Lamé coefficients. Then compute

$$J(\mathbf{q}, \mathbf{p}) \equiv \det \frac{\partial(\mathbf{x}, \mathbf{v})}{\partial(\mathbf{q}, \mathbf{p})}, \tag{9.47}$$

and show that it is the determinant of a *block diagonal matrix*. From the well-known properties of determinants and Eqs. (9.46) and (A.141), then show that $|J| = 1$. Repeat the exercise with curvilinear coordinates and velocities $\boldsymbol{\nu}$, and show that in this case

$$|J(\mathbf{q}, \boldsymbol{\nu})| = h_1 h_2 h_3 \tag{9.48}$$

(i.e., that at fixed $\mathbf{q}$ we have $d^3\mathbf{v} = d^3\boldsymbol{\nu}$). Finally, show that in action-angle variables $(\mathbf{J}, \boldsymbol{\theta})$, $d^3\mathbf{x} d^3\mathbf{v} = d^3\mathbf{J} d^3\boldsymbol{\theta} = (2\pi)^3 d^3\mathbf{J}$, where the last identity holds for integrable systems performing periodic motions. *Hint*: Consider a natural Lagangian $\mathcal{L} = T - U$ (i.e., a Lagrangian with U independent of $\dot{\mathbf{q}}$) and show that in curvilinear orthogonal coordinates

$$p_i = \frac{\partial \mathcal{L}}{\partial \dot{q}_i} = h_i^2 \dot{q}_i = h_i v_i. \tag{9.49}$$

9.3 From Eq. (9.20) and Exercise 9.2, show that the CBE in spherical coordinates is given by

$$\frac{\partial f}{\partial t} + v_r \frac{\partial f}{\partial r} + \frac{v_\vartheta}{r}\frac{\partial f}{\partial \vartheta} + \frac{v_\varphi}{r \sin\vartheta}\frac{\partial f}{\partial \varphi}$$

$$- \left(\frac{\partial \phi}{\partial r} - \frac{v_\vartheta^2 + v_\varphi^2}{r} \right) \frac{\partial f}{\partial v_r} - \frac{1}{r}\left(\frac{\partial \phi}{\partial \vartheta} + v_r v_\vartheta - \cot\vartheta\, v_\varphi^2 \right) \frac{\partial f}{\partial v_\vartheta}$$

$$- \frac{1}{r}\left(\frac{1}{\sin\vartheta}\frac{\partial \phi}{\partial \varphi} + v_r v_\varphi + \cot\vartheta\, v_\vartheta v_\varphi \right) \frac{\partial f}{\partial v_\varphi} = 0, \tag{9.50}$$

while in cylindrical coordinates

$$\frac{\partial f}{\partial t} + v_R \frac{\partial f}{\partial R} + \frac{v_\varphi}{R}\frac{\partial f}{\partial \varphi} + v_z \frac{\partial f}{\partial z}$$

$$- \left(\frac{\partial \phi}{\partial R} - \frac{v_\varphi^2}{R} \right) \frac{\partial f}{\partial v_R} - \frac{1}{R}\left(\frac{\partial \phi}{\partial \varphi} + v_R v_\varphi \right) \frac{\partial f}{\partial v_\varphi} - \frac{\partial \phi}{\partial z}\frac{\partial f}{\partial v_z} = 0 \tag{9.51}$$

(e.g., see Binney and Tremaine 2008; Ogorodnikov 1965).

9.4 Consider a spherical system with $\phi = \phi(r)$ and a phase-space DF $f(\mathcal{E})$ depending only on r, v_r, and $v_t = \sqrt{v_\vartheta^2 + v_\varphi^2}$. Show that Eq. (9.50) reduces to

$$\frac{\partial f}{\partial t} + v_r \frac{\partial f}{\partial r} - \left(\frac{\partial \phi}{\partial r} - \frac{v_t^2}{r} \right) \frac{\partial f}{\partial v_r} - \frac{v_t v_r}{r}\frac{\partial f}{\partial v_t} = 0. \tag{9.52}$$

9.5 Consider an axisymmetric system with $\phi = \phi(R, z)$ and a phase-space DF $f(\mathcal{E}, J_z)$ independent of φ. Show that Eq. (9.51) reduces to

$$\frac{\partial f}{\partial t} + v_R \frac{\partial f}{\partial R} + v_z \frac{\partial f}{\partial z} - \left(\frac{\partial \phi}{\partial R} - \frac{v_\varphi^2}{R} \right) \frac{\partial f}{\partial v_R} - \frac{v_R v_\varphi}{R}\frac{\partial f}{\partial v_\varphi} - \frac{\partial \phi}{\partial z}\frac{\partial f}{\partial v_z} = 0. \tag{9.53}$$

9.6 The diffusion coefficients in Eq. (9.25) can be further simplified by considering the so-called local approximation, i.e.,

$$\Psi(\mathbf{w}, \mathbf{w}'; t) = \delta(\mathbf{x}')\, \Psi_{\mathbf{v}}(\mathbf{v}, \mathbf{v}'; t), \tag{9.54}$$

in which one assumes that collisions change only the velocities of particles, while their positions remain unchanged. Show that Eq. (9.24) becomes

$$C[f] = -\sum_{i=1}^{3} \frac{\partial f\, D_i(\mathbf{v})}{\partial v_i} + \frac{1}{2}\sum_{i,j=1}^{3} \frac{\partial^2 f\, D_{ij}(\mathbf{v})}{\partial v_i \partial v_j} \tag{9.55}$$

(e.g., see Binney and Tremaine 2008; Spitzer 1987).

9.7 Let us consider the standard two-dimensional harmonic oscillator described by

$$\ddot{x} = -\lambda^2 x, \quad \ddot{y} = -\mu^2 y, \tag{9.56}$$

where λ and μ are two positive constants. Prove that the two isolating, global integrals of motion are the two energies, i.e.,

$$I_x = \frac{\dot{x}^2}{2} + \frac{\lambda^2 x^2}{2}, \quad I_y = \frac{\dot{y}^2}{2} + \frac{\mu^2 y^2}{2}. \tag{9.57}$$

Moreover, prove that the energy of the system is conserved, with

$$E = \frac{\dot{x}^2}{2} + \frac{\dot{y}^2}{2} + \phi = I_x + I_y. \tag{9.58}$$

9.8 Prove that the general solution of Eq. (9.56), in terms of initial conditions, is given by

$$\begin{cases} x(t) = x_0 \cos(\lambda t) + \dfrac{\dot{x}_0}{\lambda}\sin(\lambda t) = \dfrac{\sqrt{2I_x^0}}{\lambda}\cos(\lambda t - \varphi_x^0), \\[2ex] y(t) = y_0 \cos(\mu t) + \dfrac{\dot{y}_0}{\mu}\sin(\mu t) = \dfrac{\sqrt{2I_y^0}}{\mu}\cos(\mu t - \varphi_y^0), \end{cases} \tag{9.59}$$

where, for the x component,

$$\varphi_x^0 = \begin{cases} \alpha_x, & \dot{x}_0 \geq 0, \\ 2\pi - \alpha_x, & \dot{x}_0 \leq 0, \end{cases} \qquad \alpha_x = 0 \leq \mathrm{Arccos}\left(\frac{\lambda x_0}{\sqrt{2I_x^0}}\right) \leq \pi, \tag{9.60}$$

where Arccos is the principal determination of arccos. By inverting the first identity of Eq. (9.59), one obtains

$$\lambda t = \varphi_x^0 + 2\pi k - \mathrm{Arccos}\left[\frac{\lambda x(t)}{\sqrt{2I_x^0}}\right], \tag{9.61}$$

where $k = 0, \pm 1, \pm 2, \ldots$, and $t = 0$ for $k = 0$. The impossibility of a global inversion with respect to time for $-\infty < t < \infty$ is apparent. But this is only the *first* inconvenience; in fact, from Eqs. (9.59)–(9.61)

$$y(t) = \frac{\sqrt{2I_y^0}}{\mu}\cos\left\{\frac{2\pi\mu k}{\lambda} - \frac{\mu}{\lambda}\mathrm{Arccos}\left[\frac{\lambda x(t)}{\sqrt{2I_x^0}}\right] + \frac{\mu}{\lambda}\varphi_x^0 - \varphi_y^0\right\}. \tag{9.62}$$

Show that for a given $x(t)$, if μ and λ are rationally dependent, $y(t)$ assumes a finite number of values (in general more than one), while if μ and λ are rationally independent, the set of $y(t)$ corresponding to a given $x(t)$ is *dense* (e.g., see Woltjer 1967).

9.9 By using Eq. (9.15), prove by explicit computation of the Poisson brackets that for the canonical variables in an n-dimensional Hamiltonian system

$$[p_i, p_j] = [q_i, q_j] = 0, \quad [q_i, p_j] = \delta_{ij}, \quad i, j = 1, \ldots, n. \tag{9.63}$$

Moreover, prove the well-known relations obeyed by angular momentum $\mathbf{J}$ in Cartesian coordinates

$$[J_i, J_j] = \epsilon_{ijk} J_k, \quad [J_i, \|\mathbf{J}\|^2] = 0 \tag{9.64}$$

(e.g., see Goldstein et al. 2000).

9.10 Prove the Jeans theorem. *Hints*: If f is a time-independent solution of the CBE (9.26), then from Eq. (9.27) f is an integral of motion. Conversely, assume that $f = f(I_1, I_2, \ldots, I_k)$ as in Eq. (9.40) and conclude

$$\frac{Df}{Dt} = \sum_{i=1}^{k} \frac{\partial f}{\partial I_i} \frac{DI_i}{Dt} = 0, \tag{9.65}$$

because $DI_i/Dt = 0$ for $i = 1, 2, \ldots, k$.

10

The Jeans Equations and the Tensor Virial Theorem

In this chapter, we show how the (infinite) set of equations known as the *Jeans equations* are derived by considering velocity moments of the collisionless Boltzmann equation (CBE) discussed in Chapter 9. The Jeans equations are very important for physically intuitive modeling of stellar systems, and they are some of the most useful tools in stellar dynamics. In fact, while the natural domain of existence of the solution of the CBE is the six-dimensional phase space, the Jeans equations are defined over three-dimensional configuration space, allowing us to achieve more intuitive modeling of directly observable quantities. The physical meaning of the quantities entering the Jeans equations is also illustrated by comparison with the formally analogous equations of fluid dynamics. Finally, by taking the spatial moments of the Jeans equations over configuration space, the virial theorem in tensorial form is derived, complementing the more elementary discussion in Chapter 6.

10.1 The Method of Moments

As we have seen, the general solution of the CBE depends on knowledge of the properties of orbits in three-dimensional (possibly time-dependent!) potentials, a problem that is well beyond the present possibilities of mathematics (as openly admitted in Arnold 1978). Therefore, in addition to the use of numerical simulations, various techniques have been developed in order to "extract" information from the CBE. These methods can be broadly divided into (1) the method of moments and (2) the construction of particular solutions for stationary systems from the Jeans theorem; as we will see, in the common case of stationary systems, the two approaches can be used together to maximum profit.

The basic idea behind the method of moments, the subject of the present chapter, is to look for differential equations that are simpler than the CBE and describing the relations between particular functions defined as *moments* of the distribution function (DF) over velocity space (the Jeans equations) and over configuration space (the tensor and scalar virial theorems). All of the results presented in this chapter are derived in an inertial reference system S_0, and the convention of summing over repeated indices is used.

The starting point in deriving the Jeans equations for a system with a phase-space DF f is to consider a suitably choosen *microscopic* function defined over the phase space γ, which we indicate generically with $F = F(\mathbf{x}, \mathbf{v}; t)$. Intuitively, F is the mathematical expression of any physical property of interest that "knows" what a star at position $\mathbf{x}$ and with velocity $\mathbf{v}$ is doing at time t; for example, it could be the position or the velocity of the stars, their kinetic energy or angular momentum, and so on. The associated *macroscopic* function $\overline{F}$ is then naturally defined as

$$\overline{F}(\mathbf{x}; t) \equiv \frac{1}{\rho(\mathbf{x}; t)} \int_{\Re^3} F(\mathbf{x}, \mathbf{v}; t) f(\mathbf{x}, \mathbf{v}; t) d^3\mathbf{v}, \quad \rho(\mathbf{x}; t) \equiv \int_{\Re^3} f(\mathbf{x}, \mathbf{v}; t) d^3\mathbf{v}, \tag{10.1}$$

where $\rho(\mathbf{x}; t)$ is the material density of our stellar system at position $\mathbf{x}$ and at time t. From now on, a bar over a symbol will represent the operator given in Eq. (10.1); of course, if F is independent of $\mathbf{v}$, then $F = \overline{F}$. From a physical point of view, $\overline{F}$ can be interpreted as the average value (at time t) of F over all of the velocities of the stars that (at time t) contribute to the density ρ at $\mathbf{x}$.

In stellar dynamics, among all of the functions F that can be imagined, especially important are those leading to the so-called *velocity moments*. For example, up to the second order,

$$\overline{v_i}(\mathbf{x}; t) \equiv \frac{1}{\rho(\mathbf{x}; t)} \int_{\Re^3} v_i f d^3\mathbf{v}, \tag{10.2}$$

$$\overline{v_i v_j}(\mathbf{x}; t) \equiv \frac{1}{\rho(\mathbf{x}; t)} \int_{\Re^3} v_i v_j f d^3\mathbf{v}, \tag{10.3}$$

$$\sigma_{ij}^2(\mathbf{x}; t) \equiv \frac{1}{\rho(\mathbf{x}; t)} \int_{\Re^3} (v_i - \overline{v_i})(v_j - \overline{v_j}) f d^3\mathbf{v} = \sigma_{ji}^2, \tag{10.4}$$

where $i, j = 1, 2, 3$, $\overline{v_i}(\mathbf{x}; t)$ are the components of the *streaming velocity* field, and the (symmetric) tensor $\sigma_{ij}^2(\mathbf{x}; t)$ is the *velocity dispersion tensor*. Obviously, higher-order velocity moments are defined by considering higher-order products between the velocity components (e.g., see Exercise 10.6). Note that

$$\sigma_{ij}^2 = \overline{v_i v_j} - \overline{v_i}\, \overline{v_j}, \tag{10.5}$$

so that σ_{ij}^2 can be *negative* for $i \neq j$, while the diagonal terms are necessarily positive. The reason for adopting the formal notation with a square in Eq. (10.4) is as follows: given the symmetry of σ_{ij}^2 under the exchange of i and j, an orthogonal rotation matrix $\mathcal{R}(\mathbf{x}; t)$ exists so that at each point of the system the velocity dispersion tensor is diagonal in the new reference system, with only positive diagonal entries. In summary, σ_{ij}^2 is a positive definite symmetric tensor. This leads to a simple geometric interpretation of σ_{ij}^2: at each point $\mathbf{x}$ of a stellar system, it is possible to associate a *velocity dispersion ellipsoid* whose "surface" is defined as a function of the unitary versor $\mathbf{n} = n_i \mathbf{e}_i$ by

$$\sigma^2(\mathbf{x}, \mathbf{n}) = \sigma_{ij}^2 n_i n_j, \tag{10.6}$$

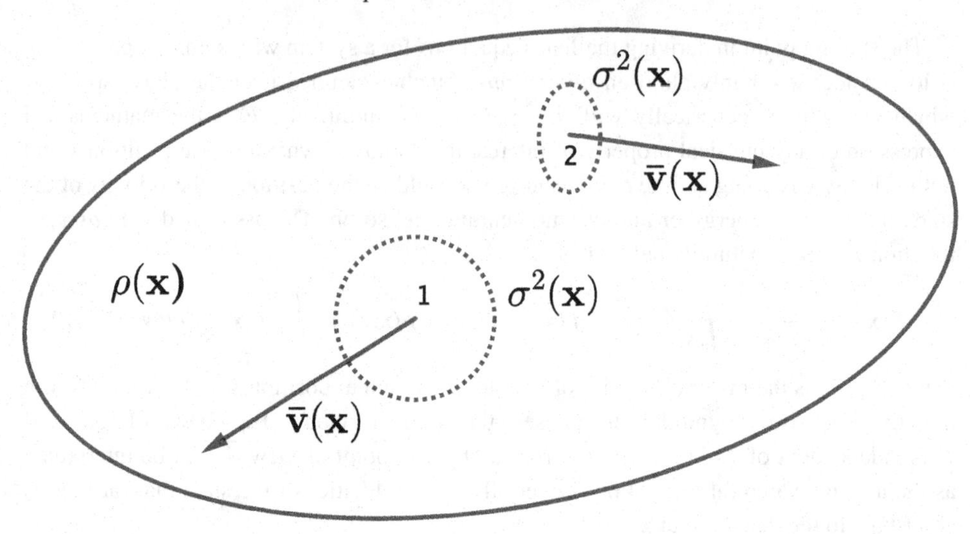

Figure 10.1 Schematic representation of the first three velocity moments of the phase-space DF f of a stellar system (i.e., the scalar density field, the streaming velocity vector field, and the velocity dispersion tensor). At Point 1 the system is isotropic, and at Point 2 some anisotropy is present.

where **n** spans the whole solid angle around **x**. In general, the orientation and the axial ratios of the velocity dispersion ellipsoid change from place to place and with time inside a system (see Figure 10.1). A highly idealized but very important class of systems is that of *isotropic* systems – systems for which, at each point in space

$$\sigma_{ij}^2(\mathbf{x};t) = \sigma^2(\mathbf{x};t)\delta_{ij}. \tag{10.7}$$

In this case, the velocity dispersion ellipsoid is everywhere a sphere. Of course, a stellar system can be isotropic just at some position, or in a limited region of space, or nowhere; in all of these cases we say that the system is *anisotropic*. As we will see, velocity anisotropy is a key concept in stellar dynamics.

10.2 The Jeans Equations

We can now move to derive the Jeans equations associated with a DF f and a total potential ϕ_T. We first prove that the general differential equation (for the moment in Cartesian coordinates for simplicity) obeyed by velocity moments of F can be written as

$$\frac{\partial \rho \overline{F}}{\partial t} + \frac{\partial \rho \overline{F v_i}}{\partial x_i} = -\rho \frac{\partial \phi_T}{\partial x_i}\overline{\frac{\partial F}{\partial v_i}} + \rho\, \overline{v_i \frac{\partial F}{\partial x_i}} + \rho\, \overline{\frac{\partial F}{\partial t}}. \tag{10.8}$$

In Eq. (10.8), $\phi_T = \phi + \phi_{\rm ext}$ is the total potential, with

$$\phi(\mathbf{x};t) = -G\int_{\Re^3} \frac{\rho(\mathbf{y};t)}{\|\mathbf{x}-\mathbf{y}\|} d^3\mathbf{y}, \quad \phi_{\rm ext}(\mathbf{x};t) = -G\int_{\Re^3} \frac{\rho_{\rm ext}(\mathbf{y};t)}{\|\mathbf{x}-\mathbf{y}\|} d^3\mathbf{y}. \tag{10.9}$$

For example, ρ could be the stellar density of a galaxy component and $\rho_{\rm ext}$ the density distribution of a dark matter halo or a massive gaseous component not described by the DF of interest. $\phi_{\rm ext}$ could also contain a term describing the effect of a supermassive central black hole (BH).

Equation (10.8) is easily obtained from the identity

$$\int_{\Re^3} \frac{Df}{Dt} F d^3\mathbf{v} = 0, \tag{10.10}$$

a direct consequence of the CBE. In fact, from Eq. (9.9), the first term in Eq. (10.10) can be rewritten as

$$\int_{\Re^3} \frac{\partial f}{\partial t} F d^3\mathbf{v} = \int_{\Re^3} \left(\frac{\partial f F}{\partial t} - f \frac{\partial F}{\partial t} \right) d^3\mathbf{v} = \frac{\partial \rho \overline{F}}{\partial t} - \rho \overline{\frac{\partial F}{\partial t}}, \tag{10.11}$$

because the time derivative and integration over $\mathbf{v}$ can be interchanged, being the coordinates $\mathbf{v}$ independent of t. The second term is

$$\int_{\Re^3} v_i \frac{\partial f}{\partial x_i} F d^3\mathbf{v} = \int_{\Re^3} v_i \left(\frac{\partial f F}{\partial x_i} - f \frac{\partial F}{\partial x_i} \right) d^3\mathbf{v} = \frac{\partial \rho \overline{F v_i}}{\partial x_i} - \rho \overline{v_i \frac{\partial F}{\partial x_i}}, \tag{10.12}$$

because the spatial derivative and integration over $\mathbf{v}$ can be interchanged, being $\mathbf{v}$ independent of $\mathbf{x}$. Finally, the third term in Eq. (10.10) is

$$\int_{\Re^3} \frac{\partial \phi_{\rm T}}{\partial x_i} \frac{\partial f}{\partial v_i} F d^3\mathbf{v} = \frac{\partial \phi_{\rm T}}{\partial x_i} \int_{\Re^3} \left(\frac{\partial f F}{\partial v_i} - f \frac{\partial F}{\partial v_i} \right) d^3\mathbf{v} = -\rho \frac{\partial \phi_{\rm T}}{\partial x_i} \overline{\frac{\partial F}{\partial v_i}}, \tag{10.13}$$

because $\phi_{\rm T}$ is independent of $\mathbf{v}$ and we assume $fF \to 0$ for $\|\mathbf{v}\| \to \infty$ (and so the integration over all of the velocity space of the exact differential with respect to v_i evaluates to 0). Equation (10.8) is now proved by adding together Eqs. (10.11)–(10.13).

The most common Jeans equations are now derived in Cartesian coordinates with some elementary algebra by choosing $F = 1$ and $F = v_j$ in Eq. (10.8):

$$\frac{\partial \rho}{\partial t} + \frac{\partial \rho \overline{v_i}}{\partial x_i} = \frac{D\rho}{Dt} + \rho \frac{\partial \overline{v_i}}{\partial x_i} = 0, \tag{10.14}$$

$$\frac{\partial \rho \overline{v_i}}{\partial t} + \frac{\partial \rho \overline{v_i v_j}}{\partial x_j} = -\rho \frac{\partial \phi_{\rm T}}{\partial x_i}, \quad (i = 1, 2, 3) \tag{10.15}$$

(e.g., see Bertin 2014; Binney and Tremaine 2008; Chandrasekhar 1942; Ciotti 2000; Ogorodnikov 1965; Saslaw 1987). Note that with some work Eq. (10.15) can also be written as[1]

$$\frac{D\overline{v_i}}{Dt} = \frac{\partial \overline{v_i}}{\partial t} + \overline{v_j} \frac{\partial \overline{v_i}}{\partial x_j} = -\frac{\partial \phi_{\rm T}}{\partial x_i} - \frac{1}{\rho} \frac{\partial \rho \sigma_{ij}^2}{\partial x_j}, \quad (i = 1, 2, 3). \tag{10.16}$$

[1] In the rewriting of the Jeans equations (10.14)–(10.16) we used the material derivative defined in Eq. (A.173), which in the present case reads $D/Dt = \partial/\partial t + \overline{v_i}\partial/\partial x_i$.

Naturally, the Jeans equations can be extended to higher orders (and sometimes are; e.g., see Amendt and Cuddeford 1994; Magorrian and Binney 1994 and references therein) by considering $F = v_i v_j$, $F = v_i v_j v_k$, etc.

Finally, a technical (but important) aspect concerning the Jeans equations is how they can be rewritten in curvilinear coordinates such as spherical, cylindrical, or ellipsoidal (often used in applications; e.g., see Chapter 13). A full discussion of the transformation of differential operators in the general context of continuum mechanics can be found in Narashiman (1993), and the student following this path will appreciate how the direct transformation of partial derivatives is not a trivial task. In particular, the student will appreciate the power of different approaches based on the use of the Hamiltonian formulation and canonical coordinates to express the CBE, or on the use of Reynold's transport theorem in curvilinear orthogonal coordinates (see Chapter 9 and Exercise 10.2).

10.3 Analogies with and Differences from Fluid Dynamics

The student with some knowledge of fluid dynamics will certainly notice the striking similarities between the Jeans equations and the fluid dynamical equations of *continuity* and *momentum* in the presence of a gravitational field and viscosity and in absence of mass, momentum, and energy sources and sinks. The analogy is not by chance. In fact, the equations of fluid dynamics can also be derived by using Reynold's transport theorem (Appendix A.10) with the fluid velocity $\mathbf{u}(\mathbf{x};t) = u_i(\mathbf{x};t)\mathbf{e}_i$ as the basic vector field.[2] In the following, we also consider source/sink terms for later use in this chapter.

We begin with the continuity equation for a fluid with density $\rho(\mathbf{x};t)$ and mass source/sink (per unit volume and time) term $\mathcal{M}(\mathbf{x};t)$, obtained by imposing the mass conservation for arbitrary volumes $\Omega(t) \subseteq \Re^3$ transported by the velocity field; in other words, by requiring that

$$\frac{d}{dt}\int_{\Omega(t)} \rho\, d^3\mathbf{x} = \int_{\Omega(t)} \mathcal{M}\, d^3\mathbf{x}, \quad \forall\, \Omega(t). \tag{10.17}$$

The time derivative is carried out under the integration with the aid of the transport theorem, the integral at the right-hand side is carried out at the left-hand side, and from the requirement that the resulting integral vanishes for arbitrary volumes, it follows immediately that

$$\frac{D\rho}{Dt} + \rho\frac{\partial u_i}{\partial x_i} = \frac{\partial \rho}{\partial t} + \frac{\partial \rho u_i}{\partial x_i} = \mathcal{M}. \tag{10.18}$$

Therefore, Eq. (10.18) shows that a fluid without sources/sinks is *incompressible* if and only if $\operatorname{div}\mathbf{u} = 0$. Incidentally, we notice that in fluid dynamics quite often one encounters the generalization of Eq. (10.17) where, instead of ρ, the product of the density and of some

[2] For remarkably clear discussions of the equations of fluid dynamics, see, for example, Aris (1989), Clarke and Carswell (2007), Currie (1993), Meyer (1982), Milne-Thomson (1996), Narashiman (1993), Pringle and King (2007), Shu (1992), and Tassoul (1978). For more physically oriented presentations, see the magnificient books of Batchelor (1967), Lamb (1945), and Landau and Lifshitz (1986).

property (per unit mass) of the fluid (temperature, velocity, internal energy, entropy, etc.) appears, and in Exercise 10.3 we derive a very useful lemma concerning these cases that is of frequent use.

We now consider the i-th component of the momentum equation (also known as the *Euler equation*) that is obtained from the Second Law of Thermodynamics

$$\frac{d}{dt}\int_{\Omega(t)} \rho u_i \, d^3\mathbf{x} = -\int_{\Omega(t)} \rho \frac{\partial \phi_T}{\partial x_i} d^3\mathbf{x} + \int_{\partial\Omega(t)} p_{ij} n_j \, d^2\mathbf{x} + \int_{\Omega(t)} \mathcal{P}_i \, d^3\mathbf{x}, \tag{10.19}$$

where the three terms at the right-hand side describe the effect of *volume* and *surface* forces over the fluid volume $\Omega(t)$, respectively, and $\mathcal{P}(\mathbf{x};t) = \mathcal{P}_i(\mathbf{x};t)\mathbf{e}_i$ is the momentum source/sink (per unit volume and time). In particular, in the surface integral $\mathbf{n} = n_i \mathbf{e}_i$ is the outward unit vector normal to the surface $\partial\Omega(t)$, and the function

$$p_{ij}(\mathbf{x};t) = -p(\mathbf{x};t)\delta_{ij} + \tau_{ij} \tag{10.20}$$

is the *stress tensor*, where $p(\mathbf{x};t)$ is the *thermodynamical pressure* and τ_{ij} is the *shear tensor*.[3] According to *Cauchy's second law* (e.g., see Truesdell 1991), p_{ij} is a symmetric tensor: note that from Eq. (10.21), in the hydrostatic case (i.e., $\mathbf{u} = 0$), the stress tensor reduces to the thermodynamical pressure only, independently of the viscosity. Using the divergence theorem, the surface integral in Eq. (10.19) can be transformed into a volume integral

$$\int_{\partial\Omega(t)} p_{ij} n_j d^2\mathbf{x} = \int_{\Omega(t)} \frac{\partial p_{ij}}{\partial x_j} d^3\mathbf{x} = -\int_{\Omega(t)} \frac{\partial p}{\partial x_i} d^3\mathbf{x} \tag{10.22}$$

(where the last identity holds for inviscid fluids). Finally, from the lemma in Eq. (10.69) applied to the left-hand side of Eq. (10.19), one obtains for $i = 1,2,3$

$$\frac{Du_i}{Dt} = \frac{\partial u_i}{\partial t} + u_j \frac{\partial u_i}{\partial x_j} = -\frac{\partial \phi_T}{\partial x_i} + \frac{1}{\rho}\frac{\partial p_{ij}}{\partial x_j} + \frac{\mathcal{P}_i - \mathcal{M} u_i}{\rho}. \tag{10.23}$$

Of course, one could write an equation for the angular momentum of the fluid analogous to Eq. (10.19) by starting from the appropriate equation of dynamics, and this is left as a (very important) exercise for the interested student (see also Exercise 10.5).

We now momentarily stop to appreciate the striking similarities between the Jeans equations (10.14)–(10.16) and the fluid dynamical equations (10.18)–(10.23). Note how the two sets of equations formally coincide in the absence of sources/sinks, provided we identify the streaming velocity of stars $\overline{v_i}$ with the fluid velocity u_i and the velocity dispersion tensor of the stellar system $\rho\sigma_{ij}^2$ with the fluid stress tensor $-p_{ij}$. Regarding the last point, note how from Eq. (10.21) the isotropic velocity dispersion tensor in Eq. (10.7) would correspond

[3] For the so-called *Newtonian fluids*,

$$\tau_{ij} = \mu\left(\frac{\partial u_i}{\partial x_j} + \frac{\partial u_j}{\partial x_i}\right) + \lambda \frac{\partial u_k}{\partial x_k}\delta_{ij}, \tag{10.21}$$

where μ and λ are the *first* and *second* viscosity coefficients, respectively (e.g., see Truesdell 1984); in incompressible fluids, the shear tensor is independent of λ.

(prove it!) to the identification $\rho\sigma^2 = p - (\lambda + 2\mu/3)\mathrm{div}\,\mathbf{u}$; from a monoatomic perfect gas, Stokes' hypothesis assumes $\lambda + 2\mu/3 = 0$, and so an isotropic velocity dispersion would be formally identified with "thermodynamic pressure."

An obvious question remains open: Can the similarities exhibited by the continuity and momentum equations of stellar dynamics and fluid dynamics be extended to the corresponding energy equations? We answer this question by deriving first the energy equation for a fluid with heat conduction and energy sinks/sources (e.g., radiative loss/heating). The continuum formulation of the First Law of Thermodynamics reads

$$\frac{d}{dt}\int_{\Omega(t)}\left(E + \rho\frac{\|\mathbf{u}\|^2}{2}\right) d^3\mathbf{x} = -\int_{\Omega(t)} \rho u_i \frac{\partial \phi_{\mathrm{T}}}{\partial x_i} d^3\mathbf{x} + \int_{\partial\Omega(t)} u_i p_{ij} n_j \, d^2\mathbf{x}$$
$$- \int_{\partial\Omega(t)} h_i n_i \, d^2\mathbf{x} + \int_{\Omega(t)} (\mathcal{E} - \mathcal{L})\, d^3\mathbf{x}, \qquad (10.24)$$

where E is the *internal energy* per unit volume, and on the right-hand side the work per unit time associated with the volume and surface forces is considered. The *heat conduction* is described by the *heat flux vector* $\mathbf{h}(\mathbf{x})$, and the minus sign in front of the surface integral is associated with the fact that $\mathbf{n}$ is directed outward from the control volume $\Omega(t)$. Finally, $\mathcal{E}(\mathbf{x})$ and $\mathcal{L}(\mathbf{x})$ are, respectively, the energy sources and sinks per unit volume and time. Repeating the treatment used for the continuity and momentum equations, and after a simplification taking into account Eq. (10.23), we finally obtain (prove it!)

$$\frac{DE}{Dt} + E\frac{\partial u_i}{\partial x_i} = \frac{\partial E}{\partial t} + \frac{\partial E u_i}{\partial x_i} = p_{ij}\frac{\partial u_i}{\partial x_j} - \frac{\partial h_i}{\partial x_i} + \mathcal{M}\frac{\|\mathbf{u}\|^2}{2} - \mathcal{P}_i u_i + \mathcal{E} - \mathcal{L}. \qquad (10.25)$$

Note how volume forces do not affect (directly) the internal energy of a fluid, an aspect that is not always appreciated; we also recall that by using the same approach it would be possible to express the Second Law of Thermodynamics in a continuum formulation, even though we will not pursue the subject further (e.g., see Currie 1993; Meyer 1982).

We now derive the energy Jeans equation for collisionless systems and we compare it with Eq. (10.25). For a system made of single stars, the quantity equivalent to the internal plus kinetic energy appearing on the left-hand side of Eq. (10.24) is given by the phase-space average of the microscopic function $F = \|\mathbf{v}\|^2/2$, and in Exercise 10.6 three important properties of $\overline{F}$ are proved; in particular, Eq. (10.72) suggests that, for a stellar system, we should identify $E = \rho\mathrm{Tr}(\sigma_{ij}^2)/2$. Moreover, from Eq. (10.8), after some algebra, and taking into accout the symmetry of σ_{ij}^2 and p_{ij}, one obtains

$$\frac{\partial E}{\partial t} + \frac{\partial E\overline{v_i}}{\partial x_i} = \frac{DE}{Dt} + E\frac{\partial \overline{v_i}}{\partial x_i} = -\rho\sigma_{ij}^2\frac{\partial \overline{v_i}}{\partial x_j} - \frac{\partial \rho\chi_i}{\partial x_i}, \qquad (10.26)$$

where the analogy with Eq. (10.25) is complete in the absence of source/sink terms and if one identifies the components of the heat flux vector $\mathbf{h} = h_i\mathbf{e}_i$ with the third-order velocity moments $\rho\chi_i$ in Eq. (10.73).

Despite this striking formal analogy, a fundamental difference between fluids and stellar systems is apparent. In fact, while the equations of fluid dynamics are a *closed* set of equations because the pressure is related to ρ and E via the *equation of state*, the concept of the equation of state[4] cannot be introduced from basic physical arguments in the Jeans equations. This means that the latter are an *infinite set* of equations. Thus, in order to solve the Jeans equations, we are forced to assume a (more or less physically motivated) *closure relation*. In applications, this is often done by specifying the properties of the velocity dispersion tensor, and due to its relevance for the modeling of stellar systems, the problem will be extensively discussed in Chapters 13 and 14. Moreover, in Exercise 10.2 we derive the generalization of Eq. (10.8) for curvilinear (orthogonal) coordinates, while in the next section we illustrate an astrophysical problem where both the Jeans equations and the equations of fluid dynamics play a fundamental role.

10.4 Stellar Dynamics and Gas Dynamics in Stellar Systems

A very broad class of astrophysical problems concerns the interesting situations where *both* stellar dynamics and fluid dynamics are simultaneously involved, such as the physics of galactic gas flows, of gaseous cold disks and hot coronae around disk galaxies, the hot intra-cluster medium in clusters of galaxies, the Bondi accretion on supermassive BHs at the center of elliptical galaxies, and so on. The link between stellar dynamics and fluid dynamics is not suprising: after all, in a stellar system, stars and gases share the same gravitational potential, as is apparent from Eqs. (10.16) and (10.23). The literature on these subjects is immense, and we cannot even attempt to give a partial covering of the most important references. Instead, we simply present a few representative cases aimed at introducing the student to such a wonderful branch of astrophysics.

[4] For example, a perfect gas is defined as any substance for which the equation of state

$$p = \frac{k_B \rho T}{\langle\mu\rangle m_p}, \quad k_B = 1.38\,10^{-16}\,\frac{\text{erg}}{K}, \quad m_p = 1.67\,10^{-24}\ \text{g} \tag{10.27}$$

holds, where $\langle\mu\rangle$ is the mean molecular weight, k_B is Boltzmann's constant, and m_p is the proton mass. The student is recommended to distinguish between the concepts of the equation of state and of *thermodynamical transformation*. Some of the most important are the *barotropic* transformations and their subset of *polytropic* transformations of index γ (the polytropic index), given respectively by

$$p = p(\rho), \quad p = \frac{p_0}{\rho_0^\gamma}\rho^\gamma, \tag{10.28}$$

where ρ_0 and p_0 are normalization values for density and pressure; nonbarotropic fluids are called *baroclinic*. A very important subset of polytropics is represented by the *reversible adiabatics*, where $\gamma = \gamma_{ad} = \mathcal{C}_p/\mathcal{C}_V$ is the ratio of the molar specific heats at constant pressure and temperature. As is well known (e.g., see eq. 25 in Chandrasekhar 1939), polytropic transformations are characterized by the constant molar specific heat

$$\mathcal{C}_{poly} = \mathcal{C}_V \frac{\gamma_{ad} - \gamma}{1 - \gamma}, \tag{10.29}$$

a *negative* number for $1 < \gamma < \gamma_{ad}$ (see also Korol et al. 2016).

10.4.1 Hydrostatic Equilibria

An important class of astrophysical problems where the interplay between fluid dynamics and stellar dynamics is particularly relevant is that of *hydrostatic equilibria* (i.e., the study of the admissible configurations of gases at equilibrium in the total potential well of stellar systems such galaxies or clusters of galaxies). This area is motivated by the hope of gaining information on the gravitational potential ϕ_T by observation of the gas properties. In fact, from Eq. (10.23), in the hydrostatic case, and in the absence of sources/sinks,

$$\nabla p = -\rho \nabla \phi_T, \tag{10.30}$$

where p is the gas pressure and ρ is the gas density. Following Poincaré (e.g., Tassoul 1978), some general properties of the solutions of Eq. (10.30) can be easily derived. First, as ∇p is parallel to $\nabla \phi_T$, it follows immediately that the gas pressure is stratified on isopotential surfaces (i.e., $p = p(\phi_T)$). Second, by taking the rot of Eq. (10.30), we deduce that $\nabla \rho \wedge \nabla \phi_T = 0$ (i.e., isodensity surfaces coincide with isopotential surfaces), and so also $\rho = \rho(\phi_T)$. Therefore, by elimination of the potential, we conclude that hydrostatic equilibria are necessarily *barotropic* (i.e., $p = p(\rho)$) independently of the specific equation of state of the gas (see Footnote 4). Finally, if our gas is *perfect*, then from Eq. (10.27) we obtain that $T/\langle\mu\rangle$ is also constant over isopotential surfaces (see also Appendix A.6). It is not a surprise then that from observations of the gas properties in systems at fiducial equilibrium, we can gain information on the total gravitational potential of the system (e.g., see Fabricant et al. 1980).

The previous considerations show that for an assigned potential, an assigned equation of state, and an assigned stratification $p(\rho)$, the integration of Eq. (10.30) can be easily performed (e.g., Ciotti and Pellegrini 2008). One of the most simple and important examples is represented by hydrostatic polytropic equilibria. In fact, it is a simple exercise (do it!) to show that by inserting the second identity of Eq. (10.28) into Eq. (10.30), one gets

$$\begin{cases} \dfrac{\rho}{\rho_0} = \left(\dfrac{\gamma - 1}{\beta_0 \gamma}\right)^{\frac{1}{\gamma-1}} (\Psi_T - \mathcal{E}_0)^{\frac{1}{\gamma-1}} \, \theta(\Psi_T - \mathcal{E}_0), \\ \dfrac{T}{T_0} = \dfrac{\gamma - 1}{\beta_0 \gamma} (\Psi_T - \mathcal{E}_0) \, \theta(\Psi_T - \mathcal{E}_0), \\ \beta_0 \equiv \dfrac{k_B T_0}{\langle\mu\rangle m_p}, \quad \mathcal{E}_0 \equiv \Psi_{T0} - \dfrac{\beta_0 \gamma}{\gamma - 1}, \end{cases} \tag{10.31}$$

where $\Psi_T \equiv -\phi_T$ and the subscript 0 indicates the values of the various physical quantities at some prescribed position $\mathbf{x}_0$ in the system. Notice that the solutions above for sufficiently low values of T_0 can be spatially truncated on the surface $\Psi_T(\mathbf{x}_t) = \mathcal{E}_0$ depending on the value of γ and of the behavior of ϕ_T. A particularly important class of solutions is obtained in the *isothermal* limit $\gamma \to 1$, when $\mathcal{E}_0 \to -\infty$, and the student is encouraged to prove that the solutions in Eq. (10.31) become untruncated, with

$$\rho = \rho_0 \, e^{\frac{\Psi_T - \Psi_0}{\beta_0}}, \qquad T = T_0. \tag{10.32}$$

We will return to these solutions in Chapter 12.

Another significant class of fluid equilibria in gravitational fields of obvious importance when considering disk galaxies is represented by axisymmetric gaseous configurations in permanent rotation (i.e., their velocity field is written in cylindrical coordinates as $\mathbf{u} = u_\varphi(R,z)\mathbf{e}_\varphi$). In this case, Eq. (10.23) reduces to (prove it!)

$$\nabla p = -\rho\nabla\phi_{\rm T} + \rho\frac{u_\varphi^2}{R}\mathbf{e}_R, \tag{10.33}$$

and again an important theorem from Poincaré states that the (rotating) equilibrium is barotropic if and only if the rotational velocity is constant on cylinders (i.e., u_φ is independent of z), and so an *effective* potential can be immediately introduced into Eq. (10.33) (see Tassoul 1978). Of course, if the density distribution and the gravitational potential are assigned, then the gas pressure is fixed by the vertical equilibrium condition and there is no reason to expect cylindrical rotation, so that the gas distribution is *baroclinic*, with the pressure not being constant over isodense surfaces. It is easy to prove (do it!) that for gas distributions that are sufficiently regular at infinity (the usual case)

$$\frac{\rho u_\varphi^2}{R} = \int_z^\infty \left(\frac{\partial\rho}{\partial R}\frac{\partial\phi_{\rm T}}{\partial z'} - \frac{\partial\rho}{\partial z'}\frac{\partial\phi_{\rm T}}{\partial R}\right) dz'. \tag{10.34}$$

Equation (10.34) can be used to study the effect of the shape of the gravitational potential of disk galaxies on the rotational field of their hot and rotating gaseous coronae, which present a clear dependence of their rotational velocity on z (e.g., see Barnabè et al. 2006 and references therein). We will return to a consideration of Eq. (10.34) in Chapter 13.

Another important combined use of the Jeans equations and fluid dynamics is obtained by elimination of the total potential from Eqs. (10.16) and (10.23) so that for the gas and stellar distributions necessarily

$$\frac{Du_i}{Dt} + \frac{1}{\rho}\frac{\partial p}{\partial x_i} = \frac{D\overline{v_i}}{Dt} + \frac{1}{\rho_*}\frac{\partial\rho_*\sigma_{ij}^2}{\partial x_j}, \tag{10.35}$$

an identity that, among other applications, can be used to probe the assumption of hydrostatic equilibrium, to gain information on anisotropy, and finally to test how departures from the assumptions can affect mass estimates of the stellar systems under consideration in a systematic way (e.g., see Ciotti and Pellegrini 2004; Mathews and Brighenti 2003b; Pellegrini and Ciotti 2006).

10.4.2 Galactic Gas Flows

An important astrophysical problem where *both* stellar dynamics and fluid dynamics are at play is that of the structure and evolution of the hot gaseous coronae observed in elliptical galaxies and in clusters of galaxies. The literature on the subject is extremely vast, and attempting of providing even of a very partial list of the results in the field devoted to the subject would be foolish at best: Kim and Pellegrini (2012) and Mathews and Brighenti (2003a) and the references therein represent very good and general overviews of the problem and of the literature (see also Sarazin 1988 for the related problem in clusters of galaxies). Here, we necessarily limit ourselves to one particular aspect: we show how the Jeans equations are needed to specify the source terms in Eqs. (10.18), (10.23), and (10.25) describing gas flows. In fact, the important information here is that the stars in a galaxy inject gas as they evolve (on the main sequence, as red giants, on the asymptotic giant branch, exploding as supernovae, etc.) and that the momentum and energy of the stellar ejecta substantially affect the structure and evolution of the resulting X-ray-emitting coronae, playing a central role in the *cooling flow* model and all of the associated variants of *galactic winds*, with or without active galactic nuclei feedback from accretion on the central BH (e.g., see Ciotti and Ostriker 2012; Ostriker and Ciotti 2005 and references therein; see also Ciotti and Pellegrini 2017, 2018).

Motivated by the previous discussion, let us now focus for simplicity on a simple stellar population in a galaxy with internal dynamics described by a phase-space DF $f(\mathbf{x}, \mathbf{v}; t)$; the generalization to the case of more than one stellar component is straightforward in the context of extended DFs. For the present purposes, it is useful to adopt the interpretation of the DF in terms of the number of stars per unit volume in phase space (i.e., we assume that $n(\mathbf{x}; t) = \int_{\Re^3} f \, d^3\mathbf{v}$). As each star will spend different amounts of time in each of the different evolutionary phases in the following, we only concentrate on some unspecified evolutionary phase of stars of a well-defined mass; as far as the whole stellar population is described by the *same* phase-space distribution, the dynamics of each population are the same and the contributions to mass, momentum, and energy injections during the different phases can simply be added to determine the total source terms. In the more general case, the extended DF formalism should be used.

In all generality, let $m = m(\mathbf{x}, \mathbf{v}, \mathbf{n}; t)$, the *mass return* per unit time and unit solid angle along the direction $\mathbf{n}$ associated with each star, where $\mathbf{n} = n_i \mathbf{e}_i$ is a unitary vector measured from the center of the star. In principle, the dependence of m on $\mathbf{n}$ would account for the possibility of *anisotropic* mass losses, while the dependence on $\mathbf{v}$ could describe some dependence of the stellar population properties on phase space (e.g., a counter-rotating disk of stars with different metallicity). The total mass return per unit time, volume, and solid angle associated with the stars at $\mathbf{x}$ is indicated as $\mu(\mathbf{x}, \mathbf{n}; t) = \int_{\Re^3} m \, f \, d^3\mathbf{v}$, so that for mass sources independent of the source velocity $\mathbf{v}$ (the obvious situation), $\mu(\mathbf{x}, \mathbf{n}; t) = nm$. Integration over the solid angle finally gives the total mass return rate per unit time and volume at $\mathbf{x}$

$$\mathcal{M}(\mathbf{x}; t) = \int_{4\pi} \mu \, d^2\mathbf{n} = 4\pi n \, m, \tag{10.36}$$

where the last expression holds for isotropic mass losses.

The *momentum return* associated with each star is given by the vectorial function $\mathbf{p} = m(\mathbf{x}, \mathbf{v}, \mathbf{n}; t)[\mathbf{v} + u_{\mathrm{s}}(\mathbf{x}, \mathbf{v}, \mathbf{n}; t)\mathbf{n}]$, where $u_{\mathrm{s}} = \|\mathbf{u}_{\mathrm{s}}\|$ is the modulus of the velocity of the ejecta along $\mathbf{n}$ measured with respect to the star's velocity $\mathbf{v}$. The total momentum injected per unit time, volume, and solid angle from the considered stellar population at $\mathbf{x}$ is then given by $\boldsymbol{\pi}(\mathbf{x}, \mathbf{n}; t) = \int_{\Re^3} \mathbf{p} \, f \, d^3\mathbf{v}$, so that for mass losses and ejection velocities independent of the specific stellar velocity $\mathbf{v}$, we have $\boldsymbol{\pi}(\mathbf{x}, \mathbf{n}; t) = nm \, (\overline{\mathbf{v}} + u_{\mathrm{s}}\mathbf{n})$, where $\overline{\mathbf{v}}$ is the streaming velocity field at position $\mathbf{x}$ in the galaxy, as is given by Eq. (10.2). Integration over the solid angle finally gives the total momentum injection rate per unit time and volume at $\mathbf{x}$

$$\mathcal{P}(\mathbf{x};t) = \int_{4\pi} \pi \, d^2\mathbf{n} = \mathcal{M}\overline{\mathbf{v}}, \qquad (10.37)$$

where again the last expression holds if m and u_s are also isotropic. In practice, *the amount of injected momentum in the case of isotropic mass losses is simply proportional to the local streaming velocity of the stellar population.*

The *energy source* associated with the stellar mass losses is made from the contributions of two distinct parts: the *internal energy source* and the *kinetic energy source*. The internal energy source (energy per unit time per unit solid angle per unit mass) is described by the function $e = e(\mathbf{x}, \mathbf{v}, \mathbf{n}; t)$. The total internal energy injected at $\mathbf{x}$ in the galaxy per unit time, volume, and solid angle is then given by $\epsilon(\mathbf{x}, \mathbf{n}; t) = \int_{\Re^3} m \, e \, f \, d^3\mathbf{v}$, so that for mass and internal energy sources independent of the stellar velocity $\mathbf{v}$, $\epsilon(\mathbf{x}, \mathbf{n}; t) = \mu e$. Integration over the solid angle finally gives the total inernal energy injection rate per unit time and volume at $\mathbf{x}$

$$\mathcal{E}_{\rm int}(\mathbf{x};t) = \int_{4\pi} \epsilon \, d^2\mathbf{n} = \mathcal{M}e, \qquad (10.38)$$

where the last expression holds if m and u_s are isotropic. The kinetic energy injection due to a given star is given by $k = \frac{1}{2} m(\mathbf{x}, \mathbf{v}, \mathbf{n}; t) \|\mathbf{v} + u_s(\mathbf{x}, \mathbf{v}, \mathbf{n}; t)\mathbf{n}\|^2$, so that the total kinetic energy per unit time, volume, and solid angle associated with the stars at $\mathbf{x}$ is $\kappa(\mathbf{x}, \mathbf{n}; t) = \int_{\Re^3} k \, f \, d^3\mathbf{v}$. In particular, for mass losses and ejection velocities independent of the stellar velocity $\mathbf{v}$, Eq. (10.72) proves that $\kappa(\mathbf{x}, \mathbf{n}; t) = nm \left(\|\overline{\mathbf{v}}\|^2 + \mathrm{Tr}\,\sigma^2 + u_s^2 + 2u_s\langle \mathbf{n}, \overline{\mathbf{v}}\rangle\right)/2$ where $\mathrm{Tr}\,\sigma^2$ is the trace of the velocity dispersion tensor in Eq. (10.4) associated with the stellar population. Finally, integrating over the whole solid angle,

$$\mathcal{E}_{\rm kin}(\mathbf{x};t) = \int_{4\pi} \kappa \, d^2\mathbf{n} = \mathcal{M}\left(\frac{\|\overline{\mathbf{v}}\|^2}{2} + \frac{u_s^2}{2} + \frac{\mathrm{Tr}\,\sigma^2}{2}\right), \qquad (10.39)$$

where the last expression holds if m and u_s are isotropic because the angular average of $\mathbf{n}$ appearing in the expression of κ vanishes.

In summary, in a stellar system where the aging stars inject (isotropically) mass, momentum, and energy, from Eqs. (10.23)–(10.25), restricting to the nonviscous case, and finally using the expressions just derived for the source terms, the momentum and energy equations for the interstellar medium are (prove it!)

$$\frac{Du_i}{Dt} = -\frac{\partial \phi_{\rm T}}{\partial x_i} - \frac{1}{\rho}\frac{\partial p}{\partial x_i} + \frac{\mathcal{M}}{\rho}(\overline{v_i} - u_i), \qquad (10.40)$$

$$\frac{DE}{Dt} + (E + p)\frac{\partial u_i}{\partial x_i} = \mathcal{M}\left(e + \frac{u_s^2}{2} + \frac{\mathrm{Tr}\,\sigma^2}{2} + \frac{\|\overline{\mathbf{v}} - \mathbf{u}\|^2}{2}\right) - \frac{\partial h_i}{\partial x_i} - \mathcal{L}, \qquad (10.41)$$

where the dependence of the source heating term from the kinematical fields σ^2 and $\overline{\mathbf{v}}$, determined by stellar dynamics, is apparent. In fact, we can illustrate the impact of stellar

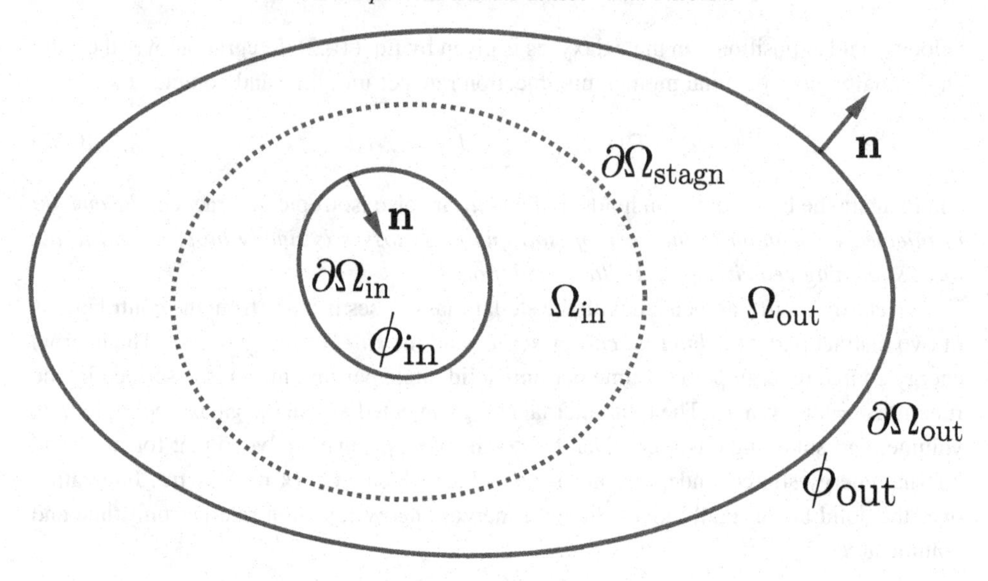

Figure 10.2 Schematic representation of stationary gas flow in the presence of stellar source terms, as described in Eq. (10.43). The dotted line represents a stagnation surface defined by $\mathbf{u} = 0$ and the gas is outflowing at $\partial\Omega_{out}$ at the constant rate $\dot{M}_{out}$ and inflowing at $\partial\Omega_{in}$ at the constant rate $\dot{M}_{in}$.

dynamics on the energetics of the flows by considering the special and very idealized case of a time-independent, nonconductive gas flow in a galaxy and integrating Eq. (10.75) over a time-independent volume Ω bounded by two isopotential surfaces $\partial\Omega_{in}$ and $\partial\Omega_{out}$ as in Figure 10.2. Moreover, we also assume that a *stagnation surface* (i.e., a surface over which $\mathbf{u} = 0$) $\partial\Omega_{stagn}$ exists, so that the gas must be on average spatially outflowing over $\partial\Omega_{out}$ and on average spatially inflowing toward the center over $\partial\Omega_{in}$ in order to ensure time independence.[5] Under these assumptions, volume integration over the two regions Ω_{out} and Ω_{in} finally shows that (prove it!)

$$\int_\Omega \mathcal{L} d^3\mathbf{x} = \int_\Omega \mathcal{M}\left(e + \frac{u_s^2}{2} + \frac{\|\overline{\mathbf{v}}\|^2}{2} + \frac{\mathrm{Tr}\,\sigma^2}{2}\right) d^3\mathbf{x}$$
$$- \int_{\partial\Omega} \left\langle \mathcal{M}\left(H + \rho\frac{\|\mathbf{u}\|^2}{2}\right)\mathbf{u}, \mathbf{n}\right\rangle d^2\mathbf{x} - L^-_{grav} + L^+_{rad}, \tag{10.43}$$

where

$$L^-_{grav} = \int_{\Omega_{out}} \mathcal{M}\,(\phi_{out} - \phi) d^3\mathbf{x}, \qquad L^+_{rad} = \int_{\Omega_{in}} \mathcal{M}\,(\phi - \phi_{in}) d^3\mathbf{x}, \tag{10.44}$$

[5] From the continuity equation and the definition of a stagnation surface

$$\dot{M}_{out} = \int_{\Omega_{out}} \mathcal{M} d^3\mathbf{x} = \int_{\partial\Omega_{out}} \rho\langle\mathbf{u}, \mathbf{n}\rangle d^2\mathbf{x}, \tag{10.42}$$

and a similar indentity holds for the inflow region Ω_{in} (see Figure 10.2).

so that it should be clear by now how the internal dynamics of stars and the gravitational potential of the galaxy deeply affect the energy balance of the gas, determining its evolution (e.g., see Ciotti and Pellegrini 1996; Ciotti et al. 1991; Negri et al. 2014a,b; Posacki et al. 2013 and references therein). As a nice exercise, the student is invited to derive the balance equation for the case of a global outflow and a global inflow.

10.5 The Tensor Virial Theorem

The Jeans equations are invaluable tools, especially when modeling stellar systems. They also contain important information that is not immediately accessibile without some additional work. In fact, after averaging the CBE over the velocity space, a further (and equally fundamental) step can be taken: as is well known, it is possible to obtain exact relations (actually, an infinite number of tensorial identities) from integration over the configuration space of the Jeans equations after multiplication with suitable functions of the spatial coordinates. The resulting set of identities is the tensor generalization of the Lagrange–Jacobi identity in Eq. (6.10); note that we already obtained an analogous result in the context of the N-body problem in Eq. (6.47). Moreover, by reducing to equilibrium configurations we will obtain the tensorial virial theorem (TVT) a generalization of the scalar virial theorem (SVT), derived in Chapter 6 (e.g., see Binney and Tremaine 2008; Chandrasekhar 1969; Ciotti 2000). In the following, we will limit ourselves to the second-order virial theorem, which is by far the most used in applications. All quantities appearing in the discussion can be time dependent in general, but for simplicity time dependence will not be indicated if it is not strictly required in order to avoid possible ambiguities.

First, for a stellar system described with the Jeans equation and with a density distribution $\rho(\mathbf{x};t)$, we introduce the symmetric *second-order mass tensor*[6]

$$I_{ij} \equiv \int_{\Re^3} \rho x_i x_j d^3\mathbf{x}, \quad I = \mathrm{Tr}(I_{ij}). \tag{10.45}$$

Second, three other manifestly symmetric second-order tensors measure the kinetic energy of the system, namely

$$K_{ij} \equiv \frac{1}{2}\int_{\Re^3} \rho\, \overline{v_i v_j} d^3\mathbf{x}, \quad T_{ij} \equiv \frac{1}{2}\int_{\Re^3} \rho\, \overline{v_i}\, \overline{v_j} d^3\mathbf{x}, \quad \Pi_{ij} \equiv \int_{\Re^3} \rho\, \sigma_{ij}^2 d^3\mathbf{x}, \tag{10.46}$$

and they are usually named as the *total, streaming*, and *velocity dispersion kinetic energy tensors*. Their trace is indicated by

$$K = \mathrm{Tr}(K_{ij}), \quad T = \mathrm{Tr}(T_{ij}), \quad \Pi = \mathrm{Tr}(\Pi_{ij}), \tag{10.47}$$

and so from Eq. (10.5)

$$K_{ij} = T_{ij} + \frac{\Pi_{ij}}{2}, \quad K = T + \frac{\Pi}{2}. \tag{10.48}$$

[6] Other second-order mass tensors that are important in stellar dynamics and that can be expressed as linear functions of I_{ij} have been encountered in Eq. (2.89).

Third, from the general results on the N-body problem illustrated in Chapter 6, we know that in the Lagrange–Jacobi identity and in the SVT, terms related to gravitational energy appear, and so we now look for a tensor generalization of this term. We recall that two scalar functions describe the gravitational energy $U_{\rm T}$ of the density distribution ρ in the total potential $\phi_{\rm T} = \phi + \phi_{\rm ext}$, where ϕ is the gravitational potential produced by the component ρ and $\phi_{\rm ext}$ is the gravitational potential produced by external densities $\rho_{\rm ext}$ (e.g., a central BH, a dark matter halo, a massive gaseous component, and so on). In particular, for a system for which we can set $\phi_{\rm T} = 0$ at infinity, the gravitational *self-energy* is

$$U = \frac{1}{2}\int_{\Re^3} \rho\phi d^3\mathbf{x} = -\frac{G}{2}\int_{\Re^6} \frac{\rho(\mathbf{x};t)\rho(\mathbf{y};t)}{\|\mathbf{x}-\mathbf{y}\|} d^3\mathbf{x} d^3\mathbf{y}, \tag{10.49}$$

while

$$U_{\rm ext} = \int_{\Re^3} \rho\phi_{\rm ext} d^3\mathbf{x} = -G\int_{\Re^6} \frac{\rho(\mathbf{x};t)\rho_{\rm ext}(\mathbf{y};t)}{\|\mathbf{x}-\mathbf{y}\|} d^3\mathbf{x} d^3\mathbf{y} \tag{10.50}$$

is the gravitational *external energy*; note that the factor $1/2$ appears only in the expression of self-gravity for the reasons illustrated in Chapter 6. In Exercise 10.9, a very elegant field expression for U is derived.

Fourth, remarkably (and unintuitively), another second-order tensor (which we call the *interaction energy tensor*) can be associated with the gravitational energy of the density ρ in the total potential $\phi_{\rm T}$ as follows:

$$W_{ij}^{\rm T} = -\int_{\Re^3} \rho x_i \frac{\partial \phi_{\rm T}}{\partial x_j} d^3\mathbf{x} = W_{ij} + W_{ij}^{\rm ext}, \tag{10.51}$$

where the meaning[7] of W_{ij} and $W_{ij}^{\rm ext}$ follows immediately from $\phi_{\rm T} = \phi + \phi_{\rm ext}$. Therefore,

$$W_{\rm T} = {\rm Tr}(W_{ij}^{\rm T}) = -\int_{\Re^3} \rho\langle\mathbf{x}, \nabla\phi_{\rm T}\rangle d^3\mathbf{x} = W + W_{\rm ext}. \tag{10.52}$$

Now, W_{ij} is related to U by a beautiful identity. In fact, in Exercise 10.10 we will prove that

$$W_{ij} = -\frac{G}{2}\int_{\Re^6} \frac{(x_i - y_i)(x_j - y_j)\rho(\mathbf{x};t)\rho(\mathbf{y};t)}{\|\mathbf{x}-\mathbf{y}\|^3} d^3\mathbf{x} d^3\mathbf{y}, \tag{10.53}$$

so that not only is W_{ij} manifestly symmetric (a property not evident from its definition), but also

$$U = {\rm Tr}(W_{ij}) = -\int_{\Re^3} \rho\langle\mathbf{x}, \nabla\phi\rangle d^3\mathbf{x}. \tag{10.54}$$

Again from Exercise 10.10, the student can convince themselves that, at variance with W_{ij}, the tensor $W_{ij}^{\rm ext}$ in general is *not* symmetric.

[7] Note how $W_{ij}^{\rm ext}$ is strictly related to the gravitational torque in Eq. (3.18), with $N_i = -\epsilon_{ijk} W_{jk}^{\rm ext}$.

We can now move to derive the Lagrange–Jacobi identity in Eq. (6.10) (see also Exercises 6.4–6.6) in tensorial form, and from that the second-order tensor and scalar virial theorems for a stellar system possibly interacting with some other mass distribution

$$\frac{1}{2}\frac{d^2 I_{ij}}{dt^2} = 2K_{ij} + W_{ij} + \frac{W_{ij}^{\text{ext}} + W_{ji}^{\text{ext}}}{2}, \quad \frac{1}{2}\frac{d^2 I}{dt^2} = 2K + W + W_{\text{ext}}. \tag{10.55}$$

This is arguably one of the most important results due to its numerous applications in theoretical and observational works in stellar dynamics. As the scalar identity is a trivial consequence of the tensorial identity obtained by its trace, and due to the fact that the associated virial theorems are obtained simply by considering the equilibrium configuration, our attention will be directed toward proving the first identity in Eq. (10.55).

First, notice that from Eqs. (10.14) and (10.45)

$$\frac{dI_{ij}}{dt} = \int_{\Re^3} \frac{\partial \rho}{\partial t} x_i x_j d^3\mathbf{x} = -\int_{\Re^3} \frac{\partial \rho \overline{v_k}}{\partial x_k} x_i x_j d^3\mathbf{x} = \int_{\Re^3} \rho\,(x_i \overline{v_j} + x_j \overline{v_i}) d^3\mathbf{x}, \tag{10.56}$$

where the last identity follows from integration by parts over x_k and assuming the vanishing of the finite term at infinity.[8] We now multiply the momentum in Eq. (10.15) by x_k and we integrate over all of the configuration space: the two terms on the left-hand side become

$$\int_{\Re^3} x_k \frac{\partial \rho \overline{v_i}}{\partial t} d^3\mathbf{x} = \frac{d}{dt}\int_{\Re^3} \rho x_k \overline{v_i}\, d^3\mathbf{x}, \tag{10.57}$$

$$\int_{\Re^3} x_k \frac{\partial \rho \overline{v_i v_j}}{\partial x_j} d^3\mathbf{x} = -\int_{\Re^3} \rho \overline{v_i v_k} d^3\mathbf{x} = -2K_{ik}, \tag{10.58}$$

while on the right-hand side we have W_{ki}^{T}. A comparison of the identity obtained from symmetrization with respect to the indices with the time derivative of Eq. (10.56) finally establishes Eq. (10.55).

In summary, we obtained the Lagrange–Jacobi identity in tensorial form for a stellar system described by the Jeans equations, and these equations in turn have been derived from the CBE (i.e., in a collisionless regime). Thus, it is legitimate to ask whether Eq. (10.55) holds only in these special circumstances. The answer to this question is provided in Exercises 6.4–6.6, where the Lagrange–Jacobi identity in tensorial (and scalar) form is proved for a system made of N point masses moving in their potential and in an external potential ϕ_{ext} by using the Newtonian equations of motion in Eq. (6.1). Therefore, Eq. (10.55) actually holds (with the kinetic energy tensor $K_{ij} = T_{ij}$) *independently* of any assumptions about collisionality.

[8] In Exercise 10.18, we elaborate on this point. In fact, the second-order mass tensor and the inertia tensors diverge for several density distributions commonly used in stellar dynamics, such as those in Eqs. (2.63), (2.83), (2.93), and (5.64), as well as many others that will be encountered in Chapter 13. For these systems, we will discuss the behavior at infinity of the discarded integrals and obtain the needed generalization of the Lagrange–Jacobi identity.

10.5.1 A Note about the Virial Theorem in Multicomponent Systems

Particular attention should be paid to avoiding the incorrect (and unfortunately not uncommon) application of the SVT to multicomponent systems. Let us imagine arbitrarily decomposing a self-gravitating system of density ρ and potential ϕ into two parts with $\rho = \rho_1 + \rho_2$ and $\phi = \phi_1 + \phi_2$; due to the additive nature of kinetic energy and the mass tensor, it follows that $K = K_1 + K_2$ and $I = I_1 + I_2$. From the second identity of Eq. (10.55), the SVT for the total system is

$$\frac{\ddot{I}}{2} = 2K + U, \tag{10.59}$$

where

$$U = \frac{1}{2}\int_{\Re^3} \rho\phi d^3\mathbf{x} = U_1 + U_2 + U_{1,2} = U_1 + U_2 + U_{2,1}, \tag{10.60}$$

and we indicate with U_1 and U_2 the gravitational self-energies of the two components, while

$$U_{1,2} = \int_{\Re^3} \rho_1\phi_2 d^3\mathbf{x}, \quad U_{2,1} = \int_{\Re^3} \rho_2\phi_1 d^3\mathbf{x} \tag{10.61}$$

are their interaction energies. Note that we also have $U_{1,2} = U_{2,1}$ from the *reciprocity theorem*, which is immediately proved from Eq. (10.50). The *correct* application of the SVT to each component is obviously $\ddot{I}_1/2 = 2K_1 + U_1 + W_{1,2}$ and $\ddot{I}_2/2 = 2K_2 + U_2 + W_{2,1}$, where

$$W_{1,2} = -\int_{\Re^3} \rho_1\langle\mathbf{x}, \nabla\phi_2\rangle d^3\mathbf{x}; \quad W_{2,1} = -\int_{\Re^3} \rho_2\langle\mathbf{x}, \nabla\phi_1\rangle d^3\mathbf{x}. \tag{10.62}$$

An *incorrect* application of the SVT would be to naively assume that $\ddot{I}_1/2 = 2K_1 + U_1 + U_{1,2}$ (i.e., that in virialized multicomponent systems twice the value of the kinetic energy of each component sums up to the negative of the gravitational energy of the component), instead of the correct expression $2K_1 = -U_1 - W_{1,2}$. The error becomes apparent if we proceed to add the two incorred relations, thus obtaining $\ddot{I}/2 = 2K_1 + 2K_2 + U_1 + U_2 + U_{1,2} + U_{2,1}$, with *twice* the contribution of the mutual gravitational energy when compared with the correct Eq. (10.60). Obviously, by adding the correct expression we recover the correct tensor virial theorem, and the student can prove this by using the same approach in Exercise 10.10 to show that

$$W_{1,2} + W_{2,1} = U_{2,1} = U_{1,2}. \tag{10.63}$$

10.5.2 The Potential Energy Tensor for Ellipsoidal Systems

We conclude this chapter, giving the expression for the tensor W_{ij} of ellipsoidally stratified density distributions $\rho = \rho_n\tilde{\rho}(m)$ as defined in Eq. (2.28). The resulting formula, albeit limited to this class of models (which certainly do not encompass the variety of structures shown by real stellar systems), has been of great importance for our understanding of elliptical galaxies, particularly for proving that in many cases elliptical galaxies are not properly modeled by classical ellipsoids flattened by rotation (Chandrasekhar 1969), but rather are systems whose shape is also determined by anisotropy of the velocity dispersion tensor (e.g., see Binney 1978; Binney and Tremaine 2008 and references therein). In addition, other important quantitative information can be obtained from W_{ij}, particularly in relation to stability problems (e.g., see Bertin 2014; Chandrasekhar 1969; see also Exercise 13.33).

In fact, it can be proved (see Roberts 1962) that $W_{ij} = 0$ for $i \neq j$, while the diagonal terms can be obtained from the quite remarkable formula

$$W_{ii} = -G\pi^2 a_1 a_2 a_3 a_i^2 w_i \int_0^\infty [\Delta\Psi(m)]^2\, dm, \tag{10.64}$$

where no summing over repeated indices is intended, $\Delta\Psi(m)$ is defined in Eq. (2.22) with $\mathbf{x}_0 = \infty$ and $m(\mathbf{x}, \tau) = m$, and the dimensionless coefficients w_i are given in Eqs. (3.12), (3.13), (3.62), and (3.63). Notice how the *ratio* of the two elements W_{ii}/W_{jj} is *independent* of the specific density profile of the ellipsoid and only depends on the axial ratio! Now, from Eq. (10.53), it is easy to prove (do it!) that for obvious symmetry reasons the off-diagonal terms of the potential energy tensor vanish identically, and in Exercise 10.19 we prove by direct computation that Eq. (10.64) in fact gives the correct result in the special case of the constant-density ellipsoid.

Finally, with the aid of the expansion in Eq. (3.13) for small flattenings and some careful algebra, we can prove (do it!) that at the linear order in the flattenings

$$U = W_{11} + W_{22} + W_{33} \sim -\frac{GM^2}{8a_1}\left(1 + \frac{\epsilon + \eta}{3}\right)\int_0^\infty [\widetilde{\Delta\Psi}(m)]^2 dm, \tag{10.65}$$

where $\widetilde{\Delta\Psi} \equiv \Delta\Psi/\rho_{\rm n}$ is the function associated with $\tilde{\rho}$ in Eq. (2.28). As a useful exercise, the student should prove the result in Eq. (10.65) by direct integration of the formulae obtained in Chapter 2 regarding homeoidal expansion up to the linear terms in the flattenings for the product $\rho\phi$.

Exercises

10.1 In the so-called *extended DF modeling*, stellar systems can be described as a sum of a finite number of different DFs (e.g., distinguished by stellar age, metallicity, kinematical properties), or even as the integral over some parameter of a DF depending on the parameter itself (e.g., see Chapter 8 for some examples). Prove that in a multicomponent system described by a discrete set of DFs f_k, the *total* macroscopic quantity $\overline{F}$ is given by the mass weighted sum of the values of each component, i.e.,

$$\overline{F} = \frac{\sum_k \rho_k \overline{F}_k}{\sum_k \rho_k}. \tag{10.66}$$

Then prove that the total streaming velocity and total velocity dispersion tensor are expressed in terms of the analogous quantities for each component f_k as

$$\overline{v_i} = \frac{\sum_k \rho_k\,(\overline{v_i})_k}{\sum_k \rho_k}, \quad \sigma_{ij}^2 = \frac{\sum_k \rho_k\,(\sigma_{ij}^2)_k}{\sum_k \rho_k} + \frac{\sum_k \rho_k\,[(\overline{v_i})_k - \overline{v_i}]\,[(\overline{v_j})_k - \overline{v_j}]}{\sum_k \rho_k}. \tag{10.67}$$

Expand and further simplify the last term in the expression for σ_{ij}^2; conclude that in multicomponent systems we can have a nonzero velocity dispersion even if the velocity dispersion of each component is zero and explain physically the result. *Hints*: In the expression of σ_{ij}^2, rewrite each term of the sum over the different phase-space components f_k as $(v_i - \overline{v_i})(v_j - \overline{v_j})\,f_k = [v_i - (\overline{v_i})_k + (\overline{v_i})_k - \overline{v_i}][v_j - (\overline{v_j})_k + (\overline{v_j})_k - \overline{v_j}]\,f_k$, expand and perform integration over the velocity space.

10.2 Derive the general Jeans equation in curvilinear orthogonal coordinates and velocities $(\mathbf{q}, \boldsymbol{v})$, analogous to Eq. (10.8). First, consider a macroscopic function $F(\mathbf{q}, \boldsymbol{v}; t)$ and a phase-space DF $f(\mathbf{q}, \boldsymbol{v}; t)$, and from Eq. (10.1) define the quantity $\overline{F}(\mathbf{q}; t)$, recalling that at fixed $\mathbf{q}$, $d^3\mathbf{v} = d^3\boldsymbol{v}$ (see Exercise 9.2). Second, repeat the treatment in Section 10.2 for the CBE in curvilinear coordinates in Eq. (9.20), with $f' = f$. Finally, consider the common case of $F(\boldsymbol{v}; t)$ and show that

$$\frac{\partial \rho \overline{F}}{\partial t} + \sum_{i=1}^{3} \frac{1}{h_i}\frac{\partial \rho \overline{F v_i}}{\partial q_i} - \sum_{i,j,k=1}^{3} \frac{\alpha_{ijk}\rho}{h_j}\overline{\frac{\partial F v_j v_k}{\partial v_i}} = -\sum_{i=1}^{3} \frac{\rho}{h_i}\frac{\partial \phi_{\mathrm{T}}}{\partial q_i}\overline{\frac{\partial F}{\partial v_i}} + \rho\overline{\frac{\partial F}{\partial t}}. \tag{10.68}$$

10.3 Show that for a fluid of density ρ, the identity

$$\frac{d}{dt}\int_{\Omega(t)} \rho\, h\, d^3\mathbf{x} = \int_{\Omega(t)} \left(\rho\frac{Dh}{Dt} + \mathcal{M}\, h\right) d^3\mathbf{x} \tag{10.69}$$

holds, where h is a generic property (per unit mass) of the fluid. *Hint*: First apply Reynold's transport theorem, use Foonote 6 in Chapter 9, and finally use Eq. (10.18).

10.4 With this exercise, we pay due hommage to Archimedes (287–212 BCE). First, from Eq. (10.69), in the absence of sources/sinks, show that for a fluid with density distribution ρ and mass $M = \int_{\Omega(t)} \rho d^3\mathbf{x}$ over some region $\Omega(t)$, the acceleration of the center of mass is given by

$$M\mathbf{A}_{\mathrm{CM}} = \frac{d^2}{dt^2}\int_{\Omega(t)} \rho\, \mathbf{x} d^3\mathbf{x} = \frac{d}{dt}\int_{\Omega(t)} \rho\, \mathbf{u} d^3\mathbf{x} = \int_{\Omega(t)} \rho\frac{D\mathbf{u}}{Dt} d^3\mathbf{x}. \tag{10.70}$$

Then consider Eq. (10.19) in the hydrostatic case and in the absence of sources/sinks, and further restrict yourself to the case of a uniform gravitational field; from the divergence theorem, prove that the total force enacted by surface (pressure) forces on $\partial\Omega$ is the opposite of the weight of M. Finally, suppose that we substitute the fluid in the volume with a body of the same shape and deduce Archimedes' law of hydrostatics.

10.5 Prove the validity of Eqs. (3.14)–(3.16) for a generic fluid system described by Eqs. (10.18) and (10.23) in the absence of source/sink terms. *Hints*: To prove the first identity in Eq. (3.14), assume $\rho = \rho_*$ in Eq. (10.70) and $\Omega(t) = \Re^3$. Then use Eq. (10.23) in the absence of sources/sinks and consider $p_{ij} = 0$ for $\|\mathbf{x}\| \to \infty$. Regarding the volume forces, a symmetrization procedure analogous to that used to prove Eq. (10.53) shows that the integral contribution of the self-force is zero from Newton's Third Law of Dynamics, so that the only remaining contribution to the integral is due to the external field $\mathbf{g} = -\nabla\phi_{\text{ext}}$, and this completes[9] the proof. For Eq. (3.16), again from Eq. (10.69) in the absence of sources/sinks applied to the first identity in Eq. (3.15), one proves immediately the second identity, where

$$\mathbf{N} = \int_{\Re^3} (\mathbf{x} - \mathbf{x}_0) \wedge \frac{D\mathbf{u}}{Dt} \rho d^3\mathbf{x}. \tag{10.71}$$

Again use Eq. (10.23) in the absence of sources/sinks: integration by parts and the Cauchy symmetry of p_{ij} proves that the contribution of surface forces vanishes. Moreover, symmetrization of the integral involving self-forces again shows that the contribution is zero, leaving external forces $\mathbf{g}$ as the only contributors to the torque, completing the proof.

10.6 This exercise is relevant for the comparison of the Jeans equations and the equations of fluid dynamics. Starting from the microscopic function $F = \|\mathbf{v}\|^2/2$, prove that

$$\overline{F} = \frac{\|\overline{\mathbf{v}}\|^2}{2} + \frac{\mathrm{Tr}(\sigma_{ij}^2)}{2}, \quad \frac{\overline{\partial F}}{\partial v_i} = \overline{v_i}, \tag{10.72}$$

$$\overline{Fv_i} = \chi_i + \sigma_{ij}^2 \overline{v_j} + \overline{v_i} \left[\frac{\mathrm{Tr}(\sigma_{ij}^2)}{2} + \frac{\|\overline{\mathbf{v}}\|^2}{2} \right], \quad \chi_i = \frac{\overline{(v_i - \overline{v_i})\|\mathbf{v} - \overline{\mathbf{v}}\|^2}}{2}. \tag{10.73}$$

10.7 By using Eq. (10.30), show that for a perfect gas at equilibrium in a spherical potential $\phi_{\mathrm{T}}(r)$ the temperature and density profiles of the gas are related to the total enclosed mass $M_{\mathrm{T}}(r)$ via Newton's second theorem as

$$M_{\mathrm{T}}(r) = -\frac{k_{\mathrm{B}} T(r) r^2}{\langle \mu \rangle m_{\mathrm{p}}} \left[\frac{d \ln \rho(r)}{dr} + \frac{d \ln T(r)}{dr} \right]. \tag{10.74}$$

10.8 With this exercise, we elaborate on the energy balance of idealized, stationary gas flows in galaxies. Show that in the stationary case Eq. (10.41) can be rewritten as

$$\mathcal{L} = -\mathrm{div} \left[\mathbf{h} + \left(H + \rho\phi + \rho \frac{\|\mathbf{u}\|^2}{2} \right) \mathbf{u} \right] + \mathcal{M} \left(e + \frac{u_{\mathrm{s}}^2}{2} + \phi + \frac{\|\overline{\mathbf{v}}\|^2}{2} + \frac{\mathrm{Tr}\,\sigma^2}{2} \right), \tag{10.75}$$

where $H = E + p = E\gamma_{\mathrm{ad}} = p\gamma_{\mathrm{ad}}/(\gamma_{\mathrm{ad}} - 1)$ is the enthalpy per unit volume of the gas and $\mathbf{h}$ is the heat conduction vector. *Hints*: Use Eqs. (10.18) and (10.40) to obtain the expression for $\langle \mathbf{u}, \nabla p \rangle$.

[9] Note that the same result holds if $\Omega(t)$ is finite *and* $p_{ij} = 0$ over $\partial\Omega(t)$ vanishes.

10.9 Let ρ be a self-gravitating density distribution of total mass M defined over a region Ω bounded by $\partial\Omega$ with normal $\mathbf{n}$ and vanishing outside. By using the Poisson equation $\Delta\phi = 4\pi G\rho$ and the identity (verify it!)

$$\phi\Delta\phi = \mathrm{div}\,(\phi\,\mathrm{grad}\,\phi) - \|\mathrm{grad}\,\phi\|^2, \tag{10.76}$$

prove that Eq. (10.49) can be rewritten as

$$U = -\frac{1}{8\pi G}\int_{\Omega}\|\nabla\phi\|^2 d^3\mathbf{x} + \frac{1}{8\pi G}\int_{\partial\Omega}\phi\langle\nabla\phi, \mathbf{n}\rangle d^2\mathbf{x} \tag{10.77}$$

(see also Jackson 1998). Moreover, if $\partial\Omega$ is an equipotential surface, prove that the surface integral evaluates to $\phi(\partial\Omega)M/2$, and finally that if $\Omega = \Re^3$ and $\phi \to 0$ for $\|\mathbf{x}\| \to \infty$, then the surface integral vanishes.

10.10 Prove that the self-interaction tensor W_{ij} can be written in the manifestly symmetric form of Eq. (10.53). *Hint*: From Eq. (10.51)

$$W_{ij} = -G\int_{\Re^3}\frac{x_i(x_j - y_j)\rho(\mathbf{x};t)\rho(\mathbf{y};t)}{\|\mathbf{x}-\mathbf{y}\|^3}d^3\mathbf{x}d^3\mathbf{y}, \tag{10.78}$$

and the exchange of $\mathbf{x}$ and $\mathbf{y}$ obviously does not affect the value of the integral. By adding the two expressions, we obtain the desired result. Note that the exchange of $\mathbf{x}$ and $\mathbf{y}$ does not affect either W_{ij}^{ext}; however, in general, W_{ij}^{ext} is *not* a symmetric tensor due to the fact that, in general, $\rho \neq \rho_{\mathrm{ext}}$. What happens if $\rho = \alpha\rho_{\mathrm{ext}}$?

10.11 Generalize Eq. (10.53) for a system of total mass M governed by additive $r^{-\alpha}$ forces as in Eq. (1.13) and show that

$$W_\alpha = \begin{cases} (\alpha - 1)U_\alpha, & \alpha \neq 1; \\ -\dfrac{GM^2}{2}, & \alpha = 1, \end{cases} \tag{10.79}$$

where U_α is the self-energy obtained from Eq. (2.126) under the assumption that $\phi(\mathbf{x}_0) = 0$ and $\mathbf{x}_0 = \infty$ for $\alpha > 1$ and $\mathbf{x}_0 = 0$ for $\alpha < 1$. Compare the result with that in Eq. (6.44) for a system of point masses and explain why W_1 in Eq. (10.79) is *not* coincident with W in Eq. (6.42) with $A = -G$, $B = 0$, and $\alpha = 1$. *Hints*: Consider the different role of the self-interaction energy for the point masses in Eq. (6.41) and for the density elements in Eq. (10.53).

10.12 Consider the deep modified Newtonian dynamics limit in Eq. (2.123) for a system of total mass M. Show that

$$W = -\frac{2}{3}\sqrt{Ga_0M^3}, \quad \sigma_{\mathrm{V}}^2 = \frac{2}{3}\sqrt{Ga_0M} \tag{10.80}$$

independently of the shape of the system so that from the Lagrange–Jacobi identity in Eq. (10.55) the changes of the moment of inertia of the system are only associated with changes in the kinetic energy, in analogy to the case of the $\alpha = 1$ force in Eq. (10.79). *Hint*: Consider Eq. (10.52) for the potential ϕ_{M} and express the density distribution in terms of the potential from Eq. (2.123). Following Nipoti et al. (2007), show that the integrand can be rewritten as

$$\mathcal{D}(\phi_{\rm M})\,\mathrm{div}\,(\|\nabla\phi_{\rm M}\|\nabla\phi_{\rm M}) = \mathrm{div}\left[\mathcal{D}(\phi_{\rm M})\|\nabla\phi_{\rm M}\|\nabla\phi_{\rm M} - \frac{\mathbf{x}\|\nabla\phi_{\rm M}\|^3}{3}\right], \quad (10.81)$$

where $\mathcal{D} \equiv \langle \mathbf{x}, \nabla \rangle$. Finally, use the divergence theorem for the integration volume extending to infinity and evaluate the surface integral from Eq. (2.125).

10.13 Consider a razor-thin disk as in Eq. (1.18) with a surface density $\Sigma(R)$ so that the self-gravitational energy U converges. From Eqs. (10.49) and (10.54), prove that

$$U = \pi \int_0^\infty \Sigma(R)\phi(R,0)R\,dR = -2\pi \int_0^\infty \Sigma(R)\frac{\partial \phi(R,0)}{\partial R}R^2\,dR$$
$$= -2\pi \int_0^\infty \Sigma(R)v_{\rm c}^2(R)R\,dR = -2\pi^2 G \int_0^\infty [\hat{\Sigma}(k)]^2\,dk. \quad (10.82)$$

The student is encouraged to prove the second identity without resorting explicitly to the expression of W. *Hint*: The last integral in Eq. (10.82) is immediately established by inserting the first identity in Eq. (2.105) evaluated at $z = 0$ to the first integral, inverting the order of integration, and evaluating the resulting Hankel transform.

10.14 Consider a galaxy with a density distribution $\rho_*(\mathbf{x})$ and total mass M_* hosting at its center a supermassive BH of mass $M_{\rm BH}$ and potential $\phi_{\rm BH}$. We focus on the stellar component so that $\phi_{\rm BH}$ acts as an external potential. Show that, independently of the galaxy shape,

$$W_{\rm BH} = -\int_{\Re^3} \rho_* \langle \mathbf{x}, \nabla\phi_{\rm BH}\rangle d^3\mathbf{x} = \int_{\Re^3} \rho_*\,\phi_{\rm BH} d^3\mathbf{x} = U_{\rm BH}, \quad (10.83)$$

therefore in this very special case $W_{\rm ext} = U_{\rm ext}$.

10.15 Consider again the stellar system of Exercise 10.14, now embedded in an external potential produced by the ellipsoidal generalization of the singular isothermal sphere in Eq. (2.63), which with abuse of language we call a *singular isothermal ellipsoid*

$$\rho_{\rm SIE} = \frac{\rho_0}{q_y q_z m^2}, \quad (10.84)$$

where we follow the convention in Eqs. (2.26)–(2.28). By using Eqs. (2.21) and (2.22), show that, independently of the shape of ρ_*,

$$W_{\rm SIE} = -\int_{\Re^3} \rho_* \langle \mathbf{x}, \nabla\phi_{\rm SIE}\rangle d^3\mathbf{x} = -2\pi G\rho_0 a_1^3 w_0(0) M_* = -v_{\rm c}^2 M_*, \quad (10.85)$$

where $w_0(0)$ is given in Eq. (2.67) and the last identity holds for the axisymmetric case ($q_y = 1$), with $v_{\rm c}$ being the constant circular velocity in the equatorial plane of $\rho_{\rm SIE}$, as obtained from Eqs. (5.64) and (5.65) evaluated for $\gamma = 2$.

10.16 Show that for a generic density distribution ρ_* embedded in an external triaxial logarithmic potential

$$\phi_{\log} = v_c^2 \ln m, \quad W_{\log} = -v_c^2 M_*, \tag{10.86}$$

where v_c^2 is a scale potential (reducing to the constant squared circular velocity in the axisymmetric case) and m is defined as in Eq. (2.8). Convince yourself (by direct calculation or by using the result from Exercise 2.9) that the potential of the singular isothermal ellipsoid is not $\phi_{\log}$, even though in the equatorial plane they coincide in the axisymmetric case (prove it!). What happens at the spherical limit? Conclude that in the last case Eqs. (10.85) and (10.86) concide with $W_{\rm SIS}$ of ρ_* embedded in the external potential of the singular isothermal sphere of rotational velocity v_c.

10.17 Consider a stellar system with density ρ_* and streaming velocity field $\overline{v_i}$. Then consider an arbitrary volume $\Omega(t)$ of the system, transported by the streaming velocity. Finally, let $\partial\Omega(t)$ be the boundary of the region and $\mathbf{n}$ its unit normal. Starting from Eqs. (10.14), (10.16), and (10.45), show that

$$\frac{\ddot{I}}{2} = 2K + W + W_{\rm ext} - \mathcal{P}, \quad \mathcal{P} = {\rm Tr}(\mathcal{P}_{ij}), \tag{10.87}$$

where all of the quantities are intended to be evaluated[10] over $\Omega(t)$, and in particular

$$\mathcal{P}_{ij} \equiv \frac{1}{2} \int_{\partial\Omega(t)} \rho_* \, (\sigma_{ik}^2 x_j + \sigma_{jk}^2 x_i) n_k d^2\mathbf{x}. \tag{10.88}$$

If $\rho_* \sigma_{ij}^2$ is constant on $\partial\Omega$, show that

$$\mathcal{P} = {\rm Volume}[\Omega(t)] \times \rho_* {\rm Tr}(\sigma_{ij}^2)\Big|_{\partial\Omega(t)}. \tag{10.89}$$

Obtain the fomulae for the case of a system with null streaming velocity and a fixed control volume and derive the SVT for a fixed region Ω in a stellar system at equilibrium (see also Exercise 10.18). Notice that the case in Eq. (10.89) happens, for example, in the case of a spherical region Ω centered on the center of a spherical stellar system such as those we will encounter in Chapters 12 and 13 (see Exercise 13.21). An important application of the virial theorem with boundary terms can be found in section 8.2 of Binney and Tremaine (2008). *Hints*: Use Reynold's transport theorem (Appendix A.10) with $\overline{v_i}$ as the velocity field to compute the second-order time derivative of I_{ij} defined over the region $\Omega(t)$ by using Eq. (10.69). From Eqs. (10.14)–(10.16) and the divergence theorem applied to volume integrals of quantities such as $x_i \partial(\rho_* \sigma_{jk}^2)/\partial x_k = \partial(\rho_* \sigma_{jk}^2 x_i)/\partial x_k - \rho_* \sigma_{ij}^2$, obtain Eqs. (10.87) and (10.88), the generalizations of Eq. (10.55). For Eq. (10.89), again use the divergence theorem to show that

[10] Of course, the potential used to compute W also contains the contribution from the part of the system outside $\Omega(t)$. If the system is included and $\partial\Omega(t)$ is the boundary of the stellar system, then $W = U$. See also Exercise 10.18, where the inner and outer contributions are made explicit.

$$\int_{\partial\Omega(t)} x_i n_k d^2\mathbf{x} = \text{Volume}[\Omega(t)]\delta_{ik}. \tag{10.90}$$

10.18 As anticipated in Footnote 7, the validity of the Lagrange–Jacobi identity as a tool to prove the virial theorem can be legitimately questioned in the case of density profiles with a divergent second-order mass tensor. With this exercise, we elaborate on this point. For a given density profile ρ, fix a time-independent closed control volume Ω and define $\rho^{\rm in}$ and $\rho^{\rm out}$ as the densities inside and outside Ω, respectively. Repeat the treatment in Section 10.3 and derive the generalization of Eq. (10.55) for the density $\rho^{\rm in}$. In particular, show that for time-independent systems

$$\frac{1}{2}\int_{\partial\Omega} \rho_* \left(x_i \overline{v_j v_k} + x_j \overline{v_i v_k}\right) d^2\mathbf{x} = 2K_{ij}^{\rm in} + W_{ij}^{\rm in} + \frac{W_{ij}^{\rm out} + W_{ji}^{\rm out}}{2} + \frac{W_{ij}^{\rm ext} + W_{ji}^{\rm ext}}{2}, \tag{10.91}$$

where $W_{ij}^{\rm out}$ and $W_{ij}^{\rm ext}$ are the interaction energy tensors of $\rho^{\rm in}$ due to the potentials $\phi^{\rm out}$ and $\phi^{\rm ext}$, respectively. The standard virial theorem can be recovered case by case, showing the existence of the limit of the various quantities for $\Omega \to \infty$ (see again Exercise 13.21).

10.19 Verify with explicit integration that for the constant-density ellipsoid obtained from Eq. (2.74) with $\rho(m) = \rho_0$, the self-interaction tensor W_{ij} in Eq. (10.64) agrees with the expression in Eq. (10.51). *Hints*: Compute the integral in Eq. (10.51) with the potential $\phi(\mathbf{x})$ at the interior points of the constant-density ellipsoid obtained from Eq. (2.75) and the potential derivatives given by Eq. (2.72) and integrate over the ellipsoidal volume. Evaluate Eq. (10.64) with $\Delta\Psi(m) = 2\int_m^{m_{\rm t}} t\rho(t)dt$ from Eq. (2.75) and compare the results.

10.20 We now use the SVT to estimate the average orbital times $\langle P_{\rm orb}\rangle$ of stars in a self-gravitating ellipsoidal galaxy in the limit of small flattenings. Of course, our definition is necessarily arbitrary (but not unreasonable), and several others analogous definitions can be easily imagined.[11] The point here is to obtain a robust order-of-magnitude estimate based on the virial theorem. First, prove from Eq. (6.26) that

$$\langle P_{\rm orb}\rangle \equiv \frac{4r_{\rm V}}{\sigma_{\rm V}} = \frac{4GM_*^{5/2}}{|U|^{3/2}} = \frac{64\sqrt{2}}{\widetilde{U}^{3/2}}\left(1 - \frac{\epsilon_* + \eta_*}{2}\right)\sqrt{\frac{a_1^3}{GM_*}}, \tag{10.92}$$

where $\widetilde{U} = \int_0^\infty [\widetilde{\Delta\Psi}(m)]^2 dm$; explain physically the dependence of $\langle P_{\rm orb}\rangle$ on the flattenings. Second, evaluate U for the ellipsoidal generalization of the γ-models, the isochrone model, the perfect sphere model, and the Plummer model and in the spherical limit verify the equivalence with the values of U calculated directly for the parent spherical models in Exercises 13.6, 13.7, and 13.17.

[11] For example, the student is invited to compute (and compare) the periods of radial and circular orbits in truncated power-law spherical systems such those in Eq. (13.29) for different values of the density slope γ.

10.21 Use Exercise 3.10 to estimate the libration frequencies of an elliptical galaxy at the center of a nonspherical cluster and compare them with the characteristic stellar orbital times inside the galaxy obtained in Exercise 10.20. For the galaxy, consider a $\gamma = 1$ ellipsoidal model with $\epsilon_* = 0.1$ and $\eta_* = 0.2$ and physical scales $M_* = 10^{11} M_\odot$ and $a_1 = 5$ kpc. For the cluster, consider $\epsilon = 0.1$ and $\eta = 0.2$ in Exercise 3.10 and estimate $\rho(0)$ of the cluster from representative values of $M_{\rm cluster} = 10^{13} M_\odot$ and a scale length of 100 kpc.

10.22 Consider a self-gravitating stellar system of density distribution $\rho_*(\mathbf{x})$, gravitational potential $\phi_*(\mathbf{x})$, and finite total mass M_*. The local *escape velocity* is given by $v_{\rm esc}(\mathbf{x}) = \sqrt{-2\phi_*(\mathbf{x})}$. Show that the mass-averaged escape velocity is related to the virial velocity dispersion of the system by the remarkable identity

$$\langle v_{\rm esc}^2 \rangle \equiv \frac{1}{M_*} \int_{\Re^3} \rho_*(\mathbf{x}) v_{\rm esc}^2(\mathbf{x}) d^3\mathbf{x} = 4\sigma_{\rm V}^2. \tag{10.93}$$

This identity plays a relvant role in the study of the gravitational evaporation of collisional systems such as globular clusters (e.g., see Chapters 6 and 7; see also Bertin 2014; Binney and Tremaine 2008; Chandrasekhar 1942; Spitzer 1987). *Hint*: Use Eq. (10.49).

10.23 Consider a spherical isotropic stellar system of density profile $\rho_*(r)$ and total stellar mass M_*. Determine the expression for the total potential $\phi_{\rm T}(r)$ so that the system is isothermal (i.e., $\sigma_{\rm r}(r) = \sigma_{\rm o}$, an assigned constant value) and show that the associated total density profile is given by

$$\rho_{\rm T}(r) = -\frac{\sigma_0^2}{4\pi G r^2} \frac{d}{dr} \left[r^2 \frac{d \ln \rho_*(r)}{dr} \right], \tag{10.94}$$

and that

$$W_{\rm T} = \frac{3\sigma_0^2 M_*}{G}. \tag{10.95}$$

Finally, assume that an isothermal (perfect) gas distribution ρ of temperature T_0 is at hydrostatic equilibrium in the total potential and show that the density profiles of the gas and stars are related as

$$\rho(r) = A \times [\rho_*(r)]^{\frac{\mu m_{\rm p} \sigma_0^2}{k_{\rm B} T_0}}, \tag{10.96}$$

where A is a constant (e.g., see Cavaliere and Fusco-Femiano 1976). *Hints*: Use the isotropic limit of Eq. (13.82) and the spherical limit of Eq. (10.30).

11
Projected Dynamics

When observed as astronomical objects, stellar systems appear projected on the plane of the sky. As a consequence, it is important to set the framework for a correct understanding of the relation between intrinsic dynamics and projected properties. Unfortunately, while it is always possible (at least in principle) to project a model and then compare the results with observational data, the operation of inversion (i.e., the recovery of three-dimensional information starting from projected properties) is generally impossible due to obvious geometric degeneracies. Spherical and ellipsoidal geometries are among the few exceptional cases that will be discussed in some depth in Chapter 13. Here, instead, the reader is provided with some of the general concepts and tools needed for projecting the most important properties of stellar systems on a projection plane.

11.1 The Projection Operator

Let us consider a given component of a (possibly) multicomponent collisionless stellar system described in an inertial orthogonal reference system $S_0 = (O; \mathbf{e}_1, \mathbf{e}_2, \mathbf{e}_3)$ by its density $\rho = \rho(\mathbf{x}; t)$. In S_0 we introduce the orthogonal (time-independent) observer's system $S' = (O'; \mathbf{f}_1, \mathbf{f}_2, \mathbf{f}_3)$, and we assume that $O \equiv O'$ at all times. Throughout this chapter, the repeated indices sum convention is used, so that $\mathbf{x} = x_i \mathbf{e}_i = x_1\mathbf{e}_1 + x_2\mathbf{e}_2 + x_3\mathbf{e}_3$ is a vector in S_0 and $\boldsymbol{\xi} = \xi_i \mathbf{f}_i$ is the same vector in S'. In S', the observer's direction is that of $\mathbf{f}_3$ (i.e., the observer's "eye" is placed at infinity on the ξ_3-axis of S'): this direction is usually referred to as the *line of sight* (hereafter *los*).

We do not use the standard Euler angles[1] to specify the orientation of S' with respect to S_0, but instead we apply a 3–2–3 rotation (see Figure 11.1). In this way, the first $(0 \leq \varphi < 2\pi)$ and second $(0 \leq \vartheta \leq \pi)$ rotation angles coincide with the polar coordinates of $\mathbf{f}_3$ in S_0, and $\mathbf{e}_3$ and $\mathbf{e}_1$ are the polar and azimuthal axes, respectively. It follows that the positions and velocities in the two systems are related as

$$\mathbf{x} = \mathcal{R}\boldsymbol{\xi}, \quad \mathbf{v} = \mathcal{R}\boldsymbol{\nu}, \quad \mathcal{R} = \mathcal{R}_3(\varphi)\mathcal{R}_2(\vartheta)\mathcal{R}_3(\psi), \tag{11.1}$$

[1] Starting from two coincident systems, the Euler angles are usually defined by a 3–1–3 rotation (i.e., by a rotation around $\mathbf{e}_3$ (φ), then around $\mathbf{e}'_1$ (ϑ), and finally around $\mathbf{e}''_3$ (ψ)). As a result, the angles φ and ϑ are not the common spherical angular coordinates of $\mathbf{e}''_3$ in S_0.

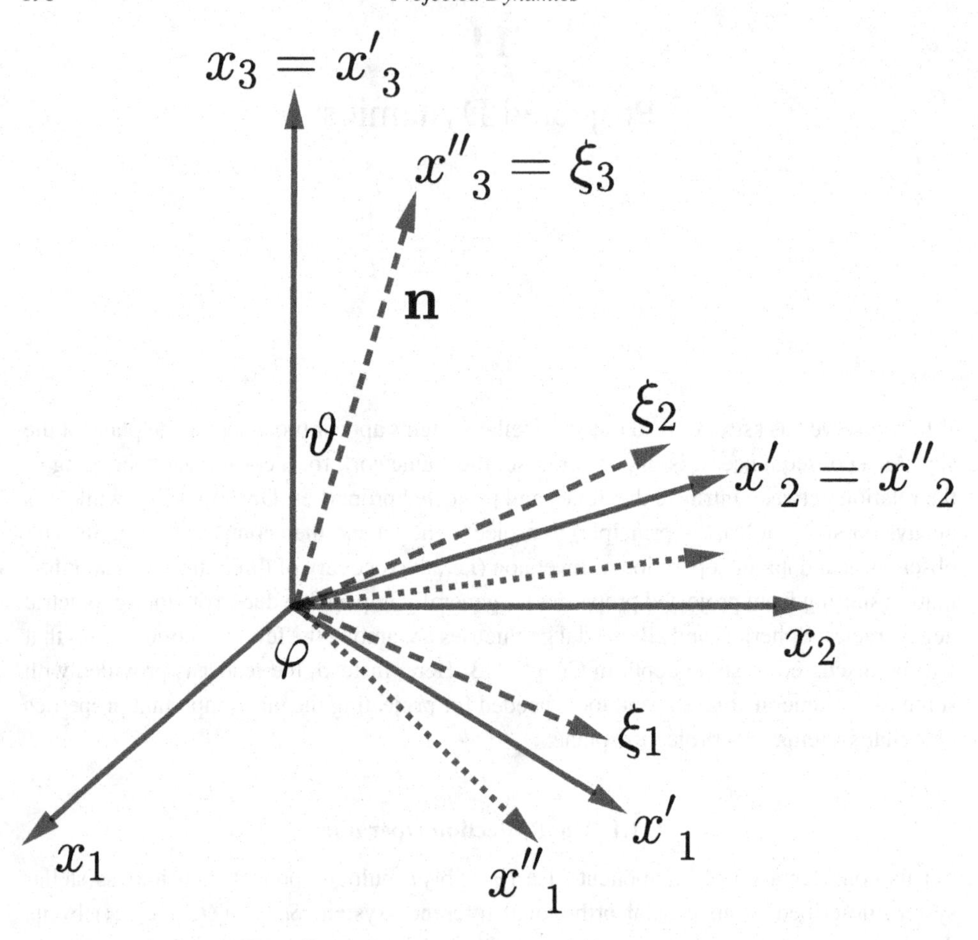

Figure 11.1 The three successive rotations of angle φ around x_3 (with $x_1 \to x_1'$), then of ϑ around x_2' (with $x_1' \to x_1''$), and finally of ψ around x_3'' (with $x_1'' \to \xi_1'$) needed to define the total rotation matrix $\mathcal{R}$ between S_0 and the observer's system S' (see Exercise 11.1).

where $\boldsymbol{v}$ is the velocity as seen from S', and the rotation matrix $\mathcal{R}$ is derived explicitly in Exercise 11.1. As a consequence, the *los* direction in S_0 is given by the unit vector $\mathbf{n}$, with

$$\mathbf{n} = \mathcal{R}\mathbf{f}_3, \quad \mathbf{n} = (\cos\varphi\sin\vartheta,\ \sin\varphi\sin\vartheta,\ \cos\vartheta). \tag{11.2}$$

Notice that from the adopted definition $\mathbf{n}$ points toward the observer. Therefore, for a given position $\mathbf{x}$, the corresponding ξ_3 coordinate is given by $\xi_3 = \langle\mathbf{n}, \mathbf{x}\rangle$. We call the *projection plane* the plane π' containing the origin and perpendicular to $\mathbf{n}$. This is given algebraically by $\xi_3 = 0$, and it can be identified with the set of two-dimensional vectors $\boldsymbol{\xi}_\perp = (\xi_1, \xi_2)$; similarly, we define the velocity vector perpendicular to the *los* as $\boldsymbol{v}_\perp = (v_1, v_2)$. It should be clear that the last rotation of angle ψ around the *los* simply reorients the projection plane; in other words, the projection is intrinsically determined only by the angles φ and ϑ.

As we have seen in Chapter 9, all of the dynamical properties of a stellar system are known if its distribution function $f = f(\mathbf{x}, \mathbf{v}; t)$ is known. In S_0, this function obeys the collisionless Boltzmann equation (9.9), and from it all of the moments[2] can be obtained. In S', the phase-space distribution function (DF) is given by

$$f'(\boldsymbol{\xi}, \boldsymbol{\nu}; t) = f(\mathcal{R}\boldsymbol{\xi}, \mathcal{R}\boldsymbol{\nu}; t). \tag{11.3}$$

Moreover, as has been discussed in Chapter 10, every macroscopic function $\overline{F}(\mathbf{x})$ is defined by integration on the velocity space of its microscopic counterpart $F(\mathbf{x}, \mathbf{v})$, using f as a weight function, and the resulting differential equation for $\overline{F}$ is given in Eq. (10.8). In the observer's reference frame S', the microscopic function F' is given by $F'(\boldsymbol{\xi}, \boldsymbol{\nu}) = F(\mathcal{R}\boldsymbol{\xi}, \mathcal{R}\boldsymbol{\nu})$. The student is invited to prove as a useful exercise that $\rho'\overline{F'} = \int f' F' d^3\boldsymbol{\nu} = \left(\int f F d^3\mathbf{v}\right)' = (\rho\overline{F})' = \rho'\overline{F}'$ by using the identity $d^3\mathbf{v} = d^3\boldsymbol{\nu}$ due to the orthogonality of $\mathcal{R}$. It follows that under our assumptions the macroscopic properties in the observer's system S' can be obtained from the corresponding property in S_0 noticing that (prove it!) $\overline{F'} = \overline{F}'$.

The *mass-weighted projection* of a given property $\overline{F} = \overline{F}(\mathbf{x})$ onto the projection plane π' is naturally defined as

$$\Sigma(\boldsymbol{\xi}_\perp)\overline{F}_{\mathrm{los}}(\boldsymbol{\xi}_\perp) \equiv \int_{-\infty}^{\infty} \rho'(\boldsymbol{\xi})\overline{F}'(\boldsymbol{\xi})d\xi_3, \tag{11.4}$$

where

$$\Sigma(\boldsymbol{\xi}_\perp) \equiv \int_{-\infty}^{\infty} \rho'(\boldsymbol{\xi})d\xi_3 \tag{11.5}$$

is the projected density of the system. It is important to note that the definition of projection above assumes implicitly that the observer is at infinite distance from the system (as all points are projected parallel to $\mathbf{n}$) *and* that the system is "transparent" (i.e., all of the information along $\mathbf{n}$ is summed in the projection plane with no superposition of two stars along the *los*; see Exercise 11.2).

We are now ready to present a simple but important result about projection: that for arbitrary function $\overline{F}$,

$$\int_{\pi'} \Sigma(\boldsymbol{\xi}_\perp)\overline{F}_{\mathrm{los}}(\boldsymbol{\xi}_\perp)d^2\boldsymbol{\xi}_\perp = \int_{\Re^3} \rho(\mathbf{x})\overline{F}(\mathbf{x})d^3\mathbf{x}. \tag{11.6}$$

The proof can be obtained by inserting Eq. (11.4) into Eq. (11.6), changing the variables according to Eq. (11.1), and using the orthogonality of $\mathcal{R}$. In particular, the result means that the mass-weighted surface integral of any projected property equals the (mass-weighted) volume integral of the spatial property, and of course this is independent of the *los* direction. Quite trivially, by assuming $\overline{F} = 1$, one obtains that the surface integral of the projected

[2] For simplicity of notation, in the following discussion, the possible dependence on time of the phase-space averages over f will not be indicated.

density equals the total mass associated with the density distribution ρ *independently* of the particular orientation of the observer:

$$\int_{\pi'} \Sigma(\boldsymbol{\xi}_\perp) d^2\boldsymbol{\xi}_\perp = \int_{\Re^3} \rho(\mathbf{x}) d^3\mathbf{x}. \tag{11.7}$$

Finally, an identical result holds if the mass density ρ is substituted with the *light* density ν when all averages in Eq. (11.6) are intended as luminosity weighted; of course, in the case of a constant mass-to-light ratio Υ_*, where $\rho(\mathbf{x}) = \Upsilon_* \nu(\mathbf{x})$, the mass-weighted and luminosity-weighted averages coincide. In Chapter 13, we will encounter some illustrative cases of projection by considering ellipsoidal (and in particular spherical) systems.

11.2 Projected Velocity Moments

In Chapter 10 it was shown that a special status within the infinite family of microscopic functions F that can be constructed is assumed by the velocity monomials of any order. In fact, these functions enter directly into the virial theorem, and here we take F to be the N-th power of the $\mathbf{v}$ component along $\mathbf{n}$. Therefore, we introduce the concept of a *phase-space velocity field of order N* as

$$v_{\mathrm{p}}^N(\mathbf{v}) \equiv \langle \mathbf{n}, \mathbf{v} \rangle^N = (n_i v_i)^N$$

$$= \sum_{K=0}^{N} \sum_{L=0}^{K} \binom{N}{K} \binom{K}{L} (n_3^{N-K} n_2^{K-L} n_1^L)(v_3^{N-K} v_2^{K-L} v_1^L), \tag{11.8}$$

where $N \geq 0$ is an integer[3] and $\binom{a}{b}$ is the standard binomial coefficient in Eq. (A.48). With $\overline{v_{\mathrm{p}}^N}(\mathbf{x})$ we indicate the mean over the velocitiy space of $v_{\mathrm{p}}^N(\mathbf{v})$, and we call this quantity the *N-th-order projected velocity field* along $\mathbf{n}$ at position $\mathbf{x}$; according to Eq. (11.4), its integration along the *los* is the *N-th-order los projected velocity field* $\overline{v_{\mathrm{p}}^N}_{\mathrm{los}}(\boldsymbol{\xi}_\perp)$. The phase-space average and integration along the *los* lead to similar expressions for $\overline{v_{\mathrm{p}}^N}(\mathbf{x})$ and $\overline{v_{\mathrm{p}}^N}_{\mathrm{los}}(\boldsymbol{\xi}_\perp)$, with the monomials $v_3^{N-K} v_2^{K-L} v_1^L$ substituted by $\overline{(v_3^{N-K} v_2^{K-L} v_1^L)}$ and $\overline{(v_3^{N-K} v_2^{K-L} v_1^L)}_{\mathrm{los}}$, respectively. In Summary, $v_{\mathrm{p}}^N(\mathbf{v})$, its average over the velocity space $\overline{v_{\mathrm{p}}^N}(\mathbf{x})$, and finally its *los* projection $\overline{v_{\mathrm{p}}^N}_{\mathrm{los}}(\boldsymbol{\xi}_\perp)$ can all be expressed as linear combinations of the corresponding velocity monomials of order N in S_0, with coefficients depending on the *los* direction, A function of great importance in observational and theoretical works is obtained for $N = 1$

$$\overline{v_{\mathrm{p}}}(\mathbf{x}) = \overline{\langle \mathbf{n}, \mathbf{v} \rangle} = n_i \overline{v_i}. \tag{11.9}$$

$\overline{v_{\mathrm{p}}}(\mathbf{x})$ is called the *projected streaming velocity field* along $\mathbf{n}$ at position $\mathbf{x}$. By construction, a positive $\overline{v_{\mathrm{p}}}$ means that the component of the projected streaming velocity along $\mathbf{n}$ points

[3] For $N = 1$, $v_{\mathrm{p}} = \langle \mathbf{n}, \mathbf{v} \rangle = \langle \mathcal{R}\mathbf{f}_3, \mathcal{R}\boldsymbol{\nu} \rangle = \nu_3$ from the orthogonality of $\mathcal{R}$.

toward the observer and the integration along the *los* of $\overline{v_{\rm p}}$ represent the *los projected streaming velocity field* $\overline{v_{\rm p}}_{\rm los}(\boldsymbol{\xi}_\perp)$. Note that if the system does not have intrinsic streaming motions (i.e., $\overline{v_i} = 0$ for $i = 1,2,3$), then $\overline{v_{\rm p}}_{\rm los} = \overline{v_{\rm p}} = 0$, but the converse is not true: for a given orientation of the *los*, the quantities $\overline{v_{\rm p}}$ and $\overline{v_{\rm p}}_{\rm los}$ may vanish, but the system can still possess streaming motions. Moreover $\overline{v_{\rm p}}_{\rm los}$ can vanish even if $\overline{v_{\rm p}} \neq 0$. The student is encouraged to imagine some illustrative cases of these possibilities.

Another important family of functions is naturally introduced: the *phase-space velocity dispersion field of order N*

$$\sigma_{\rm p}^N(\mathbf{v}) \equiv (v_{\rm p} - \overline{v_{\rm p}})^N = \sum_{K=0}^{N} \binom{N}{K} (-1)^K v_{\rm p}^{N-K}\, \overline{v_{\rm p}}^K. \tag{11.10}$$

With $\overline{\sigma_{\rm p}^N}(\mathbf{x})$ we indicate its mean over the velocity space, and we call this quantity the *N-th-order projected velocity dispersion field* along $\mathbf{n}$ at position $\mathbf{x}$; its integration along the *los* is the *N-th-order los projected velocity dispersion field* $\overline{\sigma_{\rm p}^N}_{\rm los}(\boldsymbol{\xi}_\perp)$. In practice, $\sigma_{\rm p}^N$ "measures" the dispersion of stellar motions at point $\mathbf{x}$ in the direction $\mathbf{n}$ with respect to the local average value of $\overline{v_{\rm p}}$ along the same direction, and $\overline{\sigma_{\rm p}^N}_{\rm los}$ "measures" the mass-weighted (or light-weighted) integral of these dispersions along the *los*. Similar formulae for $\overline{\sigma_{\rm p}^N}(\mathbf{x})$ and $\overline{\sigma_{\rm p}^N}_{\rm los}(\boldsymbol{\xi}_\perp)$ are obtained from Eq. (11.10) where the terms $v_{\rm p}^{N-K}\,\overline{v_{\rm p}}^K$ are replaced by $\overline{v_{\rm p}^{N-K}}\;\overline{v_{\rm p}}^K$ and $(\overline{v_{\rm p}^{N-K}\;\overline{v_{\rm p}}^K})_{\rm los}$, respectively. Of course, if $\overline{v_{\rm p}} = 0$, then $\sigma_{\rm p}^N = v_{\rm p}^N$ and so $\overline{\sigma_{\rm p}^N}_{\rm los} = \overline{v_{\rm p}^N}_{\rm los}$. Moreover, note that for $N = 1$, necessarily $\overline{\sigma_{\rm p}} = \overline{\sigma_{\rm p}}_{\rm los} = 0$. In this family, an extremely important function is obtained for $N = 2$, with

$$\overline{\sigma_{\rm p}^2}(\mathbf{x}) = n_i n_j \sigma_{ij}^2, \tag{11.11}$$

where σ_{ij}^2 is the velocity dispersion tensor introduced in Eq. (10.4); its projection is $\overline{\sigma^2_{\rm p}}_{\rm los}(\boldsymbol{\xi}_\perp)$. It should be appreciated that in general $\overline{\sigma_{\rm p}^N}_{\rm los}$ is *not* the N-th "observed" *los* velocity dispersion field. In fact, the observations give the *los* velocity dispersion centered on the *los* streaming velocity field $\overline{v_{\rm p}}_{\rm los}(\boldsymbol{\xi}_\perp)$, and not on the local $\overline{v_{\rm p}}(\mathbf{x})$ as in Eq. (11.10); nonetheless, as we will see, $\overline{\sigma_{\rm p}^N}_{\rm los}(\boldsymbol{\xi}_\perp)$ is an important projected field.

These considerations lead us to introduce a third and last function over the phase space: the *phase-space N-th-order local los velocity dispersion field*

$$\sigma_{\rm l}^N(\mathbf{v}) \equiv (v_{\rm p} - \overline{v_{\rm p}}_{\rm los})^N = \sum_{K=0}^{N} \binom{N}{K} (-1)^K v_{\rm p}^{N-K}\, (\overline{v_{\rm p}}_{\rm los})^K. \tag{11.12}$$

As with the previous quantities, with $\overline{\sigma_{\rm l}^N}(\mathbf{x})$ we indicate the mean over the velocity space of $\sigma_{\rm l}^N(\mathbf{v})$ at $\mathbf{x}$, and finally the *N-th los velocity dispersion field* is given by $\overline{\sigma_{\rm l}^N}_{\rm los}(\boldsymbol{\xi}_\perp)$. Explicit expressions for $\overline{\sigma_{\rm l}^N}(\mathbf{x})$ and $\overline{\sigma_{\rm l}^N}_{\rm los}(\boldsymbol{\xi}_\perp)$ can be obtained by substitution into Eq. (11.12) of the corresponding quantities $\overline{v_{\rm p}^{N-K}}$ and $(\overline{v_{\rm p}^{N-K}})_{\rm los}$, respectively; of course, for $N = 1$, $\overline{\sigma_{\rm l}}_{\rm los} = 0$ always. Note that if the system is characterized by $\overline{v_{\rm p}}_{\rm los} = 0$ for some $\boldsymbol{\xi}_\perp$, then

$\sigma_1^N = v_{\rm p}^N$ and $\overline{\sigma_1^N}_{\rm los} = \overline{v_{\rm p}^N}_{\rm los}$. Moreover, if $\overline{v_{\rm p}} = 0$ all along a given *los* through $\boldsymbol{\xi}_\perp$, not only do the previous simplifications hold, but in also $\sigma_1^N = \sigma_{\rm p}^N$ and $\overline{\sigma_1^N}_{\rm los} = \overline{\sigma_{\rm p}^N}_{\rm los}$. In particular, the last identity holds everywhere in the projection plane independently of the direction **n** for stellar systems with a vanishing streaming velocity field.

Before proceeding with the discussion, a comment is in order. As we have seen, the fields $\overline{\sigma_{\rm p}^N}$ and $\overline{\sigma_1^N}$ (and the associated *los* fields) can be expressed as functions of the fields $\overline{v_{\rm p}^K}$ (and their *los* projections). Is the opposite true? This is a very important question because from an observational point of view we can assume that we have access to the members of the family $\overline{\sigma_1^N}_{\rm los}$ only, together with $\overline{v_{\rm p}^N}_{\rm los}$. The following results provide the answer to this question. In fact, for $N \geq 0$, we can recast Eq. (11.8) as

$$v_{\rm p}^N(\mathbf{v}) = \sum_{K=0}^{N} \binom{N}{K} \overline{v_{\rm p}}^{N-K} \sigma_{\rm p}^K = \sum_{K=0}^{N} \binom{N}{K} (\overline{v_{\rm p}}_{\rm los})^{N-K} \sigma_1^K, \tag{11.13}$$

and the associated expressions for $\overline{v_{\rm p}^N}$ and $\overline{v_{\rm p}^N}_{\rm los}$ are obtained by considering velocity averages and projections. Similarly, Eq. (11.10) can be rewritten as

$$\sigma_{\rm p}^N(\mathbf{v}) = \sum_{K=0}^{N} \binom{N}{K} (-1)^K \sigma_1^{N-K}\, \overline{\sigma_1^1}^K, \tag{11.14}$$

and the expressions for $\overline{\sigma_{\rm p}^N}$ and $\overline{\sigma_{\rm p}^N}_{\rm los}$ are obtained by inserting in Eq. (11.14) the terms $\overline{\sigma_1^{N-K}}\, \overline{\sigma_1^1}^K$ and $(\overline{\sigma_1^{N-K}}\, \overline{\sigma_1^1}^K)_{\rm los}$, respectively (see Exercise 11.3).

In the vast majority of theoretical and observational works, the second-order moments play a special role. From the phase-space average of Eq. (11.13) with $N = 2$,

$$\overline{v_{\rm p}^2} = \overline{\sigma_{\rm p}^2} + \overline{v_{\rm p}}^2, \tag{11.15}$$

because trivially $\overline{\sigma_{\rm p}^1} = 0$, and from the phase-space and *los* projection of Eq. (11.12) with $N = 2$,

$$\overline{\sigma_1^2}_{\rm los} = \overline{v_{\rm p}^2}_{\rm los} - (\overline{v_{\rm p}}_{\rm los})^2 = \overline{\sigma_{\rm p}^2}_{\rm los} + \overline{v_{\rm p}}^2_{\rm los} - (\overline{v_{\rm p}}_{\rm los})^2, \tag{11.16}$$

where the second identity follows from the first and from the *los* projection of Eq. (11.15). Finally, notice how from Eq. (11.16) one can prove the simple but important identity $\overline{\sigma_1^2}_{\rm los} + (\overline{v_{\rm p}}_{\rm los})^2 = \overline{\sigma_{\rm p}^2}_{\rm los} + \overline{v_{\rm p}}^2_{\rm los}$; sometimes (e.g., Cappellari 2008) this latter quantity is indicated in the literature as $V_{\rm rms}^2$.

Two important conclusions relevant to obervational works can be derived from Eq. (11.16). The first is that $\overline{\sigma_1^2}_{\rm los}$ can be greater than zero even in an idealized stellar system with no intrinsic velocity dispersion (and so $\overline{\sigma_{\rm p}^2}_{\rm los} = 0$), as for example in the edge-on view of a "cold" stellar disk where all stars circulate around the disk's center in the same sense. The second is that in fully pressure-supported stellar systems (i.e., $\overline{v_i} = 0$ everywhere), $\overline{\sigma_1^2}_{\rm los} = \overline{v_{\rm p}^2}_{\rm los} = \overline{\sigma_{\rm p}^2}_{\rm los}$. In Chapter 13, we will consider in detail the construction of the first- and second-order projected velocity fields for the most simple cases routinely encountered in stellar dynamics.

11.3 Velocity and Line Profiles

A natural question with obvious implications for theoretical and observational works arises: How are the fields introduced in the previous section related to the distribution function f' referred to in the observer's frame of reference S'? We start by introducing the concept of a *velocity profile* (VP), a function defined at the generic point $\boldsymbol{\xi}_\perp$ in the projection plane as

$$\Sigma(\boldsymbol{\xi}_\perp)\mathrm{VP}(\boldsymbol{\xi}_\perp, v_\mathrm{p}) \equiv \int_{\Re^3} f'(\boldsymbol{\xi}, \boldsymbol{v}; t)\, d^2\boldsymbol{v}_\perp d\xi_3, \tag{11.17}$$

where we use the fact that $v_\mathrm{p} = v_3$ (see Footnote 3). In practice, the VP gives the distribution of the velocity component v_p of all stars along the *los*, obtained by integrating the DF over all of the velocities $\boldsymbol{v}_\perp \perp \mathbf{n}$ and then summing along ξ_3. By definition, the physical units of the VP are those of an inverse velocity; moreover, $\int \mathrm{VP}\, dv_\mathrm{p} = \left(\int \rho'\, d\xi_3\right)/\Sigma = 1$ (i.e., by construction, the integral of the VP over $-\infty < v_\mathrm{p} < \infty$ is normalized to unity (prove it!)), and of course in stellar systems the VP vanishes for sufficiently high values of $|v_\mathrm{p}|$.

It is a simple but useful exercise to show that two important projected fields that we discussed in the previous section can in fact be obtained from the VP, i.e.,

$$\int_{-\infty}^{\infty} v_\mathrm{p}\, \mathrm{VP}(\boldsymbol{\xi}_\perp, v_\mathrm{p}) dv_\mathrm{p} = \overline{v_\mathrm{p}}_{\mathrm{los}}(\boldsymbol{\xi}_\perp), \tag{11.18}$$

and

$$\int_{-\infty}^{\infty} (v_\mathrm{p} - \overline{v_\mathrm{p}}_{\mathrm{los}})^N\, \mathrm{VP}(\boldsymbol{\xi}_\perp, v_\mathrm{p}) dv_\mathrm{p} = \overline{\sigma_1^N}_{\mathrm{los}}(\boldsymbol{\xi}_\perp). \tag{11.19}$$

The proof (do it!) simply consists of substituting the definition of the VP into Eqs. (11.18) and (11.19) and performing the resulting (formal) integrations.

We now move another step toward the observational world by linking the VP to another extremely important quantity associated with spectroscopic observations of galaxies when the spectra of single (in general unresolved) stars are "summed" (luminosity weighted) along the *los*. This function is the so-called *line profile* (LP; i.e., the function of the wavelength λ describing the shape of the cumulative spectrum obtained by the superposition of the red-shifted and blue-shifted lines of all of the stars along a given *los* direction $\mathbf{n}$ and at position $\boldsymbol{\xi}_\perp$ in the projection plane). Of course, even a partial treatment of the technical aspects of the problem is well beyond the introductory nature of this book (e.g., see Binney and Tremaine 2008), yet it is important to present the basic ideas of the subject.

Let us assume for simplicity that the stellar population of our system is homogeneous with a position-independent *mass-to-light* ratio Υ_*, and that, when observed at rest, the spectral line of interest produced by a given star is described by the normalized function of wavelength λ $P_0 = P_0(\lambda)$, with $\int_0^\infty P_0(\lambda) d\lambda = 1$. The physical dimensions of P_0 are therefore the inverse of the length. In the resulting LP, each star contributes the profile P_0, red- or blue-shifted according to the sign of $v_{\rm p}$. As a result of the assumed orientation of S', a negative $v_{\rm p}$ (a star receding from the observer along the *los* $\mathbf{n}$) corresponds to a red-shift, and so for $|v_{\rm p}| \ll c$ (where c is the speed of light), a given λ in the spectrum of the star is shifted in the observer's system to $\lambda' = \lambda(1 - v_{\rm p}/c)$. It follows immediately (prove it!) that the normalized shifted spectral LP as seen in S' is

$$P_v(\lambda) = P_0\left(\frac{\lambda}{1 - v_{\rm p}/c}\right)\frac{1}{1 - v_{\rm p}/c}. \tag{11.20}$$

Therefore, by summing the spectral contributions of all of the stars along the *los*, the resulting LP is

$$\mathrm{LP}(\boldsymbol{\xi}_\perp, \lambda) \equiv \int_{-\infty}^{\infty} P_0\left(\frac{\lambda}{1 - v_{\rm p}/c}\right)\frac{\mathrm{VP}(\boldsymbol{\xi}_\perp, v_{\rm p})}{1 - v_{\rm p}/c}dv_{\rm p}, \tag{11.21}$$

where of course no problems may arise from the formal integration over velocities due to the vanishing of the VP at high velocities. Note that from the normalization of the VP in Eq. (11.20) and of P_0, the LP is also normalized to unity, with $\int_0^\infty \mathrm{LP}\, d\lambda = 1$. The simplest and idealized case of the LP is given by a perfectly monochromatic line, when P_0 is simply described by a Dirac δ-function (see Exercise 11.4).

Now let $\lambda_0 = \int_0^\infty P_0(\lambda)\lambda d\lambda$ be the characteristic wavelength (in the laboratory rest frame) associated with the P_0 profile and let $\sigma_0^2 \equiv \int_0^\infty P_0(\lambda)(\lambda - \lambda_0)^2 d\lambda$ be the measure of the line width. The same quantities can be associated with the "observed" LP, and in fact at every place in the projection plane (see Exercise 11.5)

$$\lambda_v \equiv \int_0^\infty \mathrm{LP}(\lambda)\lambda\, d\lambda = \left(1 - \frac{\overline{v_{\rm p}}_{\rm los}}{c}\right)\lambda_0, \tag{11.22}$$

$$\sigma_v^2 \equiv \int_0^\infty \mathrm{LP}(\lambda)(\lambda - \lambda_v)^2 d\lambda = \left(1 - \frac{\overline{v_{\rm p}}_{\rm los}}{c}\right)^2\sigma_0^2 + \frac{\overline{\sigma^2_{1\,\rm los}}}{c^2}(\lambda_0^2 + \sigma_0^2). \tag{11.23}$$

The identities established here show how from the knowledge of λ_0 and σ_0 of the spectral line of interest and of the observed values of λ_v and σ_v, it is possible to measure $\overline{v_{\rm p}}_{\rm los}$ and $\overline{\sigma^2_{1\,\rm los}}$ at each point in the projection plane! Of course, increasingly complicated relations between higher-order moments of the LPs and higher-order projected kinematical fields can be established following the same approach as above by considering quantities similar to those in Eq. (11.23) evaluated for $(\lambda - \lambda_v)^N$. In summary, the LPs are natural tools that connect theory to observations, and it is quite obvious that a substantial amount of theoretical work has been done to develop methods that are able to extract information from the observed LPs, with important results concerning the possibility of measuring orbital anisotropy or the amount of dark matter in a galaxy. We cannot enter into more details here, nor can we give even a partial list of the relevant references, but the student is encouraged to explore and study the literature concerning this fascinating subject (e.g., see Carollo et al. 1995a,b; Ciotti and Lanzoni 1997; Dehnen and Gerhard 1993; Gerhard 1993; Magorrian and Binney 1994; van der Marel 1994; van der Marel and Franx 1993; van der Marel et al. 1994 and references therein).

11.4 The Projected Virial Theorem

In the previous section, we showed how in principle it is possible to measure, at each point $\boldsymbol{\xi}_\perp$ in the projection plane, the values of the projected kinematics. In other words, it is possible to recover the two-dimensional maps of the relevant projected kinematical fields of a stellar system. Today, thanks to sophisticated observational techniques, this field of research is undergoing rapid evolution with fundamental improvements to our knowledge of the internal dynamics of galaxies (e.g., see Cappellari 2016 and references therein for an excellent review). Here, limiting ourselves to a very elemntary discussion, we simply describe what kind of information is contained in the projected fields, and we focus in particular on the N-th-order *projected virial theorem* (PVT; e.g., see Ciotti 1994, 2000; Kent 1990; Merrifield and Kent 1990 and references therein).

First, we obtain the moment equation for the quantity $\overline{v_{\rm p}^N}$: we specialize Eq. (10.8) to the case $F(\mathbf{v}) = v_{\rm p}^N(\mathbf{v})$ in Eq. (11.8) and we obtain immediately (prove it!)

$$\frac{\partial \rho \overline{v_{\rm p}^N}}{\partial t} + \frac{\partial \rho \overline{v_{\rm p}^N v_i}}{\partial x_i} = -N n_i \rho \overline{v_{\rm p}^{N-1}} \frac{\partial \phi_{\rm T}}{\partial x_i}. \tag{11.24}$$

Then, we generalize the definition of total kinetic energy by introducing the *los N-th-order total kinetic energy*

$$K_{\rm los}^{(N)} \equiv \frac{1}{N} \int_{\pi'} \Sigma \, \overline{v_{\rm p}^N}_{\rm los} \, d^2\boldsymbol{\xi}_\perp = \frac{1}{N} \int_{\Re^3} \rho \, \overline{v_{\rm p}^N} \, d^3\mathbf{x}, \tag{11.25}$$

where the last identity follows immediately from Eq. (11.6). In practice, $K_{\rm los}^{(N)}$ is the N-th-order generalization of the "kinetic energy" that would be measured by considering the N-th-order *los* velocity field in the projection plane, integrated over all of the "image" and weighted with the projected density. The two simplest cases are obtained for $N = 1, 2$

$$K_{\rm los}^{(1)} = n_i P_i, \quad K_{\rm los}^{(2)} = n_i n_j K_{ij}, \tag{11.26}$$

where P_i is the i-th component of the linear momentum associated with ρ, so that in the barycentric system it would be zero.

Following a similar approach to that described in Section 10.5, after multiplication of Eq. (11.24) for $n_j x_j = \langle \mathbf{n}, \mathbf{x} \rangle$, integration over the configuration space, and recognizing that $\overline{v_{\rm p}^N v_i n_i} = \overline{v_{\rm p}^{N+1}}$, we finally obtain (do it!) the N-th-order PVT

$$\frac{dI_{\rm los}^{(N)}}{dt} = (N+1) K_{\rm los}^{(N+1)} + N W_{\rm T,los}^{(N-1)}, \tag{11.27}$$

where

$$W_{\rm T,los}^{(N-1)} \equiv -n_i n_j \int_{\Re^3} \rho \overline{v_{\rm p}^{N-1}} x_j \frac{\partial \phi_{\rm T}}{\partial x_i} \, d^3\mathbf{x} = n_i n_j W_{{\rm T},ij}^{(N-1)}, \tag{11.28}$$

and

$$I_{\rm los}^{(N)} \equiv \int_{\Re^3} \langle \mathbf{n}, \mathbf{x} \rangle \rho \overline{v_{\rm p}^N} d^3\mathbf{x}. \tag{11.29}$$

In particular, from Eq. (10.56)

$$I_{\rm los}^{(1)} = \frac{1}{2} \int_{\Re^3} \rho n_i n_j (x_i \overline{v_j} + x_j \overline{v_i}) d^3\mathbf{x} = \frac{\dot{I}_{ij} n_i n_j}{2}. \tag{11.30}$$

The projection of the second-order virial theorem is then immediately obtained by setting $N = 1$ in Eq. (11.27)

$$\frac{n_i n_j \ddot{I}_{ij}}{2} = 2K_{\rm los}^{(2)} + n_i n_j \left(W_{ij} + \frac{W_{ij}^{\rm ext} + W_{ji}^{\rm ext}}{2} \right). \tag{11.31}$$

Of course, this identity is not suprising, as it is simply the double contraction of the tensorial identity in Eq. (10.55) over the *los* direction $\mathbf{n}$!

11.4.1 The Angle-Averaged PVT

As is apparent from Eq. (11.31), when applying the PVT to a system with no special symmetries, the direction of the observer *los* must be specified, but of course we cannot take for granted that in any given practical application we know the shape of the stellar system or its relative inclination $\mathbf{n}$ with respect to the observer's *los*. For this reason, even if the idea may appear initially strange, it is interesting to work out the formalism needed to compute angle-averaged analogs of the formulae derived in the previous section and to explore some consequences of the angle-averaged Jeans equations and of the PVT. Among other results, we will obtain as a bonus some exact relations holding for spherical systems. In fact, in the highly idealized (and thus frequently used) situation of spherical symmetry, the PVT cannot depend on the observation direction, and so in spherically symmetric systems the angle-averaged results simply coincide with the intrinsic identities.

Let us start by considering a fictitious "observer" that can move all around a stellar system observing some projected property $\overline{F}_{\rm los}$. In general, this property will depend on the *los* direction $\mathbf{n}$. The *angular mean* of $\overline{F}_{\rm los}$ over the solid angle, determined by our observer looking at the system from all possible directions, is naturally defined as

$$[\overline{F}_{\rm los}]_\Omega \equiv \frac{1}{4\pi} \int_{4\pi} \overline{F}_{\rm los} d^2\mathbf{\Omega} = \frac{1}{4\pi} \int_0^\pi \int_0^{2\pi} \overline{F}_{\rm los} \sin\vartheta \, d\vartheta d\varphi. \tag{11.32}$$

The last expression is determined by our choice of the orientation of S' and the two angles are the spherical coordinates angles used in Eq. (11.2). Therefore, if we consider the angular average of the PVT in Eq. (11.27), we can write

$$\frac{d[I_{\rm los}^{(N)}]_\Omega}{dt} = (N+1)[K_{\rm los}^{(N+1)}]_\Omega + N[W_{\rm T,los}^{(N-1)}]_\Omega. \tag{11.33}$$

Our task is now to obtain the expressions for the quantities involved. This can be done most easily by first showing that (do it!) at each position **x** in the system

$$\overline{[v_{\rm p}^N]_\Omega} = [\overline{v_{\rm p}^N}]_\Omega, \tag{11.34}$$

$$\overline{[n_i v_{\rm p}^N]_\Omega} = \frac{1}{N+1}\,\overline{\frac{\partial [v_{\rm p}^{N+1}]_\Omega}{\partial v_i}}, \tag{11.35}$$

$$\overline{[n_i n_j v_{\rm p}^{N-1}]_\Omega} = \frac{1}{N(N+1)}\,\overline{\frac{\partial^2 [v_{\rm p}^{N+1}]_\Omega}{\partial v_i \partial v_j}}. \tag{11.36}$$

The identities above are established with little effort by first recognizing that averages over the velocity space and angular means commute and then by using Eq. (11.8). The first identity shows the obvious fact that velocity averages and angular means commute, and the second and third identities show that angular means of special combinations of the components of **n** and $\overline{v_{\rm p}^N}$ can be computed as velocity averages of velocity derivatives of the angular means of the microscopic functions $v_{\rm p}^N(\mathbf{v})$. We will return to this point later on; for the moment, with the aid of Eqs. (11.34)–(11.36), it is a simple exercise to show that the quantities appearing in Eq. (11.33) can be written (do it!) as

$$[I_{\rm los}^{(N)}]_\Omega = \frac{1}{N+1}\int_{\Re^3} \rho x_i\, \overline{\frac{\partial [v_{\rm p}^{N+1}]_\Omega}{\partial v_i}}\, d^3\mathbf{x}, \tag{11.37}$$

$$[W_{\rm T,los}^{(N-1)}]_\Omega = -\frac{1}{N(N+1)}\int_{\Re^3} \rho x_j \frac{\partial \phi_{\rm T}}{\partial x_i}\, \overline{\frac{\partial^2 [v_{\rm p}^{N+1}]_\Omega}{\partial v_i \partial v_j}}\, d^3\mathbf{x}, \tag{11.38}$$

$$[K_{\rm los}^{(N+1)}]_\Omega = \frac{1}{N+1}\int_{\Re^3} \rho\, \overline{[v_{\rm p}^{N+1}]_\Omega}\, d^3\mathbf{x}. \tag{11.39}$$

The problem is therefore reduced to the calculation of $[v_{\rm p}^N]_\Omega$ for a generic N, and in turn from Eq. (11.8) this requires the evaluation of terms of the form $\binom{N}{K}\binom{K}{L}[n_3^{N-K} n_2^{K-L} n_1^L]_\Omega$. The algebra involved is quite boring but not difficult, and in Exercise 11.7 the curious student will find the steps needed to show that for $N \geq K \geq L \geq 0$

$$\binom{N}{K}\binom{K}{L}[n_3^{N-K} n_2^{K-L} n_1^L]_\Omega = \frac{[1+(-1)^{N-K}][1+(-1)^K][1+(-1)^L]}{8(N+1)} \times \frac{\Gamma\left(\frac{N+2}{2}\right)}{\Gamma\left(\frac{N-K+2}{2}\right)\Gamma\left(\frac{K-L+2}{2}\right)\Gamma\left(\frac{L+2}{2}\right)}. \tag{11.40}$$

From the general expression in Eq. (11.40), it follows that for any odd natural number $N = 2n + 1$

$$[v_{\mathrm{p}}^{2n+1}]_\Omega = 0. \tag{11.41}$$

This is because in Eq. (11.40) only terms with even K and L can be nonzero, but then $N - K$ is odd, and the result is proved. For $N = 2n$ instead

$$[v_{\mathrm{p}}^{2n}]_\Omega = \frac{1}{2n+1}\sum_{k=0}^{n}\sum_{l=0}^{k}\binom{n}{k}\binom{k}{l}\overline{(v_3^{n-k}v_2^{k-l}v_1^{l})^2}, \tag{11.42}$$

as can be proved (do it!) by setting $N = 2n$, $K = 2k$, and $L = 2l$ in Eq. (11.40).

In particular, for $n = 1$, $[v_{\mathrm{p}}^2]_\Omega = (\overline{v_1^2} + \overline{v_2^2} + \overline{v_3^2})/3$, and so

$$[\overline{v_{\mathrm{p}}^2}]_\Omega = \frac{\overline{v_1}^2 + \overline{v_2}^2 + \overline{v_3}^2}{3} + \frac{\sigma_{11}^2 + \sigma_{22}^2 + \sigma_{33}^2}{3}, \tag{11.43}$$

and from Eq. (11.39)

$$[K_{\mathrm{los}}^{(2)}]_\Omega = \frac{K}{3} \tag{11.44}$$

(i.e., the angular mean of the total *los* kinetic energy equals *a third* of the total kinetic energy). The same result can be stated by saying that the angular mean of the second-order projected velocity field of a stellar system, weighted over the projected density distribution Σ, is equal to a third of the squared virial velocity dispersion σ_{V}^2 of the considered component. The angular mean of the second-order PVT reads

$$2[K_{\mathrm{los}}^{(2)}]_\Omega = -\frac{U + W_{\mathrm{ext}}}{3} + \frac{\ddot{I}}{6}, \tag{11.45}$$

which is of course in perfect agreement with the expression obtained by computing the angular mean of Eq. (11.31). Clearly, for a spherical stellar system, Eq. (11.44) also holds with $[K_{\mathrm{los}}^{(2)}]_\Omega = K_{\mathrm{los}}^{(2)}$.

Exercises

11.1 Construct the rotation matrix $\mathcal{R}$ in Eq. (11.1) so that the observer's *los* is aligned along the ξ_3-axis of the rotated system and that in S_0 the direction $\mathbf{n}$ of the observer is parametrized by the usual spherical coordinates. Show that the three successive rotations are described by the matrices

$$\mathcal{R}_3(\varphi) = \begin{pmatrix} \cos\varphi & -\sin\varphi & 0 \\ \sin\varphi & \cos\varphi & 0 \\ 0 & 0 & 1 \end{pmatrix}, \tag{11.46}$$

$$\mathcal{R}_2(\vartheta) = \begin{pmatrix} \cos\vartheta & 0 & \sin\vartheta \\ 0 & 1 & 0 \\ -\sin\vartheta & 0 & \cos\vartheta \end{pmatrix}, \tag{11.47}$$

$$\mathcal{R}_3(\psi) = \begin{pmatrix} \cos\psi & -\sin\psi & 0 \\ \sin\psi & \cos\psi & 0 \\ 0 & 0 & 1 \end{pmatrix}, \tag{11.48}$$

and prove Eq. (11.2). Note that $\mathcal{R}_2(\vartheta)$ and $\mathcal{R}_3(\psi)$ also appear in Eq. (3.20).

11.2 With this exercise, we estimate the "transparency" of a stellar system (i.e., the probability that two stars are aligned along the *los*, thus in principle violating the mass (number or luminosity) conservation in Eq. (11.7)). For the galaxy toy model in Exercise 1.1, define the average covering factor as

$$\text{Covering factor} \equiv \frac{N\pi R_\odot^2}{\pi R^2}, \tag{11.49}$$

and estimate its value for an elliptical galaxy and a globular cluster. What are the conclusions you can draw from this? Notice that the stellar density in real stellar systems depends strongly on the distance from the center.

11.3 Prove the identities in Eqs. (11.13) and (11.14). *Hint*: Rewrite Eq. (11.8) as $v_{\rm p}^N = (v_{\rm p} - \overline{v_{\rm p}} + \overline{v_{\rm p}})^N = (v_{\rm p} - \overline{v_{\rm p}}_{\rm los} + \overline{v_{\rm p}}_{\rm los})^N$ and Eq. (11.10) as $\sigma_{\rm p}^N = (v_{\rm p} - \overline{v_{\rm p}}_{\rm los} + \overline{v_{\rm p}}_{\rm los} - \overline{v_{\rm p}})^N$.

11.4 Consider a perfectly monocromatic LP with $P_0 = \delta(\lambda - \lambda_0)$, so that $P_v = \delta[\lambda/(1 - v_{\rm p}/c) - \lambda_0](1 - v_{\rm p}/c)^{-1}$. By using the standard rule for the change of variable in the Dirac δ-function in Eq. (A.98), show that

$$\mathrm{LP}(\lambda) = \frac{c}{\lambda_0}\mathrm{VP}\left[\boldsymbol{\xi}_\perp, \left(1 - \frac{\lambda}{\lambda_0}\right)c\right], \tag{11.50}$$

and verify the normalization property $\int_0^\infty \mathrm{LP}(\lambda)\,d\lambda = 1$.

11.5 Prove the identities for λ_v and σ_v in Eqs. (11.22) and (11.23). What happens in the special case of a monochromatic LP with $P_0 = \delta(\lambda - \lambda_0)$? *Hints*: Substitute Eq. (11.21) into the integral expressions of λ_v and σ_v. For λ_v, invert the order of integration, integrate over λ, and then use Eq. (11.18). For σ_v, consider $(\lambda - \lambda_v)^2 = \lambda^2 - 2\lambda\lambda_v + \lambda_v^2$; from Eq. (11.22) and the normalization of the LP, the integration of the last two terms gives $-\lambda_v^2$. To integrate the λ^2 term, invert the order of integration, change the variable in the inner integral to $\lambda/(1 - v_{\rm p}/c) = t$, and set $t^2 = (t - \lambda_0 + \lambda_0)^2$, so that from the definition of P_0 and its normalization, the inner integral evaluates to $\sigma_0^2 + \lambda_0^2$. For the outer integral, write $(1 - v_{\rm p}/c)^2 = [1 - \overline{v_{\rm p}}_{\rm los}/c - (v_{\rm p} - \overline{v_{\rm p}}_{\rm los})/c]^2$, expand the integrand, and from Eq. (11.18) and Eq. (11.19) with $N = 2$ and the normalization of the VP, finally conclude.

11.6 Estimate the expected value of σ_0 for a spectral line produced by the photosphere of a star of spectral type (say) G or K and compare the result with the Doppler broadening σ_v due to the motions of the stars in a stellar system, such as an elliptical galaxy. What are the practical consequences for Eq. (11.23)?

11.7 Prove Eq. (11.40). *Hints*: Let a and b be non-negative integers and let $I_\phi(a,b) \equiv \int_0^\phi \sin^a x \cos^b x\,dx$. From Eq. (11.8) and the definition of the angular mean, first show that

$$[n_3^{N-K} n_2^{K-L} n_1^L]_\Omega = \frac{I_\pi(K+1, N-K) I_{2\pi}(K-L, L)}{4\pi}. \tag{11.51}$$

Then, by successive reduction to the first quadrant, prove that

$$I_{2\pi}(a,b) = [1+(-1)^{a+b}] I_\pi(a,b), \quad I_\pi(a,b) = [1+(-1)^b] I_{\pi/2}(a,b), \tag{11.52}$$

and from Eq. (A.53)

$$I_{\pi/2}(a,b) = \frac{1}{2}\mathrm{B}\left(\frac{a+1}{2}, \frac{b+1}{2}\right) = I_{\pi/2}(b,a), \tag{11.53}$$

where $\mathrm{B}(x,y)$ is the complete Euler beta function. Identity (11.40) is finally established by multiplication for the binomial coefficients appearing there using Eqs. (A.49) and (A.54) and by performing the needed simplifications.

11.8 Following Exercise 10.1, we now consider the relations between first, and second-order projected velocity moments, of a multi-component stellar system described by a discrete set of distribution functions f_k. Show that

$$\overline{v}_{\mathrm{Plos}} = \frac{\sum_k \Sigma_k (\overline{v}_{\mathrm{Plos}})_k}{\Sigma}, \quad \overline{\sigma_1^2}\mathrm{los} = \frac{\sum_k \Sigma_k (V^2_{\mathrm{rms}})_k}{\Sigma} - (\overline{v}_{\mathrm{Plos}})^2, \tag{11.54}$$

where $\Sigma = \sum_k \Sigma_k$ is the total surface density of the system and from Eq. (11.16) $(V^2_{\mathrm{rms}})_k \equiv (\overline{v^2_{\mathrm{Plos}}})_k = (\overline{\sigma^2_{\mathrm{P\,los}}})_k + (\overline{v}^2_{\mathrm{P\,los}})_k$. *Hints*: Consider the $N = 2$ case of Eq. (11.12), expand the square and take separately the phase-space average over each f_k, then project along the *los* and use the first identity in Eq. (11.54).

12

Modeling Techniques 1

Phase-Space Approach

Armed with the power of the Jeans theorem, we now proceed to formulate and discuss the so-called direct problem of collisionless stationary stellar dynamics. This approach is best suited for systems where empirical/dynamical arguments can lead to a plausible *ansatz* for the form of the underlying distribution function (DF), expressed in terms of the relevant integrals of motion. In the absence of such an ansatz and in the presence of specific requirements (in general motivated by observations) for the density and velocity dispersion profiles, a different and complementary approach based on the use of the Jeans equations is often followed, which is the subject of Chapter 13.

12.1 The Construction of a Galaxy Model: From f to ρ

In the most general case, the *direct* problem of collisionless stationary stellar dynamics for a system made of n different density components consists of the assignment of the functional form of n DFs f_k, dependent on a set of isolating integrals of motion of the (unknown) *total* potential and some free parameters. The choice of the integrals of motion can be based on physical arguments and/or on the phenomenological properties expected for the resulting system. The task is to determine explicitly (usually numerically, but in some very special cases analytically) the functions f_k through the determination of the total potential $\phi_{\mathrm{T}}(\mathbf{x})$. In addition to the potential $\phi(\mathbf{x})$ produced by the n components, the total potential can also contain some imposed potential (a central black hole, an external tidal field, a dark matter halo, etc.) that we will call "external" and indicate with $\phi_{\mathrm{ext}}(\mathbf{x})$. Once the total potential is determined by solving the associated Poisson equation, the properties of the system can finally be studied and compared with observations. Some care should be taken regarding the concept of a *component* of a stellar system: by "component" we mean each part (e.g., each stellar and/or dark matter component) of the system that is characterized by specific properties (chemical composition, age, mass-to-light ratio, internal kinematics, etc.). In principle, the "number" of physically distinct components can be "infinite" when the *extended* DF depends on some continuous parameter (e.g., as in the presence of a continuous distribution of metallicity/age/initial mass functions (IMFs) or other properties of stellar populations). In this latter case, the total DF is obtained by integration over such parameters, and a simple example is given in Chapter 8 (see also Exercise 10.1); in the rest of the present chapter, for simplicity we restrict ourselves to the discrete case of a finite number of components.

A basic requirement to be satisfied by the DF associated with each component is that it has to be non-negative over the phase space. When this requirement is satisfied each component is said to be *consistent*, and a *self-consistent* stellar system is a self-gravitating system for which each component is consistent. For technical reasons, at least for systems of finite total mass (or when the total potential ϕ_T can be set equal to zero at infinity; see Chapter 2), it can be convenient to work with *relative* energies and potentials. In particular, the *relative potential* Ψ_T and *relative energy* per unit mass $\mathcal{E}$ are defined as

$$\Psi_T = \Psi + \Psi_{ext} \equiv -\phi - \phi_{ext}, \qquad \mathcal{E} \equiv -E = \Psi_T - \frac{\|\mathbf{v}\|^2}{2}. \tag{12.1}$$

Thus, in general, the relative potential defines a family of isopotential surfaces enclosing more and more volume of space as the value of Ψ_T *decreases.*

The previous and qualitative considerations can be formalized as follows: let $f_k(\mathbf{x}, \mathbf{v}; t)$ be regular functions for $k = 1, \ldots, n$. From Eq. (9.9), the function

$$f = \sum_{k=1}^{n} f_k \tag{12.2}$$

is said to be the DF of a *collisionless, (self)-consistent multicomponent system* if and only if for $k = 1, \ldots, n$,

$$\frac{\partial f_k}{\partial t} + \langle \mathbf{v}, \nabla_{\mathbf{x}} f_k \rangle + \langle \nabla_{\mathbf{x}} \Psi_T, \nabla_{\mathbf{v}} f_k \rangle = 0, \tag{12.3}$$

$$f_k \geq 0, \tag{12.4}$$

$$\Delta \Psi_k = -4\pi G \rho_k, \tag{12.5}$$

where from Eq. (9.10)

$$\rho = \sum_{k=1}^{n} \rho_k = \sum_{k=1}^{n} \int_{\Re^3} f_k d^3\mathbf{v}, \quad \Psi = \sum_{k=1}^{n} \Psi_k, \quad \Delta \Psi = -4\pi G \rho. \tag{12.6}$$

Each function f_k is said to describe a *consistent component.*

The student will recognize that at the core of the problem, leaving aside all technical details, these are basically all of the important concepts encountered so far. In practice, we ask for a system where each of the components must obey the collisionless Boltzmann equation in the total potential through the global integrals of motion in the total potential, and the resulting density distribution therefore must obey the Poisson equation. Physically, we are asking something quite formidable: we are attempting to determine orbital distributions such that the resulting total density ρ produces a potential ϕ that, possibly added to an imposed external potential ϕ_{ext}, admits the adopted global integrals of motion! In practice, the problem is a sort of "bootstrap" process. As we will see, the "closure" of the problem (physically, the consistence, or self-consistence when $\phi_{ext} = 0$) is provided by the Poisson equation. Therefore, the student will appreciate that just the possibility of formulating this problem is by itself quite a beautiful piece of physics.

We are now in a position to formulate the *direct problem* of collisionless stellar dynamics for multicomponent stationary systems. For $i = 1, \ldots, m$, let $I_i(\mathbf{x}, \mathbf{v})$ be *regular* functions (i.e., at least $\mathcal{C}^{(1)}$) functionally independent over the one-particle phase space γ, with $I_1 \equiv \mathcal{E}$ given[1] in Eq. (12.1). Moreover, let

$$f_k = f_k(\boldsymbol{\lambda}_k, I_i), \quad k = 1, \ldots, n, \tag{12.7}$$

which is the DF of the k-th component. For each k, $\boldsymbol{\lambda}_k$ is a *vector parameter* "controlling" some property of f_k (the examples in the following will elucidate the meaning and role of the quite mysterious object $\boldsymbol{\lambda}_k$). In practice, f_k depends on a choice of the integrals of motion I_i of the potential Ψ_{T} of the *total* system and on a set of parameters adopted to hopefully control the properties of interest of f_k.

For a given k and $\boldsymbol{\lambda}_k$, the set of points in phase space

$$\Omega_k \equiv \{(\mathbf{x}, \mathbf{v}) \in \gamma : f_k \geq 0\} \tag{12.8}$$

is associated with f_k. In practice, for each f_k, the set Ω_k is the region of phase space where f_k is non-negative for the given choice of $\boldsymbol{\lambda}_k$. Note that from Eq. (12.7) the boundary $\partial\Omega_k$ necessarily depends on $\mathbf{x}$ and $\mathbf{v}$ through the integrals of motion I_i only. Moreover, for each k and $\boldsymbol{\lambda}_k$ and for each $\mathbf{x} \in \Omega_k$, the set

$$\Omega_{k\mathbf{x}} \equiv \{\mathbf{v} : (\mathbf{x}, \mathbf{v}) \in \Omega_k\} \tag{12.9}$$

is called the *velocity section* of Ω_k at $\mathbf{x}$. In practice, $\Omega_{k\mathbf{x}}$ is the set of admissible values of $\mathbf{v}$ at the position $\mathbf{x}$ so that f_k is positive for the given choice of $\boldsymbol{\lambda}_k$.

The mass density of each component at position $\mathbf{x}$ is naturally obtained as

$$\rho_k[\mathbf{x}, \Psi_{\mathrm{T}}(\mathbf{x})] = \int_{\Omega_{k\mathbf{x}}} f_k d^3\mathbf{v}. \tag{12.10}$$

As we will see in the following discussion, not all of the (a priori plausible) choices of $\boldsymbol{\lambda}_k$ can be physically acceptable. This is formalized by the definition of the three parameter sets

$$\Lambda_+ \equiv \bigcup_{k=1}^{n} \{\boldsymbol{\lambda}_k : \Omega_k \neq \emptyset\}, \tag{12.11}$$

$$\Lambda_\Psi \equiv \bigcup_{k=1}^{n} \{\boldsymbol{\lambda}_k : \Delta\Psi_{\mathrm{T}} = -4\pi G\rho_{\mathrm{T}} \text{ has a regular/physically acceptable solution}\}, \tag{12.12}$$

$$\Lambda_I \equiv \bigcup_{k=1}^{n} \{\boldsymbol{\lambda}_k : I_i \text{ are global regular integrals of motion for } \Psi_{\mathrm{T}}\}. \tag{12.13}$$

[1] Note that due to the time independence of Ψ_{T}, the (isolating) integral $I_1 = \mathcal{E}$ is always available.

Therefore, the direct problem is *solvable* if and only if $\Lambda_\exists \equiv \Lambda_+ \cap \Lambda_\Psi \cap \Lambda_I \neq \emptyset$. Note that for $\boldsymbol{\lambda}_k \in \Lambda_+$ the density ρ_k in Eq. (12.10) is non-negative by construction, but its dependence on the coordinates $\mathbf{x}$ is known only *after* the problem is solved, and $\Psi_{\rm T}(\mathbf{x})$ is determined by the solution of the Poisson equation in Eq. (12.6) where the total ρ is obtained by summing the functions in Eq. (12.10). In summary, if the direct problem is solvable, from the Jeans theorem the functions f_k are (stationary) DFs for each admissible $\boldsymbol{\lambda}_k$, and the stellar system described by $f = \sum_{k=1}^{n} f_k$ is a collisionless, stationary, self-consistent, multicomponent system. In applications, the starting point of the procedure is the assumption of the desired geometry for the system (i.e., for $\Psi_{\rm T}$), so that the functions I_i are usually determined by symmetry considerations. The problem then reduces to the determination of the set Λ_+, to the construction of the n density functions in Eq. (12.10) as a function of $\mathbf{x}$ and $\Psi_{\rm T}$, to the solution (numerical or analytical) of the nonlinear Poisson equation for the total potential in Eq. (12.6) with suitably assigned boundary conditions, and to a final check of $\Lambda_\exists$. Only at this stage, after the determination of $\Psi_{\rm T}(\mathbf{x})$, is the density distribution of each component finally known from Eq. (12.10).

Before proceeding with the discussion of the most elementary cases, it is important to remind ourselves that the "orbital description" sketched out above is at the basis of a very powerful numerical method that is used to (numerically) build self-consistent systems, namely the *Schwarzschild orbital superposition method* (e.g., see Richstone 1980, 1984; Schwarzschild 1979). Other quite sophisticated methods, directly based on the Hamiltonian structure of phase space and in particular on the numerical evaluation of the *actions* of given potentials (e.g., see Binney and Kumar 1993; Binney and McMillan 2016; Kaasalainen and Binney 1994; McGill and Binney 1990; Sanders and Binney 2015 and references therein; see also Binney and Spergel 1982, 1984), are today being actively studied. All of these methods are well beyond the level of the present discussion, and the interested reader is invited to consult the more advanced treatises of Bertin (2014) and Binney and Tremaine (2008) and references therein.

We now discuss some general properties of systems that are generated by elementary choices regarding the integrals of motion and functional forms for f_k in order to introduce the student to some illustrative cases that should clarify the previous quite technical but necessary preliminary discussion.

12.2 Spherical Systems

It is not a suprise that the case of spherically symmetric systems is the simplest from a mathematical point of view, and in this family we first consider the even simpler case of one-component, self-consistent spherical systems. Yet, from a physical point of view, the situation is already interesting enough to allow for a clarification of several of the aspects mentioned in Section 12.1, and for us to construct systems of great relevance for observational and theoretical works. The student should not be surprised then that the majority of models adopted in the research literature are in fact spherically symmetric.

12.2.1 Isotropic Spherical Systems

As we will now see, restricting ourselves to systems with a DF depending only on the orbital energy of stars leads us to build stellar systems with a velocity dispersion tensor that is necessarily isotropic over all of the space; notice that for simplicity of notation we avoid introducing the subscript k and we stress that for the moment we are *not* assuming spherical symmetry.

We begin by assuming $I_1 \equiv \mathcal{E}$ as defined in Eq. (12.1), and

$$f = h(\mathcal{E})\theta(\mathcal{E}) \geq 0, \tag{12.14}$$

where $h(\mathcal{E})$ is some regular function and $\theta(\mathcal{E})$ is the Heaviside step function in Eq. (A.99). The "truncation" at negative relative energies is a natural requirement for a system of finite total mass (or more generally with a potential that can be fixed to zero at infinity; see Chapter 2), but in Exercise 12.1 we will briefly address more general cases of energy truncation, or of no energy truncation.

We now prove that from Eq. (12.14) it follows that

$$\rho(\Psi_T) = 4\pi \int_0^{\Psi_T} \sqrt{2(\Psi_T - \mathcal{E})}h(\mathcal{E})d\mathcal{E}, \tag{12.15}$$

$$\overline{v_i} = 0, \quad \overline{v_i v_j} = 0, \quad i \neq j = 1,2,3, \tag{12.16}$$

and so $\sigma_{ij}^2 = 0$ for $i \neq j$. Moreover (no sum over i intended)

$$\rho(\Psi_T)\sigma_{ii}^2(\Psi_T) = \frac{4\pi}{3} \int_0^{\Psi_T} [2(\Psi_T - \mathcal{E})]^{3/2} h(\mathcal{E})d\mathcal{E}, \quad i = 1,2,3. \tag{12.17}$$

In fact, from Eq. (12.8), $\Omega_k = \{(\mathbf{x},\mathbf{v}) \in \gamma : \Psi_T - \|\mathbf{v}\|^2/2 \geq 0\}$, and according to Eq. (12.9) the velocity sections $\Omega_{k\mathbf{x}}$ are spheres of "radius" $\sqrt{2\Psi_T(\mathbf{x})}$. The natural coordinates in velocity space to be used to integrate over $\Omega_{k\mathbf{x}}$ are the spherical ones, i.e.,

$$v_x = v \sin\lambda \, \cos\mu, \quad v_y = v \sin\lambda \, \sin\mu, \quad v_z = v\cos\lambda, \tag{12.18}$$

with $0 \leq v = \|\mathbf{v}\| \leq \sqrt{2\Psi_T}$, $0 \leq \lambda < \pi$, and $0 \leq \mu < 2\pi$ (see Figure 12.1), so that[2]

$$d^3\mathbf{v} = v^2 \sin\lambda \, dv \, d\lambda \, d\mu. \tag{12.19}$$

From Eq. (12.10), one then obtains

$$\rho = \int_{\Omega_{k\mathbf{x}}} f d^3\mathbf{v} = 4\pi \int_0^{\sqrt{2\Psi_T}} hv^2 dv, \tag{12.20}$$

[2] The angles λ and μ coincide with the two angles ϑ and φ of spherical coordinates in the configuration space; however, this will not be true anymore in the next cases.

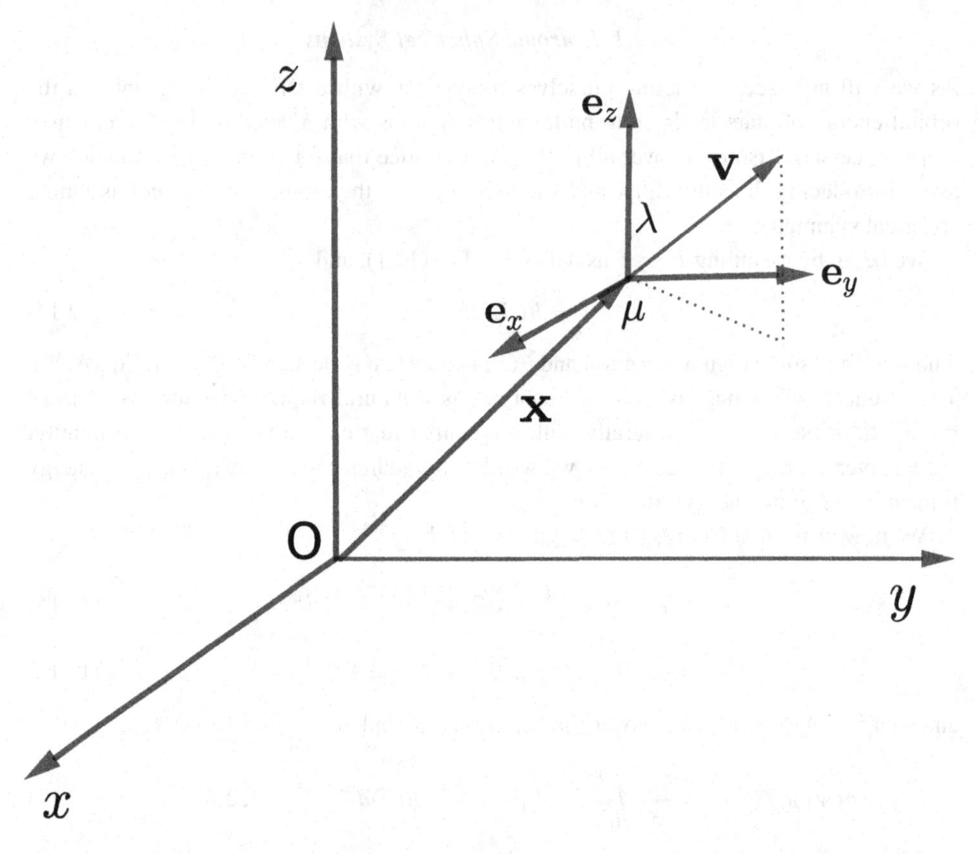

Figure 12.1 The decomposition of the velocity vector **v** adopted to integrate the DF in the isotropic case (Section 12.2.1). μ is the angle between $\mathbf{e}_x$ and the projection of **v** on the $(\mathbf{e}_x, \mathbf{e}_y)$ plane and λ is the angle between $\mathbf{e}_z$ and **v**.

because f depends only on v. Finally, by changing variable from v to $\mathcal{E}$ at fixed **x** (do it!)

$$dv = -\frac{d\mathcal{E}}{\sqrt{2(\Psi_T - \mathcal{E})}}, \tag{12.21}$$

and so Eq. (12.15) is recovered because[3] $0 \le \mathcal{E} \le \Psi_T$. With a similar procedure, it is trivial to show that Eq. (12.16) holds due to the vanishing of angular integrals. Finally

$$\rho\sigma_{ii}^2 = \int_0^{\sqrt{2\Psi_T}} hv^2 dv \int_0^{\pi} \sin\lambda \, d\lambda \int_0^{2\pi} v_i^2 d\mu, \tag{12.22}$$

and after integration over the angular variables, Eq. (12.17) is recovered. Therefore, we can conclude that under the hypothesis in Eq. (12.14), the density ρ depends on **x** only through

[3] Of course, for a generic value $\mathcal{E}_t$ of the truncation energy, the velocity sections $\Omega_{k\mathbf{x}}$ are spheres of radius $\sqrt{2[\Psi_T(\mathbf{x}) - \mathcal{E}_t]}$ and the energy integration interval is $\mathcal{E}_t \le \mathcal{E} \le \Psi_T(\mathbf{x})$. Absence of truncation means $\mathcal{E}_t = -\infty$ and $\Omega_{k\mathbf{x}} = \Re^3$; see also Exercise 12.1.

Ψ_T (i.e., $\rho = \rho(\Psi_T)$); streaming motions cannot be present and the velocity dispersion tensor is *globally isotropic* (and so $\overline{v_i^2} = \sigma_{ii}^2$ for $i = 1, 2, 3$).

As has already been stressed, up to now *we have not yet imposed spherical symmetry (i.e., the shape of Ψ_T is not specified)*, and for this reason we maintained in all generality a Cartesian decomposition of the velocity vector in Eq. (12.18). In turn, as all stationary stellar systems (independently of their shape) admit the orbital energy as a global isolating integral of motion, one could feel free to imagine the existence of (isotropic) stationary systems without special symmetries characterized by DFs that depend only on the energy (perhaps simply because other isolating integrals do not exist!). However, an important (and perhaps unexpected) consequence of global isotropy can be derived by imposing self-consistency through the Poisson equation

$$\Delta\Psi = -4\pi G\rho(\Psi_T). \tag{12.23}$$

In fact, a quite remarkable mathematical theorem (Gidas et al. 1979; see also An et al. 2017; Ciotti 2000) states that the *only* positive and nontruncated solutions for Eq. (12.23) (with $\Psi = \Psi_T$) are *spherically symmetric*, and so we conclude that *a nonspherically symmetric, spatially untruncated, one-component stellar system with no other regular integral of motion in addition to $\mathcal{E}$ cannot be stationary (and so even $\mathcal{E}$ is not an integral of motion)*. In fact, we assume the system to be stationary, so that $\mathcal{E}$ is an isolating integral of motion, and nonspherical, with no other regular integrals of motion. From the Jeans theorem, necessarily $f = f(\mathcal{E})$, but from the previous result, the Poisson equation does not admit acceptable solutions. This contradiction implies that one of the assumptions is logically inconsistent (i.e., the system cannot be stationary). As a consequence, in order to construct nonspherical, self-consistent stellar systems, at least *two* regular integrals of motion are needed.

12.2.2 Anisotropic Spherical Systems

We now proceed to explore the general consequences of assuming a *two-integral* DF for a spherical stellar system. The most natural assumption (but not the only one!) is to adopt as global integrals the orbital energy and the modulus of the orbital angular momentum.

Therefore, we assume that the total potential $\Psi_T(r)$ is spherically symmetric, with $r = \|\mathbf{x}\|$ and $I_1 \equiv \mathcal{E}$, $I_2 \equiv \|\mathbf{J}\|^2 = J^2$ (where $\mathbf{J} = \mathbf{x} \wedge \mathbf{v}$ is the angular momentum per unit mass of each star). Moreover, let

$$f = h(\mathcal{E}, J^2)\theta(\mathcal{E}) \geq 0, \tag{12.24}$$

where again we allow for a possible energy truncation. We will prove that

$$\rho = \frac{4\pi}{r^2}\int_0^{\Psi_T} d\mathcal{E}\int_0^{r\sqrt{2(\Psi_T-\mathcal{E})}} \frac{h(\mathcal{E}, J^2)J\,dJ}{\sqrt{2(\Psi_T-\mathcal{E}) - J^2/r^2}}, \tag{12.25}$$

$$\overline{v_r} = \overline{v_\vartheta} = \overline{v_\varphi} = \overline{v_r v_\vartheta} = \overline{v_r v_\varphi} = \overline{v_\vartheta v_\varphi} = 0. \tag{12.26}$$

Moreover

$$\rho\sigma_r^2 = \frac{4\pi}{r^2}\int_0^{\Psi_T} d\mathcal{E}\int_0^{r\sqrt{2(\Psi_T-\mathcal{E})}} h(\mathcal{E},J^2)\sqrt{2(\Psi_T-\mathcal{E})-\frac{J^2}{r^2}}\,JdJ, \tag{12.27}$$

$$\rho\sigma_t^2 = \frac{4\pi}{r^4}\int_0^{\Psi_T} d\mathcal{E}\int_0^{r\sqrt{2(\Psi_T-\mathcal{E})}} \frac{h(\mathcal{E},J^2)J^3dJ}{\sqrt{2(\Psi_T-\mathcal{E})-J^2/r^2}}. \tag{12.28}$$

In fact, from Eq. (12.8), $\Omega_k = \{(\mathbf{x},\mathbf{v}) \in \gamma : \Psi_T - \|\mathbf{v}\|^2/2 \geq 0\}$, and according to Eq. (12.9), the velocity sections $\Omega_{k\mathbf{x}}$ are spheres of radius $\sqrt{2\Psi_T(r)}$. In this case, the natural coordinates in configuration space are the spherical ones (r,ϑ,φ), and the associated velocity components are $(v_r, v_\vartheta, v_\varphi)$. From the geometry of $\Omega_{k\mathbf{x}}$, the natural coordinates to perform integrations over the velocity space are the spherical ones, with

$$v_\vartheta = v\sin\lambda\,\cos\mu, \quad v_\varphi = v\sin\lambda\,\sin\mu, \quad v_r = v\cos\lambda, \tag{12.29}$$

with $0 \leq v = \|\mathbf{v}\| \leq \sqrt{2\Psi_T(r)}$, $0 \leq \lambda < \pi$, and $0 \leq \mu < 2\pi$. Note that $J = rv_t$, with $v_t = \sqrt{v_\vartheta^2 + v_\varphi^2}$, is the *tangential* velocity, and $v^2 = v_r^2 + v_t^2$; thus, v_ϑ and v_φ appear in the DF only through $v_t = v\sin\lambda$ and f is idependent of μ. From Figure 12.2, the student will notice that the angles λ and μ are defined with respect to the local spherical coordinate system, so they are *not* the same as in Eq. (12.18); of course, we could have also used the present velocity decomposition in the isotropic case in Section 12.2.1 once spherical symmetry is assumed.

In order to proceed with integration, we first use Exercise 12.2 and then we change the coordinates from Eq. (12.29), finally reobtaining Eq. (12.19). After integration over μ and reducing the interval of integration of λ from $(0,\pi)$ to $(0,\pi/2)$, the velocity moments of interest are immediately written as

$$\rho = 4\pi\int_0^{\sqrt{2\Psi_T}} v^2dv\int_0^{\pi/2} h\,\sin\lambda\,d\lambda, \tag{12.30}$$

$$\rho\sigma_r^2 = 4\pi\int_0^{\sqrt{2\Psi_T}} v^4dv\int_0^{\pi/2} h\,(\cos\lambda)^2\,\sin\lambda\,d\lambda, \tag{12.31}$$

$$\rho\sigma_t^2 = 4\pi\int_0^{\sqrt{2\Psi_T}} v^4dv\int_0^{\pi/2} h\,(\sin\lambda)^3\,d\lambda. \tag{12.32}$$

With a last change of variables from (v,λ) to $(\mathcal{E},J)$ (do it!), one finally obtains

$$dvd\lambda = \frac{d\mathcal{E}dJ}{r\sqrt{2(\Psi_T-\mathcal{E})}\sqrt{2(\Psi_T-\mathcal{E})-J^2/r^2}}, \tag{12.33}$$

with $0 \leq \mathcal{E} \leq \Psi_T$, $0 \leq J \leq r\sqrt{2(\Psi_T-\mathcal{E})}$, and this completes the proof of Eqs. (12.25)–(12.28). The student is invited to evaluate the identities in this section for h independent of J^2 and to recover the corresponding indentities in Section 12.2.1.

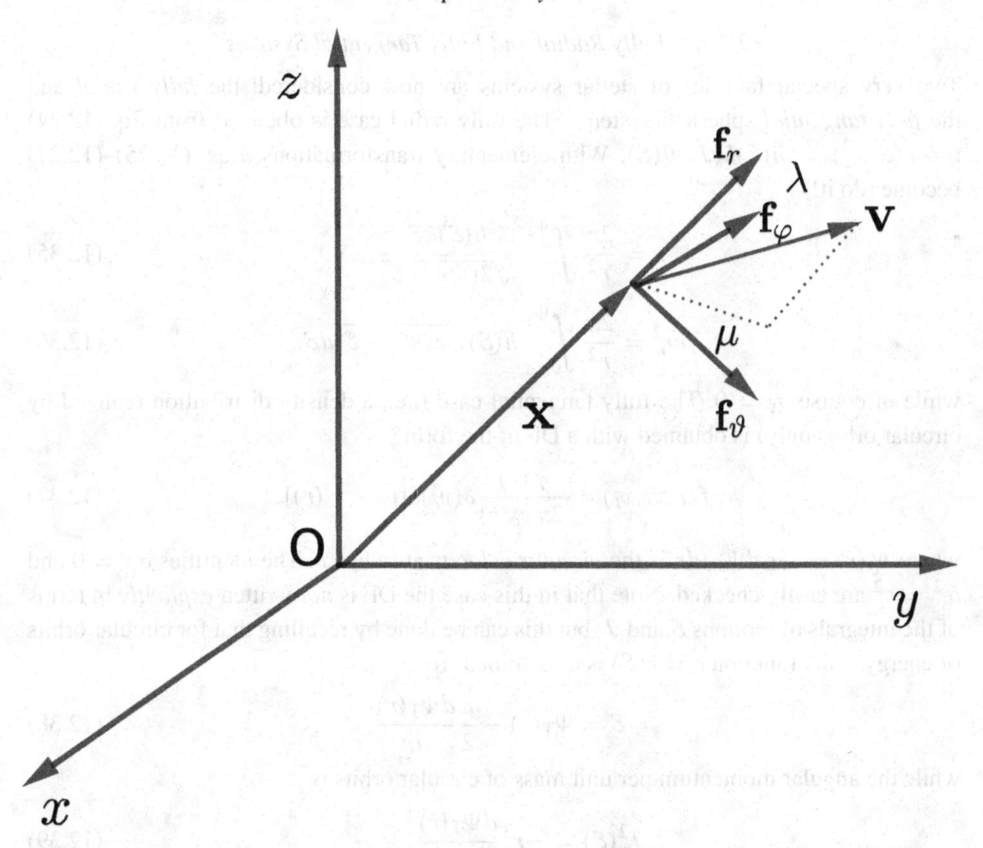

Figure 12.2 The decomposition of the velocity vector **v** adopted to integrate the DF in the two-integral case for spherical systems (Section 12.2.2). μ and λ are the usual spherical angles in the local reference system $(\mathbf{f}_r, \mathbf{f}_\vartheta, \mathbf{f}_\varphi)$. Note that we could have also used this decomposition in the isotropic case instead of Cartesian coordinates.

Therefore, at variance with the isotropic case, now the density ρ depends explicitly on both r and Ψ_T (i.e., $\rho(r, \Psi_T)$ in Eq. (12.25)); streaming motions cannot be present and $\sigma_\varphi^2 = \sigma_\vartheta^2 = \sigma_t^2/2$, where σ_t is the tangential velocity dispersion. The velocity dispersion tensor is therefore *anisotropic*, and the associated velocity dispersion ellipsoids are axisymmetric around the radial direction. In order to characterize this type of anisotropy, the function

$$\beta(r) = 1 - \frac{\sigma_t^2(r)}{2\sigma_r^2(r)} \tag{12.34}$$

is introduced (e.g., see Binney and Tremaine 2008). When $\beta(r) = 0$, the system is isotropic (at r); when $\beta(r) > 0$, the system is *radially anisotropic* (i.e., the velocity dispersion ellipsoid at r is prolate); and when $\beta(r) < 0$, the system is *tangentially anisotropic* (i.e., the velocity dispersion ellipsoid at r is oblate). In Exercise 12.3, two important and general identities about anisotropic spherical systems will be proved.

12.2.2.1 Fully Radial and Fully Tangential Systems

Two very special families of stellar systems are now considered: the *fully radial* and the *fully tangential* spherical systems. The fully radial case is obtained from Eq. (12.24) for $f(\mathcal{E}, J^2) = h(\mathcal{E})\delta(J^2)\theta(\mathcal{E})$. With elementary transformations, Eqs. (12.25)–(12.27) become (do it!)

$$\rho = \frac{2\pi}{r^2}\int_0^{\Psi_T} \frac{h(\mathcal{E})d\mathcal{E}}{\sqrt{2(\Psi_T - \mathcal{E})}}, \tag{12.35}$$

$$\rho\sigma_r^2 = \frac{2\pi}{r^2}\int_0^{\Psi_T} h(\mathcal{E})\sqrt{2(\Psi_T - \mathcal{E})}d\mathcal{E}, \tag{12.36}$$

while of course $\sigma_t = 0$. The fully tangential case (i.e., a density distribution realized by circular orbits only) is obtained with a DF of the form

$$f(r, v_r, v_t) = \frac{\rho(r)}{2\pi v_c(r)}\delta(v_r)\delta[v_t - v_c(r)], \tag{12.37}$$

where $v_c^2(r) = -r\,d\Psi_T/dr$ is the *circular velocity* at radius r. The identities $\sigma_r^2 = 0$ and $\sigma_t^2 = v_c^2$ are easily checked. Note that in this case the DF is *not* written *explicitly* in terms of the integrals of motions $\mathcal{E}$ and J, but this can be done by recalling that for circular orbits of energy $\mathcal{E}$ the function $r = r(\mathcal{E})$ is determined from

$$\mathcal{E} = \Psi_T(r) - \frac{r}{2}\frac{d\Psi_T(r)}{dr}, \tag{12.38}$$

while the angular momentum per unit mass of circular orbits is

$$J_c^2(\mathcal{E}) = -r^3\frac{d\Psi_T(r)}{dr}\bigg|_{r=r(\mathcal{E})} \tag{12.39}$$

(see Chapter 5).

12.2.2.2 Osipkov–Merritt Systems

A very important and widely studied family of spherical, anisotropic stellar systems is given by the so-called *Osipkov–Merritt* (OM) systems (named after their two independent discoverers, Merritt 1985a; Osipkov 1979). In this family, one defines the quantity

$$Q = \mathcal{E} - \frac{J^2}{2r_a^2}, \tag{12.40}$$

where r_a is a free parameter called the *anisotropy radius* (for reasons that will become clear in the following), a first example of the parameter(s) $\boldsymbol{\lambda}_k$ introduced in Section 12.1, and we assume

$$f = h(Q)\theta(Q) \geq 0. \tag{12.41}$$

We now prove that

$$\rho = 4\pi A(r)\int_0^{\Psi_T} \sqrt{2(\Psi_T - Q)}h(Q)dQ, \tag{12.42}$$

$$\rho\sigma_r^2 = 4\pi B(r) \int_0^{\Psi_T} [2(\Psi_T - Q)]^{3/2} h(Q) dQ, \tag{12.43}$$

$$\rho\sigma_t^2 = 4\pi C(r) \int_0^{\Psi_T} [2(\Psi_T - Q)]^{3/2} h(Q) dQ, \tag{12.44}$$

where the three radial factors are

$$A(r) = \frac{1}{1 + r^2/r_a^2}, \quad B(r) = \frac{A(r)}{3}, \quad C(r) = \frac{2A^2(r)}{3}. \tag{12.45}$$

The natural coordinates in configuration space are again the spherical ones. In fact, $\Omega_k = \{(\mathbf{x}, \mathbf{v}) \in \gamma : \Psi_T - \|\mathbf{v}\|^2/2 - J^2/2r_a^2 = \Psi_T - v_r^2/2 - v_t^2(1 + r^2/r_a^2)/2 \geq 0\}$, and the velocity sections $\Omega_{k\mathbf{x}}$ are rotation ellipsoids around $\mathbf{f}_r$ (see again Figure 12.2). As a consequence of the geometry of $\Omega_{k\mathbf{x}}$, the natural coordinates for integration in velocity space are the ellipsoidal ones, i.e.,

$$v_\vartheta = \frac{v \sin\lambda \cos\mu}{\sqrt{1 + r^2/r_a^2}}, \quad v_\varphi = \frac{v \sin\lambda \sin\mu}{\sqrt{1 + r^2/r_a^2}}, \quad v_r = v \cos\lambda, \tag{12.46}$$

with $0 \leq v \leq \sqrt{2\Psi_T(r)}$, $0 \leq \lambda \leq \pi$, and $0 \leq \mu < 2\pi$, so that from Exercise 12.2 we now have (prove it!)

$$d^3\mathbf{v} = \frac{v^2 \sin\lambda}{1 + r^2/r_a^2}\, d\lambda\, d\mu. \tag{12.47}$$

Note that $v_t = v \sin\lambda / \sqrt{1 + r^2/r_a^2}$. With the transformation in Eq. (12.46) f becomes independent of μ and λ, with

$$\rho = 4\pi A(r) \int_0^{\sqrt{2\Psi_T}} v^2 h dv \int_0^{\pi/2} \sin\lambda\, d\lambda, \tag{12.48}$$

$$\rho\sigma_r^2 = 4\pi A(r) \int_0^{\sqrt{2\Psi_T}} v^4 h dv \int_0^{\pi/2} \sin\lambda (\cos\lambda)^2\, d\lambda, \tag{12.49}$$

$$\rho\sigma_t^2 = 4\pi A^2(r) \int_0^{\sqrt{2\Psi_T}} v^4 h dv \int_0^{\pi/2} (\sin\lambda)^3\, d\lambda. \tag{12.50}$$

Performing the elementary integration over λ and changing the variables from v to Q, one finally obtains (do it!)

$$dv = -\frac{dQ}{\sqrt{2(\Psi_T - Q)}}, \tag{12.51}$$

and so Eqs. (12.42)–(12.44) are recovered, because $0 \leq Q \leq \Psi_T$. OM stellar systems are obviously a subset of those described in Section 12.2.2; they do not present streaming velocities, while their velocity dispersion tensor is anisotropic with two different components, σ_r^2 and σ_t^2. The relation between radial and tangential velocity dispersions in OM systems is very simple, and from Eq. (12.34) we find

$$\beta(r) = \frac{r^2}{r^2 + r_a^2}. \tag{12.52}$$

As a consequence, in OM systems, the velocity dispersion tensor is isotropic at the center *independently* of the particular value of r_a. When $r_a \to \infty$, $\beta(r) = 0\ \forall r \geq 0$ (i.e., the system is globally isotropic) due to the fact that now $Q = \mathcal{E}$. For fixed r_a the velocity dispersion tensor becomes more and more radially anisotropic for $r \to \infty$, while inside r_a the velocity dispersion tensor becomes more and more isotropic. This is the reason why the parameter r_a is called the anisotropy radius. OM models are widely used in applications due to their flexibility and simplicity of use (see Chapters 13–14).

12.2.2.3 Cuddeford Systems

An interesting generalization of OM models is given by the *Cuddeford anisotropic models* (Cuddeford 1991). Let Q be defined again by Eq. (12.40) and assume

$$f = J^{2\alpha} h(Q)\theta(Q) \geq 0, \quad \alpha > -1, \tag{12.53}$$

so that OM models are recovered for $\alpha = 0$. Then,

$$\rho = 4\pi A(r,\alpha) \int_0^{\Psi_T} [2(\Psi_T - Q)]^{\alpha+1/2} h(Q) dQ, \tag{12.54}$$

$$\rho\sigma_r^2 = 4\pi B(r,\alpha) \int_0^{\Psi_T} [2(\Psi_T - Q)]^{\alpha+3/2} h(Q) dQ, \tag{12.55}$$

$$\rho\sigma_t^2 = 4\pi C(r,\alpha) \int_0^{\Psi_T} [2(\Psi_T - Q)]^{\alpha+3/2} h(Q) dQ, \tag{12.56}$$

where now

$$A(r,\alpha) = \frac{\sqrt{\pi}\,\Gamma(\alpha+1)}{2\Gamma(\alpha+3/2)} r^{2\alpha} A^{\alpha+1}(r), \tag{12.57}$$

$$B(r,\alpha) = \frac{\sqrt{\pi}\,\Gamma(\alpha+1)}{4\Gamma(\alpha+5/2)} r^{2\alpha} A^{\alpha+1}(r), \tag{12.58}$$

$$C(r,\alpha) = \frac{\sqrt{\pi}\,\Gamma(\alpha+2)}{2\Gamma(\alpha+5/2)} r^{2\alpha} A^{\alpha+2}(r), \tag{12.59}$$

the function $A(r)$ is given in Eq. (12.45), and $\Gamma(x)$ is the complete Euler gamma function in Eq. (A.42). The proof of the identities above is similar to that of OM models: with the same change of coordinates, one obtains

$$\rho = 4\pi r^{2\alpha} A^{\alpha+1}(r) \int_0^{\sqrt{2\Psi_T}} v^{2\alpha+2} h dv \int_0^{\pi/2} (\sin\lambda)^{2\alpha+1}\, d\lambda, \tag{12.60}$$

where the convergence of the angular integral requires $\alpha > -1$. Moreover

$$\rho\sigma_r^2 = 4\pi r^{2\alpha} A^{\alpha+1}(r) \int_0^{\sqrt{2\Psi_T}} v^{2\alpha+4} h dv \int_0^{\pi/2} (\sin\lambda)^{2\alpha+1} (\cos\lambda)^2 \, d\lambda, \quad (12.61)$$

$$\rho\sigma_t^2 = 4\pi r^{2\alpha} A^{\alpha+2}(r) \int_0^{\sqrt{2\Psi_T}} v^{2\alpha+4} h dv \int_0^{\pi/2} (\sin\lambda)^{2\alpha+3} \, d\lambda. \quad (12.62)$$

After integration over λ by using Eqs. (A.53)–(A.54), the change of variable in Eq. (12.51) completes the proof.

The velocity dispersion anisotropy profile of Cuddeford models is

$$\beta(r) = \frac{r^2 - \alpha r_a^2}{r^2 + r_a^2}. \quad (12.63)$$

Therefore, for $-1 < \alpha < 0$, the function β is positive and the velocity dispersion is radially anisotropic at all radii independently of the particular (finite) value of r_a, while as expected for $\alpha = 0$ one recovers the OM anisotropy. Finally, for $\alpha > 0$, the anisotropy is tangential in the inner regions with $r < \sqrt{\alpha} r_a$ and radial for $r > \sqrt{\alpha} r_a$; in the limit $\alpha \to \infty$, the orbital structure is fully tangentially anisotropic (i.e., $\beta \to -\infty$). Note also that for $r_a \to \infty$, $\beta(r) = -\alpha \; \forall r \geq 0$ (i.e., the systems are characterized by a *constant anisotropy* profile). In practice, Cuddeford models can be seen as simple generalizations of both OM models (for $\alpha = 0$) and constant-anisotropy models (for $r_a \to \infty$), having the advantage of describing some amount of tangential anisotropy without the problems of OM tangential models (see Exercise 12.5).

12.2.2.4 *Generalized Cuddeford Systems*

We finally consider a very general family of models containing as special cases the models encountered in the previous sections. These models have been introduced in Ciotti and Morganti (2010a), and they will be used to prove some quite interesting results in Chapter 14. We refer to this family as the multicomponent, generalized Cuddeford systems, in which the DF associated with *each* density component is the sum of an arbitrary number of Cuddeford DFs in Eq. (12.53) with arbitrary positive (constant) weights w_i and possibly different anisotropy radii r_{ai}; in other words, for *each* component

$$f = J^{2\alpha} \sum_i w_i h(Q_i)\theta(Q_i), \quad Q_i = \mathcal{E} - \frac{J^2}{2r_{ai}^2}, \quad \alpha > -1. \quad (12.64)$$

Note that for a given density component, the function h and the angular momentum exponent α are fixed, but they can be different for different components; the list $(\alpha, w_1, \ldots, r_{a1}, \ldots)$ is another example of the vector parameter $\boldsymbol{\lambda}_k$. The student should prove that Eqs. (12.54)–(12.56) still hold, where now

$$A(r,\alpha) = \frac{\sqrt{\pi}\,\Gamma(\alpha+1) r^{2\alpha}}{2\Gamma(\alpha+3/2)} \sum_i \frac{w_i}{(1 + r^2/r_{ai}^2)^{\alpha+1}}, \quad (12.65)$$

$$B(r,\alpha) = \frac{\sqrt{\pi}\,\Gamma(\alpha+1)r^{2\alpha}}{4\Gamma(\alpha+5/2)} \sum_i \frac{w_i}{(1+r^2/r_{\rm ai}^2)^{\alpha+1}}, \tag{12.66}$$

$$C(r,\alpha) = \frac{\sqrt{\pi}\,\Gamma(\alpha+2)r^{2\alpha}}{2\Gamma(\alpha+5/2)} \sum_i \frac{w_i}{(1+r^2/r_{\rm ai}^2)^{\alpha+2}}. \tag{12.67}$$

Of course, the orbital anisotropy profile associated with Eq. (12.64) is in general *not* a Cuddeford one, being

$$\beta(r) = 1 - \frac{C(r,\alpha)}{2B(r,\alpha)} = 1 - (\alpha+1)\frac{\sum_i w_i/(1+r^2/r_{\rm ai}^2)^{\alpha+2}}{\sum_i w_i/(1+r^2/r_{\rm ai}^2)^{\alpha+1}}. \tag{12.68}$$

Quite general anisotropy profiles can be obtained through specific choices of the weights w_i, the anisotropy radii $r_{\rm ai}$, and the exponent α. However, near the center, $\beta(r) \sim -\alpha$ and $\beta(r) \sim 1$ for $r \to \infty$ independently of the specific values of w_i and $r_{\rm ai}$ and the number of subcomponents i. We conclude this discussion by stressing that the given examples should not be considered to be even a partial list of the models that can be found in the literature; for example, the student is invited to consult Cuddeford and Louis (1995), Gerhard (1991), and Louis (1993).

12.2.2.5 *Stellar Polytropes, King, Wilson, Michie, f_∞, and $f^{(\nu)}$ Models*

Having introduced some of the most used families of DFs for spherical stellar systems, we now move on to one of the most important cases of the direct problems of stellar dynamics: the construction of the family of so-called *spherically symmetric self-gravitating stellar polytropes*. The applications of this are numerous and far-reaching because from a mathematical point of view the same problem is encountered in the construction of self-gravitating, polytropic gaseous spheres (see Chapter 10; see also Binney and Tremaine 2008). The starting point is to consider the isotropic DF

$$f = A_n(\mathcal{E} - \mathcal{E}_{\rm t})^{n-3/2}\theta(\mathcal{E} - \mathcal{E}_{\rm t}), \tag{12.69}$$

where n is a free parameter and $\mathcal{E}_{\rm t}$ is a truncation energy. From Eq. (12.15) with $\Psi_{\rm T}(r) = \Psi(r)$ (the self-gravitating condition) and intoducing $\Phi \equiv \Psi - \mathcal{E}_{\rm t}$, one immediately obtains (prove it!)

$$\rho = B_n\Phi^n\theta(\Phi), \quad B_n = A_n\frac{(2\pi)^{3/2}\Gamma(n-1/2)}{\Gamma(n+1)}, \quad n > \frac{1}{2}, \tag{12.70}$$

so that the density vanishes at the radius $r_{\rm t}$ (if it exists!) where $\Phi(r_{\rm t}) = 0$ (i.e., where $\Psi(r_{\rm t}) = \mathcal{E}_{\rm t}$). Moreover, it is a simple exercise to show that from Eq. (12.17) each of the three identical diagonal terms of the velocity dispersion tensor is

$$\sigma_{ii}^2 = \frac{\Phi}{n+1}\theta(\Phi) \equiv \sigma^2, \tag{12.71}$$

so that the velocity dispersion also vanishes at the system edge and it is linearly proportional to $\Psi - \mathcal{E}_{\rm t}$. Notice that no sum is intended over the repeated indices.

The Poisson equation in spherical coordinates is

$$\frac{1}{r^2}\frac{d}{dr}\left(r^2\frac{d\Phi}{dr}\right) = -4\pi G B_n \Phi^n \theta(\Phi), \tag{12.72}$$

with the regular boundary conditions $\Phi(0) = \Phi_0$ and $d\Phi/dr = 0$ at $r = 0$, and the Heaviside function takes care of the fact that the system density is zero outside the truncation radius. Two physical scales can be associated with this problem, and Eq. (12.72) can be cast in dimensionless form by introducing

$$s \equiv \frac{r}{r_0}, \quad \varphi \equiv \frac{\Phi}{\Phi_0}, \quad r_0 \equiv \frac{1}{\sqrt{4\pi G B_n \Phi_0}}. \tag{12.73}$$

The resulting dimensionless equation is the well-known *Lane–Emden* equation

$$\frac{1}{s^2}\frac{d}{ds}\left(s^2\frac{d\varphi}{ds}\right) = -\varphi^n \theta(\varphi), \tag{12.74}$$

the solutions of which have also been investigated extensively in the context of stellar evolution (e.g., see Chandrasekhar 1939). It can be proved that: (1) for $n < 5$, the total mass associated with the density distribution is finite and the density vanishes at a finite truncation radius r_t; (2) for $n = 5$, the density is spatially untruncated, but the mass is still finite; and (3) for $n > 5$, the density is nontruncated and the mass is infinite. Moreover, analytical solutions exist for two values of $n > 0$, namely $n = 1$ (the linear Helmholtz equation)

$$\varphi = \begin{cases} \dfrac{\sin(s)}{s}, & \text{if } s < \pi, \\ \dfrac{\pi}{s} - 1, & \text{if } s \geq \pi, \end{cases} \tag{12.75}$$

and $n = 5$ (the Schuster solution)

$$\varphi = \frac{1}{\sqrt{1 + s^2/3}}. \tag{12.76}$$

In both cases, the total mass can be found explicitly from the Newton theorem, namely

$$M = -\lim_{r \to r_t} \frac{r^2}{G}\frac{d\Phi}{dr} = -\frac{\Phi_0 r_0}{G} \lim_{s \to s_t} s^2 \frac{d\varphi}{ds} \tag{12.77}$$

(i.e., $M_1 = \pi \Phi_0 r_0 / G$ and $M_5 = \sqrt{3}\Phi_0 r_0 / G$). In the astrophysical literature, the $n = 5$ case is also known as the *Plummer model* (Plummer 1911; see also Chapter 13).

The term *stellar polytropes* for the objects in Eq. (12.70) derives from their formal similarity to *gaseous polytropes* (e.g., see Chandrasekhar 1939; see also Section 10.4.1). In fact, the equation of hydrostatic equilibrium (10.30) for a self-gravitating gaseous sphere with polytropic stratification $p \propto \rho^\gamma$ (where γ is the *polytropic index*) reads $dp/dr = -\rho d\phi/dr$, and its solutions are given by Eq. (10.31). In particular, notice how Eqs. (12.70)–(12.71) are formally identical to Eq. (10.31) provided

$$\gamma = 1 + \frac{1}{n}, \quad \sigma^2 = \frac{k_B T}{\langle \mu \rangle m_p}, \quad B_n = \frac{\rho_0}{\sigma_0^{2n}(1+n)^n}, \quad \mathcal{E}_t = \Psi_0 - (1+n)\sigma_0^2, \tag{12.78}$$

where Ψ_0, σ_0, and ρ_0 are the values of the potential, the velocity dispersion, and the density at some distance r_0 for the center, respectively; in particular, notice that from the second of the identities in Eq. (12.78) we have the identification $\sigma_0^2 = \beta_0$ from the last identity of Eq. (10.31). Therefore, we have obtained the important result that the density and velocity dispersion radial profiles of a stellar polytrope of index n are the same as the density and temperature radial profiles of the gaseous polytrope with the same potential and with parameters as in Eq. (12.78).

A special and extremely important limiting case of polytropic systems is the so-called *isothermal sphere*, obtained from the DF

$$f = \frac{\rho_1}{(2\pi\sigma^2)^{3/2}} e^{\frac{\mathcal{E}}{\sigma^2}}, \quad \sigma_{ii}^2 = \sigma^2, \tag{12.79}$$

where ρ_1 is a density scale, no energy truncation is imposed, and for which the velocity dispersion is independent of the radius (see Exercise 12.6). In fact, if we consider the formal analogy (for suitable choices of the parameters) of stellar and gaseous polytropes and that the limit case of a isothermal gaseous equilibrium in Eq. (10.32) is obtained from the limit $\gamma \to 1$ of gaseous polytropes, we can reasonably expect that Eq. (12.79) can also be obtained as the limit of Eq. (12.69) for $n \to \infty$, after B_n and $\mathcal{E}_t$ are given by Eq. (12.78). We prove this result in Exercise 12.7.

Other isotropic DFs are obtained by truncating the isothermal sphere for $\mathcal{E} \leq \mathcal{E}_t$ (where $\mathcal{E}_t$ is a *truncation energy* mimicking the effect of escapers above the energy $-\mathcal{E}_t$; see also Section 6.2.1). Important cases, often adopted to model globular clusters (e.g., see Bertin 2014; Binney and Tremaine 2008; Spitzer 1987), are the *King* models (King 1966)

$$f_K = \frac{\rho_1}{(2\pi\sigma^2)^{3/2}} \left(e^{\frac{\mathcal{E}}{\sigma^2}} - e^{\frac{\mathcal{E}_t}{\sigma^2}} \right) \theta(\mathcal{E} - \mathcal{E}_t), \tag{12.80}$$

and the *Wilson* models (Wilson 1975)

$$f_W = \frac{\rho_1}{(2\pi\sigma^2)^{3/2}} \left(e^{\frac{\mathcal{E}}{\sigma^2}} - e^{\frac{\mathcal{E}_t}{\sigma^2}} - \frac{\mathcal{E} - \mathcal{E}_t}{\sigma^2} \right) \theta(\mathcal{E} - \mathcal{E}_t). \tag{12.81}$$

Anisotropic spherical models can be obtained by modifying the King/Wilson models. For example, the *Michie* models (Michie 1963) are obtained for

$$f_M = f_K(\mathcal{E}) \times e^{-\frac{J^2}{2r_a^2\sigma^2}}. \tag{12.82}$$

Finally, special mention is due to the f_∞ models (e.g., see Bertin 2014; Bertin and Stiavelli 1984; Stiavelli and Bertin 1985, 1987; see also Merritt et al. 1989 for the "negative temperature" version of these models), associated with a DF that in the spherical limit becomes

$$f_\infty = A\mathcal{E}^{3/2} e^{a\mathcal{E} - cJ^2} \theta(\mathcal{E}), \tag{12.83}$$

and to the $f^{(\nu)}$ models (e.g., see Bertin 2014; Bertin and Trenti 2003; Stiavelli and Bertin 1987)

$$f^{(\nu)} = A e^{a\mathcal{E} - d\left(\frac{J^2}{\mathcal{E}^{3/2}}\right)^{\nu/2}} \theta(\mathcal{E}). \tag{12.84}$$

These two last families are constructed on the basis of statistical mechanics arguments, and the associated density distributions, obtained by numerical solution of the associated Poisson equations, reproduce remarkably well the density (light) distribution of elliptical galaxies (see also Section 6.2). Other self-consistent models built to describe galactic disks coupled self-consistently with a dark matter halo or considering the effects of the galactic tidal field and internal rotation, which have been particularly successful in the modeling of globular clusters, have been recently proposed (Amorisco and Bertin 2010; Bertin and Varri 2008; Varri and Bertin 2009, 2012, and references therein). Unfortunately (but not unexpectedly), for all of these models, only numerical solutions of the Poisson equations can be found.

12.3 Two-Integral Axisymmetric Systems

We now move to consider the case of axisymmetric stellar systems, so that the total potential can be written in cylindrical coordinates (R,φ,z) as $\Psi_T = \Psi_T(R,z)$, independent of the azimuthal angle φ. We study the natural case of a DF depending on the two classical integrals of motion related to the assumed symmetry (i.e., $I_1 = \mathcal{E} = \Psi_T - \|\mathbf{v}\|^2/2$) and $I_2 = J_z$ (the component of the angular momentum along the z-axis), with

$$f = h(\mathcal{E}, J_z)\theta(\mathcal{E}). \tag{12.85}$$

In complete generality, let $f_\pm$ be the even (odd) component of the DF with respect to J_z, i.e.,

$$f_\pm(\mathcal{E}, J_z) = \frac{f(\mathcal{E}, J_z) \pm f(\mathcal{E}, -J_z)}{2}. \tag{12.86}$$

We now prove that in these systems

$$\rho = \frac{4\pi}{R}\int_0^{\Psi_T} d\mathcal{E}\int_0^{R\sqrt{2(\Psi_T-\mathcal{E})}} h_+(\mathcal{E}, J_z)\,dJ_z, \tag{12.87}$$

$$\overline{v_R} = \overline{v_z} = \overline{v_R v_z} = \overline{v_R v_\varphi} = \overline{v_z v_\varphi} = 0, \quad \sigma_R^2 = \sigma_z^2, \tag{12.88}$$

$$\rho\overline{v_\varphi} = \frac{4\pi}{R^2}\int_0^{\Psi_T} d\mathcal{E}\int_0^{R\sqrt{2(\Psi_T-\mathcal{E})}} h_-(\mathcal{E}, J_z)J_z\,dJ_z. \tag{12.89}$$

Moreover

$$\rho\overline{v_\varphi^2} = \frac{4\pi}{R^3}\int_0^{\Psi_T} d\mathcal{E}\int_0^{R\sqrt{2(\Psi_T-\mathcal{E})}} h_+(\mathcal{E}, J_z)J_z^2\,dJ_z, \tag{12.90}$$

and with the introduction of the *meridional* velocity dispersion $\sigma_m^2 = \sigma_R^2 + \sigma_z^2 = 2\sigma_z^2$

$$\rho\sigma_m^2 = \frac{4\pi}{R}\int_0^{\Psi_T} d\mathcal{E}\int_0^{R\sqrt{2(\Psi_T-\mathcal{E})}} h_+(\mathcal{E}, J_z)\left[2(\Psi_T - \mathcal{E}) - \frac{J_z^2}{R^2}\right] dJ_z. \tag{12.91}$$

In practice, in a two-integral axisymmetric system, ordered rotation can be present only around the z-axis, and only the *even* part of the DF contributes to the density. Furthermore, $\rho = \rho(R, \Psi_T)$, obtained from Eq. (12.87), depends explicitly both on R and Ψ_T. The only admissible streaming motion is along the azimuthal direction, and it is determined uniquely by the *odd* component of the DF. The velocity dispersion ellipsoids are aligned with the local basis $(\mathbf{f}_R, \mathbf{f}_\varphi, \mathbf{f}_z)$ and are rotationally symmetric around the azimuthal direction $\mathbf{f}_\varphi$, with $\sigma_\varphi^2 = \overline{v_\varphi^2} - \overline{v_\varphi}^2$ and $\sigma_R^2 = \sigma_z^2 = \sigma_\mathrm{m}^2/2$. Thus, the systems described here can be either isotropic or azimuthally anisotropic. In the isotropic case, the velocity dispersion ellipsoid is spherical and the deviation from spherical symmetry of the system is due to the ordered rotation field $\overline{v_\varphi}$, in analogy with rotating fluid systems. By contrast, when $\overline{v_\varphi} = 0$, all of the deviations from spherical symmetry are supported by the tangential velocity dispersion σ_φ. In the next chapter, we will discuss these important aspects in some detail. Finally, note that as a spherical system is also axisymmetric, in principle a collisionless stellar system could have some net rotation (Lynden-Bell 1960). Of course, this is impossible in the standard cases encountered in fluid dynamics, as the tensor pressure is isotropic (see Chapter 10).

From Eq. (12.8), $\Omega_k = \{(\mathbf{x}, \mathbf{v}) \in \gamma : \Psi_T - \|\mathbf{v}\|^2/2 \geq 0\}$, and so the velocity sections $\Omega_{k\mathbf{x}}$ are spheres of radius $\sqrt{2\Psi_T(\mathbf{x})}$. In this case, the natural coordinates in configuration space are the cylindrical ones (R, φ, z) and the associated velocity components are (v_R, v_φ, v_z). Note that in the DF v_R and v_z appear only in the combination $v_m^2 = v_R^2 + v_z^2$, because $J_z = Rv_\varphi$, while $v^2 = v_m^2 + v_\varphi^2$, where v_m is the so-called *meridional* velocity. As a direct consequence of the geometry of $\Omega_{k\mathbf{x}}$, the natural coordinates for integration in velocity space are the spherical ones, with

$$v_R = v \sin\lambda \ \sin\mu, \quad v_z = v \sin\lambda \ \cos\mu, \quad v_\varphi = v \cos\lambda, \tag{12.92}$$

with $0 \leq v = \|\mathbf{v}\| \leq \sqrt{2\Psi_T}$, $0 \leq \lambda < \pi$, and $0 \leq \mu < 2\pi$ (see Figure 12.3), so that $v_m = v \sin\lambda$ and f is independent of μ; finally (prove it!)

$$d^3\mathbf{v} = dv_R dv_z dv_\varphi = v^2 \sin\lambda \ d\lambda \ d\mu. \tag{12.93}$$

Integration of μ and reduction of the integration over λ from $(0, \pi)$ to $(0, \pi/2)$, taking into account the parity of f_+ and f_-, leads to

$$\rho = 4\pi \int_0^{\sqrt{2\Psi_T}} v^2 dv \int_0^{\pi/2} h_+ \sin\lambda \ d\lambda, \tag{12.94}$$

$$\rho\overline{v_\varphi} = 4\pi \int_0^{\sqrt{2\Psi_T}} v^3 dv \int_0^{\pi/2} h_- \sin\lambda \ \cos\lambda \ d\lambda, \tag{12.95}$$

$$\rho\overline{v_\varphi^2} = 4\pi \int_0^{\sqrt{2\Psi_T}} v^4 dv \int_0^{\pi/2} h_+ \sin\lambda \ (\cos\lambda)^2 \ d\lambda, \tag{12.96}$$

$$\rho\sigma_\mathrm{m}^2 = 4\pi \int_0^{\sqrt{2\Psi_T}} v^4 dv \int_0^{\pi/2} h_+ (\sin\lambda)^3 \ d\lambda. \tag{12.97}$$

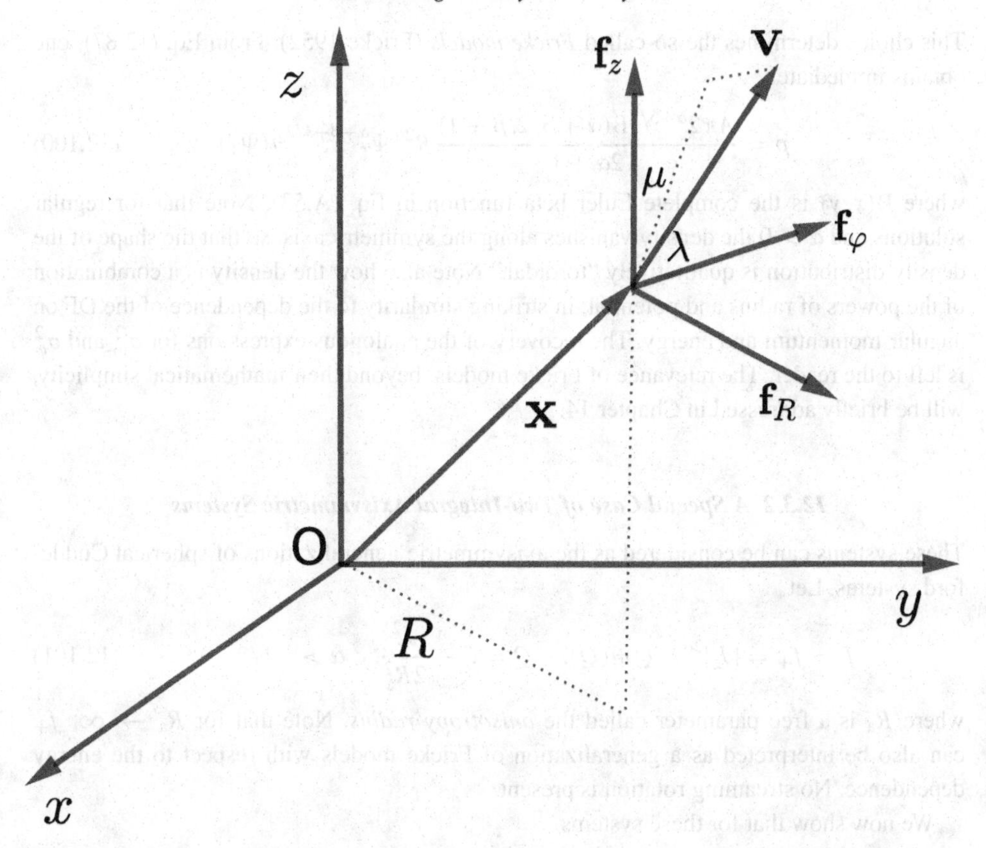

Figure 12.3 The decomposition of the velocity vector $\mathbf{v}$ adopted to integrate the DF in the two-integral case for cylindrically symmetric systems (Section 12.3). Due to the degeneracy of the velocity components in the meridional plane, the natural coordinates are the spherical ones, with the colatitude λ measured from $\mathbf{f}_\varphi$ and the longitude μ measured from $\mathbf{f}_z = \mathbf{e}_z$ in the plane $(\mathbf{f}_z, \mathbf{f}_R)$.

Finally, by changing the coordinates from (v, λ) to $(\mathcal{E}, J_z)$

$$dv d\lambda = \frac{d\mathcal{E} dJ_z}{R\sqrt{2(\Psi_T - \mathcal{E})}\sqrt{2(\Psi_T - \mathcal{E}) - J_z^2/R^2}}, \tag{12.98}$$

and so Eqs. (12.87)–(12.91) are recovered, because $0 \leq \mathcal{E} \leq \Psi_T, 0 \leq J_z \leq R\sqrt{2(\Psi_T - \mathcal{E})}$.

12.3.1 Fricke's DF

A particularly simple and useful specialization of the general identities just derived is obtained by considering the special family

$$f_+ = A|J_z|^{2\alpha}\mathcal{E}^\beta\theta(\mathcal{E}), \quad A > 0, \quad \alpha > -1/2, \quad \beta > -1. \tag{12.99}$$

This choice determines the so-called *Fricke models* (Fricke 1952). From Eq. (12.87), one obtains immediately

$$\rho = \frac{A\pi 2^{\alpha+5/2}\mathrm{B}(\alpha+3/2,\beta+1)}{2\alpha+1} R^{2\alpha}\Psi_{\mathrm{T}}^{\alpha+\beta+3/2}\theta(\Psi_{\mathrm{T}}), \tag{12.100}$$

where $\mathrm{B}(x,y)$ is the complete Euler beta function in Eq. (A.53). Note that for regular solutions and $\alpha > 0$ the density vanishes along the symmetry axis, so that the shape of the density distribution is qualitatively "toroidal." Note also how the density is a combination of the powers of radius and potential, in striking similarity to the dependence of the DF on angular momentum and energy. The recovery of the analogous expressions for σ_{m}^2 and σ_{φ}^2 is left to the reader. The relevance of Fricke models, beyond their mathematical simplicity, will be briefly addressed in Chapter 14.

12.3.2 A Special Case of Two-Integral Axisymmetric Systems

These systems can be considered as the axisymmetric generalizations of spherical Cuddeford systems. Let

$$f = f_+ = |J_z|^{2\alpha}h(Q)\theta(Q), \quad Q = \mathcal{E} - \frac{J_z^2}{2R_{\mathrm{a}}^2}, \quad \alpha > -1/2, \tag{12.101}$$

where R_{a} is a free parameter called the *anisotropy radius*. Note that for $R_{\mathrm{a}} \to \infty$, f_+ can also be interpreted as a generalization of Fricke models with respect to the energy dependence. No streaming rotation is present.

We now show that for these systems

$$\rho = 4\pi A(R,\alpha)\int_0^{\Psi_{\mathrm{T}}}[2(\Psi_{\mathrm{T}}-Q)]^{\alpha+1/2}h(Q)dQ, \tag{12.102}$$

$$\rho\sigma_{\varphi}^2 = 4\pi B(R,\alpha)\int_0^{\Psi_{\mathrm{T}}}[2(\Psi_{\mathrm{T}}-Q)]^{\alpha+3/2}h(Q)dQ, \tag{12.103}$$

$$\rho\sigma_{\mathrm{m}}^2 = 4\pi C(R,\alpha)\int_0^{\Psi_{\mathrm{T}}}[2(\Psi_{\mathrm{T}}-Q)]^{\alpha+3/2}h(Q)dQ, \tag{12.104}$$

where

$$A(R,\alpha) = \frac{R^{2\alpha}}{(2\alpha+1)(1+R^2/R_{\mathrm{a}}^2)^{\alpha+1/2}}, \tag{12.105}$$

$$B(R,\alpha) = \frac{R^{2\alpha}}{(2\alpha+3)(1+R^2/R_{\mathrm{a}}^2)^{\alpha+3/2}}, \tag{12.106}$$

$$C(R,\alpha) = \frac{2A(R,\alpha)}{2\alpha+3}. \tag{12.107}$$

The natural coordinates in configuration space are the cylindrical ones. In fact, $\Omega_k = \{(\mathbf{x},\mathbf{v}) \in \gamma : \Psi_{\rm T} - \|\mathbf{v}\|^2/2 - J_z^2/2R_{\rm a}^2 = \Psi_{\rm T} - v_m^2/2 - v_\varphi^2(1+R^2/R_{\rm a}^2)/2 \geq 0\}$, so that the velocity sections $\Omega_{k\mathbf{x}}$ are rotation ellipsoids around the azimuthal direction $\mathbf{e}_\varphi$. With f being an even function in J_z, no streaming motions can be present ($v_\varphi = 0$), and the density flattening is due to azimuthal velocity dispersion σ_φ only. As a direct consequence of the geometry of $\Omega_{k\mathbf{x}}$, the natural coordinates for integration in velocity space are the elliptical ones, i.e.,

$$v_R = v \sin\lambda\, \cos\mu, \quad v_z = v\sin\lambda\, \sin\mu, \quad v_\varphi = \frac{v\cos\lambda}{\sqrt{1+R^2/R_{\rm a}^2}}, \tag{12.108}$$

with $0 \leq v \leq \sqrt{2\Psi_{\rm T}}$, $0 \leq \lambda < \pi$, and $0 \leq \mu < 2\pi$. Note that $v_m = v\sin\lambda$ and that h depends only on v; moreover (prove it!)

$$d^3\mathbf{v} = \frac{v^2 \sin\lambda}{\sqrt{1+R^2/R_{\rm a}^2}}\, d\lambda\, d\mu. \tag{12.109}$$

Integration over μ and reduction of the integration interval of λ from $(0,\pi)$ to $(0,\pi/2)$ leads to

$$\rho = \frac{4\pi R^{2\alpha}}{(1+R^2/R_{\rm a}^2)^{\alpha+1/2}} \int_0^{\sqrt{2\Psi_{\rm T}}} h v^{2\alpha+2} dv \int_0^{\pi/2} \sin\lambda\, (\cos\lambda)^{2\alpha}\, d\lambda, \tag{12.110}$$

$$\rho\sigma_\varphi^2 = \frac{4\pi R^{2\alpha}}{(1+R^2/R_{\rm a}^2)^{\alpha+3/2}} \int_0^{\sqrt{2\Psi_{\rm T}}} h v^{2\alpha+4} dv \int_0^{\pi/2} \sin\lambda\, (\cos\lambda)^{2\alpha+2}\, d\lambda, \tag{12.111}$$

$$\rho\sigma_{\rm m}^2 = \frac{4\pi R^{2\alpha}}{(1+R^2/R_{\rm a}^2)^{\alpha+1/2}} \int_0^{\sqrt{2\Psi_{\rm T}}} h v^{2\alpha+4} dv \int_0^{\pi/2} (\sin\lambda)^3\, (\cos\lambda)^{2\alpha}\, d\lambda. \tag{12.112}$$

Finally, integration over λ by using Eqs. (A.53)–(A.54) and the change of variable

$$dv = -\frac{dQ}{\sqrt{2(\Psi_{\rm T}-Q)}}, \quad 0 \leq Q \leq \Psi_{\rm T}, \tag{12.113}$$

completes the proof. Note that for convergence $\alpha > -1/2$.

For this family of models

$$\beta(R,z) = 1 - \frac{\sigma_\varphi^2}{\sigma_{\rm m}^2/2} = \frac{R^2 - 2\alpha R_{\rm a}^2}{R^2 + R_{\rm a}^2} \tag{12.114}$$

(i.e., in these systems, the distribution of anisotropy is independent of z and of the specific form of $h(Q)$). The interpretation of anisotropy for increasing R and for various $\alpha > -1/2$ is left to the reader. As for the simpler case of Fricke models and for these models when $\alpha > 0$ and h is sufficiently regular, the associated density distribution is of toroidal shape, with ρ vanishing on the symmetry axis.

Exercises

12.1 Let us consider a system with $f = f(\mathcal{E})\theta(\mathcal{E} - \mathcal{E}_{\rm t})$ truncated at relative energies below some value $\mathcal{E}_{\rm t}$. Show that Eqs. (12.15)–(12.17) still hold with a lower limit of integration $\mathcal{E}_{\rm t}$ instead of 0, and conclude that for given $\Psi_{\rm T}$ a spatial truncation of the density will occur on the surface $\Psi_{\rm T}(\mathbf{x}_{\rm t}) = \mathcal{E}_{\rm t}$. If no energy truncation is imposed on the DF, then $\mathcal{E}_{\rm t} = -\infty$. *Hint*: See Footnote 3 of this chapter.

12.2 From the identity $\mathbf{v} = v_x\mathbf{e}_x + v_y\mathbf{e}_y + v_z\mathbf{e}_z = v_r\mathbf{f}_r + v_\vartheta\mathbf{f}_\vartheta + v_\varphi\mathbf{f}_\varphi$, show by direct evaluation of the Jacobian that

$$d^3\mathbf{v} = dv_x dv_y dv_z = dv_r dv_\vartheta dv_\varphi. \tag{12.115}$$

Moreover, show by direct evaluation that from $\mathbf{J} = \mathbf{x} \wedge \mathbf{v}$ one obtains

$$J = \|\mathbf{J}\| = r v_t, \quad v_t = \sqrt{v_\vartheta^2 + v_\varphi^2}. \tag{12.116}$$

Hint: See Exercise 9.2.

12.3 For a generic spherical, anisotropic stellar system with $f = f(\mathcal{E}, J^2)$, by using Eqs. (12.25), (12.27), and (12.28), define the radial and tangential "pressures," respectively, as

$$p_{\rm r}(r, \Psi_{\rm T}) \equiv \rho\sigma_{\rm r}^2, \quad p_{\rm t}(r, \Psi_{\rm T}) \equiv \rho\sigma_{\rm t}^2. \tag{12.117}$$

Prove that, independently of the specific $\Psi_{\rm T}(r)$ radial profile, the remarkable identities

$$\rho(r) = \frac{\partial p_{\rm r}}{\partial \Psi_{\rm T}}, \quad \beta(r) \equiv 1 - \frac{p_{\rm t}}{2p_{\rm r}} = -\frac{1}{2}\frac{\partial \ln p_{\rm r}}{\partial \ln r}, \tag{12.118}$$

hold (e.g., see Baes and van Hese 2007; Bertin et al. 1994; Dejonghe 1986; Dejonghe and Merritt 1992; Spies and Nelson 1974), where the anisotropy profile $\beta(r)$ is defined in Eq. (12.34).

12.4 This exercise introduces the *differential energy distribution*, defined so that the mass (or number) of stars with energy in the range $(\mathcal{E}, \mathcal{E} + d\mathcal{E})$ is given by $dM(\mathcal{E}) = \mathcal{N}(\mathcal{E})d\mathcal{E}$. Show that for an isotropic, spherical system in a total potential $\Psi_{\rm T}$, we have

$$\mathcal{N}(\mathcal{E}) = f(\mathcal{E})g(\mathcal{E}), \quad g(\mathcal{E}) = 16\pi^2 \int_0^{r(\mathcal{E})} r^2\sqrt{2(\Psi_{\rm T} - \mathcal{E})}dr, \tag{12.119}$$

where $\Psi_{\rm T}[r(\mathcal{E})] = \mathcal{E}$ (for interesting properties of this function, see Binney 1982b; Binney and Tremaine 2008; Ciotti 1991). *Hints*: In the expression for the total mass $M = 4\pi \int_0^\infty r^2\rho(r)dr$, insert the expression for ρ in Eq. (12.15) and invert the order of integration in the plane $(r, \mathcal{E})$.

12.5 The family of *tangentially anisotropic* OM models is obtained from Eq. (12.41) with

$$Q = \mathcal{E} + \frac{J^2}{r_{\rm a}^2}. \tag{12.120}$$

Show that

$$\beta(r) = -\frac{r^2}{r_a^2 - r^2}. \tag{12.121}$$

Note that the velocity dispersion tensor is still isotropic in the center, but now is increasingly tangentially anisotropic with increasing radius. Note also that, at variance with the radially anisotropic case, Eq. (12.121) can be applied only to spatially *truncated* systems, provided that $r_a > r_t$, where r_t is the truncation radius (i.e., $\rho(r) = 0$ for $r \geq r_t$). Because of this limitation, these models are not of much use.

12.6 Consider the DF of the isothermal sphere in Eq. (12.79). By integration over the velocity space, show that

$$\rho = \rho_1 e^{\frac{\Psi}{\sigma^2}}, \quad \sigma_{ii}^2 = \sigma^2, \tag{12.122}$$

and then recast the Poisson equation (12.72) in terms of the density as

$$\frac{1}{r^2}\frac{d}{dr}\left(r^2 \frac{d\ln\rho}{dr}\right) = -\frac{4\pi G}{\sigma^2}\rho. \tag{12.123}$$

Finally, use $\rho = \rho_1 (r/r_1)^{-\alpha}$ as a trial solution and prove by substitution into Eq. (12.123) that a self-gravitating singular solution exists for $\alpha = 2$ and $\rho_1 r_1^2 = \sigma^2/(2\pi G)$, coincident with the singular isothermal sphere in Eqs. (2.63)–(2.64) for $v_c^2 = 2\sigma^2$. Of course, a *regular* solution of Eq. (12.123) also exists, but it cannot be expressed in terms of elementary functions (e.g., see Binney and Tremaine 2008; Chandrasekhar 1939).

12.7 Show that the DF of the isothermal sphere in Eq. (12.79) can be obtained as the limit for $n \to \infty$ of the DF of the polytropic sphere (12.69), once Eq. (12.78) is used. *Hints*: Write A_n in terms of B_n from Eq. (12.70)

$$A_n = \frac{\rho_0}{(2\pi)^{3/2}\sigma_0^{2n}} \frac{\Gamma(n+1)}{\Gamma(n-1/2)(n+1)^n}, \tag{12.124}$$

and fix $\mathcal{E}_t$ as in Eq. (12.78). Compute the limit of the resulting Eq. (12.69) for $n \to \infty$ by using Stirling expansion in Eq. (A.47) and show that Eq. (12.79) is recovered, with $\rho_1 = \rho_0 e^{\Psi_0/\beta_0}$ and $\sigma = \sigma_0$. As a consistency check, also compute the limits for $n \to \infty$ of the density and velocity dispersions in Eqs. (12.70) and (12.71) and reobtain Eq. (12.122).

12.8 Consider the King DF in Eq. (12.80). Integrate Eq. (12.15) and show that in the self-consistent case

$$\rho_K = \rho_1 e^{\tilde{\mathcal{E}}_t}\left[e^{\tilde{\Phi}}\,\mathrm{Erf}(\sqrt{\tilde{\Phi}}) - \frac{2\sqrt{\tilde{\Phi}}}{\sqrt{\pi}}\left(1 + \frac{2\tilde{\Phi}}{3}\right)\right]\theta(\tilde{\Phi}), \tag{12.125}$$

where $\tilde{\mathcal{E}}_t \equiv \mathcal{E}_t/\sigma^2$, $\tilde{\Phi} \equiv \Psi/\sigma^2 - \tilde{\mathcal{E}}_t$ and Erf is the error function in Eq. (A.57). Obtain the analogous expression for the velocity dispersion.

12.9 Show that, with the change of coordinates in Eq. (12.92) suggested by Figure 12.3, Eq. (12.93) holds, and from computation of the Jacobian, prove Eq. (12.98).

12.10 By using Eqs. (9.41) and (9.42), show that from the identity $\mathbf{v} = v_x\mathbf{e}_x + v_y\mathbf{e}_y + v_z\mathbf{e}_z = v_r\mathbf{f}_r + v_\vartheta\mathbf{f}_\vartheta + v_\varphi\mathbf{f}_\varphi$, the Cartesian and spherical phase-space velocity components are related as

$$\begin{cases} v_x = v_r \sin\vartheta\cos\varphi + v_\vartheta\cos\vartheta\cos\varphi - v_\varphi\sin\varphi, \\ v_y = v_r\sin\vartheta\sin\varphi + v_\vartheta\cos\vartheta\sin\varphi + v_\varphi\cos\varphi, \\ v_z = v_r\cos\vartheta - v_\vartheta\sin\vartheta. \end{cases} \tag{12.126}$$

Then consider a spherical stellar system with $f = f(\mathcal{E}, J^2)$, recall that $\sigma_\vartheta^2 = \sigma_\varphi^2$, and from averaging over the velocity space deduce that

$$\begin{cases} \overline{v_x} = \overline{v_y} = \overline{v_z} = 0, \\ \sigma_{xx}^2 = \sigma_r^2\sin^2\vartheta\cos^2\varphi + \sigma_\varphi^2(\cos^2\vartheta\cos^2\varphi + \sin^2\varphi), \\ \sigma_{yy}^2 = \sigma_r^2\sin^2\vartheta\sin^2\varphi + \sigma_\varphi^2(\cos^2\vartheta\sin^2\varphi + \cos^2\varphi), \\ \sigma_{zz}^2 = \sigma_r^2\cos^2\vartheta + \sigma_\varphi^2\sin^2\vartheta, \\ \sigma_{xy}^2 = (\sigma_r^2 - \sigma_\varphi^2)\sin^2\vartheta\sin\varphi\cos\varphi, \\ \sigma_{xz}^2 = (\sigma_r^2 - \sigma_\varphi^2)\sin\vartheta\cos\vartheta\cos\varphi, \\ \sigma_{yz}^2 = (\sigma_r^2 - \sigma_\varphi^2)\sin\vartheta\cos\vartheta\sin\varphi. \end{cases} \tag{12.127}$$

12.11 By using Eqs. (9.41)–(9.43), show that from the identity $\mathbf{v} = v_x\mathbf{e}_x + v_y\mathbf{e}_y + v_z\mathbf{e}_z = v_R\mathbf{f}_R + v_\varphi\mathbf{f}_\varphi + v_z\mathbf{f}_z$, the Cartesian and cylindrical phase-space velocity components are related as

$$v_x = v_R\cos\varphi - v_\varphi\sin\varphi, \quad v_y = v_R\sin\varphi + v_\varphi\cos\varphi, \quad v_z = v_z. \tag{12.128}$$

Then consider an axisymmetric stellar system with $f = f(\mathcal{E}, J_z)$, recall that $\sigma_R^2 = \sigma_z^2$, and from averaging over the velocity space deduce that

$$\begin{cases} \overline{v_x} = -\overline{v_\varphi}\sin\varphi, \quad \overline{v_y} = \overline{v_\varphi}\cos\varphi, \quad \overline{v_z} = 0, \\ \sigma_{xx}^2 = \sigma_R^2\cos^2\varphi + \sigma_\varphi^2\sin^2\varphi, \\ \sigma_{yy}^2 = \sigma_R^2\sin^2\varphi + \sigma_\varphi^2\cos^2\varphi, \\ \sigma_{zz}^2 = \sigma_R^2, \\ \sigma_{xy}^2 = (\sigma_R^2 - \sigma_\varphi^2)\sin\varphi\cos\varphi, \\ \sigma_{xz}^2 = \sigma_{yz}^2 = 0. \end{cases} \tag{12.129}$$

12.12 Show that the spherical and cylindrical phase-space velocity components of $\mathbf{v} = v_r\mathbf{f}_r + v_\vartheta\mathbf{f}_\vartheta + v_\varphi\mathbf{f}_\varphi = v_R\mathbf{f}_R + v_\varphi\mathbf{f}_\varphi + v_z\mathbf{f}_z$ are related as

$$v_r = v_R \sin\vartheta + v_z \cos\vartheta, \quad v_\vartheta = v_R \cos\vartheta - v_z \sin\vartheta, \quad v_\varphi = v_\varphi. \tag{12.130}$$

Consider again an axisymmetric stellar system with $f = f(\mathcal{E}, J_z)$ and show that

$$\begin{cases} \overline{v_r} = \overline{v_\vartheta} = 0, \quad \overline{v_\varphi} = \overline{v_\varphi}, \\ \sigma_{rr}^2 = \sigma_{\vartheta\vartheta}^2 = \sigma_R^2 = \sigma_z^2 \\ \sigma_{\varphi\varphi}^2 = \sigma_\varphi^2, \\ \sigma_{r\vartheta}^2 = \sigma_{r\varphi}^2 = \sigma_{\vartheta\varphi}^2 = 0. \end{cases} \tag{12.131}$$

12.13 An important model that is often used to describe self-gravitating planar distributions (both in stellar dynamics and in fluid dynamics) is the so-called *isothermal sheet* (Bertin 2014; Binney and Tremaine 2008; Spitzer 1942). It is obtained (see the precautionary note in Chapter 1!) by assuming a density distribution stratified on z and independent of x and y, with the DF (12.79) and $\mathcal{E} = \Psi(z) - v^2/2$. First, show that the density can be written as in Eq. (12.122) and that the Poisson equation for $\phi = -\Psi$ in the self-gravitating case reads

$$\frac{d^2\phi}{dz^2} = 4\pi G\rho_1 e^{-\phi/\sigma^2}, \quad \phi(0) = 0, \quad \left.\frac{d\phi(z)}{dz}\right|_{z=0} = 0, \tag{12.132}$$

where the two conditions at $z = 0$ are fixed so that the density value ρ_1 in the plane is given as a boundary condition and the vanishing of the derivative imposes regularity. Integrate Eq. (12.132) or show by substitution that the self-gravitating solution is

$$\rho(z) = \rho_1 \cosh^{-2}\left(\frac{\sqrt{2\pi G\rho_1}z}{\sigma}\right), \quad \phi(z) = -\sigma^2 \ln\frac{\rho(z)}{\rho_1}. \tag{12.133}$$

12.14 An interesting family of power-law DFs discussed in Ciotti et al. (2004) is obtained from the models in Section 12.3.2, assuming a power-law as a special form of the function h (i.e., $h(Q) = AQ^\beta$; see also Rowley 1988), so that

$$f = f_+ = A|J_z|^{2\alpha} Q^\beta \theta(Q), \quad \alpha > -1/2, \quad \beta > -1. \tag{12.134}$$

From Eq. (12.102), show that

$$\rho = \frac{A\pi 2^{\alpha+5/2}\mathrm{B}(\alpha+3/2, \beta+1)}{(2\alpha+1)(1+R^2/R_\mathrm{a}^2)^{\alpha+1/2}} R^{2\alpha} \Psi_\mathrm{T}^{\alpha+\beta+3/2}\theta(\Psi_\mathrm{T}), \tag{12.135}$$

and obtain the analogous expressions for σ_m and σ_φ. What happens for $\beta = n$ positive integers? *Hint*: Expand f with the binomial theorem and consider a sum of Fricke DFs.

13

Modeling Techniques 2

Moments Approach

In this chapter, we discuss the complementary approach to that presented in Chapter 12 for the construction of stationary, multicomponent collisionless stellar systems. The Abel inversion theorem is introduced, and then a selection of density–potential pairs of spherical, axisymmetric, and triaxial shapes commonly used in modeling/observational works are presented. We finally discuss the solution of the Jeans equations for spherical and axisymmetric systems, and among other things we show how to compute the various quantities entering the virial theorem. For illustrative purposes, we use some of the derived results to investigate the possible physical interpretations of the fundamental plane of elliptical galaxies.

13.1 The Construction of a Galaxy Model: Starting with the Jeans Equations

One difficulty with the f-to-ρ approach illustrated in Chapter 12 is that there is little control over the resulting profiles of density and over the other dynamical quantities, even though when available these solutions are of great interest because they represent physically acceptable equilibrium models (which are not necessarily stable) of stationary, collisionless stellar systems.

An alternative (and, when possible, complementary) approach to the construction of models describing stellar systems, based on the Jeans equations (see Chapter 10), is very useful for its ability to relate observationally accessible quantities, such as the streaming velocity and velocity dispersion, to other interesting but directly unaccessible quantities (e.g., the potential and the total density field of a given galaxy); we refer to this approach as the ρ-to-f approach. However, before proceeding with the illustration of the method, the student should fully appreciate that in the ρ-to-f approach, in general the underlying distribution function (DF) cannot be determined uniquely, and even when it is (under quite restrictive assumptions), its numerical recovery is an example of an "ill-conditioned" problem. In fact, as discussed in Chapter 10, the macroscopic properties of a stellar system described by the Jeans equations are *moments* over the velocity space of the DF of the system, and from a mathematical point of view, physically acceptable moments can also be originated by an unphysical (i.e., somewhere negative) DF. Thus, in general, even though the approach described in this chapter is easier to apply than the f-to-ρ approach, the validity of the results obtained is not guaranteed a priori. Fortunately, the situation is not desperate; in fact, under special circumstances, it is possible to obtain sufficient information about the DF associated with the system and to test its properties, which is the subject of Chapter 14.

Qualitatively, we can summarize the main logical steps of the ρ-to-f approach as follows:

(1) Guided by theoretical/numerical/observational indications, a specific form for the density ρ (or for the projected density Σ) of the component of interest is assumed. Different density components can be allowed in order to simulate the presence of a dark matter halo, of a central supermassive black hole (BH) or of other stellar components with different ages, chemical compositions, mass-to-light ratios, and so on. In general, the mass-to-light ratio Υ_* of each density component is assumed to be independent of position, but its numerical value can be different for each density component.
(2) For each density component, the associated Jeans equations are solved either analytically or, more often, numerically, and their solutions are projected using the techniques presented in Chapter 11. The *closure* of the Jeans equations is in general obtained by imposing specific *ansatz* (hopefully physically motivated), such as with the task of reproducing the main features of observed (projected) velocity fields.
(3) When possible, the underlying DF is recovered and its positivity is investigated. In case of negative values, the DF cannot be accepted, and then the model obtained must be discarded, even if it is successful at reproducing observations. The recovery of the DF is usually the least trivial step of the ρ-to-f approach and, when ignored, can leave the physical significance of the entire process under serious concern.

In this chapter, we focus on Points (1) and (2) above, restricting ourselves for simplicity to some of the most common spherical and axisymmetric models. Point (3) will be discussed in Chapter 14.

13.2 The Choice of the Density Distribution

13.2.1 Spherical Models

Spherically symmetric models (single-component or multicomponent) are the simplest models used in stellar dynamics to describe stellar systems. They are usually sufficiently simple to allow an almost complete analytical study, and so they can be used as useful starting points before embarking on time-consuming investigations based on more realistic models. In general, the first step in the construction of a model is the choice of a density distribution reproducing the relevant properties of the projected/spatial density (luminosity) profile of the system of interest, so that the relation between spatial and projected density for spherical systems needs to be clarified. This is accomplished by using a very general result that comes from Abel (1802–1829). Let $g_1(x)$ and $g_2(x)$ be two sufficiently regular functions defined over the interval $x_{\rm m} \leq x \leq x_{\rm M}$. For $0 < \alpha < 1$, the following identities hold:

$$f_1(x) = \int_{x_{\rm m}}^{x} \frac{g_1(t)dt}{(x-t)^\alpha}, \quad g_1(x) = \frac{\sin(\pi\alpha)}{\pi}\frac{d}{dx}\int_{x_{\rm m}}^{x} \frac{f_1(t)dt}{(x-t)^{1-\alpha}}, \tag{13.1}$$

$$f_2(x) = \int_{x}^{x_{\rm M}} \frac{g_2(t)dt}{(t-x)^\alpha}, \quad g_2(x) = -\frac{\sin(\pi\alpha)}{\pi}\frac{d}{dx}\int_{x}^{x_{\rm M}} \frac{f_2(t)dt}{(t-x)^{1-\alpha}}, \tag{13.2}$$

and the verification of the inversion formulae above is given in Exercise 13.1. This important theorem has innumerable applications in mathematics and physics (e.g., see Gorenflo and Vessella 1991; chapter 3 of Landau and Lifshitz 1969), and even restricting ourselves just to astronomy (e.g., see Ciotti et al. 1995; see also Chapter 14), this is vital not only in stellar dynamics, but also in areas such as the modeling of hot gaseous atmospheres of elliptical galaxies and clusters of galaxies (e.g., see Ciotti and Pellegrini 2008).

Let us now consider a spherically symmetric stellar system described by a mass density $\rho(r)$ stratified on the spherical radius r; we indicate with $\Sigma(R)$ the *projected mass density*, where R is the radius on the projection plane (see Chapter 11), and for simplicity, in the following, we only consider the case of a spatially constant mass-to-light ratio Υ_*, so that $\rho(r)$ and $\Sigma(R)$ are related to the *light density* $\nu(r)$ and to the *surface brightness profile* $I(R)$ by the elementary $\rho = \Upsilon_* \nu$ and $\Sigma = \Upsilon_* I$, respectively.[1] Abel's theorem can be used to relate ρ and Σ (and ν and I) as follows: assume

$$\rho_{\rm t} = \rho(r)\theta(r_{\rm t} - r), \quad \Sigma_{\rm t} = \Sigma(R)\theta(R_{\rm t} - R), \tag{13.3}$$

where $r_{\rm t} = R_{\rm t}$ is the so-called *truncation radius*, and when $r_{\rm t} = R_{\rm t} = \infty$, the system is said to be (spatially) *nontruncated.* From Figure 13.1, simple geometric considerations show that in a *transparent system* (i.e., a system where no two stars are exactly aligned along the line of sight (*los*); see also Exercise 11.2)

$$\Sigma(R) = 2\int_R^{R_{\rm t}} \frac{\rho(r) r dr}{\sqrt{r^2 - R^2}}, \quad R_{\rm t} = r_{\rm t}, \tag{13.4}$$

and from Exercise 13.2

$$\rho(r) = -\frac{1}{\pi}\int_r^{r_{\rm t}} \frac{d\Sigma(R)}{dR}\frac{dR}{\sqrt{R^2 - r^2}} + \frac{\Sigma(R_{\rm t})}{\pi\sqrt{R_{\rm t}^2 - r^2}}. \tag{13.5}$$

A check can obtained by inserting Eq. (13.5) into Eq. (13.4), inverting the order of integration, and finally using the Euler integral (13.133) with $\alpha = 1/2$. Note that even if $\rho(r_{\rm t}) > 0$ (in practice, the system is spatially truncated with a finite "jump" at $r_{\rm t}$), necessarily $\Sigma(R_{\rm t}) = 0$. It is then not a surprise that if $\Sigma(R_{\rm t}) > 0$, then the associated spatial density must be (weakly) divergent at the boundary of the system, $\rho(r) \sim (r_{\rm t}^2 - r^2)^{-1/2}$;

[1] In the case of stellar systems with a spatially dependent mass-to-light ratio $\Upsilon_*(\mathbf{x})$, such as multicomponent systems with different stellar populations, caution is needed when relating intrinsic and projected density/luminosity fields (e.g., see Cappellari 2002; Caravita et al. 2021).

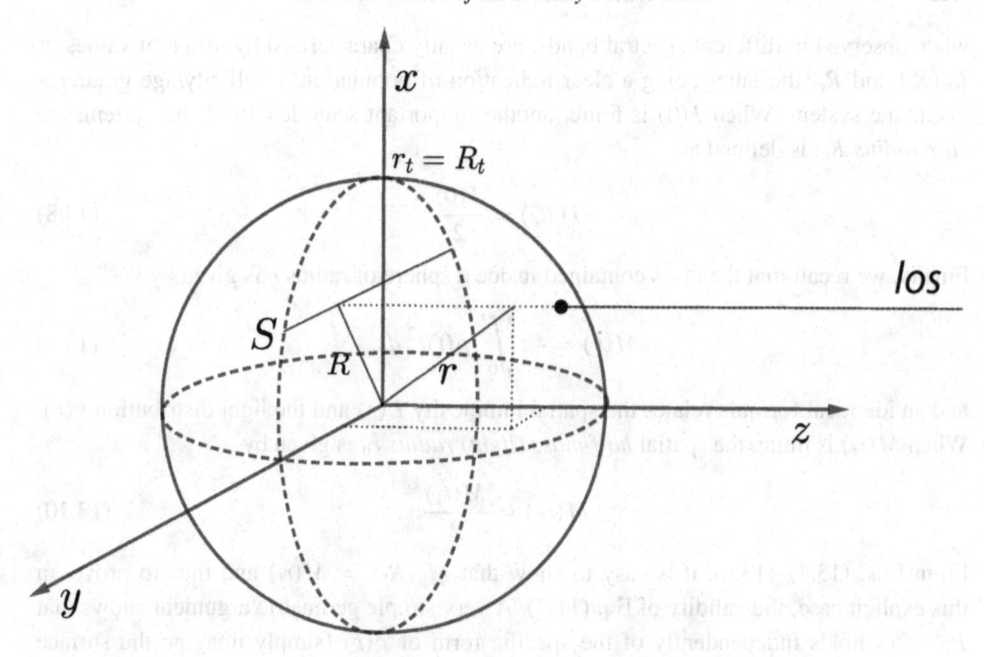

Figure 13.1 Geometry of the projection of a transparent spherical system. The *los* is directed along the z-axis, and so $r^2 = R^2 + z^2$, where $R^2 = x^2 + y^2$. By definition of projection, $\Sigma(R) = 2\int_0^{\sqrt{R_t^2 - R^2}} \rho(\sqrt{R^2 + z^2})dz$, so that Eq. (13.4) is obtained with a change of the integration coordinate from z to r, at fixed R. Integration along the horizontal line S defines the strip brightness $S(x)$ in Eq. (13.15).

the student is invited to explore the behavior of $\Sigma(R)$ near R_t for $\rho(r) \sim (r_t^2 - r^2)^\alpha$ as a function of α.

Before listing some of the most used spherical models that are encountered in the literature, we recall the relevant quantities that characterize a spherical density profile. The *projected mass* M_p inside a circle of radius R on the projection plane of a system with surface density Σ is

$$M_p(R) = 2\pi \int_0^R \Sigma(t)t\,dt, \tag{13.6}$$

and an identical expression relates the *projected luminosity* L_p and the surface brightness profile I when considering the spatial luminosity density ν instead of ρ.

When $L_p(R_t)$ is finite, the *effective radius* R_e of the system is defined as

$$L_p(R_e) = \frac{L_p(R_t)}{2} \tag{13.7}$$

(i.e., R_e is the radius of a circle in the projection plane containing half of the total luminosity of the density component under investigation). It is important to recall that real galaxies, when observed in different spectral bands, are usually characterized by different values of $L_{\rm p}(R_{\rm t})$ and $R_{\rm e}$, the latter being a clear indication of population/metallicity/age gradients inside the system. When $I(0)$ is finite, another important scale length of the system, the *core radius* $R_{\rm c}$, is defined as

$$I(R_{\rm c}) = \frac{I(0)}{2}. \tag{13.8}$$

Finally, we recall that the mass contained inside a sphere of radius r is given by

$$M(r) = 4\pi \int_0^r \rho(t)t^2\,dt, \tag{13.9}$$

and an identical formula relates the spatial luminosity $L(r)$ and the light distribution $\nu(r)$. When $M(r_{\rm t})$ is finite, the spatial *half-mass (light) radius* $r_{\rm h}$ is given by

$$M(r_{\rm h}) = \frac{M(r_{\rm t})}{2}. \tag{13.10}$$

From Eqs. (13.4)–(13.6), it is easy to show that $M_{\rm p}(R_{\rm t}) = M(r_{\rm t})$ and thus to prove, in this explicit case, the validity of Eq. (11.7). A very simple geometric argument shows that $R_{\rm e} \leq r_{\rm h}$ holds independently of the specific form of $\rho(r)$ (simply imagine the surface integral of the projected density inside the radius R as given by the sum of three volume integrals: a sphere of radius R and the two remaining parts of the projection cylinder once of the sphere is removed, then integrate up to $R = r_{\rm h}$ and deduce the statement; see also Exercise 13.3 for an explicit example). It should be fairly obvious that in the case of a *spatially constant* Υ_*, the quantities $R_{\rm e}$, $R_{\rm c}$, and $r_{\rm h}$ can be defined indifferently over the mass or over the light profile, being $M_{\rm p} = \Upsilon_* L_{\rm p}$.

In the construction of a model for a stellar system, potential and gravitational energies are quantities of fundamental interest. We have already encountered them in general cases in Chapters 2 and 10, and here, in preparation for the following discussion, we give their expressions for spherical systems in presence of an external potential $\phi_{\rm ext}(r)$. The gravitational potential produced by the density distribution in Eq. (13.3) is obtained from Eq. (2.5) by considering $r_0 \to \infty$, so that

$$\phi(r) = -\begin{cases} \dfrac{GM(r)}{r} + 4\pi G \displaystyle\int_r^{r_{\rm t}} \rho(r)r\,dr, & 0 \leq r \leq r_{\rm t}, \\ \dfrac{GM(r_{\rm t})}{r}, & r \geq r_{\rm t}, \end{cases} \tag{13.11}$$

and from Eqs. (10.49)–(10.54) the associated self-gravitational energy can be written as

$$U = 2\pi \int_0^{r_{\rm t}} \rho(r)\phi(r)r^2 dr = -4\pi G \int_0^{r_{\rm t}} M(r)\rho(r)r\,dr = W, \tag{13.12}$$

where the last identity derives from Eq. (10.54). If, in addition, a spherically symmetric external potential $\phi_{\rm ext}(r)$ is present (e.g., that of a dark matter halo or of a central BH), then an integration by parts of Eq. (10.50) with respect to the enclosed mass shows that

$$U_{\text{ext}} = 4\pi \int_0^{r_t} \rho(r)\phi_{\text{ext}}(r)r^2 dr = M(r_t)\phi_{\text{ext}}(r_t) - G\int_0^{r_t} \frac{M(r)M_{\text{ext}}(r)}{r^2} dr, \quad (13.13)$$

while the interaction energy entering the virial theorem is obtained from Eq. (10.52) as

$$W_{\text{ext}} = -4\pi G \int_0^{r_t} M_{\text{ext}}(r)\rho(r) r dr. \quad (13.14)$$

In spherical systems, a quite remarkable identity relates the projected mass (luminosity) distribution with the potential and self-gravitational energy. Following Schwarzschild (1954), we the introduce the concept of *strip brightness*. Let $I(R)$ be the surface brightness profile associated with the light distribution $\nu(r)$, where $\rho = \Upsilon_* \nu$, $R = \sqrt{x^2 + y^2}$, and Υ_* is constant. From Figure 13.1, the strip brightness $S(x)$ is defined as the integral of I along y at distance x from $R = 0$, and in Exercise 13.4 we show that the following relations hold:

$$S(x) \equiv 2\int_0^{\sqrt{R_t^2 - x^2}} I(R) dy = 2\int_x^{R_t} \frac{I(R)R}{\sqrt{R^2 - x^2}} dR = 2\pi \int_x^{r_t} \nu(r) r dr, \quad (13.15)$$

$$\nu(r) = -\frac{1}{2\pi r}\frac{dS(r)}{dr}, \quad M(r) = 2\Upsilon_* \left[\int_0^r S(x)dx - r\, S(r)\right], \quad (13.16)$$

$$U = -2G\Upsilon_*^2 \int_0^{R_t} S^2(x) dx. \quad (13.17)$$

Note that $S(R_t) = S(r_t) = 0$, so that for $r \geq r_t$ the second identity of Eq. (13.16) gives $M(r_t)$. Moreover

$$\phi(r) = -\frac{2G\Upsilon_*}{r}\int_0^r S(x)dx, \quad \phi(0) = -4G\Upsilon_* \int_0^{R_t} I(R) dR \quad (13.18)$$

(a result apparently unnoticed in Schwarzschild 1954; see Ciotti 1991 and Exercise 13.5; see also exercise 2.15 in Binney and Tremaine 2008). Finally, a quite interesting expression of U in terms of $\Sigma(R)$ can be obtained with some work from Eq. (13.17), as is shown in Exercise 13.8. The interesting point of these results is that in some cases they allow us to obtain explicit expressions for dynamical quantities in systems defined by the surface density profile, when Abel inversion cannot be performed in terms of elementary functions.

13.2.1.1 A Selection of Spherical Models of Common Use

Here, we list some of the most used density and surface brightness profiles of spherically symmetric stellar systems. These models are relatively simple, yet they can represent reasonably well the density profiles of (early-type) galaxies or other spheroidal stellar systems. Of course, the following must not be considered by any means complete, and the student is strongly encouraged to dedicate some time to the exploration of the literature. We begin with models defined by $\Sigma(R)$ or $I(R)$.

- *Sersic $R^{1/m}$ law* (Sersic 1968)

$$I(R) = I(0)\text{e}^{-b(m)\eta^{1/m}}, \quad \eta = \frac{R}{R_e}, \quad (13.19)$$

where R_e is the effective radius and $b(m)$ is determined below. It has no free parameters and depends on two physical scales, the effective radius R_e and the central surface brightness $I(0)$. For $m = 4$, it reduces to the famous de Vaucouleurs (1948) $R^{1/4}$ law, with $b(4) \simeq 7.67$. For $m = 1$, the formula becomes the exponential profile adopted to model the face-on surface brightness of disk galaxies. Due to the relevance of Sersic law in the observational works of early-type galaxies, some additional information is presented. From Eq. (13.6), the projected and total luminosities are given by

$$L_p(R) = I(0)R_e^2 \frac{2\pi m}{b^{2m}} \gamma(2m, b\eta^{1/m}), \quad L = I(0)R_e^2 \frac{2\pi m}{b^{2m}} \Gamma(2m), \tag{13.20}$$

where $\gamma(\alpha, x)$ is the incomplete gamma function in Eq. (A.51). From Eq. (13.7), it follows that $b(m)$ is the solution of the transcendental equation

$$\gamma(2m, b) = \frac{\Gamma(2m)}{2}, \tag{13.21}$$

and standard asymptotic analysis (e.g., Bender and Orszag 1978; Bleistein and Handelsman 1986; de Bruijn 1958) shows that for $m \to \infty$,

$$b(m) \sim 2m - \frac{1}{3} + \frac{4}{405m} + \frac{46}{25515m^2} + \cdots \tag{13.22}$$

(see Ciotti and Bertin 1999; see also Prugniel and Simien 1997). A slightly more difficult asymptotic analysis (Baes and Ciotti 2019a) shows that for $m \to 0$

$$b(m)^{2m} \sim m\Gamma(2m) = \frac{1}{2} - \gamma m + \left(\frac{\pi^2}{6} + \gamma^2\right) m^2 + \cdots \tag{13.23}$$

where $\gamma = 0.577\ldots$ is the Euler–Mascheroni constant; therefore, $b(m) \sim 1/(\sqrt{2})^{1/m} \to 0$ for $m \to 0$. The luminosity density ν is given by Eq. (13.5) and unfortunately it cannot be expressed in terms of elementary functions for generic values of m. However, it can be proved (Ciotti 1991) that for $r \to \infty$

$$\nu(r) \sim \frac{I(0)}{R_e} \sqrt{\frac{b}{2\pi m}} e^{-bs^{1/m}} s^{(1-2m)/2m}, \quad s \equiv \frac{r}{R_e}, \tag{13.24}$$

while for $r \to 0$

$$\nu(r) \sim \frac{I(0)}{R_e} \times \begin{cases} \dfrac{b^m \Gamma(1-m)}{\pi}, & 0 < m < 1, \\[2ex] \dfrac{b(1)}{\pi} \ln \dfrac{2}{b(1)s^{1/2}}, & m = 1, \\[2ex] \dfrac{\mathrm{B}[1/2, (m-1)/2m]}{2mb^{m-1}} e^{-bs^{1/m}} s^{(1-m)/m}, & m > 1. \end{cases} \tag{13.25}$$

For $m > 1$, the density diverges at the origin as $r^{(1-m)/m}$, and therefore the divergence of ρ is worse for higher-m models: for $m = 4$, ρ diverges as $r^{-3/4}$ for $r \to 0$. Several dynamical properties have been computed (e.g., see Baes and van Hese 2011; Binney

1982b; Ciotti and Lanzoni 1997; Graham and Driver 2005; Poveda et al. 1960; Young 1976 and references therein). The central potential is obtained from Eq. (13.18) as

$$\phi(0) = -G\Upsilon_* I(0) R_e \frac{4\Gamma(m+1)}{b^m}, \tag{13.26}$$

and a quite surprising integral expression for U is finally given in Exercise 13.8. We conclude by mentioning the *Einasto models* (Einasto 1965; Graham et al. 2006 and references therein), the three-dimensional analogs of the Sersic law, which are found to reproduce very well the structure of dark matter halos obtained from numerical simulations of structure formation.

- *Hubble (1930)–Reynolds (1913)*

$$I(R) = \frac{I(0)}{(1 + R/R_H)^2}, \tag{13.27}$$

one of the first families of profiles used to describe the surface brightness profiles of elliptical galaxies.

- *King (1962)*

$$I(R) = K \times \left(\frac{1}{\sqrt{R^2 + R_K^2}} - \frac{1}{\sqrt{R_t^2 + R_K^2}} \right) \theta(R_t - R), \tag{13.28}$$

where the coefficient K can written in terms of $I(0)$. This family has been used to describe the surface brightness profiles of globular clusters.

The following models are instead built starting from $\rho(r)$, so that $\Sigma(R)$ is a derived quantity. For a few we give also some additional formulae that can be useful in applications, such as the relative potential $\Psi(r) = -\phi(r)$, while the deduction of other formulae (that can be found in the given references) is left to the student. We indicate with $s \equiv r/r_c$ the spherical radius in units of the scale length r_c of the density profile, and with $\eta \equiv R/r_c$ the radius in the projection plane in units of r_c.

- *Power-law* (untruncated, $0 < \gamma < 3$)

$$\rho(r) = \frac{\rho_c}{s^\gamma}, \quad M(r) = \frac{M_c\, s^{3-\gamma}}{3-\gamma}, \quad M_c \equiv 4\pi \rho_c r_c^3. \tag{13.29}$$

For mass convergence near the center $\gamma < 3$ and for $\gamma = 2$, the density profile is also known as the *singular isothermal sphere* (SIS; see also Exercise 2.1). The potential is obtained from Eq. (2.5) with $r_t = \infty$ and $r_0 = \infty$ for $2 < \gamma < 3$ or $r_0 = 0$ for $0 < \gamma < 2$

$$\phi(r) = \frac{GM_c}{r_c} \frac{s^{2-\gamma}}{(3-\gamma)(2-\gamma)}. \tag{13.30}$$

For $\gamma > 1$, the projected density and projected mass profiles are

$$\Sigma(R) = \rho_c r_c \mathrm{B}\left(\frac{1}{2}, \frac{\gamma-1}{2}\right) \eta^{1-\gamma}, \quad M_p(R) = \frac{M_c}{2(3-\gamma)} \mathrm{B}\left(\frac{1}{2}, \frac{\gamma-1}{2}\right) \eta^{3-\gamma}, \tag{13.31}$$

where $B(x, y)$ is the complete beta function in Eq. (A.53). Note how Σ is shallower than ρ, a general consequence of projection. For other properties of these models, see also Exercise 3.1.

- *Quasi-isothermal sphere* (untruncated)

$$\rho(r) = \frac{\rho(0)}{s^2+1}, \quad M(r) = 4\pi\rho(0)r_c^3(s - \arctan s), \tag{13.32}$$

$$\phi(r) = 4\pi G\rho(0)r_c^2\left(\ln\frac{\sqrt{1+s^2}}{e} + \frac{\arctan s}{s}\right), \tag{13.33}$$

where the potential is obtained from Eq. (2.5) with $r_0 = 0$ and imposing $\phi(0) = 0$. This model is characterized by a constant circular velocity at large radii, with the value given by (prove it!) $v_c^2 = 4\pi G\rho(0)r_c^2$. As an exercise, compare the behavior of the potential for $s \to \infty$ with that of the SIS in Eq. (2.64) and comment on the result (consider what happens to this model for $r_c \to 0$ at fixed v_c).

- *King (1972)* (untruncated)

$$\rho(r) = \frac{\rho(0)}{(s^2+1)^{3/2}}, \quad M(r) = 4\pi\rho(0)r_c^3\left(\operatorname{arcsinh} s - \frac{s}{\sqrt{s^2+1}}\right), \tag{13.34}$$

$$\Psi(r) = 4\pi G\rho(0)r_c^2\frac{\operatorname{arcsinh} s}{s}. \tag{13.35}$$

Note that the total mass of the profile diverges; however, from the discussion in Chapter 2, it follows that the potential can be fixed to zero at infinity (for additional properties of this model, see Exercise 3.2).

- *Plummer (1911)*

$$\rho(r) = \frac{3M}{4\pi r_c^3}\frac{1}{(s^2+1)^{5/2}}, \quad M(r) = \frac{Ms^3}{(s^2+1)^{3/2}}, \quad \Psi(r) = \frac{GM}{r_c}\frac{1}{\sqrt{s^2+1}}. \tag{13.36}$$

This is one of the most important models encountered in stellar dynamics due to its numerous relevant properties (see Chapter 12 for the relation with the polytrope of index $n = 5$ and see Exercises 13.3, 13.17, 13.18, and 14.2).

- *β-models* ($\beta > 1$)

$$\rho(r) = \frac{\rho(0)}{(s^2+1)^{\beta/2}}, \quad \Sigma(R) = \frac{\rho(0)r_c B(1/2, \beta/2 - 1/2)}{(\eta^2+1)^{(\beta-1)/2}}. \tag{13.37}$$

The cumulative mass $M(r)$ and the potential $\Psi(r)$ can in general be expressed (do it!) for generic values of β in terms of the incomplete beta function in Eq. (A.55). Note that the quasi-isothermal sphere in Eq. (13.32), the King (1972) model in Eq. (13.34), the spherical limit of the perfect ellipsoid in Eq. (2.93), and the Plummer (1911) model in Eq. (13.36) can all be seen as special cases of β-models, obtained respectively for $\beta = 2, 3, 4$, and 5.

- γ*-models* ($0 \le \gamma < 3$) (Dehnen 1993; Tremaine et al. 1994)

$$\rho(r) = \frac{(3-\gamma)M}{4\pi r_c^3 s^\gamma (s+1)^{4-\gamma}}, \quad M(r) = \frac{M s^{3-\gamma}}{(s+1)^{3-\gamma}}, \quad r_h = \frac{r_c}{2^{\frac{1}{3-\gamma}} - 1}, \tag{13.38}$$

$$\Psi(r) = \frac{GM}{r_c} \begin{cases} \dfrac{1}{(2-\gamma)}\left[1 - \left(\dfrac{s}{s+1}\right)^{2-\gamma}\right], & \gamma \neq 2 \\ \ln \dfrac{s+1}{s}, & \gamma = 2. \end{cases} \tag{13.39}$$

It is an interesting exercise (do it!) to show that at $r_c \to 0$ the potential of the γ-models reduces to that of a point of mass M. From Eq. (13.38), it follows that the density of γ-models is a power-law $r^{-\gamma}$ at small radii (i.e., for $r < r_c$) and a power-law r^{-4} at large radii, and due to their simplicity and flexibility, these models are routinely used when modeling the density distribution of early-type galaxies. Special cases of great importance in theoretical and numerical works are the Jaffe (1983) and Hernquist (1990) models, obtained respectively for $\gamma = 2$ and $\gamma = 1$, with $R_e \simeq 0.75\, r_c$ and $R_e \simeq 1.82\, r_c$: these two profiles reproduce the $R^{1/4}$ profile in projection remarkably well. Generalizations of these models can be found in Burkert (1995), Evans et al. (2015), and Zhao (1996). For other properties, see Exercices 3.3 and 13.6.

- *Navarro–Frenk–White (NFW) profile* (Navarro et al. 1997)

$$\rho(r) = \frac{M}{4\pi r_c^3 f(c) s (s+1)^2}, \quad M(r) = M\frac{f(s)}{f(c)}, \quad \Psi(r) = \frac{GM}{r_c}\frac{\ln(s+1)}{s f(c)}, \tag{13.40}$$

where M is the mass contained inside the radius r_t and so

$$f(s) = \ln(s+1) - \frac{s}{s+1}, \quad c \equiv \frac{r_t}{r_c}. \tag{13.41}$$

Note that the total mass of the profile diverges; however, from the discussion in Chapter 2, it follows that the potential can be fixed to zero at infinity. This profile is found to reproduce the density radial trend of the cosmological dark matter halos obtained in numerical simulations.

- *Hénon (1959) isochrone*

$$\Psi(r) = \frac{GM}{r_c}\frac{1}{1+\sqrt{s^2+1}}, \tag{13.42}$$

where M is the total mass, and from Newton's second theorem

$$M(r) = \frac{M s^3}{\sqrt{s^2+1}(1+\sqrt{s^2+1})^2}, \quad \rho(r) = \frac{M}{4\pi r_c^3}\frac{2s^2 + 3(1+\sqrt{s^2+1})}{(s^2+1)^{3/2}(1+\sqrt{s^2+1})^3}. \tag{13.43}$$

This is a very important model (most easily introduced starting from the potential) due to the fact that the period of orbits is independent of the orbital angular momentum and is only dependent on orbital energy (see Binney and Tremaine 2008 and, in particular, Binney 2014 for a very clear and complete discussion of the model).

13.2.2 Axisymmetric and Triaxial Ellipsoidal Models

Along the lines adopted to introduce the main structural properties of spherical models, we now consider models departing from spherical symmetry, starting with the general family of triaxial and axisymmetric ellipsoidal models. In the axisymmetric case, the natural coordinates for these systems are the cylindrical ones (R, φ, z), with all of the model properties independent of the azimuthal angle φ. After ellipsoidal models, we will describe other families of nonspherical models of use in stellar dynamics, but the student is warned that many more families can be found in the literature.

We already encountered ellipsoidal models in Chapter 2 (see in particular Section 2.2 and related exercises). As we have seen, a very simple procedure that can be adopted in order to generate families of triaxial (or axisymmetric) models starting from a spherical density distribution $\rho(r) = \rho_0\tilde{\rho}(r/a_1)$ is based on the substitution (2.26). This is equivalent to the transformation of the spherical radius to

$$r \mapsto \sqrt{x^2 + \frac{y^2}{q_y^2} + \frac{z^2}{q_z^2}}, \qquad r \mapsto \sqrt{R^2 + \frac{z^2}{q_z^2}}, \tag{13.44}$$

where $q_y = a_2/a_1$ and $q_z = a_3/a_1$ are the ratios of the semiaxes of the isodensity ellipsoidal surfaces. In general, for $q_y = q_z$, the model is axisymmetric around the x-axis, for $q_z = 1$ around the y-axis, and for $q_y = 1$ around the z-axis. The standard axisymmetric case is obtained for $q_y = 1$, and q_z is responsible for the model axial flattening. For $0 < q_z < 1$, the density is stratified on *oblate ellipsoids*, for $q_z = 1$ on spheres, and for $q_z > 1$ on *prolate ellipsoids*. Usually (but not always), we will follow the convention in Eq. (2.27) (i.e., the shortest axis of the ellipsoidal surfaces is along the z-axis, the intermediate axis along the y-axis, and the long axis along the x-axis). With this substitution, the density of our system is given by Eqs. (2.28) and (2.74), i.e.,

$$\rho(\mathbf{x}) = \frac{\rho_0\tilde{\rho}(m)}{q_y q_z}\theta(m_t - m), \qquad m = \frac{1}{a_1}\sqrt{x^2 + \frac{y^2}{q_y^2} + \frac{z^2}{q_z^2}}, \tag{13.45}$$

where m_t fixes the truncation surface $m = m_t$; for $m_t = \infty$, the model is untruncated.

In analogy with Section 13.2.1, we now consider the most important properties concerning the projection of the ellipsoidal density in Eq. (13.45). For the sake of generality, we do not require that the mass M of the model is finite, so that ρ_0 is simply a density scale not necessarily equal to $M/(4\pi a_1^3)$. The astrophysical relevance of the problem should be clear, as quite often early-type galaxies can be satisfactorily described as ellipsoidal bodies. Several treatments of the projection of ellipsoidal (transparent) stellar systems are available (see, among others, Binney 1985; Lanzoni and Ciotti 2003; Stark 1977, and, in particular, Franx 1988a,b for the relevant problem of the projection of Stäckel stellar systems); however, as different conventions are adopted to define the projection angles, for consistency here we follow Chapter 11. To project the galaxy density on the plane of the sky (the *projection plane* π'), we start from the system $S_0 = (O; \mathbf{e}_1, \mathbf{e}_2, \mathbf{e}_3)$ with the origin at the center of the density distribution and with the axes aligned with the ellipsoidal axes. In S_0, the generic position vector is written as $\mathbf{x} = x_i \mathbf{e}_i = x_1\mathbf{e}_1 + x_2\mathbf{e}_2 + x_3\mathbf{e}_3$. The *los* direction in S_0 is indicated by the unitary vector $\mathbf{n}$ pointing *toward* the observer, and it is specified by using standard spherical angles ϑ and φ, as in Eq. (11.2). The observer system $S' = (O'; \mathbf{f}_1, \mathbf{f}_2, \mathbf{f}_3)$ is nonrotating and $O \equiv O'$ at all times. The position vector in S' is $\boldsymbol{\xi} = \xi_i \mathbf{f}_i$ and $\mathbf{n}$ coincides with $\mathbf{f}_3$ (i.e., the observer's "eye" is placed at infinity on the ξ_3-axis). The projection plane π' then contains the origin, and it is perpendicular to $\mathbf{n}$. This is given algebraically by $\xi_3 = 0$, and it can be identified with the set of two-dimensional vectors $\boldsymbol{\xi}_\perp = (\xi_1, \xi_2)$. From Chapter 11, it follows that the orientation of S' with respect to S_0 is given by a 3–2–3 rotation; the first $(0 \leq \varphi < 2\pi)$ and second $(0 \leq \vartheta \leq \pi)$ rotation angles coincide with the spherical angles of $\mathbf{n}$ in S_0, and the resulting rotation matrix $\mathcal{R} = \mathcal{R}_3(\varphi)\mathcal{R}_2(\vartheta)\mathcal{R}_3(\psi)$ is given by Eqs. (11.46)–(11.48), with $\mathbf{x} = \mathcal{R}\boldsymbol{\xi}$ and $\mathbf{n} = \mathcal{R}\mathbf{f}_3$. Note that the rotation $\mathcal{R}_3(\psi)$ around $\mathbf{f}_3$ simply rotates the projected image in the plane π', and so for the present purposes here we fix $\psi = 0$; in other words, the projected image is intrinsically determined only by the angles φ and ϑ.

The integral defining the density *projection* $\Sigma(\boldsymbol{\xi}_\perp)$ is given in Eq. (11.5) and, as already remarked, this definition assumes that the observer is at infinite distance from the system and that the system is "transparent" (i.e., all of the information along $\mathbf{n}$ is summed at $\boldsymbol{\xi}_\perp$ in the projection plane; see Exercise 11.2). Moreover, according to Eq. (11.7), the integration of Σ over π' equals the total mass (luminosity) of the projected system independently of the *los* direction. In order to proceed with the projection, it is convenient to introduce the diagonal matrix

$$D = \begin{pmatrix} \frac{1}{a_1^2} & 0 & 0 \\ 0 & \frac{1}{a_2^2} & 0 \\ 0 & 0 & \frac{1}{a_3^2} \end{pmatrix} = \frac{1}{a_1^2} \begin{pmatrix} 1 & 0 & 0 \\ 0 & \frac{1}{q_y^2} & 0 \\ 0 & 0 & \frac{1}{q_z^2} \end{pmatrix}, \tag{13.46}$$

so that from Eq. (2.26)

$$m^2 = \langle D\mathbf{x}, \mathbf{x}\rangle = \langle E\boldsymbol{\xi}, \boldsymbol{\xi}\rangle, \quad E \equiv \mathcal{R}^{\mathrm{T}} D \mathcal{R}, \tag{13.47}$$

where the matrix E depends on the projection angles and is apparently symmetric and definite positive. We then write $\boldsymbol{\xi} = \boldsymbol{\xi}_\perp + \xi_3\mathbf{f}_3$, where for the moment we suspend the convention $\boldsymbol{\xi}_\perp = (\xi_1, \xi_2)$ and we assume instead $\boldsymbol{\xi}_\perp = (\xi_1, \xi_2, 0)$, so that Eq. (13.47) becomes

$$\begin{aligned} m^2 &= \alpha^2\xi_3^2 + 2\langle E\mathbf{f}_3, \boldsymbol{\xi}_\perp\rangle\xi_3 + \langle E\boldsymbol{\xi}_\perp, \boldsymbol{\xi}_\perp\rangle \\ &= \left(\alpha\xi_3 + \frac{\langle E\mathbf{f}_3, \boldsymbol{\xi}_\perp\rangle}{\alpha}\right)^2 + \ell^2, \end{aligned} \tag{13.48}$$

where (prove it!)

$$\alpha^2 = \langle E\mathbf{f}_3, \mathbf{f}_3\rangle = \langle D\mathbf{n}, \mathbf{n}\rangle$$
$$= \frac{1}{a_1^2}\left(\sin^2\vartheta\cos^2\varphi + \frac{\sin^2\vartheta\sin^2\varphi}{q_y^2} + \frac{\cos^2\vartheta}{q_z^2}\right), \tag{13.49}$$

and

$$\ell^2 = \langle E\boldsymbol{\xi}_\perp, \boldsymbol{\xi}_\perp\rangle - \frac{\langle E\mathbf{f}_3, \boldsymbol{\xi}_\perp\rangle^2}{\alpha^2} \equiv \langle H\boldsymbol{\xi}_\perp, \boldsymbol{\xi}_\perp\rangle \tag{13.50}$$

is a positive definite[2] quadratic form. Notice that at fixed $\boldsymbol{\xi}_\perp$, from Eq. (13.48) m^2 attains the minimum value ℓ^2 at

$$\xi_3 = \xi_3^{\min} \equiv -\frac{\langle E\mathbf{f}_3, \boldsymbol{\xi}_\perp\rangle}{\alpha^2}. \tag{13.51}$$

In the last term in Eq. (13.50) we reverted to the convention in Chapter 11 of $\boldsymbol{\xi}_\perp = (\xi_1, \xi_2)$. H is the definite positive symmetric 2×2 matrix

$$H = \frac{1}{\alpha^2}\begin{pmatrix} \alpha^2 E_{11} - E_{13}^2 & \alpha^2 E_{12} - E_{13}E_{23} \\ \alpha^2 E_{21} - E_{13}E_{23} & \alpha^2 E_{22} - E_{23}^2 \end{pmatrix}, \tag{13.52}$$

and its entries can be easily computed for any choice of the *los* orientation. Therefore, *curves of constant ℓ are concentric and coaxial ellipses in the projection plane.*

We can now proceed with the projection of the density distribution in Eq. (13.45). First, we express the density in the observer's coordinates $\boldsymbol{\xi}$ by using Eq. (13.48) and by fixing $\boldsymbol{\xi}_\perp$; we indicate the new function as $\rho_{\mathrm{T}}'(\boldsymbol{\xi})$. Then we consider the general expression of the projection operator in Eq. (11.5) and we evaluate the minimum (ξ_3^-) and maximum (ξ_3^+) values of the integration variable setting $m = m_{\mathrm{t}}$ in Eq. (13.48), obtaining

$$\xi_3^{\pm} = \pm\frac{\sqrt{m_{\mathrm{t}}^2 - \ell^2}}{\alpha} + \xi_3^{\min}. \tag{13.53}$$

Of course, for untruncated models, the integration domain is $(-\infty, \infty)$. We then split the integration domain in $(\xi_3^-, \xi_3^{\min})$ and $(\xi_3^{\min}, \xi_3^+)$ and change the integration variable in the two integrals as $t = \alpha\xi_3 - \alpha\xi_3^{\min}$. Recognizing that the two integrals coincide, we finally obtain

$$\begin{cases} \Sigma(\boldsymbol{\xi}_\perp) = \displaystyle\int_{\xi_3^-}^{\xi_3^+} \rho'(\boldsymbol{\xi})d\xi_3 = \frac{\rho_0}{q_y q_z \alpha}\tilde{\Sigma}(\ell), \\ \tilde{\Sigma}(\ell) \equiv 2\displaystyle\int_0^{\sqrt{m_{\mathrm{t}}^2-\ell^2}} \tilde{\rho}(\sqrt{t^2+\ell^2})dt = 2\int_\ell^{m_{\mathrm{t}}} \frac{\tilde{\rho}(m)m}{\sqrt{m^2-\ell^2}}dm. \end{cases} \tag{13.54}$$

[2] From the positivity of m^2 in Eq. (13.48), it follows that the determinant of the quadratic equation in ξ_3 is negative, and this proves that $\ell^2 \geq 0$.

Therefore, we proved that the projection of the triaxial ellipsoids in Eq. (13.45) produces surface densities stratified on ellipses $\ell = const.$, and that the dimensionless function $\tilde{\Sigma}(\ell)$ is the same as that appearing in the projection of the parent spherical case, as can be immediately recognized by considering the case $q_y = q_z = 1$ in Eq. (13.54). Incidentally, this also implies that the deprojection formula for the ellipsoidal case, recovering $\rho(m)$ from $\Sigma(\ell)$, can be again obtained from Abel's theorem as in the spherical case, once the *los* direction and the two intrinsic flattenings are known (for more complicated cases of deprojection, see, among others, Binney and Gerhard 1996; Gerhard and Binney 1996).

A few comments are in order. First, it is easy to check that in the triaxial case, and for *los* inclinations not aligned with the axes of the ellipsoid, the elliptical isodense $\ell = const.$ in the projection plane is rotated with respect to the $(\mathbf{f}_1, \mathbf{f}_2)$ axes, as the nondiagonal terms of the H matrix are in general different from zero. Second, being H symmetric, a rotation of the remaining angle ψ around the $\mathbf{f}_3$ axis, leading to a diagonal matrix H', is always possible, so that the phenomenon known as *isophotal twisting* (i.e., the systematic rotation of the isophotal principal axes moving from one isodense to another; e.g., see Binney and Merrifield 1998) for the considered ellipsoids is impossibile. In Exercise 13.9, we show that in the axisymmetric case H is independent of φ and already in diagonal form for $\psi = 0$. Third, when H is in diagonal form, the diagonal elements of H' are the two eigenvalues $\lambda_+ \geq \lambda_-$ of H, given by the two (positive) solutions of the characteristic polynomial

$$\lambda^2 - \mathrm{Tr}(H)\lambda + \det(H) = 0. \tag{13.55}$$

Therefore, from Eq. (13.50), the semimajor and the semiminor axes of the isodensity ellipse $\ell = const.$ in the projection plane are (prove it!) $a = \ell/\sqrt{\lambda_-}$ and $b = \ell/\sqrt{\lambda_+}$, and so the (constant) axial ratio of the projected isodensities is given by $q \equiv b/a = \sqrt{\lambda_-/\lambda_+}$. Not only is isophotal twisting absent in our class of objects, but also the ellipticity of the isodenses is constant and depends only on the intrinsic flattenings and on the *los* orientation. Moreover, the area encircled by the ellipse $\ell = const.$ is given by $\pi ab = \pi\ell^2/\sqrt{\lambda_-\lambda_+}$, and as is well known from matrix algebra $\lambda_-\lambda_+ = \det(H)$, where from Eq. (13.52)

$$\frac{1}{\sqrt{\det(H)}} = a_1 a_2 a_3 \alpha = a_1^2 q_y q_z \sqrt{\sin^2\vartheta\cos^2\varphi + \frac{\sin^2\vartheta\sin^2\varphi}{q_y^2} + \frac{\cos^2\vartheta}{q_z^2}}. \tag{13.56}$$

From Eq. (13.56), it is possible to recover without diagonalization, as a function of the *los* direction, the explicit expression for the *circularized radius* $\langle R\rangle$, defined as $\pi\langle R\rangle^2 \equiv \pi ab$. Geometrically, $\langle R\rangle$ is the radius of a circle in the projection plane with the same area as an ellipse ℓ, and so

$$\langle R\rangle = a_1\ell\sqrt{q_y q_z \alpha a_1}. \tag{13.57}$$

Note that the expression under the square root is independent of a_1. The encircled mass in the projected ellipse ℓ is then obtained as

$$M_{\mathrm{p}}(\ell) = \int_{\ell'\leq\ell} \Sigma(\boldsymbol{\xi}_\perp) d^2\boldsymbol{\xi}_\perp = 2\pi a_1^3 \rho_0 \tilde{M}_{\mathrm{p}}(\ell), \quad \tilde{M}_{\mathrm{p}}(\ell) = \int_0^\ell \tilde{\Sigma}(\ell')\ell' d\ell'. \tag{13.58}$$

This formula is obtained by first rotating the coordinates in the projection plane so that H is in diagonal form, with $d^2\xi_\perp = d^2\xi'_\perp$, and then changing to polar coordinates $\xi'_\perp = (\ell' \cos\theta/\sqrt{\lambda_-}, \ell' \sin\theta/\sqrt{\lambda_+})$. From Eqs. (13.54) and (13.58), the student should prove that for $\ell \to \infty$, $M_P(\ell) \to M$. Moreover, in perfect analogy with the spherical case, we can define the *effective ellipse* ℓ_e as

$$M_p(\ell_e) = \frac{M}{2}, \quad \text{i.e.,} \quad \tilde{M}_p(\ell_e) = \frac{1}{2}. \tag{13.59}$$

Of course, the value of ℓ_e is independent of q_y, q_z, and the *los* inclinantion and depends *only* on the dimensionless function $\tilde{\rho}(m)$. Therefore, from Eq. (13.57), the *circularized effective radius* $\langle R_e \rangle$ is given by

$$\langle R_e \rangle = R_e \sqrt{q_y q_z \alpha a_1}, \tag{13.60}$$

where R_e is the effective radius of the parent spherical model. Note that the axisymmetric case $q_y = 1$ is particularly interesting, as from Exercise 13.9 it is elementary to prove that

$$q = \sqrt{\cos^2\vartheta + q_z^2 \sin^2\vartheta} = q_z \alpha a_1, \quad \langle R_e \rangle = R_e \sqrt{q}, \tag{13.61}$$

so that $q = 1$ in the *face-on* view and $q = q_z$ in the *edge-on* view.

13.2.3 Additional Families of Axisymmetric and Triaxial Models

Of course, other relatively simple axisymmetric systems not described by ellipsoidal distributions are known. Among the models in significant use (this list is by no means complete), we may recall

- *Miyamoto–Nagai models* (Miyamoto and Nagai 1975; Nagai and Miyamoto 1976)

$$\Psi(R,z) = \frac{GM}{\sqrt{R^2 + \zeta^2}}, \quad \zeta = a + \sqrt{z^2 + b^2}, \tag{13.62}$$

$$\rho(R,z) = \frac{Mb^2}{4\pi} \frac{aR^2 + (\zeta + 2\sqrt{z^2 + b^2})\zeta^2}{(R^2 + \zeta^2)^{5/2}(z^2 + b^2)^{3/2}}, \tag{13.63}$$

  with a characteristic disk-like shape (e.g., Binney and Tremaine 2008). From inspection of Eq. (13.62), notice how these models reduce for $a = 0$ to the Plummer sphere and for $b = 0$ to the Kuzmin razor-thin disk in Eq. (13.69).
- *Satoh models* (Satoh 1980)

$$\Psi(R,z) = \frac{GM}{\zeta}, \quad \zeta = \sqrt{R^2 + z^2 + a(a + 2\sqrt{z^2 + b^2})}, \tag{13.64}$$

$$\rho(R,z) = \frac{Mab^2}{4\pi\zeta^3(z^2 + b^2)} \left[\frac{1}{\sqrt{z^2 + b^2}} + \frac{3}{a}\left(1 - \frac{R^2 + z^2}{\zeta^2}\right) \right], \tag{13.65}$$

  another family of disk-like stellar systems.

- *Binney's logarithmic potential* (Binney 1981; Binney and Tremaine 2008)

$$\phi(R,z) = \frac{v_\infty^2}{2} \ln\left(1 + \frac{R^2 + z^2/q^2}{a^2}\right), \tag{13.66}$$

$$\rho(R,z) = \frac{v_\infty^2}{4\pi G} \frac{(1+2q^2)a^2 + R^2 + (2 - 1/q^2)z^2}{q_2(a^2 + R^2 + z^2/q^2)^2}, \tag{13.67}$$

where a is a scale length and v_∞ is the asymptotic (constant) circular velocity in the equatorial plane (prove it!), and for positivity of the density $q \leq 1/\sqrt{2}$. For values of q near the positivity limit, the density distribution presents a characteristic torus-like structure.

- *Evans' power-law models* (Evans 1994; Evans and de Zeeuw 1994)

$$\Psi(R,z) = \Psi(0,0)\left(1 + \frac{R^2 + z^2/q^2}{a^2}\right)^{-\beta/2}, \tag{13.68}$$

where a is a scale length, and $\beta \neq 0$, and the associated density distribution can be easily calculated. In fact, the Laplacian of potentials stratified on ellipsoidal surfaces, such as those in Eqs. (13.66)–(13.68), can be easily obtained following Exercise 2.9; note that the potential in Eq. (13.68) is the power-law analog of Eq. (13.66), and in fact for these models the density distribution also presents a characteristic torus-like structure for q near the positivity limit.

- *Kuzmin disk* (Kuzmin 1956; Toomre 1963)

$$\Psi(R,z) = \frac{GM}{\sqrt{R^2 + (a + |z|)^2}}, \quad \rho(R,z) = \frac{M\, a\, \delta(z)}{2\pi(R^2 + a^2)^{3/2}}, \tag{13.69}$$

where the student should prove that $R_h = \sqrt{3}a$ is the half-mass radius and verify the potential–density pair of this razor-thin disk obeys the Poisson equation by using the Gauss theorem.

As anticipated, several others families of nonspherical models are available. The study of the properties of these systems – sometimes built using quite advanced mathematical techniques – is beyond the task of an introductory book, and the student is strongly encouraged to explore the literature. Among these models, we can briefly mention the *scale-free models* (Evans et al. 1997), the *flattened isochrone* (Evans et al. 1990), the axisymmetric and triaxial models built using the homeoidal expansion technique described in Section 2.2.1 (e.g., see Ciotti and Bertin 2005; Ciotti et al. 2020; Riciputi et al. 2005), the models built with variants of the Miyamoto–Nagai technique (see An and Evans 2019; Evans and Bowden 2014 and references therein), and a few other illustrative cases described in the following sections.

13.2.3.1 Stäckel Models

A very important family of models (in general triaxial) is obtained by starting from the assignment of a potential expressed in ellipsoidal coordinates (λ, μ, ν) and then proceeding with the evaluation of the density from the Poisson equation (A.157). In this approach, the ellipsoidal coordinates are usually defined using a different convention than that adopted in Eq. (2.9): now the unitary ellipsoid is

$$\frac{x^2}{\alpha+\tau}+\frac{y^2}{\beta+\tau}+\frac{z^2}{\gamma+\tau}=1, \tag{13.70}$$

where $\alpha \leq \beta \leq \gamma \leq 0$ and $0 \leq -\gamma \leq \nu \leq -\beta \leq \mu \leq -\alpha \leq \lambda$. Surfaces of constant λ are again confocal ellipsoids, but now with the *short* axis along x and the long axis along z. Surfaces of constant μ are hyperboloids of one sheet, and surfaces of constant ν are hyperboloids ot two sheets. No difficulties are encountered in determining the new relations between Cartesian and ellipsoidal coordinates and the expressions of the differential operators (see Appendix A.8), and the student is encouraged to work out the corresponding formulae. Many interesting models have been constructed in this way (e.g., see de Zeeuw 1985a; de Zeeuw and Lynden-Bell 1985; de Zeeuw and Pfenniger 1988; de Zeeuw et al. 1986; Hunter and de Zeeuw 1992; Lynden-Bell 1962b and references therein). Among the models that can be generated with this approach, a prominent place (due to their special orbital properties) is occupied by the *Stäckel separable models*, originating from potentials of the family

$$\phi(\lambda,\mu,\nu)=-\frac{F(\lambda)}{(\lambda-\mu)(\lambda-\nu)}-\frac{F(\mu)}{(\mu-\lambda)(\mu-\nu)}-\frac{F(\nu)}{(\nu-\lambda)(\nu-\mu)}, \tag{13.71}$$

where $F(\tau)$ is an arbitrary function. In addition to the separability of the Hamilton–Jacobi equation in ellipsoidal coordinates, and so allowing for a complete description of stellar orbits in these models (see also Section 9.3.2), several interesting properties have been proved to hold for Stäckel models, the most famous being due to Kuzmin (see de Zeeuw 1985b; Kuzmin 1956). Briefly stated, for the potentials in Eq. (13.71): (1) the potential along the z-axis (the long axis of ellipsoidal coordinates) is related to the density profile along the same axis by a linear second-order ordinary differential equation (ODE; the *Kuzmin property*); (2) for the assigned density profile along the z-axis, the Kuzmin ODE can be integrated, and the solution determines *uniquely* the potential (and so the density) over the whole space – the density elsewhere is related to that on the z-axis by the so-called *Kuzmin formula*, and usually the z-axis is the short axis of the resulting density distribution; (3) the Kuzmin formula shows that the density is everywhere non-negative if this holds along the z-axis (the *Kuzmin theorem*); and (4) the Kuzmin formula also shows that density profiles that fall off along the z-axis faster than z^{-3} lead to stellar systems of finite mass, density profiles that fall off less steeply than z^{-4} become spherical at large radii, while for a density falloff $\propto z^{-4}$ or steeper the models have finite flattening at large radii. In particular, density profiles that fall off faster than z^{-4} lead to density distributions that in all other directions fall off as r^{-4}, so they are of quasi-toroidal shape. We will not discuss this subject further, but we conclude by recalling that Stäckel separable models are not the *only* models for which Kuzmin's results hold: other nonseparable models in ellipsoidal coordinates also share the same properties. Moreover, it has been shown

(Ciotti et al. 2012) that the Kuzmin property and a Kuzmin-like formula hold for Stäckel density–potential pairs in the deep modified Newtonian dynamics (MOND) regime (see Exercise 2.34); this is quite surprising, as the differential operator in Eq. (2.123) is not the Laplacian but the nonlinear p-Laplacian. For these separable MOND systems, the density profile at large radii is in general not spherically symmetric and declines as $\ln r/r^5$ instead of $1/r^4$.

13.2.3.2 Complex Shifted Systems

A method to construct axisymmetric (and triaxial) density–potential pairs from spherical density–potential pairs based on the linearity of the Poisson equation has been presented in Ciotti and Giampieri (2007) (see also Ciotti and Marinacci 2008; Vogt and Letelier 2009b and references therein). The method generalizes to extended mass distributions the concept of a *holomorphic Coulomb field* of a point charge introduced in electromagnetism (e.g., see Appell 1887; Lynden-Bell 2004; Newman 1973 and references therein).

Let us assume that the (nowhere negative) density distribution $\rho(\mathbf{x})$ satisfies the Poisson equation (2.6) with potential $\phi(\mathbf{x})$. The associated *complexified density–potential pair* with shift $\mathbf{a}$ is defined as

$$\rho_{\rm c}(\mathbf{x}) \equiv \rho(\mathbf{x} - i\mathbf{a}), \quad \phi_{\rm c}(\mathbf{x}) \equiv \phi(\mathbf{x} - i\mathbf{a}), \tag{13.72}$$

where $i^2 = -1$ is the imaginary unit and $\mathbf{a} \in \Re^3$. It is elementary to verify that

$$\Delta\phi_{\rm c} = 4\pi G\rho_{\rm c} \quad \forall \mathbf{a} \in \Re^3, \tag{13.73}$$

and so, if we separate $\rho_{\rm c}$ and $\phi_{\rm c}$ into their real and imaginary components, then we obtain at once *two* real density–potential pairs. Of course, in the gravitational case, the resulting pairs are physically acceptable provided they do not change sign as a function of $\mathbf{x}$. In fact, in Exercise 13.12 we will prove that $\Im(\rho_{\rm c})$ must necessarily change sign and so cannot be used to model a stellar system. Instead, the sign of $\Re(\rho_{\rm c})$ depends on the specific parent density–potential pair and on the shift vector $\mathbf{a}$, and can be everywhere positive, as illustrated in the following example.

For simplicity, we restrict ourselves to a parent spherically symmetric density–pontential pair $\rho(r)$ and $\phi(r)$. The complexification rule is that the "norm" of the shifted positions must be evaluated formally[3] as

$$r^2(a) \equiv \|\mathbf{x} - i\mathbf{a}\|^2 = r^2 - a^2 - 2i\langle\mathbf{x},\mathbf{a}\rangle, \quad a^2 \equiv \|\mathbf{a}\|^2. \tag{13.74}$$

Without loss of generality, we assume $\mathbf{a} = (0,0,a)$, so that $r^2(a) = r^2 - 2i\,a\,z - a^2$ and

$$\rho_{\rm c} = \rho\left(\sqrt{r^2 - 2iaz - a^2}\right), \quad \phi_{\rm c} = \phi\left(\sqrt{r^2 - 2iaz - a^2}\right). \tag{13.75}$$

Note that for a spherically symmetric (ρ,ϕ) pair, $\Re(\rho_{\rm c})$ and $\Im(\rho_{\rm c})$ can be obtained: (1) from evaluation of the Laplace operator applied to $\Re(\phi_{\rm c})$ and $\Im(\phi_{\rm c})$; (2) by expansion of

[3] If one adopts the standard inner product over the complex field, one would obtain $\|\mathbf{x} - i\mathbf{a}\|^2 = r^2 + \|\mathbf{a}\|^2$.

the complexified density in Eq. (13.75); or finally (3) by considering that for spherically symmetric systems $\rho = \rho(\phi)$, and so the real and imaginary parts of $\rho_c(\phi_c)$, can be expressed (at least in principle) as functions of the real and imaginary parts of the shifted potential. We now consider the Plummer model in Eq. (13.36), and for ease of notation, we use densities normalized to M/r_c^3, potentials to GM/r_c, and lengths to r_c (so that the vector shift $\mathbf{a}$ must also be expressed in units of r_c). It follows that $\Psi = -\phi = 1/\sqrt{1+r^2}$, and finally

$$\rho_c = \frac{3}{4\pi}[\Re(\Psi_c) + i\Im(\Psi_c)]^5, \quad \Psi_c = \frac{1}{\zeta_c} = \frac{\bar{\zeta}_c}{|\zeta_c|^2}, \tag{13.76}$$

where from Eq. (13.75)

$$\begin{cases} \zeta_c^2 = 1 + r^2(a) \equiv d e^{i\varphi}, \\ d \equiv |\zeta_c|^2 = \sqrt{(1 - a^2 + r^2)^2 + 4a^2z^2}, \\ \cos\varphi = \dfrac{1 - a^2 + r^2}{d}, \quad \sin\varphi = -\dfrac{2az}{d}. \end{cases} \tag{13.77}$$

Note that $\cos\varphi > 0$ everywhere for $a < 1$, and in the following discussion we restrict ourselves to this case. The square root $\zeta_c = \sqrt{d}e^{\pi ki + \varphi i/2}$ is made to be a single-valued function of (r, z) by cutting the complex plane along the negative real axis (which is never touched by ζ_c^2) and assuming $k = 0$, so that the model equatorial plane is mapped onto the line $\varphi = 0$. With this choice, the principal determination of Ψ_c reduces to Ψ when $a = 0$, and simple algebra shows that

$$\cos\frac{\varphi}{2} = \frac{\sqrt{1+\cos\varphi}}{\sqrt{2}}, \quad \sin\frac{\varphi}{2} = -\frac{\sqrt{2}az}{d\sqrt{1+\cos\varphi}}, \tag{13.78}$$

so that

$$\Re(\Psi_c) = \frac{\sqrt{d + 1 - a^2 + r^2}}{\sqrt{2}\,d}, \quad \Im(\Psi_c) = \frac{az}{d^2\Re(\Psi_c)}. \tag{13.79}$$

From Eq. (13.76), we now obtain the expressions of the (normalized) axisymmetric densities

$$\Re(\rho_c) = \frac{3\Re(\Psi_c)}{4\pi}\left[\Re(\Psi_c)^4 - \frac{10a^2z^2}{d^4} + 5\Im(\Psi_c)^4\right], \tag{13.80}$$

$$\Im(\rho_c) = \frac{3\Im(\Psi_c)}{4\pi}\left[5\Re(\Psi_c)^4 - \frac{10a^2z^2}{d^4} + \Im(\Psi_c)^4\right]. \tag{13.81}$$

It can be easily verified (do it!) that the axisymmetric density–potential pair $\Re(\rho_c) - \Re(\Psi_c)$ satisfies the Poisson equation, and that for $a \leq 1$ the shifted density $\Re(\rho_c)$ is nowhere negative. An idea of the shapes of the density–potential pairs produced with the complex shift method can be obtained from Ciotti and Giampieri (2007), where the complexification was also applied to the isochrone sphere in Eq. (13.43); from Ciotti and Marinacci (2008), where the complex shift was applied in the equatorial plane of the Miyamoto–Nagai disk in Eqs. (13.62)–(13.63), obtaining structures with three reflection planes of symmetry; or from Vogt and Letelier (2009b), where the shift was applied to the Kuzmin disk in Eq. (13.69). For sufficiently large shifts, the shifted densities are characterized by torus-like shapes, reminding us of structures such as the Lynden-Bell (1962a) flattened Plummer model, the density associated with the Binney logarithmic potential in Eq. (13.67), the Evans (1994) power-law models, the Toomre (1982) tori, the Ciotti and Bertin (2005) tori obtained from homeoidal expansion, and the Ciotti et al. (2020) galaxy models obtained using the same method.

13.3 The Solution of the Jeans Equations

13.3.1 Spherical Systems

After choosing a density distribution, the next step in the construction of a model for a stellar system is the solution of the associated Jeans equations (e.g., with respect to the velocity dispersion). In this case, their closure (see Chapter 12) can be obtained by assuming that the (unknown) associated DF depends on certain isolating integrals of motion. For example, in spherical models, if we assume $f = f(\mathcal{E}, J^2)$ as in Section 12.2.2, the relevant second-order Jeans equation to be solved for the density component ρ is (prove it!)

$$\frac{d\rho(r)\sigma_r^2(r)}{dr} + \frac{2\beta(r)}{r}\rho(r)\sigma_r^2(r) = -\rho(r)\frac{d\phi_{\rm T}(r)}{dr}, \tag{13.82}$$

where the definition of $\beta(r)$ is given in Eq. (12.34) and Eq. (13.82) is obtained from Eq. (10.68) expressed in spherical coordinates as in Eq. (9.52). For a given choice of the functions $\rho(r)$, $\phi_{\rm T}(r)$, and $\beta(r)$, the result of the integration is the function $\sigma_r^2(r)$. Then, $\sigma_{\rm t}^2(r)$ is recovered as

$$\sigma_{\rm t}^2(r) = 2\sigma_r^2(r)[1 - \beta(r)]. \tag{13.83}$$

The natural boundary condition for Eq. (13.82) is the vanishing of radial "pressure" at $r_{\rm t}$ (see Chapter 12), i.e.,

$$\rho(r_{\rm t})\sigma_r^2(r_{\rm t}) = 0, \tag{13.84}$$

because, by definition of the truncation radius, no stars can cross $r_{\rm t}$ (=∞ in most cases). Because of the linearity of Eq. (13.82), singular points for the solution may appear only at singular points of the coefficients (i.e., singular points of β/r and $\rho d\phi_{\rm T}/dr$), and in physically acceptable models these points may occur only at the center and/or at $r_{\rm t}$. Remarkably, the solution of Eq. (13.82) can be cast in explicit form. In fact, this is a Bernoulli first-order inhomogeneous linear ODE (e.g., Ince 1927; Lomen and Mark 1988), and its solution (Binney and Mamon 1982) is given by

$$\rho(r)\sigma_r^2(r) = \int_r^{r_{\rm t}} \rho(x)\frac{d\phi_{\rm T}(x)}{dx}{\rm e}^{-2\int_x^r \frac{\beta(y)dy}{y}}\, dx. \tag{13.85}$$

Due to its importance, we now show that Eq. (13.85) is the solution of Eq. (13.82) with the appropriate boundary condition. The simplest way to do this is to define the new function $Y(r) = \rho(r)\sigma_r^2(r)$ and to solve the resulting equation with the boundary condition $Y(r_t) = 0$. From the theory of ODEs, the solution of the resulting ODE is given by $Y = Y_0 + Y_1$, where Y_0 is the general solution of the associated homogeneous ODE and Y_1 is a particular solution of the inhomogeneous ODE. Let us determine Y_0. The integration is trivial (do it!), and one obtains

$$Y_0(r) = Y_0(r_0)\mathrm{e}^{-2\int_{r_0}^{r}\frac{\beta(y)dy}{y}}, \tag{13.86}$$

where $r_0 \neq r_t$ is for the moment left unspecified. The function Y_1 is now determined using the method of variation of constants; in other words, we set

$$Y_1(r) = Y_0(r)Y_2(r). \tag{13.87}$$

After substitution of Y_1 into the inhomogeneous equation, we are left with an ODE for Y_2 with solution (prove it!)

$$Y_2(r) = Y_2(r_0) - \frac{1}{Y_0(r_0)}\int_{r_0}^{r}\rho(x)\frac{d\phi_T(x)}{dx}\mathrm{e}^{2\int_{r_0}^{x}\frac{\beta(y)dy}{y}}dx, \tag{13.88}$$

where $Y_2(r_0)$ is an arbitrary constant. A rearrangement of $Y_0(r) + Y_1(r)$ shows that, for the choice $Y_2(r_0) = -1$, a cancellation of $Y_0(r_0)$ happens, and it is now possible to choose $r_0 = r_t$ so that the boundary condition in Eq. (13.84) is satisfied.

For example, in the case of anisotropic Osipkov–Merritt (OM) spherical systems from Eqs. (12.52)–(13.85)

$$\begin{aligned}\rho(r)\sigma_r^2(r) &= \frac{G}{r^2+r_a^2}\int_r^{r_t}\rho(x)M_T(x)\left(1+\frac{r_a^2}{x^2}\right)dx\\ &= \frac{A(r)+r_a^2 I(r)}{r^2+r_a^2} = I(r) + \frac{A(r)-r^2 I(r)}{r^2+r_a^2},\end{aligned} \tag{13.89}$$

where the two functions $A(r)$ and $I(r)$ correspond to the purely radial and to the isotropic velocity dispersion profiles, obtained for $r_a = 0$ and $r_a \to \infty$, respectively.

More complicated systems can be constructed. For example, in the case of anisotropic Cuddeford spherical systems, from Eqs. (12.63) and (13.85)

$$\rho(r)\sigma_r^2(r) = \frac{Gr^{2\alpha}}{(r^2+r_a^2)^{\alpha+1}}\int_r^{r_t}\rho(x)M_T(x)\left(1+\frac{r_a^2}{x^2}\right)^{\alpha+1}dx. \tag{13.90}$$

As we know from Chapter 12, when $\alpha = 0$, this reduces to the OM solution, while for $r_a \to \infty$ this reduces to the constant-anisotropy case.

We now derive the projection formula for the second-order velocity dispersion profile associated with spherical models with $f = f(\mathcal{E}, J^2)$ and anisotropy profile $\beta(r)$. Because in these systems the streaming velocity field vanishes identically, from Chapter 11 $\overline{\sigma_l^N}_{\rm los} = \overline{\sigma_p^N}_{\rm los} = \overline{v_p^N}_{\rm los}$. In particular, for the moment of order 2, in Exercise 13.14 we show that

$$\Sigma(R)\sigma_{\rm P}^2(R) = 2\int_R^{R_{\rm t}} \left[1-\beta(r)\frac{R^2}{r^2}\right]\frac{\rho(r)\sigma_r^2(r)r}{\sqrt{r^2-R^2}}dr, \tag{13.91}$$

where for simplicity we adopted the standard notation $\sigma_{\rm P}^2 = \overline{v_{\rm P_{los}}^2}$; note how the projection integral for isotropic models is similar to that of the density projection. In Exercise 13.15, we specialize Eq. (13.91) to the case of OM systems. At the center of the image in the projection plane, when the integral in Eq. (13.91) converges, we have

$$\Sigma(0)\sigma_{\rm P}^2(0) = 2\int_0^{R_{\rm t}} \rho(r)\sigma_r^2(r)\,dr. \tag{13.92}$$

The "observed" $\sigma_\circ$ does not correspond to $\sigma_{\rm P}(0)$, but rather to the luminosity (mass in the case of constant Υ_*) average over the aperture used for the spectrographic observations

$$\sigma_{\rm a}^2(R_{\rm ap}) = \frac{2\pi}{M_{\rm p}^*(R_{\rm ap})}\int_0^{R_{\rm ap}} \Sigma(R)\,\sigma_{\rm P}^2(R)\,R\,dR, \tag{13.93}$$

where $M_{\rm p}^*(R_{\rm ap})$ is the projected stellar mass inside $R_{\rm ap}$. Usually, one identifies this quantity with $\sigma_\circ$.

With the aid of this obtained formula, it is an interesting exercise to verify directly the projected virial theorem for spherical systems introduced in Chapter 11. In fact, in Exercise 13.16 we show by direct integration that the general identity in Eq. (11.44) reduces for nonrotating spherical systems and generic β profiles to

$$[K_{\rm los}^{(2)}]_\Omega = K_{\rm los}^{(2)} = \pi\int_0^{R_{\rm t}} \Sigma(R)\sigma_{\rm P}^2(R)R\,dR = \frac{2\pi}{3}\int_0^{r_{\rm t}} \rho(r)[\sigma_r^2(r)+\sigma_{\rm t}^2(r)]r^2dr = \frac{K}{3}. \tag{13.94}$$

13.3.1.1 Power-Law Spherical Models

As an example of highly idealized spherical systems, for which many dynamical quantities can be obtained explicitly yet retaining some qualitative features of more realistic models, we consider the untruncated one-component power-law models in Eq. (13.29) with OM internal dynamics and with $2 < \gamma < 3$. A central BH and an additional external power-law density component can be easily added, keeping the level of the presentation simple (see Exercise 13.19; see also the spherical limits of the models in Ciotti and Bertin 2005; Riciputi et al. 2005). The student is also invited to repeat the treatment of this section for OM models and spatially truncated power-law OM models, also adding a BH at their center, and to discuss the full range of slopes $0 \leq \gamma < 3$ (see also Exercise 13.21).

The radial component of the velocity dispersion is obtained from Eq. (13.89)

$$\sigma_r^2(r) = \frac{GM_{\rm c}}{r_{\rm c}}\frac{s^{2-\gamma}}{2(3-\gamma)(s^2+s_{\rm a}^2)}\left(\frac{s^2}{\gamma-2}+\frac{s_{\rm a}^2}{\gamma-1}\right), \quad s_{\rm a} \equiv \frac{r_{\rm a}}{r_{\rm c}}, \tag{13.95}$$

where $s = r/r_c$, and the isotropic and purely radial anisotropic cases are recovered for $s_{\rm a} \to \infty$ and $s_{\rm a} = 0$, respectively

$$\begin{cases} [\sigma_r^2(r)]_{\rm iso} = \dfrac{GM_{\rm c}}{r_{\rm c}} \dfrac{s^{2-\gamma}}{2(3-\gamma)(\gamma-1)} = \dfrac{\Psi(r)}{2} \dfrac{\gamma-2}{\gamma-1}, \\ [\sigma_r^2(r)]_{\rm rad} = \dfrac{GM_{\rm c}}{r_{\rm c}} \dfrac{s^{2-\gamma}}{2(3-\gamma)(\gamma-2)} = \dfrac{\Psi(r)}{2}, \end{cases} \tag{13.96}$$

where we used $\Psi = -\phi$ from Eq. (13.30). Therefore

$$\frac{[\sigma_r^2(r)]_{\rm rad}}{[\sigma_r^2(r)]_{\rm iso}} = \frac{\gamma-1}{\gamma-2} > 1, \tag{13.97}$$

and σ_r in the radially anisotropic case is larger than that in the case of orbital isotropy, as expected. The projected velocity dispersion can be obtained from Eq. (13.91), and in the two relevant cases

$$\begin{cases} [\sigma_{\rm P}^2(R)]_{\rm iso} = \dfrac{GM_{\rm c}}{r_{\rm c}} \dfrac{\eta^{2-\gamma}}{2(3-\gamma)(\gamma-1)} \dfrac{{\rm B}(1/2,\gamma-3/2)}{{\rm B}(1/2,\gamma/2-1/2)}, \\ [\sigma_{\rm P}^2(R)]_{\rm rad} = \dfrac{GM_{\rm c}}{r_{\rm c}} \dfrac{\eta^{2-\gamma}}{2(3-\gamma)(\gamma-2)} \dfrac{{\rm B}(3/2,\gamma-3/2)}{{\rm B}(1/2,\gamma/2-1/2)}, \end{cases} \tag{13.98}$$

where $\eta = R/r_{\rm c}$, so that

$$\frac{[\sigma_{\rm P}^2(R)]_{\rm rad}}{[\sigma_{\rm P}^2(R)]_{\rm iso}} = \frac{1}{2(\gamma-2)}. \tag{13.99}$$

Finally, the aperture velocity dispersion defined in Eq. (13.93) converges at the center for $\gamma < 5/2$, and the student is invited to prove that

$$[\sigma_{\rm a}^2(R)]_{\rm iso} = \frac{3-\gamma}{5-2\gamma}[\sigma_{\rm P}^2(R)]_{\rm iso}, \quad [\sigma_{\rm a}^2(R)]_{\rm rad} = \frac{3-\gamma}{5-2\gamma}[\sigma_{\rm P}^2(R)]_{\rm rad}. \tag{13.100}$$

A comment is in order regarding the behavior of $\sigma_r(r)$ for $r \to 0$ in isotropic spherical models that can be studied with the aid of power-law models. In fact, it turns out (e.g., see Bertin et al. 2002; Binney and Ossipkov 2001) that in the central regions of spherical systems with density approximated by $\rho \sim 1/r^\gamma$, the corresponding isotropic $\sigma_r(r)$ stays constant for $\gamma = 0$, decreases to zero for $0 < \gamma < 2$, is again a positive constant for $\gamma = 2$, and diverges for $\gamma > 2$. A simple explanation of this nonmonotonic behavior of the central σ_r with the local value of γ is that in a self-gravitating system, by imposing the density profile, we also impose the gravitational potential (i.e., the gravitational field that the system must balance with the "pressure" $\rho\sigma_r^2$ to be at equilibrium). But ρ is fixed, and so the "temperature" σ_r^2 remains determined by the relative profiles of ρ and the gravitational field. Of course, for all of the values of γ, the "pressure" $\rho\sigma_r^2$ is a decreasing function of r, in accordance with the (isotropic) Eq. (13.82), and needless to say, an exact parallel of this phenomenon happens in fluid dynamics when imposing a potential and a gas density profile and searching for the temperature profile (see Section 10.4.1). It should remarked that several of the models encountered in the literature (e.g., the γ-models or the Sersic $R^{1/m}$ profiles) are characterized by a sizable central depression in their $\sigma_{\rm P}$, with quite important implications for the interpretation of observational data (e.g., see Bertin et al. 2002; Binney 1980; Ciotti 1991; Ciotti and Lanzoni 1997; Ciotti and Pellegrini 1992; Ciotti et al. 1996).

We conclude this section by reminding the student that they will find in the literature numerous examples of more complicated (realistic) multicomponent spherical stellar systems where the same calculations can be more or less easily carried out analytically (e.g., see the *very* incomplete list Ciotti 1996, 1999; Ciotti and Ziaee Lorzad 2018; Ciotti et al. 2009, 2019; Dehnen 1993; Hernquist 1990; Hiotelis 1994; Jaffe 1983; Łokas and Mamon 2001; Tremaine et al. 1994 and references therein).

13.3.2 Axisymmetric Systems

As with the spherical case, after the density distribution is assigned, the next step in the construction of a model for an axisymmetric stellar system is the integration of the associated Jeans equations. If we assume that the underlying DF is of the form $f = f(\mathcal{E}, J_z)$, then from Chapter 12 $\sigma_R^2 = \sigma_z^2$, the velocity dispersion ellipsoid is rotationally symmetric in the meridional plane (R, z), and from Eq. (10.68) in cylindrical coordinates the stationary Jeans equations (prove it!) are obtained from Eq. (9.53)

$$\begin{cases} \dfrac{\partial \rho\sigma_z^2}{\partial z} = -\rho \dfrac{\partial \phi_{\mathrm{T}}}{\partial z}, \\ \dfrac{\partial \rho\sigma_z^2}{\partial R} - \dfrac{\rho\Delta_\sigma}{R} = -\rho \dfrac{\partial \phi_{\mathrm{T}}}{\partial R}, \quad \Delta_\sigma \equiv \overline{v_\varphi^2} - \sigma_z^2 = \overline{v_\varphi}^2 + \sigma_\varphi^2 - \sigma_z^2. \end{cases} \tag{13.101}$$

Note that in the *isotropic* case $\Delta_\sigma = \overline{v_\varphi}^2$ (i.e., it reduces to the squared streaming velocity field), in complete (formal) analogy with the equations of fluid dynamics of (in general baroclinic) self-gravitating rotating structures, where the term $\rho\sigma_z^2$ acts as a "pressure" and the streaming velocity field $\overline{v_\varphi}$ acts as the fluid velocity (see Section 10.4.1). In stellar dynamics, this special but important class of models is known as the family of *isotropic rotators*: the deviation from sphericity of their density distributions can only be due to the azimuthal streaming velocity.

Before proceeding with the integration of Eq. (13.101), a discussion of the boundary condition for the velocity dispersion is needed. If the system has an infinite extent, then the natural boundary condition is $\rho\sigma_z^2 \to 0$ for $\|\mathbf{x}\| \to \infty$. Instead, systems spatially truncated by self-gravitation require that everywhere on the system boundary the normal component of the "pressure" $\rho\sigma_{\mathrm{n}}^2 = \rho\sigma_{ij}^2 n_i n_j$ vanishes, and in Exercise 13.23 we show that in our case this is equivalent to having $\sigma_R^2 = \sigma_z^2 = 0$, independently of the shape of the truncation surface $S(R, z) = 0$. For assigned ρ and ϕ_{T}, the general solution of Eq. (13.101) is easily obtained by integrating the first equation at fixed R

$$\rho\sigma_z^2 = \int_z^{z_{\mathrm{t}}(R)} \rho \frac{\partial \phi_{\mathrm{T}}}{\partial z'} dz' = \int_r^{r_{\mathrm{t}}(R)} \rho \frac{\partial \phi_{\mathrm{T}}}{\partial r'} dr', \tag{13.102}$$

where the function $z_{\mathrm{t}}(R)$ is given implicitly by $S[R, z_{\mathrm{t}}(R)] = 0$ (and for spatially untruncated systems $z_{\mathrm{t}} = \infty$), and the second expression is useful when the density and potential are expressed in terms of r and R instead of z and R (e.g., see Exercise 13.32), with $r_{\rm t}^2(R) = R^2 + z_{\rm t}^2(R)$. The quantity Δ_σ (and $\overline{v_\varphi^2}$) is then recovered from the second equation by calculating the radial derivatives, and so we conclude that actually *only one* integration is required to solve the system in Eq. (13.101). However, this approach tends to obscure some important properties of the solutions, and in fact an alternative expression for Δ_σ can be obtained following an approach adopted in fluid dynamics to treat the case of axisymmetric rotating fluids (e.g., see Barnabè et al. 2006 and references therein; see also Tassoul 1978). An integration by parts, after substitution of Eq. (13.102) into the second equation on Eq. (13.101), shows that (prove it!)

$$\begin{aligned}\frac{\rho\Delta_\sigma}{R} &= \rho(R, z_{\rm t})\frac{d\phi(R, z_{\rm t})}{dR} + \int_z^{z_{\rm t}(R)} \left(\frac{\partial\rho}{\partial R}\frac{\partial\phi_{\rm T}}{\partial z'} - \frac{\partial\rho}{\partial z'}\frac{\partial\phi_{\rm T}}{\partial R}\right) dz' \\ &= \rho(R, z_{\rm t})\frac{d\phi(R, z_{\rm t})}{dR} + [\rho, \phi_{\rm T}]\end{aligned} \tag{13.103}$$

(e.g., see Hunter 1977; Smet et al. 2015), in perfect analogy with Eq. (10.34) for untruncated isotropic rotators. Therefore, for a system with density vanishing on the boundary (or untruncated), the boundary term vanishes and only the *commutator* term $[\rho, \phi_{\rm T}]$ survives; in Exercises 13.24 and 13.25, some revelant properties of the commutator are derived. In particular, in case of spherical symmetry of the density and potential, the commutator vanishes and so $\Delta_\sigma = 0$. Of course, for assigned density and potential, the model must certainly be discarded if $\Delta_\sigma + \sigma_z^2 = \overline{v_\varphi^2} < 0$, a manifestation of phase-space inconsistency.

When $\Delta_\sigma \geq 0$ everywhere,[4] the phenomenological *Satoh decomposition* is often adopted to break the degeneracy between ordered and random motions along the azimuthal direction shown by Eq. (13.101), where only the sum $\overline{v_\varphi^2} = \overline{v_\varphi}^2 + \sigma_\varphi^2$ appears

$$\overline{v_\varphi} = k\sqrt{\Delta_\sigma}, \quad |k| \leq 1 \tag{13.104}$$

(Satoh 1980), where k is a free parameter. From Eq. (13.104), one then obtains

$$\sigma_\varphi^2 = \overline{v_\varphi^2} - \overline{v_\varphi}^2 = (1 - k^2)\Delta_\sigma + \sigma_z^2 = (1 - k^2)\overline{v_\varphi^2} + k^2\sigma_z^2. \tag{13.105}$$

For $k = 0$, no ordered rotation is present (i.e., $\sigma_\varphi^2 = \overline{v_\varphi^2}$), and the model's asphericity is due to *tangential anisotropy* only, while for $k = 1$, the velocity dispersion tensor is isotropic (i.e., $\sigma_\varphi^2 = \sigma_R^2 = \sigma_z^2$). Finally, note that Satoh systems in the spherical limit are necessarily isotropic due to the vanishing of the left-hand side terms in Eq. (13.105). Of course, k can also be a function of (R, z), with the only requirement being the non-negativity of σ_φ. As is shown in Ciotti and Pellegrini (1996), this limit case is obtained by imposing the *maximum* amount of ordered rotation (i.e., for $\sigma_\varphi = 0$ everywhere) when the velocity dispersion

[4] See Barnabè et al. (2006) and Smet et al. (2015) for some general results regarding the positivity of Δ_σ based on the use of the commutator.

ellipsoid reduces to a zero-thickness disk. For a given model, the maximum bound of $|k|$ at each point is therefore obtained once the solution of the Jeans equations is known, and it is given by

$$k^2 \leq k_{\max}^2(R,z) = 1 + \frac{\sigma_z^2}{\Delta_\sigma}. \tag{13.106}$$

By using this approach, counter-rotation can also be imposed on the Jeans solutions (Negri et al. 2014a). Other more complicated phenomenological decompositions can be made, such as that proposed by Cappellari (2008, 2020) with $\sigma_R \neq \sigma_z$, with the implicit assumption of a three-integral DF. Explicit examples of solutions are the two-component Miyamoto–Nagai models in Ciotti and Pellegrini (1996) and the fully analytical solution for Miyamoto–Nagai models embedded in the Binney logarithmic potential (Smet et al. 2015).

Obviously, the Satoh decomposition and its generalization can be applied only provided $\Delta_\sigma \geq 0$ everywhere. However, we can also consider the case of $\Delta_\sigma \leq 0$ somewhere provided $\overline{v_\varphi^2} \geq 0$ everywhere. In this case, a possible decomposition of $\overline{v_\varphi^2}$ is

$$\overline{v_\varphi} = k\sqrt{\Delta_\sigma + \sigma_z^2}, \quad \sigma_\varphi^2 = (1-k^2)(\Delta_\sigma + \sigma_z^2), \qquad |k| \leq 1. \tag{13.107}$$

For $k = 0$, the model in the tangential direction is fully velocity-dispersion supported, while it is fully supported by ordered rotation for $k = 1$ (see Caravita et al. 2021 for the multicomponent case). Note that, at variance with the Satoh decomposition, isotropy cannot be realized when $\Delta_\sigma < 0$, because by imposing $\sigma_\varphi^2 = \sigma_z^2$ in the second identity of Eq. (13.107), we would have (prove it!) $0 \geq \Delta_\sigma = \overline{v_\varphi}^2$, which is obviously impossible.

Once the azimuthal decomposition is fixed, the next step in the model construction is the projection of the density and the velocity fields along a chosen *los*, fixed by the unit vector $\mathbf{n}$. Following Chapter 11, in order to obtain the projected velocity fields at position $\boldsymbol{\xi}_\perp$ in the projection plane, we must integrate along the whole *los* their projected component on $\mathbf{n}$. In the case of a two-integral axisymmetric systems, we can assume without loss of generality $\varphi = 0$, $\theta = i$, and $\psi = 0$ in the rotation matrix $\mathcal{R}$ in Eq. (11.1), so that along the *los*,

$$x = \xi_1 \cos i + \xi_3 \sin i, \quad y = \xi_2, \quad z = -\xi_1 \sin i + \xi_3 \cos i, \tag{13.108}$$

and $\mathbf{n} = (\sin i, 0, \cos i)$ from[5] Eq. (11.2). In practice, the observer system S' is determined by a rotation of ϑ around the axis $y = \xi_2$, and for fixed ξ_1 and ξ_2 in the projection plane the projection is an integration along ξ_3. Let us now consider two of the most used projected velocity fields in applications, namely the projected streaming velocity field and the projected velocity dispersion field in Eqs. (11.9)–(11.11). Before integration along ξ_3,

[5] Recall that $\mathbf{n}$ points *toward* the observer, so that a receding velocity is *negative*.

as dictated by Eq. (11.4), we transform these quantities from cylindrical to Cartesian coordinates as in Eqs. (12.129) and (12.130)

$$\begin{cases} \overline{v_{\rm p}}(\mathbf{x}) = -\overline{v_\varphi}(\mathbf{x}) \sin\varphi \sin i, \\ \overline{\sigma^2_{\rm p}}(\mathbf{x}) = \sigma^2_z(\mathbf{x}) + [\sigma^2_\varphi(\mathbf{x}) - \sigma^2_z(\mathbf{x})] \sin^2\varphi \sin i. \end{cases} \tag{13.109}$$

Of course, in Eq. (13.109) i is fixed along the *los*, while the angle φ (not to be confused with the angle $\varphi = 0$ in the rotation matrix!) is given by

$$\cos\varphi = \frac{x}{R}, \quad \sin\varphi = \frac{y}{R}. \tag{13.110}$$

From Chapter 11, the corresponding (mass-weighted) *los* projected fields are obtained by changing the coordinates in Eqs. (13.109) to those from Eq. (13.108) and then integrating on ξ_3

$$\overline{v_{\rm p}}_{\rm los}(\boldsymbol{\xi}_\perp) = \frac{1}{\Sigma(\boldsymbol{\xi}_\perp)} \int_{-\infty}^{\infty} \rho \overline{v_{\rm p}} d\xi_3, \tag{13.111}$$

and from Eq. (11.16)

$$\overline{\sigma^2_{\rm l}}_{\rm los}(\boldsymbol{\xi}_\perp) = \frac{1}{\Sigma(\boldsymbol{\xi}_\perp)} \int_{-\infty}^{\infty} \rho \, \overline{(v_{\rm p} - \overline{v_{\rm p}}_{\rm los})^2} d\xi_3 = \overline{\sigma^2_{\rm p}}_{\rm los} + \overline{v_{\rm p}}^2_{\rm los} - (\overline{v_{\rm p}}_{\rm los})^2, \tag{13.112}$$

where

$$\overline{\sigma^2_{\rm p}}_{\rm los}(\boldsymbol{\xi}_\perp) = \frac{1}{\Sigma(\boldsymbol{\xi}_\perp)} \int_{-\infty}^{\infty} \rho \overline{\sigma^2_{\rm p}} d\xi_3, \quad \overline{v_{\rm p}}^2_{\rm los}(\boldsymbol{\xi}_\perp) = \frac{1}{\Sigma(\boldsymbol{\xi}_\perp)} \int_{-\infty}^{\infty} \rho \overline{v_{\rm p}}^2 d\xi_3. \tag{13.113}$$

Note that, independently of the *los* orientation i, on the axis ξ_1 (where, by definition, $\sin\varphi = 0$, and usually corresponding to the short axis of the isophotal ellipse), $\overline{v_{\rm p}}(\mathbf{x}) = 0$ and $\overline{\sigma^2_{\rm p}}(\mathbf{x}) = \sigma^2_R(\mathbf{x})$, so that $\overline{v_{\rm p}}_{\rm los} = 0$ and $\overline{\sigma^2_{\rm l}}_{\rm los} = \overline{\sigma^2_{\rm p}}_{\rm los}$. In addition, the last identity holds everywhere when observing the galaxy face on ($i = 0$) or in the case $k = 0$. In Exercise 13.26, we prove a simple but useful property of second-order projected fields.

Since the observed velocity dispersion is always measured within a given aperture, we finally integrate $\overline{\sigma^2_{\rm l}}_{\rm los}$ over the isophotes (even though $\overline{\sigma^2_{\rm l}}_{\rm los}$ in general is not constant over isophotes)

$$M_{\rm p}\sigma^2_{\rm a} \equiv \int_{\rm Aperture} \Sigma(\boldsymbol{\xi}_\perp) \overline{\sigma^2_{\rm l}}_{\rm los}(\boldsymbol{\xi}_\perp) d^2\boldsymbol{\xi}_\perp. \tag{13.114}$$

13.3.2.1 Ferrers Axisymmetric Ellipsoids

An interesting (albeit quite artificial) family of *triaxial* ellipsoidal systems is represented by the *Ferrers ellipsoids* (Binney and Tremaine 2008; Chandrasekhar 1969; Ferrers 1877). Following Eq. (2.74), they are defined as

$$\rho_{\rm t}(m) = \rho(0) \times (1 - m^2)^n \theta(1 - m), \tag{13.115}$$

where $\rho(0)$ is the central density and $n \geq 0$ is usually an integer number: the $n = 0$ case corresponds to the homogeneous ellipsoid. In the axisymmetric case around the z-axis, a_1 is the semimajor axis of the truncation surface $m_t = 1$ in the equatorial plane, while the flattening is given by $0 < q_z = 1 - \eta \leq 1$. The interesting aspect of these models is that many of their properties can be computed analytically, which is quite unusual for ellipsoidal systems (e.g., see de Zeeuw and Pfenniger 1988; Lanzoni and Ciotti 2003). The mass within m, and the total mass of the model are given by

$$M(m) = 2\pi\rho(0)a_1^3 q_y q_z \mathrm{B}\left(\frac{3}{2}, n+1; m^2\right), \quad M = 2\pi\rho(0)a_1^3 q_y q_z \mathrm{B}\left(\frac{3}{2}, n+1\right), \tag{13.116}$$

where the Euler beta function is given in Eqs. (A.53)–(A.55); for integer n, the expressions in Eq. (13.116) reduce to simple algebraic form. Notice that the last identity in Eq. (13.116) can be used to express $\rho(0)$ in terms of $\rho_n = M/(4\pi a_1^3)$, and so to recast Eq. (13.115) in the form of Eq. (2.28), and finally to determine $\tilde{\rho}(m)$. From Eqs. (13.54) and (13.58), the surface density and the projected mass profiles are given by (prove it!)

$$\Sigma(\ell) = \frac{\rho(0)}{\alpha}\mathrm{B}\left(\frac{1}{2}, n+1\right)(1-\ell^2)^{n+\frac{1}{2}}, \; M_p(\ell) = M\left[1 - (1-\ell^2)^{n+\frac{3}{2}}\right], \tag{13.117}$$

so that from Eqs. (13.57) and (13.59) it follows that

$$\ell_e = \sqrt{1 - 2^{-1/(n+3/2)}}, \quad \langle R_e \rangle = a_1 \ell_e \sqrt{q_y q_z \alpha a_1}. \tag{13.118}$$

Note that Eq. (13.115) in the axisymmetric case with n integer can be rewritten for $m \leq 1$ as

$$\rho(m) = \rho_0 \sum_{i=0}^{n}\sum_{j=0}^{i} \rho_{nij}(q_z)\, \tilde{R}^{2j}\tilde{z}^{2(i-j)}, \quad \rho_{nij}(q_z) \equiv \frac{(-1)^i}{q_z^{2(i-j)}}\binom{n}{i}\binom{i}{j}, \tag{13.119}$$

where $\tilde{R} \equiv R/a_1$, $\tilde{z} \equiv z/a_1$. For this distribution, the potential in Eq. (2.75) can be easily calculated, because $\Delta\Psi$ reduces to a finite sum, and after some algebra (prove it!)

$$\begin{cases} \phi(R,z) = -\pi G\rho(0)a_1^2 \displaystyle\sum_{i=0}^{n+1}\sum_{j=0}^{i} \phi_{nij}(q_z)\tilde{R}^{2(i-j)}\tilde{z}^{2j}, \\ \phi_{nij}(q_z) \equiv \dfrac{q_z(-1)^i}{(n+1)(i+1/2)}\dbinom{n+1}{i}\dbinom{i}{j}\, {}_2\mathrm{F}_1\left(j+\dfrac{1}{2}, i+\dfrac{1}{2}, i+\dfrac{3}{2}; 1-q_z^2\right), \end{cases} \tag{13.120}$$

where ${}_2\mathrm{F}_1$ is the standard hypergeometric function in Eq. (A.69). Note that ${}_2\mathrm{F}_1(a,b,c;0) = 0$, and that for i and j integers ${}_2\mathrm{F}_1$ in Eq. (13.120) simplifies to a combination of elementary functions. Moreover, note that in the special case of the constant-density ellipsoid, from Eq. (2.67) we have $\phi_{000} = a_1 q_z w_0(0)$ and from Eq. (3.11) we have $\phi_{010}(q_z) = -w_1 = -w_2$ and $\phi_{011}(q_z) = -w_3$, in accordance with Eqs. (3.62) and (3.63); for the quasi-spherical limit, see Eq. (3.13). The components of the gravitational energy tensor W_{ij} can be easily computed by using Eq. (10.64) specialized to the truncated case discussed in Exercise 2.8. In the axisymmetric case, $W_{22} = W_{11}$ and

$$W_{11} = -G\rho(0)^2 a_1^5 \frac{\pi^2 q_z w_1}{2(n+1)^2} \mathrm{B}\left(\frac{1}{2}, 2n+3\right),$$

$$W_{33} = -G\rho(0)^2 a_1^5 \frac{\pi^2 q_z^3 w_3}{2(n+1)^2} \mathrm{B}\left(\frac{1}{2}, 2n+3\right). \tag{13.121}$$

Finally, it is obvious that the Jeans equations (13.101) can be easily integrated, and in particular in Eq. (13.102) we have $z_t \equiv q_z\sqrt{a_1^2 - R^2}$. In fact, thanks to the finite expansions of ρ and ϕ, the quantities $\rho\sigma_z^2$ and $\rho\Delta_\sigma$ in Eqs. (13.102) and (13.103) can also be obtained as polynomials. All of the steps of the computations, up to the projection of the resulting velocity dispersion fields and integration over isophotal ellipses in the projection plane, can be found in Lanzoni and Ciotti (2003). The resulting formulae are quite cumbersome and not illuminating, and so we do not report them here: however, the special case of the constant-density ellipsoid is remarkably simple (see Exercise 13.27).

13.3.2.2 Homeoidally Expanded Axisymmetric Two-Tntegral Systems

A quite general family of axisymmetric models, whose two-integral Jeans equations are amenable to an interesting formal solution, are those derived from homeoidal expansion described in Chapter 2. We recall that the obtained density–potential pairs can be interpreted in two different ways: as the limit of ellipsoidal systems for small flattenings *or* as exact pairs with finite (and not necessarily small) flattenings. In the latter case, when evaluating products of the expanded functions (as required by the Jeans equations), all of the flattening terms up to the second order inclusive must be retained, while in the first case only linear terms in the flattenings matter. Of course, in both cases no special difficulties are encountered, but here for simplicity we restrict ourselves to the first case. Worked-out examples of solutions of the Jeans equations (also in the case of multicomponent systems with a central BH) can be found in Ciotti and Bertin (2005), Ciotti et al. (2020), and Riciputi et al. (2005). We first consider the solution of the vertical Jeans equation (13.102), and we recognize that the density–potential pair in Eqs. (2.29)–(2.31), when specialized to the *oblate* axisymmetric case ($\epsilon = 0$), is *not* in the most useful form to be integrated, as it explicitly contains the variable z. However, with the substitution $z^2 = r^2 - R^2$, we can rewrite the dimensionless density–potential pair associated with $\rho = \rho_n\tilde{\rho}(m)$ and $\phi = -\Psi_n\tilde{\phi}(\mathbf{x})$ as

$$\frac{\tilde{\rho}(m)}{(1-\eta)} \sim \tilde{\rho}_0(s) + \eta\tilde{\rho}_1(s) + \eta\tilde{R}^2\tilde{\rho}_2(s), \quad \tilde{\phi}(\mathbf{x}) \sim \tilde{\phi}_0(s) + \eta\tilde{\phi}_1(s) + \eta\tilde{R}^2\tilde{\phi}_2(s), \tag{13.122}$$

where the meaning of the six spherical functions is obvious from Eqs. (2.29)–(2.31). Equation (13.102) is evaluated by expanding the integrand and retaining the terms up to the linear order in the flattening and changing the variable from z' to the spherical radius r' at fixed as R, obtaining

$$\rho\sigma_z^2 = \int_s^\infty \rho(s')\frac{\partial\phi(s',R)}{\partial s'}\,ds' \sim \frac{GM\rho_0}{a_1}[A(s)+\eta B(s)+\eta\tilde{R}^2C(s)], \qquad (13.123)$$

where the three radial integrals are

$$\begin{cases} A(s) = -\displaystyle\int_s^\infty \tilde{\rho}_0(s')\tilde{\phi}_0'(s')\,ds', \\ B(s) = -\displaystyle\int_s^\infty [\tilde{\rho}_0(s')\tilde{\phi}_1'(s') + \tilde{\rho}_1(s')\tilde{\phi}_0'(s')]\,ds', \\ C(s) = -\displaystyle\int_s^\infty [\tilde{\rho}_0(s')\tilde{\phi}_2'(s') + \tilde{\rho}_2(s')\tilde{\phi}_0'(s')]\,ds'. \end{cases} \qquad (13.124)$$

Equation (13.103) is solved similarly by using the rule established in Eq. (13.172), and after integration by parts one obtains the remarkable expression

$$\rho\Delta_\sigma = \frac{GM\rho_0}{a_1}2\eta\tilde{R}^2\left[C(s) - \tilde{\rho}_0(s)\tilde{\phi}_2(s)\right]. \qquad (13.125)$$

Note how the quantity in Eq. (13.125) vanishes in case of a spherical model, as is expected on general grounds for two-integral systems, and how the function $C(s)$ appears the same in the vertical velocity dispersion. Finally, the student is invited to show that the self-gravitational energy can be written as

$$U = -\frac{GM^2}{a_1}(\tilde{U}_0 + \eta\tilde{U}_1), \qquad (13.126)$$

and to write $\tilde{U}_0$ and $\tilde{U}_1$ in terms of the functions appearing in the homeoidally expanded density and potential (see Ciotti et al. 2020).

A few comments are in order before concluding this section. First, perfectly analogous formulae hold in the *prolate* case, obtained from Eqs. (2.29)–(2.31) with $\epsilon = \eta$, and so $R^2 = y^2 + z^2$, and x is the symmetry axis. Of course, now the transformation $z^2 = r^2 - R^2$ is not required, and $(1-\eta)^2$ appears at the denominator of $\tilde{\rho}(m)$, so that the radial functions in the expanded density–potential pair are not the same as in Eq. (13.122), but they can be worked out (do it!) without much difficulty. Therefore, the structure of Eqs. (13.123) and (13.126) is unchanged, even though the functions B and C change. Second, it is not difficult to prove (do it!) that in general $A \geq 0$, and that if $\tilde{\rho}' \leq 0$, then $C \geq 0$ in the oblate case, while $C \leq 0$ in the prolate case; this means that we can have a (low-flattening) self-gravitating oblate isotropic rotator, but this is impossible in the prolate case. As we will see in Exercise 13.29, this result actually holds even for finite flattenings.

We conclude this section by reminding the student that the solutions of the Jeans equations can also be obtained for genuinely triaxial systems or for axisymmetric models with three integrals of motion, but these technical problems are well beyond the level of this introductory book (e.g., see de Zeeuw et al. 1996; Dejonghe and de Zeeuw 1988; Evans and Lynden-Bell 1989; van de Ven et al. 2003 and references therein).

13.4 The Fundamental Plane of Elliptical Galaxies and the Virial Theorem

A very specific but far-reaching application of the Jeans-based modeling techniques is the physical interpretation of a remarkable *empirical* relation obeyed by stellar spheroids (i.e., early-type galaxies and bulges): the so-called *fundamental plane* (FP; e.g., see Djorgovski and Davis 1987; Dressler et al. 1987; see also Cappellari 2016; Cimatti et al. 2019). We cannot describe all of the observational and theoretical issues surrounding the problem, as an immense literature is available, so we simply focus on the most elementary dynamical aspects (for a more extended overview, see, among others, Ciotti 2009; D'Onofrio et al. 2016 and references therein). In particular, we will attempt to clarify some of the confusion that unfortunately is present in the literature and that can be quite dangerous for the student (i.e., the idea that, after all, the FP is nothing more than the virial theorem in disguise, and so that it is possible to "deduce" the existence of the FP just from the virial theorem).

Stellar spheroids are characterized by three main observables: the circularized effective radius $\langle R\rangle_{\rm e}$ defined in Eq. (13.60), the luminosity-weighted projected velocity dispersion $\sigma_\circ$ measured within some prescribed aperture (e.g., $R_{\rm a} = \langle R\rangle_{\rm e}/8$) defined in Eq. (13.93), and the mean effective surface brightness $\langle I\rangle_{\rm e} \equiv L/(2\pi\langle R\rangle_{\rm e}^2)$, where L is the total luminosity of the galaxy in the adopted band. Therefore, each galaxy is represented by a point in the three-dimensional parameter space $(\langle I\rangle_{\rm e}, \langle R\rangle_{\rm e}, \sigma_\circ)$ or $(L, \langle R\rangle_{\rm e}, \sigma_\circ)$. Statistical studies reveal that the parameter space is not populated uniformly; rather, the data points are confined to the vicinity of a narrow logarithmic plane, the FP

$$\log\langle R\rangle_{\rm e} = \alpha\log\sigma_\circ + \beta\log\langle I\rangle_{\rm e} + \gamma. \tag{13.127}$$

The coefficients α, β, and γ depend slightly on the considered photometric band. For example (see Jorgensen et al. 1996), by measuring $\langle R\rangle_{\rm e}$ in kpc, $\sigma_\circ$ in km s_{-1}, and $\langle I\rangle_{\rm e}$ in $L_\odot/\text{pc}^2$, reported values in the Gunn r band for Coma Cluster galaxies (with $H_0 = 50$ km s^{-1} Mpc^{-1}) are $\alpha \simeq 1.24$, $\beta \simeq -0.82$, and $\gamma \simeq 0.182$. Moreover, the FP presents a strikingly small and nearly constant scatter; $\langle R\rangle_{\rm e}$ presents a scatter around the best-fit FP at fixed $\sigma_\circ$ and $\langle I\rangle_{\rm e}$ in the order of 15% or less. As is well known, the famous scaling laws of Faber and Jackson (1976) and Kormendy (1977) and $D_n - \sigma$ (Dressler et al. 1987) are the natural consequences of the distribution of data points on the FP and of its projections on the three coordinate planes of the parameter space.

No definite/universally accepted interpretation of the FP and its origin has been found yet, while it is sometimes stated that – after all – the FP is "nothing more" that the virial theorem. As we will see, if interpreted literally this is a wrong statement, akin to claiming that the main sequence in the Hertzsprung–Russell diagram is nothing more than the hydrostatic equilibrium of gas spheres! In fact, we will see that even though the virial theorem is a *necessary* condition for the existence of the FP, it is by no means *sufficient*. We remark that, for the following discussion, the "exact" values of the FP coefficients α, β, and γ are irrelevant – what is fundamental is that they are robustly measured and that the data points have a small scatter around the FP.

The first consideration comes from the fact that stellar spheroids are much older than their characteristic dynamical times $P_{\rm dyn} \simeq 2R_{\rm e}/\sigma_\circ \approx 10^8$ yrs, so that these systems, in the absence of strong perturbations, are essentially at equilibrium (Lynden-Bell 1967; see also Section 6.2); this conclusion is reinforced by the observed large-scale regularity of their density profiles. It is then plausible to assume that the vast majority of stellar spheroids are virialized systems. Now, for a galaxy of total stellar mass M_* embedded in a dark matter halo of total mass $M_{\rm h} = \mathcal{R}M_*$, the virial theorem can be written in full generality (see Chapters 6 and 10) as

$$\sigma_{\rm V}^2 \equiv \frac{2K_*}{M_*} = \frac{|W_{**}| + |W_{*\rm h}|}{M_*} = \frac{GM_*}{r_*} \times (|\widetilde{W}_{**}| + \mathcal{R}|\widetilde{W}_{*\rm h}|), \tag{13.128}$$

where $\sigma_{\rm V}$ and K_* are the (three-dimensional) virial velocity dispersion and total kinetic energy of the stellar component, respectively, $\Upsilon_* = M_*/L$ is the *stellar* mass-to-light ratio in the specific band used to measure L and $\langle R\rangle_{\rm e}$, and finally r_* is a scale length of the stellar component. From Eq. (10.52)

$$W_{**} = -\int_{\Re^3} \rho_* \langle \mathbf{x}, \nabla\phi_*\rangle d^3\mathbf{x}, \quad W_{*\rm h} = -\int_{\Re^3} \rho_* \langle \mathbf{x}, \nabla\phi_{\rm h}\rangle d^3\mathbf{x}, \tag{13.129}$$

where ϕ_* and $\phi_{\rm h}$ are the gravitational potentials generated by stars and dark matter, respectively. As an example, in Exercise 13.17 we compute these functions for the two-component Plummer model.

We now relate the previous identity with observational quantities. From Chapter 11, we know that $\langle R\rangle_{\rm e}$ and $\sigma_\circ^2$ are related, respectively, to r_* and $\sigma_{\rm V}^2$ by dimensionless coefficients that depend on projection effects combined with the intrinsic shape of the stellar distribution, on the dark matter amount and distribution, and on the specific internal dynamics (rotation, orbital anisotropy, etc.). For each galaxy, we can then write

$$\begin{cases} r_* = C_{\rm S}(\text{structure, projection}) \times \langle R\rangle_{\rm e}, \\ \sigma_{\rm V}^2 = C_{\rm D}(\text{structure, anisotropy, projection}) \times \sigma_\circ^2, \end{cases} \tag{13.130}$$

where in the presence of color and/or metallicity gradients in the galaxy, the coefficients will also depend on the adopted wavelength of observations. It follows that Eq. (13.128) can be rewritten for each galaxy as

$$L = \frac{K_{\rm V}}{G\Upsilon_*}\langle R\rangle_{\rm e}\sigma_\circ^2, \quad K_{\rm V} \equiv \frac{C_{\rm S}C_{\rm D}}{|\widetilde{W}_{**}| + \mathcal{R}|\widetilde{W}_{*\rm h}|}. \tag{13.131}$$

Equations (13.127) and (13.131) are all that we need. In fact, their combination (they must be *both* true!) implies that in real spheroids, no matter how complex their structure is, $\Upsilon_*/K_{\rm V}$ is a well-defined function of any two of the three quantities $(L, R_{\rm e}, \sigma_\circ)$; in fact, by eliminating (for example) $\sigma_\circ$ from the two identities, one concludes that on the FP

$$\frac{\Upsilon_*}{K_{\rm V}} \propto \langle R\rangle_{\rm e}^{(2+4\beta+\alpha)/\alpha} L^{-(2\beta+\alpha)/\alpha}. \tag{13.132}$$

The dependence of the ratio Υ_*/K_V on galaxy properties is commonly referred to as the "FP tilt." The physical content of Eq. (13.132) is truly remarkable: *all* stellar systems (real or hypothetical) described by Eq. (13.131) are virialized, but *only* those for which Υ_*/K_V scales according to Eq. (13.132) correspond to real galaxies. In other words, the very existence of the (thin) FP dictates that the *structural/dynamical* (K_V) and *stellar population* (Υ_*) properties of stellar spheroids are strictly connected, possibly as a consequence of their formation process; understanding the origin of the FP tilt is thus of the utmost importance for the understanding of galaxy formation. Of course, this argument conclusively shows that *the virial theorem by itself does not imply any FP*. In fact, different galaxies, all virialized, in principle could have very different K_V and Υ_* and so be scattered everywhere in the (R_e, L, σ_o) space; perhaps the simplest concrete example is given by spiral galaxies, which are certainly virialized, but they do not define any FP. Note that by using values for α and β such as those reported above, there is only a mild dependence of Υ_*/K_V on $\langle R \rangle_e$, and in the hypothetical case of $\beta = -1$ and $\alpha = 2$, from Eq. (13.132) we would deduce $\Upsilon_*/K_V = const.$ (i.e., that stellar spheroids would follow *strong homology*). The fact that in nature the values of α and β almost correspond to strong homology led sometimes to the erroneous identification of the FP with the virial theorem, while it should be clear by now that the correct conclusion is that stellar spheroids are almost homologous systems, and that a remarkable *fine-tuning* between Υ_* and K_V is required to produce the FP tilt and yet preserve its surprising tightness. Note that this conclusion would be the same even if α and β were found to be very different from their current values, as far as the FP would be thin.

Therefore, the central problem posed by the existence of the FP is to determine what are the physical ingredients responsible for the tilt. A very large body of literature has been produced in this area, and here we necessarily restrict ourselves to some selected examples, without any pretense of completeness. Perhaps the simplest approach to extracting information from Eq. (13.132) is to focus on the variation of a *single* galaxy property among the plethora in principle appearing in the quantity Υ_*/K_V, while fixing all of the others to some prescribed value. For instance, one can explore the possibility that a systematic variation of Υ_* with L is at the origin of the FP tilt, while considering the galaxies as structurally and dynamically strongly homologous systems with the same K_V. In this case, it is elementary to conclude that the FP would be due to a systematic increase of Υ_* with the galaxy L, and the study then moves to the field of stellar evolution and stellar populations (e.g., see Bender et al. 1992; Prugniel and Simien 1996; Renzini and Ciotti 1993; van Albada et al. 1995).

The other extreme possibility is to assume a constant Υ_* among all galaxies and so to require that the galaxy density profiles, dark matter content and distribution, stellar orbital distribution, and so on vary systematically with $\langle R \rangle_e$ and L, so that K_V obeys Eq. (13.132). This possibility is called *weak homology*. The modeling of weak homology is a natural field for the use of Jeans modeling: the usual approach is to build two-component models (stars plus dark matter), solve the Jeans equations, project them and obtain the K_V coefficient, and change the structural/dynamical parameters until Eq. (13.132) is satisfied (e.g., see Bertin et al. 2002; Caon et al. 1993; Ciotti and Pellegrini 1992; Ciotti et al. 1996;

Graham 1998; Graham and Colless 1997; Hjorth and Madsen 1995; Prugniel and Simien 1997; Renzini and Ciotti 1993). In general, one sees that the relative amount/distribution of dark matter versus the stellar component should increase along the FP or (for example) have a Sersic index m of the stellar distribution that increases for increasing L, in broad agreement with observations. Another possibility for systematically changing (decreasing) K_V at increasing L is that more and more luminous galaxies are more and more radially anisotropic due to obvious projection effects (e.g., see Eq. (13.162)), with the consequent increase of $\sigma_\circ$ at fixed σ_V. However, it is difficult to reconcile there being a major role of this effect in explaining the FP tilt with the expected onset of the radial orbit instability (see Section 6.2) in the resulting anisotropic models (Ciotti and Lanzoni 1997; Nipoti et al. 2002).

Three final points are worth mentioning at the end of this brief discussion. The first is that galaxies are not spherical, and so we must understand what happens to a representative point of a galaxy in the parameter space $(L, \sigma_\circ, \langle R \rangle_e)$ as a function of its relative orientation with the observer, because both $\sigma_\circ$ and $\langle R \rangle_e$ change while L stays constant (see Chapter 11). By using analytical and numerical models (e.g., see Bertin et al. 2002; González-García and van Albada 2003; Jorgensen et al. 1993; Lanzoni and Ciotti 2003; Nipoti et al. 2002, 2003; Saglia et al. 1993; van Albada et al. 1995), it is found that projection effects move models not exactly parallel to the edge-on FP by an amount that can be comparable with the observed FP thickness, and so the *intrinsic* scatter of the FP is smaller than the observed one, making the problem of fine-tuning even more interesting. The second point relates to the effects of galaxy merging, which, of course, if they occur, must preserve the FP tilt and thickness, and this is not obvious at all (see Section 6.2.2 and references therein, in particular Ciotti et al. 2007). Third, a very interesting problem, which we only mention briefly here, is posed by the comparison of the scaling laws followed by the dark matter halos predicted by cosmological simulations with the observed FP of galaxies (Lanzoni et al. 2004).

Exercises

13.1 By changing the order of integration, verify the Abel inversion theorem in Eqs. (13.1) and (13.2) by using the Euler integral

$$\int_x^y \frac{dt}{(t-x)^{1-\alpha}(y-t)^\alpha} = \mathrm{B}(\alpha, 1-\alpha) = \frac{\pi}{\sin(\pi\alpha)}, \quad 0 < \alpha < 1. \tag{13.133}$$

Note how the value of this remarkable integral (a complete beta function in disguised form) is independent of x and y. *Hint*: Use the second identity of Eq. (A.46) and

Eq. (A.54).

13.2 Deduce the deprojection formula in Eq. (13.5) for transparent spherical systems. *Hints*: In Eq. (13.4), define $x = r^2$, $X = R^2$, $X_t = R_t^2$, $U(X) = \Sigma(\sqrt{X})$, $u(x) = \rho(\sqrt{x})$. Show that $U(X) = \int_X^{X_t} u(x)(x-X)^{-1/2} dx$ and invert it by using the

Abel formula. After restoring the original variables and conducting an integration by parts and a successive differentiation with respect to r, complete the proof.

13.3 For the Plummer sphere in Eq. (13.36), show that the projected density and the spatial half-mass radius are given by

$$\Sigma(R) = \frac{Mr_c^2}{\pi(R^2+r_c^2)^2}, \quad r_h = \frac{1+2^{1/3}}{\sqrt{3}}r_c, \tag{13.134}$$

in agreement with Eq. (13.37) for β-models evaluated for $\beta = 5$. Moreover, show that for the family of β-models

$$M = 2\pi\rho(0)r_c^3 \mathrm{B}\left(\frac{\beta-3}{2}, \frac{3}{2}\right), \quad \beta > 3, \tag{13.135}$$

and that, under the assumption of a constant mass-to-light ratio Υ_*, the core radius (for $\beta > 1$) and the effective radius (for $\beta > 3$) are given by

$$R_c = \sqrt{2^{\frac{2}{\beta-1}} - 1}\, r_c, \quad R_e = \sqrt{2^{\frac{2}{\beta-3}} - 1}\, r_c. \tag{13.136}$$

13.4 Prove Eqs. (13.15)–(13.17). *Hints*: The first two identities in Eq. (13.15) are immediate consequences of the definition of S, while for the last identity use Eq. (13.4), invert the order of integration to obtain

$$S(x) = \int_x^{R_t} \nu(r)dr \int_x^r \frac{4RdR}{\sqrt{(R^2-x^2)(r^2-R^2)}}, \tag{13.137}$$

and finally evaluate the inner integral to 2π from Eq. (13.133). The identities in Eq. (13.16) are now elementary. Finally, Eq. (13.17) is proved by inserting into the last identity of Eq. (13.12) the two expressions in Eq. (13.16) and performing integration by parts with the conditions $S(r_t) = 0$ and $rS(r) \to 0$ for $r \to 0$.

13.5 Prove the identities in Eq. (13.18). *Hints*: For the first identity, use Eqs. (13.11) and (13.16). The second identity is obtained from the first considering the limit for $r \to 0$ under the assumption that $S(0) = 2\int_0^{R_t} I(R)dR$ is finite and then using the mean value theorem for integrals (or de l'Hôpital's rule).

13.6 Compute the self-gravitational energy for γ-models in Eq. (13.38) and show that

$$U = W = -\frac{GM^2}{r_c}\frac{1}{2(5-2\gamma)}, \quad 0 \le \gamma < 5/2. \tag{13.138}$$

Explain with physical arguments what happens for $\gamma \ge 5/2$. *Hints*: Use the last identity of Eq. (13.12) and change the variable to $t = r/(1+r)$ in the resulting Tchebishev binomial integral.

13.7 By using Eq. (13.12), compute the self-gravitational energy for the King (1972) model in Eq. (13.34), for the NFW model in Eq. (13.40), for the Hénon (1959) isochrone in Eq. (13.43), and finally for the spherical limit of the perfect ellipsoid in Eq. (2.93), and show that

$$U = W = -G\left[4\pi^3\rho(0)^2 r_c^5, \frac{M^2}{2f(c)^2 r_c}, \frac{M^2(3\pi - 8)}{12 r_c}, \frac{M^2}{2\pi r_c}\right], \tag{13.139}$$

respectively.

13.8 Here, we prove an interesting identity concerning the self-gravitational energy U of a spherical system expressed in terms of the projected density profile $\Sigma(R)$. First, show that the triple integral in Eq. (13.17) can be rewritten as

$$U = -16G \int_0^{\pi/4} \mathbf{K}(\tan\varphi) f(\varphi) \sin\varphi \, d\varphi, \tag{13.140}$$

where $\mathbf{K}$ is the complete elliptic integral of the first kind in Eq. (A.65) and

$$f(\varphi) \equiv \int_0^\infty \Sigma(R\cos\varphi)\Sigma(R\sin\varphi) R^2 dR. \tag{13.141}$$

Then, show that in the case of the Sersic profile in Eq. (13.19), the function f evaluates to

$$f(\varphi) = \frac{\Sigma(0)^2 R_e^3 \, m\Gamma(3m)}{[b(m)\Omega(\varphi)]^{3m}}, \quad \Omega(\varphi) = (\cos\varphi)^{1/m} + (\sin\varphi)^{1/m}, \tag{13.142}$$

so that from $x = \tan\varphi$ for the Sersic profile

$$U = -G\Sigma(0)^2 R_e^3 \frac{16 m\Gamma(3m)}{b(m)^{3m}} \int_0^1 \frac{\mathbf{K}(x) x}{(1 + x^{1/m})^{3m}} dx. \tag{13.143}$$

Hint: For a proof, see Baes and Ciotti (2019b); see also Ciotti (2019).

13.9 Show that for an axisymmetric ellipsoidal system with $a_1 = a_2$ and $q_z = a_3/a_1$, independently of φ

$$E = \frac{1}{a_1^2}\begin{pmatrix} \cos^2\vartheta + \dfrac{\sin^2\vartheta}{q_z^2} & 0 & \left(1 - \dfrac{1}{q_z^2}\right)\sin\vartheta\cos\vartheta \\ 0 & 1 & 0 \\ \left(1 - \dfrac{1}{q_z^2}\right)\sin\vartheta\cos\vartheta & 0 & \sin^2\vartheta + \dfrac{\cos^2\vartheta}{q_z^2} \end{pmatrix}, \tag{13.144}$$

$$\alpha^2 = \frac{1}{a_1^2}\left(\sin^2\vartheta + \frac{\cos^2\vartheta}{q_z^2}\right), \quad \langle E\mathbf{f}_3, \boldsymbol{\xi}_\perp\rangle = \frac{\xi_1}{a_1^2}\left(1 - \frac{1}{q_z^2}\right)\sin\vartheta\cos\vartheta, \tag{13.145}$$

and that the matrix H is diagonal with

$$H = \frac{1}{a_1^2}\begin{pmatrix} \dfrac{1}{\cos^2\vartheta + q_z^2\sin^2\vartheta} & 0 \\ 0 & 1 \end{pmatrix}, \quad \ell^2 = \langle H\boldsymbol{\xi}_\perp, \boldsymbol{\xi}_\perp\rangle. \tag{13.146}$$

Finally, prove that in the spherical case, $\alpha = 1/a_1$, H and E are independent of ϑ, and $\ell = R/a_1$, where R is the radius in the projection plane.

13.10 With this exercise, we compute the self-gravitational energy of some of the most famous axisymmetric razor-thin disks of finite mass. From the identities in Exercise 10.13, prove that for the Kuzmin disk in Eq. (13.69), for the exponential disk in Eq. (2.110), for the truncated constant-density disk in Eq. (5.47), for the Maclaurin disk in Eq. (5.49), for the (truncated) Mestel disk in Eq. (2.115), and finally for the finite Mestel disk in Eq. (5.51), one has

$$U = -GM^2 \times \left[\frac{1}{4R_\mathrm{d}}, \frac{3\pi}{32R_\mathrm{d}}, \frac{8}{3\pi R_\mathrm{t}}, \frac{3\pi}{10R_\mathrm{t}}, \frac{4\mathbf{G}}{\pi R_\mathrm{t}}, \frac{\pi}{2R_\mathrm{t}}\right], \tag{13.147}$$

respectively, where $\mathbf{G} \simeq 0.916$ is the *Catalan constant*. For the Kuzmin disk, repeat the integration by directly using the density–potential pair. Evaluate numerically the coefficients above and relate their behavior with the disk concentration. *Hints*: In the following, we use Gradshteyn et al. (2007). For the Kuzmin disk from Gradshteyn et al.'s eq. (6.554.4)

$$\hat{\Sigma}(k) = \frac{M}{2\pi}\mathrm{e}^{-\lambda}, \quad \lambda = k\,a, \tag{13.148}$$

and the result is now elementary. For the exponential disk, use Eq. (2.114), so that the resulting integral is elementary. For the truncated constant-density disk and for the Maclaurin disk, use Eqs. (5.53) and (5.54), respectively, and calculate U from Gradshteyn et al.'s eq. (6.575.2). The (truncated) Mestel disk case is most easily solved by evaluation of U in terms of v_c in Eq. (5.45) and from Gradshteyn et al.'s eq. (6.141.1). The case of the finite Mestel disk is more difficult: use Eq. (5.55) and integrate twice by parts, with the appropriate limits at 0 and ∞ and with the aid of Gradshteyn et al.'s eqs. (3.741.3) and (6.254.2), obtaining

$$\int_0^\infty \frac{\mathrm{Si}^2(\lambda)}{\lambda^2}\,d\lambda = 2\int_0^\infty \frac{\sin^2\lambda}{\lambda^2}\,d\lambda = \pi. \tag{13.149}$$

13.11 Evaluate the self-gravitational energy of the Miyamoto–Nagai models in Eqs. (13.62) and (13.63) and show that

$$U(s) = -\frac{GM^2}{8b}\left[\frac{1-2s^2}{s(1-s^2)} - \frac{\pi}{2s^2} + \frac{F(s)}{s^2(1-s^2)}\right], \quad s = \frac{a}{b}, \tag{13.150}$$

$$F(s) = \begin{cases} \dfrac{\arccos(s)}{\sqrt{1-s^2}}, & 0 \le s < 1, \\[2ex] \dfrac{\mathrm{arccosh}(s)}{\sqrt{s^2-1}}, & s > 1, \end{cases} \tag{13.151}$$

with $F(1) = 1$. Then prove that $U(1) = -(GM^2/b)(1/3 - \pi/16)$, $U(0) = -(GM^2/b)3\pi/32$ (the Plummer sphere), and $U(\infty) = -GM^2/(4a)$ (the Kuzmin disk), in agreement with previous results. *Hints*: Normalize all lengths to b. Integrate first over R, changing the variable to $R^2 = x$. The resulting expression is rational in the funtion $\sqrt{1+z^2}$. Change the variable to $z = \sinh y$ and obtain a rational function in $\cosh y$. Perform a standard Hermite partial fraction decomposition, transform $\cosh y$ in terms of the exponential, change the variable to $e^y = t$, and integrate the rational expressions; see also Ciotti and Pellegrini (1996).

13.12 With a change of integration variables, prove that the gravitational energy of the complexified system coincides with the gravitational energy U of the parent density, i.e.,

$$U_c \equiv \frac{1}{2}\int_{\Re^3} \rho_c \phi_c d^3\mathbf{x} = \frac{1}{2}\int_{\Re^3} \rho\phi d^3\mathbf{x} = U, \tag{13.152}$$

and deduce that $\Im(U_c) = 0$. Then prove the identity

$$-G\int_{\Re^6} \frac{\Re[\rho_c(\mathbf{x})]\Im[\rho_c(\mathbf{x}')]}{\|\mathbf{x}-\mathbf{x}'\|} d^3\mathbf{x}d^3\mathbf{x}' = \Im(U_c), \tag{13.153}$$

so that the integral vanishes. Along the same lines of reasoning, show that the total mass of the complexified distribution $M_c = \int \rho_c d^3\mathbf{x}$ coincides with the total (real) mass of the seed density distribution $M = \int \rho d^3\mathbf{x}$, so that

$$\int_{\Re^3} \Im(\rho_c) d^3\mathbf{x} = 0. \tag{13.154}$$

13.13 What happens to Eq. (13.82) in the case of purely tangential orbits when $\sigma_r = 0$ at all radii? Show that in this case $\sigma_t^2(r) = v_c^2(r)$, the circular velocity of the model. *Hint*: Consider the behavior of the anisotropy parameter β in Eq. (12.34) for a vanishing radial velocity dispersion.

13.14 By direct integration, prove Eq. (13.91). *Hints*: $\Sigma\overline{v^2_{\rm p_{los}}} = \int \rho\overline{\langle \mathbf{n}, \mathbf{v}\rangle^2} d\xi_3$. Given the spherical symmetry, we can assume without loss of generality $\xi_1 = x$, $\xi_2 = y$, and $\xi_1 = z$. As a consequence, $r = \|\boldsymbol{\xi}\|$, $R^2 = x^2 + y^2$, and $\mathbf{n} = (0,0,1)$, and so $\Sigma\overline{v^2_{\rm p_{los}}} = \int \rho\overline{v_z^2} dz = 2\int_R^{R_t} \rho(r)\overline{v_z^2}(r)(r^2 - R^2)^{-1/2} dr$. Expressing v_z in terms of the velocity components in spherical coordinates and recalling that $\overline{v_r v_\vartheta} = 0$, one obtains $\overline{v_z^2} = \overline{(v_r\cos\vartheta - v_\vartheta\sin\vartheta)^2} = \overline{v_r^2}\cos^2\vartheta + \overline{v_\vartheta^2}\sin^2\vartheta = \overline{v_r^2}(1 - \beta\sin^2\vartheta)$, and from the identity $\sin^2\vartheta = R^2/r^2$ the result is proved.

13.15 In numerical studies, the evaluation of Eq. (13.91) usually requires the evaluation of a double integral. By using integration by parts, show that in the case of OM anisotropy (with the exception of the special case $R = 0$ and $r_a = 0$), the integral can in fact be reduced to a single integration

$$\Sigma(R)\sigma_p^2(R) = G\int_R^{R_t} K(r)\rho(r)M_T(r)\left(1 + \frac{r_a^2}{r^2}\right) dr, \tag{13.155}$$

where

$$K(r) = \frac{2r_a^2 + R^2}{(r_a^2 + R^2)^{3/2}} \arctan\sqrt{\frac{r^2 - R^2}{r_a^2 + R^2}} - \frac{R^2\sqrt{r^2 - R^2}}{(r_a^2 + r^2)(r_a^2 + R^2)}. \tag{13.156}$$

Discuss the case for $R = 0$ and $r_a = 0$.

13.16 By direct integration, prove Eq. (13.94). *Hints*: Use Eq. (13.91) and invert the order of integration, so that

$$\pi \int_0^{R_t} \Sigma(R)\sigma_p^2(R) = 2\pi \int_0^{r_t} \rho(r)\sigma_r^2(r) r dr \int_0^r \frac{1-\beta(r)R^2/r^2}{\sqrt{r^2-R^2}} R dR$$
$$= \frac{2\pi}{3} \int_0^{r_t} \rho(r)\sigma_r(r)^2[3-2\beta(r)]r^2 dr. \quad (13.157)$$

Finally, use Eq. (12.34).

13.17 Consider a stellar Plummer model of mass M_* and scale length r_* in Eq. (13.36). First show that

$$U = W = -\frac{GM_*^2}{r_*}\frac{3\pi}{32}, \quad \sigma_r^2(r) = \frac{\Psi(r)}{6}, \quad (13.158)$$

where the second identity holds for the isotropic case. Then embed the stellar model in a dark matter Plummer sphere of total mass $M_h = \mathcal{R} M_*$ and scale length $r_h = \xi r_*$, and from Eqs. (13.13) and (13.14) show that

$$U_{ext} = -\frac{GM_*^2 \mathcal{R}}{r_*} \frac{(\xi^2+1)\mathbf{E}(\sqrt{1-\xi^2}) - 2\xi^2 \mathbf{K}(\sqrt{1-\xi^2})}{(\xi^2-1)^2}, \quad (13.159)$$

$$W_{ext} = -\frac{GM_*^2 \mathcal{R}}{r_*} \frac{\xi^2(3\xi^2+5)\mathbf{K}(\sqrt{1-\xi^2}) - (7\xi^2+1)\mathbf{E}(\sqrt{1-\xi^2})}{(\xi^2-1)^3}, \quad (13.160)$$

where **K** and **E** are the complete elliptic integrals of the first and second kinds in Eq. (A.65). Finally, solve the Jeans equations in the OM case by using Eq. (13.89) and show that σ_r can be obtained in explicit algebraic form, and that at the center

$$\sigma_r^2(0) = \frac{GM_*}{6r_*}\left\{1 + \frac{2\mathcal{R}(\xi+3)}{\xi(1+\xi)^3} + \frac{1}{s_a^2}\left[\frac{1}{2} + \frac{4\mathcal{R}}{3(\xi+1)^3}\right]\right\}, \quad (13.161)$$

where $s_a \equiv r_a/r_*$ (see also Renzini and Ciotti 1993). Discuss as a function of ξ the limits of a dominant dark matter halo and the purely isotropic and radial cases.

13.18 The projection of the OM velocity dispersion in the two-component Plummer model of Exercise 13.17 in general cannot be expressed in terms of elementary functions. However, in two important cases the projected central velocity dispersion can be obtained quite easily. Show that in the purely stellar OM case

$$\sigma_p^2(0) = \frac{GM_*}{r_*}\frac{\pi(3s_a^3+6s_a^2+5s_a+4)}{64 s_a(s_a+1)^2}. \quad (13.162)$$

What happens in the isotropic limit? What happens at decreasing s_a? Explain the behavior by considering projection effects. Compare this with Eq. (13.161) with $\mathcal{R} = 0$ and explain why $\sigma_p(0) < \sigma_r(0)$. Another simple case is the purely isotropic two-component model with dark matter. Show that in this case

$$\sigma_{\rm p}^2(0) = \frac{GM_*}{r_*}\left[\frac{3\pi}{64} + \mathcal{R}\frac{(7+\xi^2)\mathbf{E}(\sqrt{1-\xi^2}) - (5\xi^2+3)\mathbf{K}(\sqrt{1-\xi^2})}{2(\xi^2-1)^3}\right]. \tag{13.163}$$

Draw a plot as a function of ξ.

13.19 Consider a supermassive BH of mass $M_{\rm BH}$ at the center of a power-law galaxy as in Eq. (13.29). Solve the OM Jeans equations and show that for $\gamma > 1$ the contribution of the BH gravitational field to the velocity dispersion profile is

$$\sigma_r^2(r) = \frac{GM_{\rm BH}}{r_{\rm c}}\frac{1}{s(s^2+s_{\rm a}^2)}\left(\frac{s^2}{\gamma-1} + \frac{s_{\rm a}^2}{\gamma+1}\right), \quad s_{\rm a} \equiv \frac{r_{\rm a}}{r_{\rm c}}. \tag{13.164}$$

Moreover, from Eq. (13.91), show that the associated isotropic and completely radially anisotropic projected velocity dispersion profiles are

$$\begin{cases} [\sigma_{\rm P}^2(R)]_{\rm iso} = \dfrac{GM_{\rm BH}}{r_{\rm c}\eta}\dfrac{\mathrm{B}(1/2,\gamma/2)}{(\gamma+1)\mathrm{B}(1/2,\gamma/2-1/2)}, \\ [\sigma_{\rm P}^2(R)]_{\rm rad} = \dfrac{GM_{\rm BH}}{r_{\rm c}\eta}\dfrac{\mathrm{B}(3/2,\gamma/2)}{(\gamma-1)\mathrm{B}(1/2,\gamma/2-1/2)}. \end{cases} \tag{13.165}$$

Finally, from Eq. (13.93), show that a finite-aperture velocity dispersion in the presence of a central BH requires $\gamma < 2$, and in this case

$$[\sigma_{\rm a}^2(R)]_{\rm iso} = \frac{3-\gamma}{2-\gamma}[\sigma_{\rm P}^2(R)]_{\rm iso}, \quad [\sigma_{\rm a}^2(R)]_{\rm rad} = \frac{3-\gamma}{2-\gamma}[\sigma_{\rm P}^2(R)]_{\rm rad}. \tag{13.166}$$

13.20 Consider the SIS in Eqs. (2.63) and (13.29) with a constant circular velocity $v_{\rm c}$. Solve the Jeans equations in the isotropic case and show that

$$\sigma_r^2(r) = \frac{v_{\rm c}^2}{2}, \quad \sigma_{\rm t}^2(r) = v_{\rm c}^2 \tag{13.167}$$

(see also Exercise 12.6). Then, by using the considerations after Eq. (12.63), integrate Eq. (13.90) for $r_{\rm a} \to \infty$, thus obtaining the constant-anisotropy solution for $\beta = -\alpha$, with the isotropic case corresponding to $\alpha = 0$. Show that $\sigma_r^2(r) = 0.5v_{\rm c}^2/(\alpha+1)$, while $\sigma_{\rm t}^2 = v_{\rm c}^2$ independently of α.

13.21 With these (quite artificial) models, we elaborate on the peculiar behavior of $\sigma_{\rm V}$ and $r_{\rm V}$ for some model with infinite total mass (see Footnote 6 in Chapter 6). Consider the families of spherical density distributions

$$\rho(r) = \frac{A\theta(r_{\rm t}-r)}{r^\gamma}, \quad \rho(r) = \frac{A{\rm e}^{-r/r_{\rm t}}}{r^\gamma}. \tag{13.168}$$

Evaluate their total masses, solve the OM Jeans equations (also in the presence of a central BH), and finally compute the self-gravitational energy $U = W$. Determine the limit on γ for the convergence of U (see also Exercise 13.6). Discuss the behavior of $\sigma_{\rm V}$ and $r_{\rm V}$ as a function of γ for $r_{\rm t} \to \infty$. What happens for $\gamma = 2$? Is the result consistent with the expectations regarding the SIS that can be obtained from Eq. (13.167)? *Hints*: Consider Exercise 10.17 and discuss the behavior of the virial theorem surface pressure term for an SIS with a spherical control volume Ω of radius r, for $r \to \infty$.

13.22 This exercise addresses a question that emerges quite often in discussions with students (and not only them!). From Newton's second theorem and Eq. (13.82), it follows that only the mass inside the radius r determines the variation of velocity dispersion at r. However, from Eq. (13.85), σ_r appears to be determined by an integral extending from r to r_{t}. Explain why this is the case. *Hints*: For simplicity, consider the isotropic case and consider a stellar system with density profile $\rho(r)$ in the gravitational field of a shell of mass M_{s} and radius r_{s}, ignoring the self-gravitational field of the system. From the solution of the Jeans equation, show that for $r \geq r_{\mathrm{s}}$ the radial "pressure" $p_r = \rho\sigma_r^2$ is the same as that for a central mass M_{s}, while for $r \leq r_{\mathrm{s}}$ the pressure is constant $p_r(r) = p_r(r_{\mathrm{s}})$, even if there is no gravitational field inside the shell. Discuss the analogous case of a gas in equilibrium in the field of the shell and conclude that you need a pressure inside the shell to support the gas outside the shell.

13.23 Show that the natural boundary condition on σ_z for the vertical Jeans equation (13.102) of an axisymmetric stellar system with $f = f(\mathcal{E}, J_z)$ and spatially truncated on the surface $S(R,z) = 0$ is $\sigma_R = \sigma_z = 0$ on S. *Hint*: At any point on the boundary, the unit vector is given by

$$\mathbf{n} = (n_x, n_y, n_z) = \frac{(S_x, S_y, S_z)}{\sqrt{S_x^2 + S_y^2 + S_z^2}} = \frac{(S_R \cos\varphi, S_R \sin\varphi, S_z)}{\sqrt{S_R^2 + S_z^2}}, \tag{13.169}$$

where $S_x = \partial S/\partial x$, etc., and $S_R = \partial S/\partial R$. Use Eq. (12.129) to show that

$$\sigma_{\mathrm{n}}^2 = \sigma_{ij}^2 n_i n_j = \frac{S_R^2\sigma_R^2 + S_z^2\sigma_z^2}{S_R^2 + S_z^2} = \sigma_z^2, \tag{13.170}$$

and conclude.

13.24 In the same spirit as the second identity of Eq. (13.102), show that if the density and potential are assigned in terms of r and R, then the commutator in Eq. (13.103) can be written as

$$\int_z^{z_{\mathrm{t}}(R)} \left(\frac{\partial\rho}{\partial R}\frac{\partial\phi_{\mathrm{T}}}{\partial z'} - \frac{\partial\rho}{\partial z'}\frac{\partial\phi_{\mathrm{T}}}{\partial R} \right) dz' = \int_r^{r_{\mathrm{t}}(R)} \left(\frac{\partial\rho}{\partial R}\frac{\partial\phi_{\mathrm{T}}}{\partial r'} - \frac{\partial\rho}{\partial r'}\frac{\partial\phi_{\mathrm{T}}}{\partial R} \right) dr'. \tag{13.171}$$

13.25 Given three generic functions of R and z, show that the commutator introduced in Eq. (13.103) obeys the rules of *Lie algebra* (see also Section 9.3.1). Moreover, from Exercise 13.24, show that for generic spherically symmetric functions $u(r)$ and $v(r)$ and generic $f(R)$, in the untruncated case

$$[f(R)u(r), v(r)] = f'(R)\int_r^{\infty} u(x)v'(x)dx, \tag{13.172}$$

where $v'(r) = dv(r)/dr$. Finally, show that for $f(m)$ with $m^2 = R^2/a^2 + z^2/(q^2a^2)$,

$$[f(m), v(r)] = \left(1 - \frac{1}{q^2}\right)\frac{R}{a^2}\int_r^\infty \frac{f'(m)}{m} v'(x)dx, \tag{13.173}$$

where the integral is evaluated at fixed R and m in the integrand is intended to be expressed in terms of the spherical radius x (see also Exercise 13.30).

13.26 For an axisymmetric stellar system with $f = f(\mathcal{E}, J_z)$, by using Eq. (13.109) and the definition of Δ_σ in Eq. (13.101), show that

$$\overline{\sigma_{\rm p}^2}(\mathbf{x}) + \overline{v_{\rm p}}^2(\mathbf{x}) = \sigma_z^2(\mathbf{x}) + \Delta_\sigma(\mathbf{x})\sin^2\varphi\sin^2 i \tag{13.174}$$

(i.e., the sum in Eq. (13.174) is *independent* of the specific decomposition adopted to break the degeneracy of azimuthal motions $v_\varphi^2 = \overline{v_\varphi}^2 + \sigma_\varphi^2$). Finally, rewrite Eq. (13.109) in the case of the Satoh decomposition shown in Eqs. (13.104) and (13.105).

13.27 The (axisymmetric) constant-density ellipsoid of flattening q_z is obtained from Eq. (13.115) with $n = 0$. From Eqs. (13.59), (13.116), and (13.117), first show that

$$M = \frac{4\pi q_z \rho(0)a_1^3}{3}, \quad \Sigma(\ell) = \frac{2\rho(0)}{\alpha}\sqrt{1-\ell^2}, \quad \ell_{\rm e} \simeq 0.608, \tag{13.175}$$

where α and ℓ are given in Eqs. (13.145) and (13.146). Then, from Eq. (13.120), show that the potential inside the ellipsoid is

$$\phi(R,z) = -\pi G\rho(0)a_1^2\left[a_1 q_z w_0(0) - w_1\tilde{R}^2 - w_3\tilde{z}^2\right], \tag{13.176}$$

where $w_0(0)$, $w_1 = w_2$, and w_3 are given in Eqs. (2.67), (3.62), and (3.63). By integration of Eqs. (13.102) and (13.103), show that

$$\sigma_z^2 = \pi G\rho(0)a_1^2q_z^2w_3(1-m^2), \quad \Delta_\sigma = 2\pi G\rho(0)a_1^2(w_1 - q_z^2w_3)\tilde{R}^2, \tag{13.177}$$

and verify that $w_1 - q_z^2w_3 > 0$ for $0 < q_z \le 1$, with $\Delta_\sigma = 0$ for $q_z = 1$. For the projected fields in Eqs. (13.109) and (13.114), prove that in the Satoh decomposition

$$\begin{cases} \overline{v_{\rm P\,los}}(\xi_\perp) = -k\sqrt{2\pi G\rho(0)a_1^2(w_1 - q_z^2w_3)}\,\tilde{\xi}_2\sin i, \\ \overline{\sigma^2_{\rm P\,los}}(\xi_\perp) = 2\pi G\rho(0)a_1^2\left[\dfrac{q_z^2w_3(1-\ell^2)}{3} + (1-k^2)(w_1 - q_z^2w_3)\tilde{\xi}_2^{\,2}\sin^2 i\right], \end{cases} \tag{13.178}$$

and explain why $\overline{v^2_{\rm P\,los}} = (\overline{v_{\rm P\,los}})^2$, so that from Eq. (13.112) also $\overline{\sigma^2_{\rm I\,los}} = \overline{\sigma^2_{\rm P\,los}}$. Finally, evaluate the aperture velocity dispersion over a generic isophotal ellipse ℓ from Eq. (13.114), and with the help of Eq. (13.117) show that

$$\sigma_a^2 = \frac{\pi G\rho(0)a_1^2}{5} \frac{A_1 + (1-k^2)\, A_2 \sin^2 i}{1-(1-\ell^2)^{3/2}}, \tag{13.179}$$

where

$$A_1 = 2q_z^2 w_3[1-(1-\ell^2)^{5/2}],\ A_2 = (w_1 - q_z^2 w_3)\left[2-(2+\ell^2-3\ell^4)\sqrt{1-\ell^2}\right]. \tag{13.180}$$

See also Lanzoni and Ciotti (2003).

13.28 With this exercise, we illustrate the importance of the relative shape of the density and potential on the positivity of Δ_σ in Eq. (13.103). Consider for simplicity an untruncated stellar system with density profile $\rho_*(m_*)$ and $\rho_*'(m_*) = d\rho_*/dm_* \le 0$. Let $\phi_h(m_h)$ be an ellipsoidal potential (see also Exercise 2.9) with $\phi_h'(m_h) = d\phi_h(m_h)/dm_h \ge 0$, and finally

$$m_*^2 = \frac{R^2}{a_*^2} + \frac{z^2}{a_*^2 q_*^2}, \quad m_h^2 = \frac{R^2}{a_h^2} + \frac{z^2}{a_h^2 q_h^2}. \tag{13.181}$$

Show that

$$\rho_*\Delta_\sigma = \left(\frac{1}{q_*^2} - \frac{1}{q_h^2}\right) \frac{R^2}{a_*^2 a_h^2} \int_z^\infty \frac{|\rho_*'(m_*)|\phi_h'(m_h) z}{m_* m_h} dz, \tag{13.182}$$

and discuss the sign of Δ_σ as a function of the two flattenings q_* and q_h (see Barnabè et al. 2006 for further discussion in the framework of fluid dynamics). Obtain the analogous expression for $\rho_*\sigma_z^2$.

13.29 With this exercise, we address the problem of the sign of Δ_σ in Eq. (13.103) for self-gravitating (axisymmetric) ellipsoidal models. Let $\rho = \rho(m)$, with $m^2 = R^2/a^2 + z^2/(a^2q^2)$ being an untruncated ellipsoidal model, so that the oblate case is obtained for $0 < q < 1$ and the prolate case is obtained for $q > 1$. Show that in the self-gravitating case

$$\rho_*\Delta_\sigma = \frac{2\pi G a(1-q^2)R^2}{q} \int_z^\infty \frac{|\rho_*'(m)| z\, dz}{m} \int_0^\infty \frac{\tau \rho_*(m_\tau)\, d\tau}{\sqrt{\Delta(\tau)}(a^2+\tau)(a^2q^2+\tau)}, \tag{13.183}$$

and obtain the analogous expression for $\rho_*\sigma_z^2$. Conclude that it is possible to have an oblate isotropic rotator, but that this is impossible in the prolate case.

13.30 Show that for the axisymmetric Jaffe stellar density profile in Eq. (2.83) with $\gamma = 2$, $a_1 = a_*$, $M = M_*$, $\rho_0 = M_*/(4\pi a_*^3)$, and in the spherical Jaffe potential in Eq. (13.39) with $\gamma = 2$, $M = \mathcal{R} M_*$, $r_c = \xi a_*$, $\Psi_n = GM_*/a_*$, Eqs. (13.102) and (13.103) can be expressed in terms of elementary functions. In the oblate case

$$\rho_*\sigma_z^2 = \frac{\rho_0 \Psi_n \mathcal{R}\, Y^5}{q_*^2} \int_{\mathrm{arccosh}\,\zeta}^\infty \frac{dt}{(X+\mathrm{ch}\, t)\,(Y+\mathrm{sh}\, t)^2\, \mathrm{sh}\, t\, \mathrm{ch}\, t}, \tag{13.184}$$

$$\rho_* \Delta_\sigma = \frac{\rho_0 \Psi_n 2\mathcal{R} Y^5}{q_*^2} \int_{\text{arccosh}\zeta}^{\infty} \frac{(Y + 2\,\text{sh}\, t)dt}{(X + \text{ch}\, t)\,(Y + \text{sh}\, t)^3\, \text{sh}^3 t\, \text{ch}\, t}, \quad (13.185)$$

where

$$X \equiv \frac{\xi}{\sqrt{1 - q_*^2 \tilde{R}}}, \quad Y \equiv \frac{q_*}{\sqrt{1 - q_*^2 \tilde{R}}}, \quad \zeta \equiv \frac{s}{\sqrt{1 - q_*^2 \tilde{R}}}, \quad (13.186)$$

$$m^2 = \frac{s^2}{q_*^2} - \left(\frac{1}{q_*^2} - 1\right) \tilde{R}^2, \quad s \equiv \frac{r}{a}, \quad \tilde{R} \equiv \frac{R}{a}, \quad (13.187)$$

and $\zeta \geq 1$ over all of the space. Three interesting limit solutions are obtained for $q_* \to 1$ (the isotropic spherical limit; see Ciotti and Ziaee Lorzad 2018; Ciotti et al. 2020), the BH-dominated case ($\xi \to 0$ and $\mathcal{R}M_* = M_{\text{BH}}$), and the SIS potential[6] ($\mathcal{R} = v_c^2 \xi / \Psi_n$ and $\xi \to \infty$). *Hints*: From the previous exercises, we limit ourselves to the oblate case. Express the ellipsoidal coordinate m in terms of r as in Eq. (13.187); notice that this transformation holds for generic ellipsoidal models in a spherical potential. Then use Eq. (13.173). In order to obtain the explicit form of the integrals, a standard approach would be to transform the hyperbolic functions in their exponential representations to reduce the integrand to a rational function and to use the Hermite partial fraction decomposition.

13.31 Consider the stellar oblate axisymmetric power-law ellipsoid in Eq. (5.64) with $q_z = q_*$ and $m = m_*$. Following Exercises 13.28–13.30, show that the two-integral Jeans equations can be solved in closed form in the case of a potential dominated by a central BH, or by the SIS of circular velocity v_c in Eq. (2.63), with

$$\rho_* \sigma_z^2 = \rho_0 \times \begin{cases} \dfrac{\mu \Psi_n Y^{\gamma+1}}{2q_*^2} \text{B}\left(\dfrac{\gamma+1}{2}, \dfrac{2-\gamma}{2}; \dfrac{1}{\zeta^2}\right), & \text{(BH)} \\[2ex] \dfrac{v_c^2 Y^\gamma}{2q_*} \text{B}\left(\dfrac{\gamma}{2}, \dfrac{2-\gamma}{2}; \dfrac{1}{\zeta^2}\right), & \text{(SIS)} \end{cases} \quad (13.188)$$

$$\rho_* \Delta_\sigma = \rho_0 \times \begin{cases} \dfrac{\mu \Psi_n Y^{\gamma+1} \gamma}{2q_*^2} \text{B}\left(\dfrac{\gamma+3}{2}, -\dfrac{\gamma}{2}; \dfrac{1}{\zeta^2}\right), & \text{(BH)} \\[2ex] \dfrac{v_c^2 Y^\gamma \gamma}{2q_*} \text{B}\left(\dfrac{\gamma+2}{2}, -\dfrac{\gamma}{2}; \dfrac{1}{\zeta^2}\right), & \text{(SIS)} \end{cases} \quad (13.189)$$

where $\text{B}(a; x, y)$ is the incomplete beta function in Eq. (A.55) and Y and ζ are defined in Eq. (13.186). The nearly spherical case can be obtained by expansion for $q_* \to 1$ (i.e., $\zeta \to \infty$) and Eq. (A.56), or by mass-unconstrained homeoidal expansion of the power-law ellipsoid (see also Ciotti and Bertin 2005; Riciputi et al. 2005).

[6] The sum of these two limit cases will give the solution of the Jeans equations for the ellipsoidal Jaffe model in a total potential of the SIS plus a central BH. Note that analogous explicit solutions can be obtained for the ellipsoidal Hernquist model.

13.32 Solve the Jeans equations for the power-law torus in Exercise 2.14 under the assumption of an $f(\mathcal{E}, Jz)$. Show that

$$\sigma_z^2 = 4\pi G\rho_n r_*^2 \frac{\tilde{r}^{2-\alpha}}{7-\alpha}\left[\frac{2\tilde{r}^2}{(\alpha-2)^2(5-\alpha)} + \frac{\tilde{R}^2}{2(\alpha-1)}\right], \tag{13.190}$$

$$\Delta_\sigma = 4\pi G\rho_n r_*^2 \frac{2\tilde{r}^{2-\alpha}}{(\alpha-2)(7-\alpha)}\left[\frac{2\tilde{r}^2}{(\alpha-2)(5-\alpha)} - \frac{\tilde{R}^2}{\alpha-1}\right]. \tag{13.191}$$

What happens in the isotropic case? Compare v_c^2 and v_φ^2 in the equatorial plane (see Ciotti and Bertin 2005 and Exercise 5.7). Repeat the exercise adding a central BH of mass M_{BH}.

13.33 We recall three of the most important stability criteria for stellar systems, mentioned at the end of Section 6.2; the inequalities mark *stability*. The *radial orbit instability* indicator for spherical stellar systems can be written as

$$\Xi \equiv \frac{2K_{rad}}{K_{tan}} = -\frac{4}{2 + W_T/K_{rad}} \lesssim 1.7, \tag{13.192}$$

where K_{rad} and K_{tan} are, respectively, the kinetic energies associated with the radial and tangential components of the velocity dispersion tensor of the density component of interest, and the last expression is obtained from the virial theorem (prove it!). The (local) *Toomre stability criterion* for self-gravitating stellar disks can be written as

$$Q(R) \equiv \frac{\sigma_R \kappa_R}{3.36 G\Sigma} > 1, \tag{13.193}$$

where σ_R is the radial velocity dispersion in the disk, κ_R the radial epicyclic frequency, and finally Σ is the disk surface density at radius R. The *Ostriker–Peebles stability criterion* for axisymmetric systems is given by

$$t \equiv \frac{K_{str}}{|W_T|} \lesssim 0.14, \tag{13.194}$$

where K_{str} is the kinetic energy associated with the streaming motions (sometimes called "ordered" motions) of the density component of interest.

14
Modeling Techniques 3
From ρ to f

In this last chapter, we discuss a final theoretical step of the moments approach illustrated in Chapter 13: under the assumption that the macroscopic profiles (e.g., density and velocity dispersion) of each component are known, there is a possibility of recovering the phase-space distribution function (DF) of a model and checking its positivity (i.e., verifying the model consistency). The problem of recovering the DF is in general a technically difficult *inverse problem*, and even when it is doable, unicity of the recovered DF is not guaranteed, so that a simple consistency analysis is quite problematic. Fortunately, there are special cases when (in principle) the DF can be obtained analytically (generally in integral form), and in these cases a few general and useful consistency conditions can be proved, such as the so-called *global density slope–anisotropy inequality* (GDSAI). The student is warned that this chapter is somewhat more technical than the others; however, the additional effort needed for its study will be well repaid by the understanding of some nontrivial results allowing for the construction of phase-space consistent collisionless stellar systems.

14.1 Recovering the DF

As we have seen in Chapter 13, in the study of stellar systems based on the moments approach, the density distribution is given and specific assumptions on the internal dynamics of the model are made. In practice, one usually starts from the density profile and then solves the Jeans equations due to more or less motivated closure relations. Unfortunately, the fact that the Jeans equations admit a solution that is not trivially unacceptable[1] is not sufficient to guarantee that the model is viable: the minimum requirement to be met is that the DF of *each* physically distinct component must be positive definite. A model satisfying this requirement is called a *consistent* model. If the DF can be recovered, a positivity check should be performed, and as discussed in Chapter 9, in case of negative values the model must be discarded as unphysical, even if the kinematical profiles look satisfactory.

Remarkably, under special assumptions about the geometry/internal dynamics of the models, inversion formulae exist so that the DF can be obtained, usually in integral form or

[1] For instance, a somewhere negative squared velocity field; for example, see Eq. (13.103) and the subsequent comments.

as a series expansion (see Binney and Tremaine 2008; see also Cuddeford 1991; Dejonghe 1986, 1987a; Eddington 1916; Fricke 1952; Hunter and Qian 1993; Lynden-Bell 1962a; Merritt 1985a; Osipkov 1979 and references therein). However, in these fortunate cases, the difficulties inherent in the operation of recovering analytically the DF also prevent in general a simple consistency analysis, and numerical inspection of the inversion integral is required. Fortunately, at least in some special cases, criteria for phase-space consistency that can be applied without an explicit recovery of the DF are known. Finally, it is important to realize that phase-space consistency is a much weaker requirement than model stability (see Section 6.2), and consistent but unstable models should not be accepted as viable equilibria to describe real stellar systems.

In the first section of this chapter, some of the known inversion formulae that can be used to recover the DF in special cases are illustrated, and in the second section the associated consistency criteria are proved and discussed. All of the presented results (if not explicitly stated) apply to *each* density component ρ_k of multicomponent systems, and we maintain as far as possible the nomenclature introduced in Chapter 12; however, in order to avoid clumsy notation, the subscript "k" used to label each component is not reported.

14.1.1 Multicomponent Isotropic Spherical Models

We start by presenting the seminal result from Eddington (1916), namely the inversion formula for the DF of spherically symmetric, globally isotropic stellar systems. Following Chapter 12, in a multicomponent spherical system, let

$$f = h(\mathcal{E})\theta(\mathcal{E} - \mathcal{E}_{\mathrm{t}}) \tag{14.1}$$

be the DF of the isotropic k-th density component $\rho = \rho_k(r)$ immersed in a total potential $\Psi_{\mathrm{T}}(r)$, where θ is the Heaviside step function in Eq. (A.99) and $\mathcal{E}_{\mathrm{t}}$ is the component's *truncation energy*. Then

$$\begin{aligned} h(\mathcal{E}) &= \frac{1}{\sqrt{8}\pi^2}\frac{d}{d\mathcal{E}}\int_{\mathcal{E}_{\mathrm{t}}}^{\mathcal{E}} \frac{d\rho}{d\Psi_{\mathrm{T}}}\frac{d\Psi_{\mathrm{T}}}{\sqrt{\mathcal{E}-\Psi_{\mathrm{T}}}} \\ &= \frac{1}{\sqrt{8}\pi^2}\int_{\mathcal{E}_{\mathrm{t}}}^{\mathcal{E}} \frac{d^2\rho}{d\Psi_{\mathrm{T}}^2}\frac{d\Psi_{\mathrm{T}}}{\sqrt{\mathcal{E}-\Psi_{\mathrm{T}}}} + \frac{d\rho/d\Psi_{\mathrm{T}}|_{\Psi_{\mathrm{T}}=\mathcal{E}_{\mathrm{t}}}}{\sqrt{8}\pi^2\sqrt{\mathcal{E}-\mathcal{E}_{\mathrm{t}}}}, \end{aligned} \tag{14.2}$$

where it is assumed that the density ρ of the k-th component is written as a function of the total potential Ψ_{T} (i.e., $\rho = \rho(\Psi_{\mathrm{T}})$). In principle, this is always possible, because from Newton's second theorem in spherical systems the potential is a monotonic function of the radius, being $d\Psi_{\mathrm{T}}/dr = GM_{\mathrm{T}}(r)/r^2$.

The proof of Eq. (14.2) is simple. From Exercise 12.1, $\rho(\Psi_{\mathrm{T}}) = 4\pi\int_{\mathcal{E}_{\mathrm{t}}}^{\Psi_{\mathrm{T}}}\sqrt{2(\Psi_{\mathrm{T}}-\mathcal{E})}\, h(\mathcal{E})d\mathcal{E}$; after differentiation with respect to Ψ_{T}, the integral is cast in a form that is suitable for the Abel integral inversion in Eq. (13.1), and so the first identity is proved. The second identity is obtained by integrating by parts from the first identity with respect to Ψ_{T}, and then by differentiation with respect to $\mathcal{E}$. Note that for untruncated regular systems of finite total mass (or more generally when the total potential at infinity can be set equal to zero; see Chapter 2), by fixing $\mathcal{E}_t = 0$, the last term in Eq. (14.2) vanishes (see also Exercise 14.1).

14.1.2 Multicomponent Osipkov–Merritt Spherical Models

A generalization of Eq. (14.2) holds in the case of multicomponent spherical systems with radial Osipkov–Merritt (OM) orbital anisotropy (Section 12.2.2). In this case, we assume that the k-th component of a multicomponent spherical system is supported by a DF of the form

$$f = h(Q)\theta(Q - Q_t), \quad Q = \mathcal{E} - \frac{J^2}{2r_a^2}, \tag{14.3}$$

where Q_t is a truncation value; note that in principle different components of the system can have different values of the anisotropy radius and different functions h, and consistency requires that each h is nowhere negative. It is easy to prove that

$$\begin{aligned} h(Q) &= \frac{1}{\sqrt{8}\pi^2} \frac{d}{dQ} \int_{Q_t}^{Q} \frac{d\varrho}{d\Psi_T} \frac{d\Psi_T}{\sqrt{Q - \Psi_T}} \\ &= \frac{1}{\sqrt{8}\pi^2} \int_{Q_t}^{Q} \frac{d^2\varrho}{d\Psi_T^2} \frac{d\Psi_T}{\sqrt{Q - \Psi_T}} + \frac{d\varrho/d\Psi_T|_{\Psi_T = Q_t}}{\sqrt{8}\pi^2 \sqrt{Q - Q_t}}, \end{aligned} \tag{14.4}$$

where from Eq. (12.45)

$$\varrho(r) \equiv \frac{\rho(r)}{A(r)} = \left(1 + \frac{r^2}{r_a^2}\right)\rho(r). \tag{14.5}$$

We call this function the OM *augmented density*. In fact, the proof of Eq. (14.4) is identical to that of the isotropic case once we introduce the augmented density into Eq. (12.42). For some explicit examples of one- and two-component systems that can be solved analytically, as well as in the presence of a central black hole (BH), see, for example, Exercise 14.2; see also Binney and Tremaine (2008), Ciotti (1996, 1999), Ciotti and Ziaee Lorzad (2018), Ciotti et al. (2009), Hernquist (1990), and Jaffe (1983).

14.1.3 Multicomponent Cuddeford and Generalized Cuddeford Spherical Models

In the case of a density component $\rho = \rho_k$ in a multicomponent spherical Cuddeford (1991) system, from Eq. (12.53) the associated DF is given by

$$f = J^{2\alpha} h(Q)\theta(Q - Q_t), \quad \alpha > -1, \tag{14.6}$$

where again we consider the possibility of truncation; in general, the values of α, r_a, and Q_t will be different for different components.

From Eq. (12.54), the augmented density can be written as

$$\varrho(r) \equiv \frac{\rho(r)}{A(r,\alpha)} = \frac{2^{1-\alpha}}{\sqrt{\pi}} \frac{\Gamma(\alpha+3/2)}{\Gamma(\alpha+1)} \left(1+\frac{r^2}{r_{\rm a}^2}\right)^{\alpha+1} \frac{\rho(r)}{r^{2\alpha}}, \tag{14.7}$$

and an inversion formula (reducing to the radial OM case for $\alpha = 0$) can be obtained as follows: after

$$m = \mathrm{int}\left(\alpha + \frac{1}{2}\right) + 1 \tag{14.8}$$

differentiations of Eq. (12.54) with respect to $\Psi_{\rm T}$, the resulting identity (prove it!) can be Abel inverted.[2] In practice, one must perform enough differentiations as to produce a negative exponent $\alpha + 1/2 - m > -1$ in the power-law kernel; note that, as $\alpha > -1$, we have $m \geq 0$.

When $\alpha + 1/2$ is not integer

$$h(Q) = \frac{(-1)^{m+1}\cos\alpha\pi}{2\sqrt{8}\pi^2} \frac{\Gamma(\alpha+3/2-m)}{\Gamma(\alpha+3/2)} \times \frac{d}{dQ}\int_0^Q \frac{d^m\varrho}{d\Psi_{\rm T}^m} \frac{d\Psi_{\rm T}}{(Q-\Psi_{\rm T})^{\alpha+3/2-m}}, \tag{14.9}$$

and in the particular case of untruncated systems with finite total mass, an integration by parts show that

$$\frac{d}{dQ}\int_0^Q \frac{d^m\varrho}{d\Psi_{\rm T}^m} \frac{d\Psi_{\rm T}}{(Q-\Psi_{\rm T})^{\alpha+3/2-m}} = \int_0^{\Psi_{\rm T}} \frac{d^{m+1}\varrho}{d\Psi_{\rm T}^{m+1}} \frac{d\Psi_{\rm T}}{(Q-\Psi_{\rm T})^{\alpha+3/2-m}}. \tag{14.10}$$

As expected, Eqs. (14.9) and (14.10) reduce to the OM case for $\alpha = 0$. When $\alpha = n - 3/2$ with n being an integer ≥ 1, then $m = n$, and the solution of the resulting Volterra integral in Eq. (14.9) is given by

$$h(Q) = \frac{1}{2\sqrt{8}\pi(m-1)!} \left.\frac{d^m\varrho}{d\Psi_{\rm T}^m}\right|_{\Psi_{\rm T}=Q}, \tag{14.11}$$

and so in this very special case the DF is recovered analytically while avoiding integration![3] As a nice exercise, the student is encouraged to work out the explicit expressions of the augmented density and of the inversion formula for a generic component of multicomponent generalized Cuddeford spherical systems, whose DF is given in Eq. (12.64).

14.1.4 Purely Radial Spherical Models

We finally discuss the inversion problem for the special but conceptually relevant case of spherical systems made of *purely radial orbits*. In this case, we consider a density component $\rho = \rho_k$ with the DF

$$f = h(\mathcal{E})\delta(J^2)\theta(\mathcal{E} - \mathcal{E}_{\rm t}), \tag{14.12}$$

[2] $\mathrm{int}(x)$ is the largest integer $\leq x$. For example, $\mathrm{int}(1/2) = 0$, and so $m = 1$ for OM models, while $m = 0$ for Cuddeford models with $-1 < \alpha < -1/2$.

[3] In eq. (30) of Cuddeford (1991), the $(m-1)!$ in the denominator is missing. See also eqs. (49) and (51) of Baes and Dejonghe (2002).

where the Dirac δ-function guarantees that only radial orbits are present. The integral equation to be inverted is given by an equation analogous to Eq. (12.35), where $\mathcal{E}_t$ is now the lower limit of integration. Noticing that, at variance with the Eddington, OM, and Cuddeford inversions, the preparatory differentiation with respect to the potential is not required in order to apply the Abel inversion theorem, and from Eq. (13.1), we immediately obtain (prove it!)

$$
\begin{aligned}
h(\mathcal{E}) &= \frac{1}{\sqrt{2}\pi^2}\frac{d}{d\mathcal{E}}\int_{\mathcal{E}_t}^{\mathcal{E}}\frac{\varrho}{\sqrt{\mathcal{E}-\Psi_T}}\,d\Psi_T \\
&= \frac{1}{\sqrt{2}\pi^2}\int_{\mathcal{E}_t}^{\mathcal{E}}\frac{d\varrho}{d\Psi_T}\frac{d\Psi_T}{\sqrt{\mathcal{E}-\Psi_T}} + \frac{\varrho(\mathcal{E}_t)}{\sqrt{2}\pi^2\sqrt{\mathcal{E}-\mathcal{E}_t}}
\end{aligned}
\tag{14.13}
$$

(e.g., see Ciotti and Ziaee Lorzad 2018; Oldham and Evans 2016; Richstone and Tremaine 1984), where again the augmented density $\varrho = r^2\rho$ is intended to be expressed in terms of the total potential Ψ_T. As we will see later on, the seemingly minor difference of the missing preparatory differentiation has quite important consequences for system consistency.

14.2 Axisymmetric and Triaxial Models

The problem of recovering the DF for axisymmetric systems under the assumption of $f(\mathcal{E}, J_z)$ (i.e., the inversion of Eqs. (12.87)–(12.89)) is much more difficult than that of the spherically symmetric case. In fact, note that in all of the previous formulae the density must be expressed in terms of the total potential. As remarked above, in principle this is always possible in spherical systems. In axisymmetric systems with the vertical component of the gravitational field attracting everywhere toward the equatorial plane, it is instead possible to eliminate z between $\rho(R,z)$ and $\Psi_T(R,z)$, and so an expression of the form $\rho = \rho(R, \Psi_T)$ can be derived. Basically, the known methods of inversion can be listed (in a broad sense) as follows:

(1) *Fricke method*: If $\rho(R, \Psi_T)$ is expressed as a *finite* or *infinite* series of terms $R^a\Psi_T^b$, then Eq. (12.100) can be used to determine the coefficient A for each term of the sum (Fricke 1952; see also Kalnajs 1976b; Miyamoto 1971, 1974, 1975; Nagai and Miyamoto 1976). The resulting DF is given by a sum or a series of functions of the type in Eq. (12.99). The proof of the convergence of the obtained series is usually the most difficult step of this procedure, and sometimes it may even require analytic continuation. An example of the application of the method is given by the power-law torus with $\alpha = 3$ presented in Ciotti and Bertin (2005) in the study of homeoidal expansion (see also Exercise 2.14), with a two-integral DF expressed with hypergeometric function; we also recall the family of power-law tori studied by Toomre (1982), one of the extremely few cases of axisymmetric systems with elementary DFs (however, this was not discovered by using the Fricke method). Finally, a different method, but still based on a suitable expansion of the density, is given in Dehnen and Gerhard (1994).

(2) *Integral transforms*: The inversion procedure is based on classical integral transforms, and here we recall the Lynden-Bell (1962a) method based on Laplace transforms, the Hunter (1975) method based on Stieltjes transforms, and the Dejonghe (1986) method based on Laplace–Mellin transforms (the interested reader is invited to consult Dejonghe (1986) for a very detailed account of integral transform methods). All of these tools require the analytical continuation in the complex plane of the function $\rho = \rho(R, \Psi_T)$; therefore, they are not meant to be directly applied to observational data.[4] Very important examples of the Laplace transform method are the recovery of the elementary two-integral DF for the flattened Plummer sphere presented in the seminal paper by Lynden-Bell (1962a), and by the unexpected discovery made by Evans (1993) of the surprisingly simple two-integral DF of the Binney (1981) logarithmic potential in Eq. (13.66), successively extended in Evans (1994) to the power-law models in Eq. (13.68).

(3) *Hunter–Quian method*: This method is based on the theorem of residues for functions of complex variables. The requirements on the function $\rho(R, \Psi_T)$ to be continued to the complex plane are much weaker than those for the integral transforms. The papers by Hunter and Qian (1993) and Qian et al. (1995) not only give clear expositions of the method, but also provide quite complete summaries and several additional references of the inversion methods to recover two-integral DFs.

(4) *Stäckel models:* A special class of systems amenable to inversion (as well as in the three-dimensional case; i.e., when the DF depends on three isolating integrals of motion) is represented by separable Stackel models (see Section 13.2.3), and for the inversion formula, see, for example, Dejonghe (1987a).

All of these results are quite technical and well beyond the introductory level of this book; for a much more complete discussion of the inversion methods (as well as considering quite sophisticated numerical techniques), the student is directed to consult Binney and Tremaine (2008).

14.3 Testing the Consistency

As was stressed at the beginning of this chapter, in some special cases, it is possible to investigate the positivity of the DF without explicitly recovering the DF itself. In the following, we present a summary of the main results regarding the consistency of spherical models that are simple enough to be used without major effort.

14.3.1 Multicomponent OM Models

We begin by discussing in detail a criterion (Ciotti and Pellegrini 1992) that allows us to check whether the DF of a multicomponent spherical system immersed in a total potential

[4] From Exercise 2.9, the student should prove that for a potential stratified on ellipsoidal surfaces $\phi(m)$ with $\phi'(m) \neq 0$, in the axisymmetric case the corresponding density can be always written as $\rho = A(\phi) + R^2 B(\phi)$.

Ψ_T, for which the orbital anisotropy of each component is of the radial OM form, is indeed positive *without* an explicit calculation of it. Of course, the isotropic case is included in the following results when considering the limiting case of $r_a \to \infty$ (the analogous discussion of the less natural tangentially anisotropic OM models is left as an useful exercise for the student). As in the previous sections, if not strictly needed, we do not indicate explicitly the subscript k in the density of the specific density conponent ρ_k, in the associated anisotropy radius r_a, and in its potential Ψ_k, while $\Psi_T = \sum_k \Psi_k$ is the total potential.[5]

First, we show that a *necessary condition* (NC) for $h \geq 0$ in Eq. (14.4) is

$$\frac{d\varrho}{d\Psi} \geq 0, \quad \mathcal{Q}_t \leq \Psi \leq \Psi(0). \tag{14.14}$$

In fact, from Eq. (12.42), we have $d\varrho/d\Psi_T = 4\pi \int_{\mathcal{Q}_t}^{\Psi_T} h(\mathcal{Q}) d\mathcal{Q}/\sqrt{2(\Psi_T - \mathcal{Q})}$, and if $h \geq 0$, one obtains the inequality above, being $d\varrho/d\Psi_T = (d\varrho/d\Psi)(d\Psi_T/d\Psi)^{-1}$, and the last factor is necessarily positive from Newton's second theorem.[6]

For systems with $\mathcal{Q}_t = 0$ (the common situation for spatially untruncated systems of finite total mass) for which the NC is satisfied, a *strong sufficient condition* (SSC) for $h \geq 0$ is

$$\frac{d}{d\Psi}\left[\frac{d\varrho}{d\Psi}\left(\frac{d\Psi_T}{d\Psi}\right)^{-1}\sqrt{\Psi_T}\right] \geq 0, \quad 0 \leq \Psi \leq \Psi(0), \tag{14.15}$$

and a *weak sufficient condition* (WSC) for $h \geq 0$ is

$$\frac{d}{d\Psi}\left[\frac{d\varrho}{d\Psi}\left(\frac{d\Psi_T}{d\Psi}\right)^{-1}\right] \geq 0, \quad 0 \leq \Psi \leq \Psi(0). \tag{14.16}$$

The proof of the WSC is obtained directly from the second identity in Eq. (14.4) for spatially untruncated systems, while the proof of the SSC is slightly more complex, and it is deferred to Exercise 14.3.

As nice as the previous results may be, we must accept that only for *very few* special systems it is possible to obtain the explicit function $\varrho(\Psi)$; fortunately, we can reformulate Eqs. (14.14)–(14.16) in terms of the radius instead of the potential, and in Exercise 14.4 the corresponding formulae are derived. Some remarks are also in order. The first is that the violation of the NC is related only to the radial behavior of ρ and the value of r_a, and so this condition applies independently of whether any other component is added to the model. This condition is only necessary; thus, f can be negative even for values of model parameters allowed by the NC. A model failing the NC is *certainly* inconsistent, while a model satisfying the NC *may be* consistent. Similarly, a model satisfying the WSC (or the more restrictive SSC) is *certainly* consistent, while a model failing the WSC/SSC *may be* consistent. As a result, *the consistency of a model satisfying the NC and failing the*

[5] The possible presence of an external potential Ψ_{ext} is considered for simplicity in the sum.

[6] The function $\Psi_T(\Psi)$ can always be obtained (in principle) by elimination of the radius between $\Psi_T(r)$ and $\Psi(r)$ because the potential in spherical systems is monotonically decreasing and so invertible.

WSC/SSC can be proved only by direct inspection of its DF. The second remark particularly concerns the limitations on r_a. In fact, for each component of OM systems, one can write

$$h(Q) = h_i(Q) + \frac{h_a(Q)}{r_a^2}, \tag{14.17}$$

where the meaning of the two functions h_i and h_a is obvious from Eqs. (14.4) and (14.5). Importantly, note that a similar functional decomposition holds for the NC and the WSC, so that all of the following arguments can be repeated for these two conditions with the variable Q substituted by r, and h substituted by the augmented density ϱ.

Let us now discuss the positivity of $h(Q)$ in Eq. (14.17) in terms of r_a, the anistropy radius. We indicate with A_+ the set of values of Q such that $h_i > 0$ (i.e., the set of values of Q so that the isotropic component of the DF is positive definite). Then, from Eq. (14.17)

$$r_a \geq r_a^- \equiv \sqrt{\max\left\{0, \sup\left[-\frac{h_a(Q)}{h_i(Q)}\right]_{Q\in A_+}\right\}} \tag{14.18}$$

is the condition to be satisfied in order to have a positive definite DF over A_+. Obviously, when $h_i > 0$ over all of the phase space (e.g., when the isotropic component is consistent), A_+ coincides with the total range of variation for Q and r_a^- is the lower bound for the anisotropy radius; this is the common situation encountered in the vast majority of one-component systems. However, when the set A_- (complementary to A_+) is not empty (i.e., $h_i < 0$ over some region of the accessible phase space), a second inequality, derived from the request of $h(Q) \geq 0$ in Eq. (14.17), must be verified:

$$r_a \leq r_a^+ \equiv \sqrt{\inf\left[-\frac{h_a(Q)}{h_i(Q)}\right]_{Q\in A_-}}. \tag{14.19}$$

It follows that if A_- is not empty and $h_a < 0$ for some $Q \in A_-$, or $r_a^+ < r_a^-$, then the density component under investigation is inconsistent. In summary, the allowed region for phase-space consistency for each component of single-component or multicomponent spherically symmetric and radially anisotropic OM models is $r_a^- \leq r_a \leq r_a^+$.

As remarked above, formally identical considerations can also be repeated for the NC, SSC, and WSC, where of course the necessary or sufficient nature of the obtained limitations must now be taken into account (see Exercise 14.5; see also Ciotti 1996, 1999; Ciotti and Lanzoni 1997; Ciotti and Morganti 2009; Ciotti and Pellegrini 1992; Ciotti and Ziaee Lorzad 2018; Ciotti et al. 1996, 2009, 2019; Tremaine et al. 1994 and references therein for applications of the NC, WSC, and SSC conditions to single-component and multicomponent OM galaxy models also in the presence of a central BH).

Two important consequences derive from the NC radial condition. First, in the *isotropic case* ($r_a \to \infty$), the NC for consistency in Eq. (14.38) reveals that ρ must be a radially decreasing function; for instance, it is impossible to have isotropic spherically symmetric systems with a "hole" or a "depression" in the center. Second, in the fully radial case ($r_a \to 0$), the NC requires that the function $r^2\rho(r)$ must be nonincreasing with increasing radius. However, as we have seen in Section 14.1.4, the fully radial case can be also discussed directly from its inversion integral, and Eq. (14.13) shows that

$$\frac{d[r^2\rho(r)]}{dr} \leq 0, \quad 0 \leq r < \infty, \tag{14.20}$$

is also a sufficient condition for consistency. Therefore, we have reached the interesting conclusion that the necessary and sufficient condition for a spherical model to be made by purely radial orbits is that its density declines everywhere with the radius at least as fast as $1/r^2$. Again in relation to purely radial models, it is useful to mention an explicit example illustrating some of the subtleties that can occur when considering phase-space inversion. In fact, Ciotti and Ziaee Lorzad (2018) showed that, in the OM case, the (analytical) DF of Jaffe models (e.g., see Binney and Tremaine 2008; Merritt 1985b), for r_a below a (quite small) critical value, is negative for some admissible value of Q, and so one would feel authorized to conclude that a fortiori the purely radial ($r_a = 0$) Jaffe model is inconsistent. On the other hand, the Jaffe density profile (being everywhere steeper than $1/r^2$) obeys the sufficient condition for consistency of purely radial models in Eq. (14.20), as is confirmed by inspection of its nowhere negative analytical DF (Ciotti and Ziaee Lorzad 2018; Evans et al. 2015; Merritt 1985b) obtained (do it!) from Eq. (14.13). This fact clearly illustrates that the purely radial case obtained from OM anisotropy can be a *singular limit*; in practice, if the limit for $r_a \to 0$ of a given OM model is consistent, then it is possible to conclude that the purely radial model exists, but the opposite is not necessarily true.

14.3.2 Multicomponent Generalized Cuddeford Models

From the results in Section 14.1.3, it should be clear that the same arguments used to derive the necessary and sufficient consistency conditions for OM models can be repeated for multicomponent generalized Cuddeford systems. However, as m preparatory differentiations with respect to Ψ_T must be performed before inversion of the integral[7] in Eq. (12.54), with each differentiation producing an integral identity, it is easy to see that we now obtain m necessary consistency conditions, plus a sufficient condition (Ciotti and Morganti 2010a).

In fact (prove it!) each density component in a consistent multicomponent generalized Cuddeford system with $\alpha > -1$ obeys m necessary conditions (NC_l)

$$\frac{d^l \varrho}{d\Psi_T^l} \geq 0, \quad l = 1, \ldots, m, \tag{14.21}$$

where m is given by Eq. (14.8) and $\varrho(r) = \rho(r)/A(r,\alpha)$ is the augmented density. Moreover, under the assumptions of Eq. (14.9), a sufficient condition for the non-negativity of the DF in the untruncated ($Q_t = 0$) case is

$$\frac{d^{m+1}\varrho}{d\Psi_T^{m+1}} \geq 0. \tag{14.22}$$

[7] Of course, now the function $A(r,\alpha)$ is given in Eq. (12.65).

Of course, for $\alpha = n - 3/2$ and integer $n \geq 1$ (when $m = n$), the first $m - 1$ inequalities in Eq. (14.21) are still NCs, but

$$\frac{d^m \varrho}{d\Psi_T^m} \geq 0 \tag{14.23}$$

is the necessay *and* sufficient condition for consistency, as follows from the closed-form inversion formula in Eq. (14.11); moreover, the student is invited to prove that the NC and the WSC for OM systems in Eqs. (14.14)–(14.16) are reobtained as very special cases for $\alpha = 0$ (i.e., $m = 1$).

In applications, it can be useful to express the family of Eqs. (14.21) in terms of the radius instead of the potential (see Exercise 14.4); for example, NC_1 and NC_2 coincide with Eqs. (14.38)–(14.40), where of course now ϱ is the augmented density of the generalized Cuddeford component. When expressing the conditions in terms of radius, notice that NC_1 is the sole condition in which only the augmented density profile of the specific density component appears, while in the higher-order NC_l the total mass profile $M_T(r)$ is also involved (see Exercise 14.6). Also notice that from the monotonically decreasing nature of Ψ_T with the radius, the sign of the NC_l inequality changes to $(-1)^l$.

Finally, we mention a result that (at the present stage of discussion) seems to be nothing more than a mathematical curiosity, and that was obtained by direct computation in Ciotti and Morganti (2009, 2010a) (see also Exercise 14.7): that for each component ρ of multicomponent generalized Cuddeford systems, if we define the logarithmic density slope $\gamma(r) \equiv -d \ln \rho / d \ln r$, then the NC_1 holds *if and only if* $\forall r$

$$\gamma(r) \geq 2\beta(r), \tag{14.24}$$

where $\beta(r)$ is the anisotropy profile in Eq. (12.68). We will refer to the inequality in Eq. (14.24) as the GDSAI.

14.4 The GDSAI

From the GDSAI inequality (14.24), it follows that, at least for the quite large family of multicomponent generalized Cuddeford systems (containing as very special cases the isotropic, constant-anisotropy, Cuddeford, and OM models), the density slope of each consistent component is related to the maximum amount of anisotropy that can be supported at each radius by the component itself. A similar and complementary result, the *central cusp–anisotropy theorem* (An and Evans 2006; de Bruijne et al. 1996), was in fact already known: this theorem states that Eq. (14.24) holds (1) in all consistent *constant-anisotropy* systems with $\beta \leq 1/2$ and (2) asymptotically at the center (i.e., for $r \to 0$) of any consistent spherical system with a *generic* anisotropy profile. Therefore, Eq. (14.24) unexpectedly revealed that the GDSAI holds not only at the center, but at all radii in a quite large class of spherical systems, whenever their DF is positive. When considering the two previous results together, it is natural to ask whether the GDSAI is necessarily obeyed by generic spherically symmetric systems with positive DFs. This possibility is reinforced by the fact that the end products of N-body numerical simulations appear to follow interesting correlations between β and γ (e.g., see Hansen and Moore 2006).

Quite interestingly, a (partial) answer to this conjecture can be given. In fact, Ciotti and Morganti (2010b) considered the special family of stellar systems in which the radial pressure p_r defined in Eq. (12.117) is a factorized function of the radius and of the total potential, so that from the first identity of Eq. (12.119)

$$\rho(r, \Psi_T) = \mathcal{A}(r)\mathcal{B}(\Psi_T). \tag{14.25}$$

Of course, while the function $\mathcal{B}$ in the expression above is the derivative of the potential dependent factor in p_r, the radial function $\mathcal{A}$ is the same (see also Baes and Dejonghe 2002; Baes and van Hese 2007). We note that the multicomponent generalized Cuddeford models (and therefore also the constant-anisotropy, OM, and Cuddeford models) belong to such a family of systems. From the second identity of Eq. (12.119), it follows that, independently of the specific radial dependence of Ψ_T,

$$2\beta(r) = -\frac{d\ln\mathcal{A}}{d\ln r}. \tag{14.26}$$

Moreover, it is a simple exercise (do it!) to show from the first identity of Eq. (12.119) and from Eq. (14.26) that the logarithmic slope of the density profile in Eq. (14.25) can be written as

$$\gamma(r) = -\frac{d\ln\rho}{d\ln r} = -\frac{d\ln\mathcal{B}}{d\ln\Psi_T}\frac{d\ln\Psi_T}{d\ln r} + 2\beta(r), \tag{14.27}$$

so that we can express the quantity $\gamma(r) - 2\beta(r)$ in terms of the logarithmic derivative of the $\mathcal{B}$ function. Now, since from Newton's theorem $\Psi_T(r)$ is a monotonically decreasing function of the radius, Eq. (14.27) proves that *in all spherical systems whose radial pressure is a separable function of the radius and total potential, so that $\rho = A(r)\mathcal{B}(\Psi_T)$, the global GDSAI holds if and only if $\mathcal{B}(\Psi_T)$ is a monotonically increasing function of* Ψ_T. Therefore, if one is able to show that in all factorized consistent systems the $\mathcal{B}$ function is necessarily monotonically increasing, then the GDSAI is established at once for all of these systems, without the need for lengthy algebraic calculations. We will now illustrate the power and elegance of this result with some explicit examples.

14.4.1 The GDSAI for Generalized Multicomponent Cuddeford Systems Again

Following Ciotti and Morganti (2010b), we begin by considering again the case of generalized multicomponent Cuddeford models, where for $\alpha \geq -1/2$ we obtained the GDSAI in Eq. (14.24) by explicit computation (see Exercise 14.7). From Eqs. (12.65) and (14.25)

$$\mathcal{A}(r) = (2\pi)^{3/2}\frac{\Gamma(\alpha+1)r^{2\alpha}}{\Gamma(\alpha+3/2)}\sum_i \frac{w_i}{(1+r^2/r_{\rm ai}^2)^{\alpha+1}}, \tag{14.28}$$

$$\mathcal{B}(\Psi_{\rm T}) = \int_0^{\Psi_{\rm T}} (\Psi_{\rm T} - Q)^{\alpha+1/2} h(Q) dQ, \tag{14.29}$$

where differentiation shows that $\mathcal{B}$ is a monotonically increasing function of $\Psi_{\rm T}$ whenever $h > 0$ and $\alpha \geq -1/2$, proving again in a very direct way the validity of the GDSAI for this class of models. We are left with the case $-1 < \alpha < -1/2$, where Eq. (14.22) with $m = 0$ shows that the GDSAI is only a sufficient condition for consistency. Indeed, for $-1 < \alpha < -1/2$, it is not possible to compute the derivative of Eq. (14.29) directly. However, this problem can be circumvented by first integrating by parts and then performing differentiation, obtaining (prove it!)

$$\frac{d\mathcal{B}}{d\Psi_{\rm T}} = \Psi_{\rm T}^{\alpha+1/2} h(0) + \int_0^{\Psi_{\rm T}} (\Psi_{\rm T} - Q)^{\alpha+1/2} h'(Q) dQ. \tag{14.30}$$

From Eq. (14.30), we conclude that the GDSAI is again necessary for phase-space consistency in all generalized multicomponent Cuddeford systems with $-1 < \alpha < -1/2$ *and* $h'(Q) > 0$. In summary, for $\alpha > -1/2$ the GDSAI is necessary for consistency, for $\alpha = -1/2$ it is equivalent to consistency (i.e., necessary and sufficient), and for $-1 < \alpha < -1/2$ it is just sufficient, but also necessary if $h'(Q) > 0$.

14.4.2 The GDSAI for the Baes and van Hese (2007) Anisotropic Models

Again from Ciotti and Morganti (2010b), we now show that the GDSAI is obeyed by another quite large family of consistent models not belonging to the family of generalized Cuddeford models. Baes and van Hese (2007) considered the family of density profiles

$$\rho = \rho_0 \left(\frac{r}{r_{\rm a}}\right)^{-2\beta_0} \left(1 + \frac{r^{2\delta}}{r_{\rm a}^{2\delta}}\right)^{\frac{\beta_0 - \beta_\infty}{\delta}} \left(\frac{\Psi}{\Psi_0}\right)^p \left(1 - \frac{\Psi^s}{\Psi_0^s}\right)^q, \tag{14.31}$$

where $\delta > 0$, $q \leq 0$, $s > 0$, $0 \leq \Psi \leq \Psi_0$, and Ψ_0 is the value of the central relative potential. From Eq. (14.26), the anisotropy profile is

$$\beta(r) = \frac{\beta_0 + \beta_\infty (r/r_{\rm a})^{2\delta}}{1 + (r/r_{\rm a})^{2\delta}}, \tag{14.32}$$

so that β_0 and β_∞ are the values of the orbital anisotropy parameter at small and large radii, respectively; since by construction $\beta(r) \leq 1$, both β_0 and β_∞ are ≤ 1. Note that the requirement of finite total mass for the density distribution (14.31) (i.e., $\Psi(r) \sim 1/r$ for $r \to \infty$) translates into the condition $p + 2\beta_\infty > 3$, and from the limitation on β_∞, it follows that $p \geq 1$. Now, it is easy to show that $d\mathcal{B}(\Psi)/d\Psi > 0$ for $0 \leq \Psi \leq \Psi_0$, when $s > 0$, $q \leq 0$, and $p \geq 1$, and thus the GDSAI is necessarily obeyed by these profiles as well. The analytical DF associated with these systems (Baes and van Hese 2007) is given by

$$f = \sum_k \left(\frac{\mathcal{E}}{\Psi_0}\right)^{p+ks-3/2} \theta(\mathcal{E}) g_k \left(\frac{J^2}{2r_{\rm a}^2 \mathcal{E}}\right), \tag{14.33}$$

where the g_k are expressed in terms of the Fox H functions (e.g., see Mathai et al. 2009). The anisotropic Hernquist model considered in Baes and Dejonghe (2002) also belongs to this family.

14.4.3 The GDSAI for the Multicomponent Generalized Cuddeford and Louis (1995) Anisotropic Polytropes

We finally consider the multicomponent generalization of the family of anisotropic polytropes (Cuddeford and Louis 1995; Louis 1993; see also Dejonghe 1987b for the anisotropic Plummer model in the family) considered in Ciotti and Morganti (2010b). The DF of each density component is given by

$$f(\mathcal{E}, J^2) = \mathcal{E}^{q-2}\theta(\mathcal{E})\sum_i w_i h(k_i), \quad k_i \equiv \frac{J^2}{2r_{\rm ai}^2\mathcal{E}}, \tag{14.34}$$

with different anisotropy radii $r_{\rm ai}$ and positive weights w_i. In Exercise 14.9, we prove that

$$\rho = 2^{3/2}\pi \mathrm{B}\left(q, \frac{1}{2}\right)\Psi_{\rm T}^{q-1/2}\sum_i w_i\eta_i^{q-1}\int_0^\infty \frac{h(k)dk}{(k+\eta_i)^q}, \quad \eta_i \equiv \frac{r^2}{r_{\rm ai}^2}, \tag{14.35}$$

$$p_{\rm r} = 2^{5/2}\pi \mathrm{B}\left(q, \frac{3}{2}\right)\Psi_{\rm T}^{q+1/2}\sum_i w_i\eta_i^{q-1}\int_0^\infty \frac{h(k)dk}{(k+\eta_i)^q}. \tag{14.36}$$

Note that the convergence of the density integral requires $q > 0$ near $\mathcal{E} = 0$. As $p_{\rm r}$ and ρ are in factorized form and $\mathcal{B} \propto \Psi_{\rm T}^{q-1/2}$, the GDSAI is satisfied whenever $q \geq 1/2$. The situation is less straightforward in the interval $0 < q < 1/2$. In fact, if a consistent model exists in this interval, then it will represent a case of violation of the GDSAI, similarly to the case of the models in Section 14.4.1 for $-1 < \alpha < -1/2$. Therefore, it is natural to ask whether it is possible to construct a consistent dynamical model with $q > 0$, but violating the GDSAI ($q < 1/2$). Note that in the self-consistent case of finite total mass (when $\Psi_{\rm T} = \Psi \sim 1/r$ for $r \to \infty$), volume integration of Eq. (14.35) obtained with inversion of the order of integration shows that $q > 3/2$ is necessary in order to have a finite mass for the component under scrutiny. Therefore, the GDSAI is a necessary consistency condition for anisotropic polytropes in generic external potential when $q \geq 1/2$, and for all finite-mass self-consistent models.

The possibility remains open for the existence of infinite-mass, self-gravitating, and consistent anisotropic polytropes with $0 < q < 1/2$, or of Cuddeford models with angular momentum exponents in the range $-1 < \alpha < -1/2$, which would violate the GDSAI. Therefore, we can conclude that *there is* the possibility of the existence of consistent models with factorized densities as in Eq. (14.25) but nonmonotonic $\mathcal{B}$ functions that violate the GDSAI. In fact, recently it has been proved that these (quite peculiar) systems do exist, and for recent progress on this interesting problem, the student is invited to consult An et al. (2012), Barber and Zhao (2014), and Van Hese et al. (2011). Of course, much less is known for models with a nonseparable density profile.

Exercises

14.1 The interested reader should prove the following statements about the Eddington inversion of spherical, isotropic systems: (1) for all spatially untruncated systems of finite total mass, $\lim_{\Psi_T\to 0} d\rho_k/d\Psi_T = 0$, and so one term in the second identity in Eq. (14.2) is absent; and (2) if $h_k \sim (\mathcal{E}-\mathcal{E}_t)^\alpha$ for $\mathcal{E}\to\mathcal{E}_t$, then $\alpha > -1$ to ensure the convergence of the density ρ_k at the edge of the system. When $\alpha > -1$, then $\lim_{\Psi_T\to\mathcal{E}_t}\rho_k = 0$. Moreover, $d\rho_k/d\Psi_T \sim (\Psi_T-\mathcal{E}_t)^{\alpha+1/2}$, and so $\lim_{r\to r_t} = d\rho_k/dr = -\infty$ (for $-1<\alpha<-1/2$), $d\rho_k/dr_t = -const.$ (for $\alpha=-1/2$), and finally $d\rho_k/dr_t = 0$ (for $\alpha > -1/2$; i.e., the mass density goes smoothly to zero at the edge of the system).

14.2 With this illustrative exercise, we recover one of the simplest explicit cases of the OM phase-space DF. Consider the single-component Plummer model in Eq. (13.36) and define $M_n = M/r_c^3$ and $\Psi_n = GM/r_c$. Integrate Eq. (14.4) in the untruncated case ($Q_t=0$) and show that $f=(M_n/\Psi_n^{3/2})h(q)$, where

$$h(q)=\frac{24\sqrt{2}q^{7/2}}{7\pi^3}+\frac{3q^{3/2}(7-16q^2)}{7\sqrt{2}\pi^3 s_a^2},\quad s_a\equiv\frac{r_a}{r_c},\quad 0\le q\equiv\frac{Q}{\Psi_n}\le 1. \tag{14.37}$$

14.3 Prove the SSC in Eq. (14.15) for the k-th component of a multicomponent OM system. *Hint*: Let $I(Q)\equiv\int_0^Q g(x)(Q-x)^{-1/2}dx$, with $x=\Psi_T$ and $g(x)=d\varrho_k/d\Psi_T$; we look for a sufficient condition to have $I(Q)'=dI(Q)/dQ\ge 0$. Change the integration variable to x/Q and, after differentiation with respect to Q, restore the original integration variable, obtaining $I(Q)'=Q^{-1}\int_0^Q[g(x)/2+xg(x)'](Q-x)^{-1/2}dx = Q^{-1}\int_0^Q[g(x)\sqrt{x}]'\sqrt{x}(Q-x)^{-1/2}dx$. The SSC is proved by requiring positivity of the derivative in the integrand and finally setting $\Psi_T=\Psi_T(\Psi_k)$ (see Footnote 6).

14.4 By changing variable from potential to radius, show that Eqs. (14.14)–(14.16) can be rewritten as functions of the sperical radius r, respectively, as

$$\frac{d\varrho(r)}{dr}\le 0,\quad 0\le r\le r_t, \tag{14.38}$$

$$\frac{d}{dr}\left[\frac{d\varrho(r)}{dr}\frac{r^2\sqrt{\Psi_T(r)}}{M_T(r)}\right]\ge 0,\quad 0\le r\le\infty, \tag{14.39}$$

$$\frac{d}{dr}\left[\frac{d\varrho(r)}{dr}\frac{r^2}{M_T(r)}\right]\ge 0,\quad 0\le r\le\infty. \tag{14.40}$$

14.5 With this exercise, we elaborate further on the Plummer model, and we compute explicitly the critical values of the anisotropy radius for consistency in the case of OM anisotropy. By using the DF in Eq. (14.37) and the conditions in Eqs. (14.38)–(14.40) or in Eqs. (14.14)–(14.16), show that the *minimum* values of $s_a = r_a/r_c$ obtained from the NC, the DF (the true limit), the SSC, and finally the WSC are

$$s_{\rm a}^{-}({\rm NC}) = \sqrt{\frac{2}{5}} < s_{\rm a}^{-}({\rm DF}) = \frac{3}{4} < s_{\rm a}^{-}({\rm SSC}) = \sqrt{\frac{7}{10}} < s_{\rm a}^{-}({\rm WSC}) = \sqrt{\frac{2}{3}}. \quad (14.41)$$

14.6 Quite often the consistency is studied for multicomponent systems (e.g., a stellar distribution embedded in a dark matter halo and/or with a central BH). Show that in OM models a sufficient condition for the validity of the WSC($\rho, \Psi_{\rm T}$) for a specific density component in the total potential $\Psi_{\rm T}$ is to have the WSC(ρ, Ψ_k) verified *separately* in the potential Ψ_k of *each* component. *Hints*: Write $M_{\rm T}(r) = \sum_k M_k(r)$ in Eq. (14.40), evaluate the outer derivative, and show that

$$\mathrm{WSC}(\rho, \Psi_{\rm T}) = \sum_k \frac{M_k^2(r)}{M_{\rm T}^2(r)} \mathrm{WSC}(\rho, \Psi_k), \quad 0 \le r \le \infty. \quad (14.42)$$

14.7 Prove by direct evaluation that the GDSAI in Eq. (14.24) holds for each component of generalized multicomponent Cuddeford systems with $\alpha \ge -1/2$. *Hint*: First notice that for $\alpha \ge -1/2$ from Eq. (14.8) $m \ge 1$, and so NC$_1$ exists. Then construct the augmented density $\varrho = \rho(r)/A(r,\alpha)$ by using Eq. (12.65) and show that NC$_1$ in Eqs. (14.21) and (14.38) can be rewritten as

$$0 \ge \frac{d\varrho}{dr} = \frac{\varrho}{r}\frac{d\ln(\rho/A)}{d\ln r} = -\frac{\varrho}{r}\left[\gamma(r) + \frac{d\ln A(r,\alpha)}{d\ln r}\right]. \quad (14.43)$$

Finally, from Eqs. (12.65) and (12.68), show by direct computation that for $\alpha > -1$

$$-\frac{d\ln A(r,\alpha)}{d\ln r} = 2\left[1 - \frac{C(r,\alpha)}{2B(r,\alpha)}\right] = 2\beta(r). \quad (14.44)$$

14.8 Starting from the Jeans equation (13.82) for generic spherically symmetric and anisotropic systems, show that the GDSAI in Eq. (14.24) can be rewritten as

$$\gamma(r) - 2\beta(r) = \frac{r}{\sigma_{\rm r}^2}\left(\frac{d\sigma_{\rm r}^2}{dr} - \frac{d\Psi_{\rm T}}{dr}\right) \ge 0. \quad (14.45)$$

14.9 By integration over the velocity space of Eq. (14.34) using Eqs. (12.25) and (12.27), show that Eqs. (14.35) and (14.36) hold. Then, deduce that

$$\sigma_{\rm r}^2 = \frac{p_{\rm r}}{\rho} = \frac{2\Psi_{\rm T}}{2q+1}, \quad (14.46)$$

and from Eq. (14.45) prove again that the DGSAI is satisfied for $q > -1/2$.

Appendix

Mathematical Background

We summarize here the most important mathematical results used in the text (and useful for solving some of the proposed exercises). Overall, this appendix should be intended more as a selection of mathematical results relevant to stellar dynamics than an organic presentation of definitions and theorems. A good knowledge of linear algebra and calculus at the undergraduate level is assumed, and the interested reader is encouraged to refer to some of the excellent classical treatises of mathematical physics such as Arfken and Weber (2005), Bender and Orszag (1978), Courant and Hilbert (1989), Dennery and Krzywicki (1967), Ince (1927), Jeffrey and Jeffrey (1950), Kahn (2004), and Morse and Feshbach (1953). To help with further study, in the following sections more specific references are sometimes also provided.

A.1 Identities of Vector Calculus

The literature on vector calculus and linear algebra is immense. Here, we only mention three books that stand out for clarity and rigor, namely Aris (1989), Borisenko and Tarapov (1979), and Cullen (1990). In the following, we recall the main identities of vector calculus and we restrict ourselves to real vector spaces over $\Re^n$, whose defining axioms are assumed to be known. The *inner product* obeys the standard rules of *commutativity*

$$\langle \mathbf{a}, \mathbf{b} \rangle = \langle \mathbf{b}, \mathbf{a} \rangle, \tag{A.1}$$

linearity

$$\langle \mathbf{a}, \lambda \mathbf{b} + \mu \mathbf{c} \rangle = \lambda \langle \mathbf{a}, \mathbf{b} \rangle + \mu \langle \mathbf{a}, \mathbf{c} \rangle, \tag{A.2}$$

where λ and μ are two real scalars, and finally *positive definitness*, i.e.,

$$\langle \mathbf{a}, \mathbf{a} \rangle \geq 0, \quad \text{and} \quad \langle \mathbf{a}, \mathbf{a} \rangle = 0 \quad \Longleftrightarrow \mathbf{a} = 0. \tag{A.3}$$

The *norm* (the length of a vector) is introduced as

$$\|\mathbf{a}\| \equiv \sqrt{\langle \mathbf{a}, \mathbf{a} \rangle}, \tag{A.4}$$

and geometrically

$$\langle \mathbf{a}, \mathbf{b} \rangle = \|\mathbf{a}\| \, \|\mathbf{b}\| \, \cos\theta, \quad 0 \leq \theta \leq \pi, \tag{A.5}$$

where θ is the angle between the two vectors, so that the *Cauchy–Schwarz* inequality

$$|\langle \mathbf{a}, \mathbf{b} \rangle| \leq \|\mathbf{a}\| \, \|\mathbf{b}\| \tag{A.6}$$

follows. Two vectors with zero inner product are called *orthogonal*, and a vector with unitary norm is called *versor*. The norm obeys the property of *homogeneity*

$$\|\lambda \mathbf{a}\| = |\lambda| \, \|\mathbf{a}\|, \tag{A.7}$$

to *triangular inequality*

$$\|\mathbf{a} + \mathbf{b}\| \leq \|\mathbf{a}\| + \|\mathbf{b}\|, \tag{A.8}$$

and finally, if two vectors are orthogonal, the Pythagorean theorem holds

$$\|\mathbf{a}\|^2 + \|\mathbf{b}\|^2 = \|\mathbf{a} + \mathbf{b}\|^2. \tag{A.9}$$

By using the norm, it is possible to introduce in a natural way the *distance* between two points by defining $d(\mathbf{a}, \mathbf{b}) \equiv \|\mathbf{a} - \mathbf{b}\|$; it is easy to prove that the formal properties of distance are in fact verified by this function.

In linear algebra, it is proved that n orthogonal vectors (versors) in $\Re^n$,

$$\langle \mathbf{e}_i, \mathbf{e}_j \rangle = \delta_{ij} \tag{A.10}$$

(where $\delta_{ij} = 1$ if $i = j$ and 0 otherwise is the Kronecker symbol), are a *basis* for the vector space over $\Re^n$, so that a generic vector can be expressed as a linear combination of the basis as $\mathbf{a} = a_i \mathbf{e}_i$ with the convention of repeated indices.[1] Note that $a_i = \langle \mathbf{a}, \mathbf{e}_i \rangle$. From the linearity of the inner product and orthogonality of the basis, it follows immediately that

$$\langle \mathbf{a}, \mathbf{b} \rangle = a_i b_j \delta_{ij} = a_i b_i, \tag{A.11}$$

where $\mathbf{b} = b_i \mathbf{e}_i$; the expression in Eq. (A.11) is also known as a *dot product*. Moreover, for a generic $n \times n$ square matrix $\mathcal{R}$ over $\Re^n$, with the expression $\mathcal{R}\mathbf{a}$ we intend the usual row-by-column product of linear algebra, and it is a trivial exercise to show that

$$\langle \mathcal{R}\mathbf{a}, \mathbf{b} \rangle = \langle \mathbf{a}, \mathcal{R}^{\mathrm{T}} \mathbf{b} \rangle, \tag{A.12}$$

where $\mathcal{R}^{\mathrm{T}}$ is the *transpose* of $\mathcal{R}$ (i.e., $(\mathcal{R}^{\mathrm{T}})_{ij} = (\mathcal{R})_{ji}$). It follows that if $\mathcal{R}$ is *orthogonal* (i.e., $\mathcal{R}\mathcal{R}^{\mathrm{T}} = \mathcal{R}^{\mathrm{T}}\mathcal{R} = I$, where I is the identity matrix), then

$$\langle \mathcal{R}\mathbf{a}, \mathcal{R}\mathbf{b} \rangle = \langle \mathbf{a}, \mathbf{b} \rangle, \tag{A.13}$$

a formalization of the intuitive geometric fact that a rigid rotation of space does not change the length and angle between two vectors. The *cross* (or *vector*) product – an operation

[1] For typographic reasons, vectors are written in the text as *row vectors*, even though they should be intended as *column vectors* for consistency with the rules of matrix multiplication. The student will also notice that we do not introduce the concept of covariant and controvariant bases (e.g., see Narashiman 1993).

restricted to vectors of $\Re^3$ only – is here indicated (with an abuse of notation) with the symbol of *exterior product* $\wedge$ instead of $\times$, so that we write $\mathbf{a} \wedge \mathbf{b}$. We recall that

$$\|\mathbf{a} \wedge \mathbf{b}\| = \|\mathbf{a}\| \, \|\mathbf{b}\| \, \sin\theta = \sqrt{\|\mathbf{a}\| \, \|\mathbf{b}\|^2 - \langle \mathbf{a}, \mathbf{b} \rangle^2}, \quad 0 \leq \theta \leq \pi, \tag{A.14}$$

where $\mathbf{a} \wedge \mathbf{b}$ is perpendicular to the plane containing $\mathbf{a}$ and $\mathbf{b}$ and directed so that the three vectors $\mathbf{a}$, $\mathbf{b}$, and $\mathbf{a} \wedge \mathbf{b}$ (in this order) are positively oriented. Geometrically, the cross product can be visualized as the oriented area of the parallelogram with the two vectors $\mathbf{a}$ and $\mathbf{b}$ as sides. The main properties of the cross product are *anticommutativity*

$$\mathbf{a} \wedge \mathbf{b} = -\mathbf{b} \wedge \mathbf{a}, \tag{A.15}$$

so that $\mathbf{a} \wedge \mathbf{a} = 0$, *linearity*

$$\mathbf{a} \wedge (\lambda \mathbf{b} + \mu \mathbf{c}) = \lambda \mathbf{a} \wedge \mathbf{b} + \mu \mathbf{a} \wedge \mathbf{c}, \tag{A.16}$$

and the *Jacobi identity*

$$\mathbf{a} \wedge (\mathbf{b} \wedge \mathbf{c}) + \mathbf{b} \wedge (\mathbf{c} \wedge \mathbf{a}) + \mathbf{c} \wedge (\mathbf{a} \wedge \mathbf{b}) = 0. \tag{A.17}$$

It is a simple exercise to show that in any positively oriented Cartesian reference system, where

$$\mathbf{e}_x \wedge \mathbf{e}_y = \mathbf{e}_z, \quad \mathbf{e}_z \wedge \mathbf{e}_x = \mathbf{e}_y, \quad \mathbf{e}_y \wedge \mathbf{e}_z = \mathbf{e}_x, \tag{A.18}$$

the following formal expansion holds:

$$\mathbf{a} \wedge \mathbf{b} = \begin{vmatrix} \mathbf{e}_x & \mathbf{e}_y & \mathbf{e}_z \\ a_1 & a_2 & a_3 \\ b_1 & b_2 & b_3 \end{vmatrix} = \begin{pmatrix} 0 & -a_3 & a_2 \\ a_3 & 0 & -a_1 \\ -a_2 & a_1 & 0 \end{pmatrix} \mathbf{b}. \tag{A.19}$$

Equation (A.19) reveals an important property of the cross product: its relation with *antisymmetric* matrices in $\Re^3$. In practice, the action of a generic antisymmetric[2] matrix $\mathcal{R}^{\mathrm{T}} = -\mathcal{R}$ on a vector $\mathbf{x} \in \Re^3$ can be also represented as a cross product with a vector $\mathbf{a} \in \Re^3$ related to the elements of $\mathcal{R}$ as in Eq. (A.19), and vice versa. Finally, an important property of the cross product – intuitively obvious from its geometric interpretation – is its behavior under the action of a proper rotation of matrix $\mathcal{R}$ (with $\det \mathcal{R} = 1$), i.e.,

$$(\mathcal{R}\mathbf{a}) \wedge (\mathcal{R}\mathbf{b}) = \mathcal{R}(\mathbf{a} \wedge \mathbf{b}). \tag{A.21}$$

[2] Other families of (real) matrices $\mathcal{R}$ of fundamental importance in dynamics are the *symmetric* matrices $\mathcal{R}^{\mathrm{T}} = \mathcal{R}$; the *antisymmetric* matrices $\mathcal{R}^{\mathrm{T}} = -\mathcal{R}$; the $2n \times 2n$ *symplectic* matrices $\mathcal{R}^{\mathrm{T}} J \mathcal{R} = J$, where

$$J = \begin{pmatrix} 0 & I \\ -I & 0 \end{pmatrix}, \quad J^{\mathrm{T}} = -J, \tag{A.20}$$

and I is the $n \times n$ identity matrix; the $2n \times 2n$ *Hamiltonian* matrices $(J\mathcal{R})^{\mathrm{T}} = J\mathcal{R}$; and the *normal* matrices $[\mathcal{R}, \mathcal{R}^{\mathrm{T}}] = \mathcal{R}\mathcal{R}^{\mathrm{T}} - \mathcal{R}^{\mathrm{T}}\mathcal{R} = 0$ (see also Chapter 9).

With the introduction of the Levi–Civita tensor ϵ_{ijk}, we can write $\mathbf{a} \wedge \mathbf{b} = \epsilon_{ijk}\mathbf{e}_i a_j b_k$, and recalling that $\epsilon_{ijk}\epsilon_{klm} = \delta_{il}\delta_{jm} - \delta_{im}\delta_{jl}$, the student can prove that the *triple vecor product* can be written as

$$\mathbf{a} \wedge (\mathbf{b} \wedge \mathbf{c}) = \mathbf{b}\langle \mathbf{a}, \mathbf{c}\rangle - \mathbf{c}\langle \mathbf{a}, \mathbf{b}\rangle. \tag{A.22}$$

Moreover, the *mixed product* $\langle \mathbf{a}, \mathbf{b} \wedge \mathbf{c}\rangle$ is geometrically interpreted as the (oriented) volume of the parallelepiped with sides given by the three vectors, and from Eq. (A.19) it is given by

$$\langle \mathbf{a}, \mathbf{b} \wedge \mathbf{c}\rangle = \det(\mathbf{a}, \mathbf{b}, \mathbf{c}) = \begin{vmatrix} a_1 & a_2 & a_3 \\ b_1 & b_2 & b_3 \\ c_1 & c_2 & c_3 \end{vmatrix}. \tag{A.23}$$

From a well-known property of determinants, the vectors can be cyclically rotated in the mixed product, so that

$$\langle \mathbf{a}, \mathbf{b} \wedge \mathbf{c}\rangle = \langle \mathbf{b}, \mathbf{c} \wedge \mathbf{a}\rangle = \langle \mathbf{c}, \mathbf{a} \wedge \mathbf{b}\rangle. \tag{A.24}$$

A.1.1 Change of Reference Systems

The mathematical treatment of the change of reference system is of obvious importance in physics. Here, we follow the elegant geometric approach adopted by Arnold (1978).

We consider an inertial system S_0 and a second system S', so that in general

$$\mathbf{x} = \mathbf{R}_0 + \mathcal{R}\mathbf{x}', \qquad \mathbf{x}' = \mathcal{R}^{\mathrm{T}}(\mathbf{x} - \mathbf{R}_0), \tag{A.25}$$

where $\mathbf{R}_0(t)$ is the instantaneous position of the origin of S' in S_0, $\mathbf{x}$ is the vector position in S_0, and $\mathbf{x}'$ is the vector position in S'. In general, $\mathcal{R}(t)$ is a time-dependent orthogonal matrix describing the instantaneous orientation of S' with respect to S_0, and so $\mathcal{R}\mathcal{R}^{\mathrm{T}} = \mathcal{R}^{\mathrm{T}}\mathcal{R} = I$. The time derivative of this identity shows that $\mathcal{R}^{\mathrm{T}}\dot{\mathcal{R}}$ is an antisymmetric operator, and from Eq. (A.19)

$$\mathcal{R}^{\mathrm{T}}\dot{\mathcal{R}} = \begin{pmatrix} 0 & -\omega_3 & \omega_2 \\ \omega_3 & 0 & -\omega_1 \\ -\omega_2 & \omega_1 & 0 \end{pmatrix}, \quad (\mathcal{R}^{\mathrm{T}}\dot{\mathcal{R}})\mathbf{x}' = \boldsymbol{\omega} \wedge \mathbf{x}'. \tag{A.26}$$

$\boldsymbol{\omega}$ is the instantaneous angular velocity of S' with respect to S_0, as seen in S', and it is a simple exercise (do it!) to show that the angular velocity of S' observed in S_0 is given by

$$\boldsymbol{\Omega} = \mathcal{R}\boldsymbol{\omega}. \tag{A.27}$$

We are now in a position to derive the transformation formulae for the velocity in the two-coordinate systems: from Eqs. (A.25) and (A.26)

$$\dot{\mathbf{x}} = \mathbf{V}_0 + \mathcal{R}(\boldsymbol{\omega} \wedge \mathbf{x}' + \dot{\mathbf{x}}'), \qquad \dot{\mathbf{x}}' = \mathcal{R}^{\mathrm{T}}(\dot{\mathbf{x}} - \mathbf{V}_0) - \boldsymbol{\omega} \wedge \mathbf{x}', \tag{A.28}$$

where $\mathbf{V}_0 = \dot{\mathbf{R}}_0$ is the instantaneous velocity of the origin of S' as seen from S_0. The transformation of accelerations is obtained from the time derivative of Eq. (A.28).

$$\begin{cases} \ddot{\mathbf{x}} = \mathbf{A}_0 + \mathcal{R}\left[\ddot{\mathbf{x}}' + \dot{\boldsymbol{\omega}} \wedge \mathbf{x}' + 2\boldsymbol{\omega} \wedge \dot{\mathbf{x}}' + \boldsymbol{\omega} \wedge (\boldsymbol{\omega} \wedge \mathbf{x}')\right], \\ \ddot{\mathbf{x}}' = \mathcal{R}^{\mathrm{T}}(\ddot{\mathbf{x}} - \mathbf{A}_0) - \dot{\boldsymbol{\omega}} \wedge \mathbf{x}' - 2\boldsymbol{\omega} \wedge \dot{\mathbf{x}}' - \boldsymbol{\omega} \wedge (\boldsymbol{\omega} \wedge \mathbf{x}'), \end{cases} \tag{A.29}$$

where $\mathbf{A}_0 = \dot{\mathbf{V}}_0$ is the instantaneous acceleration of S' with respect to S_0. In the second identity in Eq. (A.29), the last three terms on the right-hand side are the *Euler's force*, the *Coriolis' force*, and the *centrifugal force*, respectively. Of course, in the inertial system S_0, Newton's Second Law of Dynamics dictates that for a point mass $\ddot{\mathbf{x}} = \mathbf{g}$, where $\mathbf{g}$ is the the sum of all of the forces (per unit mass) acting at $\mathbf{x}$.

Three comments are in order here. The first is that $\mathbf{g}' = \mathcal{R}^{\mathrm{T}}\mathbf{g}$ is the total (physical) force per unit mass expressed in S' so that, in the particular case of physical forces derived from a potential $\phi(\mathbf{x})$ in S_0, it follows that

$$\mathbf{g}' = -\mathcal{R}^{\mathrm{T}}\nabla_{\mathbf{x}}\phi(\mathbf{x})|_{\mathbf{x}=\mathbf{R}_0+\mathcal{R}\mathbf{x}'} = -\nabla_{\mathbf{x}'}\phi'(\mathbf{x}'), \qquad \phi'(\mathbf{x}') \equiv \phi(\mathbf{R}_0 + \mathcal{R}\mathbf{x}'), \tag{A.30}$$

as can be proved by direct evaluation; notice that ϕ' can depend explicitly on time through $\mathbf{R}_0$ even if ϕ is time independent. The second comment relates to angular accelerations: it is an useful exercise (do it!) to prove from Eq. (A.27) the not-so-obvious transformation identity

$$\dot{\boldsymbol{\Omega}} = \mathcal{R}\dot{\boldsymbol{\omega}}, \tag{A.31}$$

and to show that higher-order time derivatives of the angular velocity in general do not obey similar identities. The third and final comment relates to the centrifugal force term that from Eqs. (A.14), (A.22), and (A.24) obeys the following easily proved identities:

$$\begin{cases} \boldsymbol{\omega} \wedge (\boldsymbol{\omega} \wedge \mathbf{x}') = \langle \boldsymbol{\omega}, \mathbf{x}' \rangle \boldsymbol{\omega} - \|\boldsymbol{\omega}\|^2 \mathbf{x}' = -\nabla_{\mathbf{x}'} \dfrac{\|\boldsymbol{\omega} \wedge \mathbf{x}'\|^2}{2}, \\ \|\boldsymbol{\omega} \wedge \mathbf{x}'\|^2 = \|\boldsymbol{\omega}\|^2 \, \|\mathbf{x}'\|^2 - \langle \boldsymbol{\omega}, \mathbf{x}' \rangle^2, \end{cases} \tag{A.32}$$

so that the centrifugal force can be derived from a *centrifugal potential.*

With the aid of Eqs. (A.25), (A.28), and (A.29), we are now in a position to derive the transformation formulae for all of the important mechanical quantities of particles and systems of particles, such as kinetic energy, momentum, angular momentum, torque, and so on. In the following, we will only discuss a couple of applications of great relevance, and the student is invited to work out as an instructive exercise the remaining cases and then specialize the results for $S' = S_{\mathrm{CM}}$ (i.e., to assume[3] $\mathbf{R}_0(t) = \mathbf{R}_{\mathrm{CM}}(t)$).

[3] See Chapter 6 for the even more special case of pure translation (i.e., $\mathbf{R}_0 = \mathbf{R}_{\mathrm{CM}}$ and $\mathcal{R} = I$).

A.1.1.1 Angular Momentum, Torque, and Kinetic Energy Transformation

From the definition of angular momentum in a generic inertial reference system S_0 with reduction center $\mathbf{x}_0$ for a system of masses m_i and total mass $M = \sum_i m_i$, from Eqs. (A.21) and (A.22) we obtain the well-known identity

$$\mathbf{J} = \sum_i (\mathbf{x}_i - \mathbf{x}_0) \wedge m_i \mathbf{v}_i = (\mathbf{R}_{\rm CM} - \mathbf{x}_0) \wedge M\mathbf{V}_{\rm CM} + \mathcal{R}(\mathbf{J}' + \Im' \boldsymbol{\omega}), \tag{A.33}$$

where $\mathbf{J}' = \sum_i \mathbf{x}_i' \wedge m_i \mathbf{v}_i'$ is the system's angular momentum in the barycentric reference system $S_{\rm CM}$ and $\Im'$ is the barycentric (discrete) version of the inertia tensor in Eq. (2.89). Notice that the last term in Eq. (A.33) can be rewritten as

$$\mathcal{R}\Im'\boldsymbol{\omega} = \mathcal{R}\Im'\mathcal{R}^{\rm T}\boldsymbol{\Omega} = \Im\boldsymbol{\Omega}, \tag{A.34}$$

where $\Im$ is the (in general time-dependent) barycentric inertia tensor of the system as seen from S_0 and $\boldsymbol{\Omega}$ is the angular velocity of $S_{\rm CM}$ as seen from S_0. A particularly important case is obtained when the system of particles is a *rigid body* and $S_{\rm CM}$ is fixed in it, so that $\mathbf{J}' = 0$ and $\boldsymbol{\Omega}$ is the rigid body angular velocity; in this case

$$\mathbf{J} = (\mathbf{R}_{\rm CM} - \mathbf{x}_0) \wedge M\mathbf{V}_{\rm CM} + \Im\boldsymbol{\Omega}. \tag{A.35}$$

Similarly for the torque

$$\mathbf{N} = \sum_i (\mathbf{x}_i - \mathbf{x}_0) \wedge \mathbf{F}_i = (\mathbf{R}_{\rm CM} - \mathbf{x}_0) \wedge M\mathbf{A}_{\rm CM} + \mathcal{R}\mathbf{N}', \tag{A.36}$$

where $\mathbf{F}_i$ is the total force acting on the particle i, $\mathbf{N}' = \sum_i \mathbf{x}_i' \wedge \mathbf{F}_i'$, and $\mathbf{F}_i' = \mathcal{R}^{\rm T}\mathbf{F}_i$.

Another important transformation formula is that of the system's kinetic energy, and following the same treatment adopted for the angular momentum we now get

$$T = \frac{1}{2}\sum_i \|\mathbf{v}_i\|^2 = M\frac{\|\mathbf{V}_{\rm CM}\|^2}{2} + T' + \langle \mathbf{J}', \boldsymbol{\omega}\rangle + \frac{\langle \Im'\boldsymbol{\omega}, \boldsymbol{\omega}\rangle}{2}, \tag{A.37}$$

where $T' = \sum_i \|\mathbf{v}_i'\|^2/2$ is the system's kinetic energy in $S_{\rm CM}$ and $\mathbf{J}'$ and $\Im'$ are the quantities introduced above; notice that the identity $\langle \Im'\boldsymbol{\omega}, \boldsymbol{\omega}\rangle = \langle \Im\boldsymbol{\Omega}, \boldsymbol{\Omega}\rangle$ holds. Again, if the system of particles is a rigid body and $S_{\rm CM}$ is fixed within it, then $T' = 0$ and $\mathbf{J}' = 0$, and one recovers the well-known identity

$$T = M\frac{\|\mathbf{V}_{\rm CM}\|^2}{2} + \frac{\langle \Im\boldsymbol{\Omega}, \boldsymbol{\Omega}\rangle}{2}. \tag{A.38}$$

A.1.1.2 The Jacobi Integral

We conclude this discussion with an important result concerning a conserved quantity along the orbit of *each* particle (the so-called *Jacobi integral*) in potentials rotating with *uniform* angular velocity around some *fixed* point; the relevance of this conservation law in the modeling of rotating stellar systems should be obvious, even though we will not pursue the subject further here (for applications, see, among others, Bertin 2014; Binney and Tremaine 2008; Chandrasekhar 1942; Ogorodnikov 1965; Szebehely 1967). With the previous assumptions, consider the case of uniform rotation (i.e., $\dot{\boldsymbol{\omega}} = 0$) of the system S' with a fixed origin $\mathbf{R}_0 = 0$, and the motion of a particle under the action of a force field derived from a potential $\phi(\mathbf{x})$ in S_0, so that the transformed potential $\phi'(\mathbf{x}')$ does not depend explicitly on time.[4] Under these assumptions, consider the second identity of Eq. (A.29), where the Euler's force vanishes because $\dot{\boldsymbol{\omega}} = 0$, the centrifugal force is expressed as in Eq. (A.32), and the physical forces are given by Eq. (A.30)

$$\ddot{\mathbf{x}}' = -\nabla_{\mathbf{x}'}\phi'_{\mathrm{R}}(\mathbf{x}') - 2\boldsymbol{\omega} \wedge \dot{\mathbf{x}}', \quad \phi'_{\mathrm{R}} = \phi'(\mathbf{x}') - \frac{\|\boldsymbol{\omega} \wedge \mathbf{x}'\|^2}{2}, \tag{A.39}$$

where $\phi'_{\mathrm{R}}(\mathbf{x}')$ is known in celestial mechanics as the *Roche potential*. Following the standard treatment, we now multiply both sides of the obtained identity by $\dot{\mathbf{x}}'$, and we use the fact that Coriolis' forces do not work. It follows that in S' the *Jacobi integral* of a test particle

$$C_{\mathrm{J}} = T' + m\phi'_{\mathrm{R}} = E - \langle \mathbf{J}, \boldsymbol{\Omega} \rangle \tag{A.40}$$

is conserved. In the first expression in Eq. (A.40), T' is the kinetic energy of the test mass in S'. Remarkably, as is shown by the last expression in Eq. (A.40), the conserved quantity C_{J} can *also* be written as a combination of the particle energy $E = T + m\phi$ and angular momentum $\mathbf{J}$ in S_0 (where in general these two quantities are *not* conserved!), as can be proved with a little algebraic work by using Eqs. (A.25) and (A.28). Of course, in S' we can introduce again the concept of zero-velocity surfaces (see Chapter 5), which in the present case are known as *Hill's surfaces*, and conclude that in S' the motion of the test particle is limited to the region(s)

$$\phi'_{\mathrm{R}}(\mathbf{x}') \leq \frac{C_{\mathrm{J}}}{m}, \tag{A.41}$$

for assigned values of the Jacobi constant, while the equilibrium points coincide with the critical points of the Roche potential. Therefore, for a given rotating potential (e.g., that obtained with the methods described in Chapter 2), it is possible to study the motion of tracers (e.g., gas flows) in rotating stellar systems such as the *bars* at the center of disk galaxies.

We cannot resist mentioning that in the special case of two particles in mutual circular orbit, the positions of the resulting five critical points of $\phi'_{\mathrm{R}}(\mathbf{x}')$ give the five *Lagrange libration points* of equilibrium ($L_{1,2,3}$ corresponding to the Euler collinear stationary solution and $L_{4,5}$ corresponding to the Lagrange equilateral stationary solution; see also Chapter 6 and Figure A.1). We also recall that $L_{1,2,3}$ are unstable saddle points for $\phi'_{\mathrm{R}}(\mathbf{x}')$, while $L_{4,5}$

[4] Consider, for example, the potential produced by two point masses in circular orbit around their center of mass in the restricted three-body problem, or the potential produced by a triaxial ellipsoid rotating around its center of mass with constant angular velocity parallel to one of its principal axes.

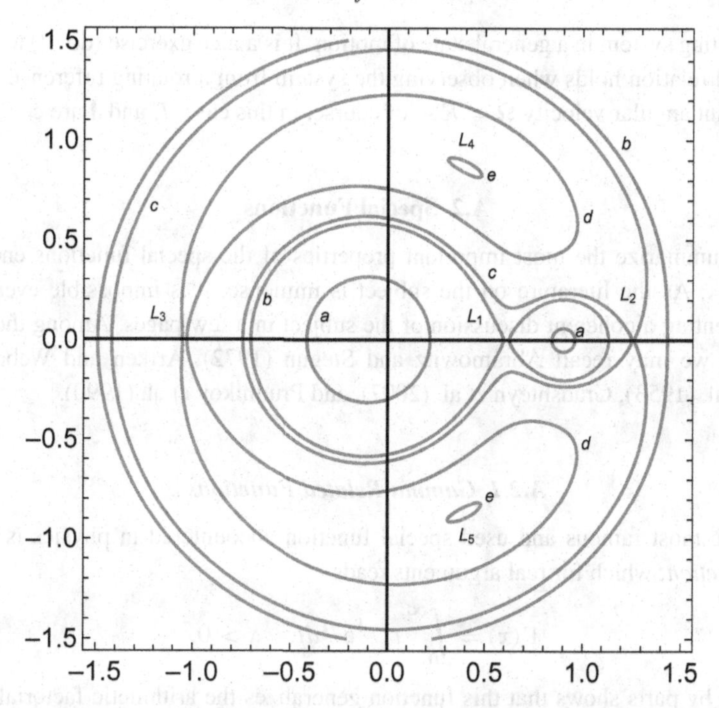

Figure A.1 Hill's surfaces for the restricted three-body problem for masses m_1 and m_2 on the x'-axis. All lengths are normalized to $d = x'_2 - x'_1$, masses to $M = m_1 + m_2$, and potentials to GM/d; the dimensionless mass ratios are $m_2/M = \lambda$ and $m_1/M = 1 - \lambda$, so that from Eq. (4.7) $x'_1/d = -\lambda$ and $x'_2/d = 1 - \lambda$, and finally $\omega^2 = GM/d^3$ from Eq. (4.31). In this figure, we adopted $\lambda = 1/10$, and the normalized values of the Jacobi constant along the zero-velocity Hill's curves are $a = -3$, $b = -1.8$, $c = -1.7335$, $d = -1.55$, and $e = -1.456$; the five Lagrange libration proints are indicated. Recall that for fixed C_J the motion of a test particle is possible only where $\phi'_R \leq C_J/m$ (e.g., see the remarkably clear figure 8.2 in Danby 1962).

are maximum points for the Roche potential, even though they can be stabilized (for suitable values of the mass ratios) by the effect of Coriolis' forces. For detailed discussions of these fascinating aspects of celestial mechanics, the student is invited to consult Binney and Tremaine (2008), and in particular specialized treatises such as Abraham and Marsden (1978), Binney and Tremaine (2008), Boccaletti and Pucacco (1996), Danby (1962), Hagihara (1970), Meyer and Hall (1992), Roy (2005), Szebehely (1967), and Valtonen and Karttunen (2006).

We conclude this section by noticing that Eq. (A.40) holds for a single particle in a time-independent external potential in the uniformly rotating reference system S'. A natural question arises as to whether something similar holds for an N-body system. Of course, a basic case is represented by a system simply described as the sum of N particles. In this case, the global Jacobi constant is just the sum of the Jacobi constant of each particle, where E and $\mathbf{J}$ are in general not conserved. A more interesting case is represented by a self-gravitating system in a general state of motion. It is a nice exercise (do it!) to show that an identical relation holds when observing the system from a rotating reference system S' with constant angular velocity $\boldsymbol{\Omega} = \mathcal{R}\boldsymbol{\omega}$; of course, in this case, E and $\mathbf{J}$ are conserved.

A.2 Special Functions

Here, we summarize the most important properties of the special functions encountered in this book. As the literature on the subject is immense, it is impossible even to think about presenting a coherent discussion of the subject in a few pages. Among the classical references, we may recall Abramowitz and Stegun (1972), Arfken and Weber (2005), Erdélyi et al. (1953), Gradshteyn et al. (2007), and Prudnikov et al. (1990).

A.2.1 Gamma-Related Functions

Perhaps the most famous and used special function encountered in physics is the *Euler gamma function*, which for real arguments reads

$$\Gamma(x) = \int_0^\infty t^{x-1} \mathrm{e}^{-t} dt, \quad x > 0. \tag{A.42}$$

Integration by parts shows that this function generalizes the arithmetic factorial function, with the recursive relation

$$\Gamma(x+1) = x\Gamma(x), \tag{A.43}$$

and for integers $n \geq 0$

$$n! = \Gamma(n+1), \quad (2n)!! = 2^n \Gamma(n+1), \quad (2n+1)!! = \frac{2^{n+1}\Gamma(n+3/2)}{\sqrt{\pi}}. \tag{A.44}$$

Special values are

$$\Gamma(1) = \Gamma(2) = 1, \quad \Gamma\left(\frac{1}{2}\right) = \sqrt{\pi}. \tag{A.45}$$

Three important relations obeyed by the Gamma function are the duplication formula and the reflection[5] formula

$$\frac{\Gamma(x)}{\Gamma(x/2)} = \frac{2^{x-1}}{\sqrt{\pi}}\Gamma\left(\frac{x+1}{2}\right), \quad \Gamma(1-x)\Gamma(x) = \frac{\pi}{\sin \pi x}, \tag{A.46}$$

and finally the *Stirling expansion*

$$\Gamma(x) \sim \sqrt{2\pi}\, x^{x-1/2}\, \mathrm{e}^{-x}, \quad x \to \infty. \tag{A.47}$$

[5] The Gamma function is not defined for negative integers; notice, however, the important alternative expression of the reflection formula that is obtained from Eq. (A.46) for $x = 1/2 - y$, and the resulting identity in the special case of integer y.

The *binomial coefficients* appearing in the well-known identity

$$(a+b)^n = \sum_{k=0}^{n} \binom{n}{k} a^{n-k} b^k, \quad \binom{n}{k} = \frac{n!}{k!\,(n-k)!}, \tag{A.48}$$

can be generalized to real variables by using the Gamma function as

$$\binom{x}{y} = \frac{\Gamma(x+1)}{\Gamma(x-y+1)\Gamma(y+1)}, \tag{A.49}$$

and in particular Newton's binomial theorem can be formally written as

$$(1+z)^\alpha = \sum_{k=0}^{\infty} \binom{\alpha}{k} z^k = 1 + \alpha z + \frac{\alpha(\alpha-1)}{2!} z^2 + \ldots \quad |z| < 1, \tag{A.50}$$

Two functions are associated with the Gamma function and are quite often encountered in problems of stellar dynamics: the *incomplete* left and right Gamma functions

$$\gamma(a,x) = \int_0^x t^{a-1} \mathrm{e}^{-t} dt, \quad \Gamma(a,x) = \int_x^\infty t^{a-1} \mathrm{e}^{-t} dt. \tag{A.51}$$

Of course, $a > 0$ for the existence of $\gamma(a,x)$, and in this case $\gamma(a,x) + \Gamma(a,x) = \Gamma(a)$. They obey the important recursion relations of easy proof

$$\gamma(a+1,x) = a\gamma(a,x) - x^a \mathrm{e}^{-x}, \quad \Gamma(a+1,x) = a\Gamma(a,x) + x^a \mathrm{e}^{-x}. \tag{A.52}$$

The *Euler Beta function* is defined for real arguments $x > 0$ and $y > 0$ as

$$\mathrm{B}(x,y) = \int_0^1 t^{x-1}(1-t)^{y-1} dt = \int_0^{\pi/2} (\sin z)^{2x-1} (\cos z)^{2y-1} dz. \tag{A.53}$$

From the first integral, it is a trivial exercise to show that $\mathrm{B}(x,y) = \mathrm{B}(y,x)$, while the equivalence with the trigonometric integral can be obtained with the change of variable $t = \sin^2 z$. A not-so-trivial result is the link with the Gamma function

$$\mathrm{B}(x,y) = \frac{\Gamma(x)\Gamma(y)}{\Gamma(x+y)}, \tag{A.54}$$

and, as for the Gamma function, an incomplete case is also considered for the Beta function

$$\mathrm{B}(p,q;x) = \int_0^x t^{p-1}(1-t)^{q-1} dt, \tag{A.55}$$

where obviously $\mathrm{B}(p,q;1) = \mathrm{B}(p,q)$, and another common notation is $\mathrm{B}(p,q;x) = \mathrm{B}_x(p,q)$. Note that for generic values of x, $\mathrm{B}(p,q;x)$ *is not* symmetric in the exchange between p and q; the expansion

$$\mathrm{B}(p,q;x) \sim \frac{x^p}{p} + \frac{(1-q)x^{1+p}}{1+p}, \quad x \to 0, \tag{A.56}$$

is of frequent use.

The *error function*

$$\mathrm{Erf}(x) = \frac{2}{\sqrt{\pi}} \int_0^x \mathrm{e}^{-t^2} dt = \frac{\gamma(x^2, 1/2)}{\sqrt{\pi}}, \tag{A.57}$$

is a special case of the right incomplete Gamma function, where $\mathrm{Erf}(\infty) = 1$ from Eq. (A.45). Strictly related to the error function are the two *Dawson functions*

$$\mathrm{F}_+(x) = \mathrm{e}^{-x^2} \int_0^x \mathrm{e}^{t^2} dt, \quad \mathrm{F}_-(x) = \mathrm{e}^{x^2} \int_0^x \mathrm{e}^{-t^2} dt. \tag{A.58}$$

Finally, we mention a few other special functions sometimes encountered in stellar dynamics: the *exponential integral*

$$\mathrm{E}_k(z) = \int_1^\infty t^{-k} e^{-tz} dt = z^{k-1}\Gamma(1-k, z), \tag{A.59}$$

$$\mathrm{E}_k(z) = \frac{\mathrm{e}^{-z}}{k-1} - \frac{z\mathrm{E}_{k-1}(z)}{k-1} = \frac{\mathrm{e}^{-z}}{z} - \frac{k\mathrm{E}_{k+1}(z)}{z}, \quad \frac{d\mathrm{E}_k(z)}{dz} = -\mathrm{E}_{k-1}(z) \tag{A.60}$$

(with the related function Ei) and the *Eulerian logarithmic integral*, *sine integral*, *cosine integral*, and the *polylogarithm*

$$\mathrm{Li}(z) = \int_2^z \frac{dt}{\ln t}, \quad \mathrm{Si}(z) = \int_0^z \frac{\sin t}{t} dt, \quad \mathrm{Ci}(z) = -\int_z^\infty \frac{\cos t}{t} dt, \quad \mathrm{Li}_s(z) = \sum_{k=1}^\infty \frac{z^k}{k^s} \tag{A.61}$$

(with the related functions li, si, and ci). As usual, the polylogarithm is given as a series over the unitary disk (see in particular Lewin 1986; Prudnikov et al. 1990).

A.2.2 Elliptic Integrals and Hypergeometric Functions

Elliptic integrals of the first, second, and third kinds in the Legendre–Jacobi form are given by

$$F(\varphi, k) = \int_0^\varphi \frac{d\vartheta}{\sqrt{1 - k^2 \sin^2 \vartheta}} = \int_0^{\sin\varphi} \frac{dt}{\sqrt{(1-t^2)(1-k^2t^2)}}, \tag{A.62}$$

$$E(\varphi, k) = \int_0^\varphi \sqrt{1 - k^2 \sin^2 \vartheta}\, d\vartheta = \int_0^{\sin\varphi} \sqrt{\frac{1-k^2t^2}{1-t^2}} dt, \tag{A.63}$$

$$\Pi(\varphi, n, k) = \int_0^\varphi \frac{d\vartheta}{(1 - n\sin^2\vartheta)\sqrt{1 - k^2\sin^2\vartheta}} = \int_0^{\sin\varphi} \frac{dt}{(1-nt^2)\sqrt{(1-t^2)(1-k^2t^2)}}, \tag{A.64}$$

where k is the *modulus* and $-\pi/2 \leq \varphi \leq \pi/2$. The *complete* elliptic integrals are also frequently encountered in applications, and they are defined as

$$\mathbf{K}(k) \equiv F(\pi/2, k), \quad \mathbf{E}(k) \equiv E(\pi/2, k), \quad \Pi(n, k) \equiv \Pi(\pi/2, n, k), \tag{A.65}$$

so that $\mathbf{K}(0) = \mathbf{E}(0) = \pi/2$, while $\mathbf{K}(k) \sim -\ln\sqrt{1-k}$ and $\mathbf{E}(k) \sim 1$ for $k \to 1^-$, and

$$\frac{d\mathbf{K}(k)}{dk} = \frac{\mathbf{E}(k)}{k(1-k^2)} - \frac{\mathbf{K}(k)}{k}, \quad \frac{d\mathbf{E}(k)}{dk} = \frac{\mathbf{E}(k)}{k} - \frac{\mathbf{K}(k)}{k}. \tag{A.66}$$

Note that the (standard) definition above also leads to real values of the integrals for purely imaginary values of k (i.e., when $k^2 < 0$). We stress that different definitions may be found in the literature (and in computer algebra systems) for the parameters k and n, so special care must be exercised in applications. Elliptic integrals satisfy an enormous number of quite amazing identities, and the student should be prepared to spend some time with tables (e.g., see Byrd and Friedman 1971; Gradshteyn et al. 2007; Prudnikov et al. 1990).

Elliptic functions were discovered and studied in depth (among others) by Abel (1802–1829), Jacobi (1804–1851), and Weierstrass (1815–1897). They play a fundamental role in several area of physics and in particular in mechanics; perhaps the simplest application is the motion of a pendulum beyond the harmonic approximation (e.g., see Lichtenberg and Lieberman 1992; Meyer and Hall 1992). Here, we simply recall that for a given value of the variable u, the angle φ obtained by inverting $u = F(\varphi, k)$ leads to the definition of (Jacobian) *amplitude*, *sine-amplitude*, *cosine-amplitude* and *δ-amplitude*, respectively

$$\varphi = \mathrm{am}\, u, \quad \sin\varphi = \mathrm{sn}\, u, \quad \cos\varphi = \mathrm{cn}\, u, \quad \sqrt{1-k^2\sin^2\varphi} = \mathrm{dn}\, u. \tag{A.67}$$

Note that for $k = 0$ the amplitudes reduce to identity and the elliptic functions reduce to trigonometric functions.

Among the special functions, the Gauss (1777–1855) *hypergeometric functions* occupy a preeminent role. They were introduced as the power series with convergence radius 1

$${}_2\mathrm{F}_1(a,b;c;z) = 1 + \frac{ab}{c}z + \frac{a(a+1)b(b+1)}{c(c+1)}\frac{z^2}{2!} + \ldots, \quad c \neq 0, -1, -2, \ldots, \tag{A.68}$$

${}_2\mathrm{F}_1(a,b;c;0) = 0$, and of course ${}_2\mathrm{F}_1(a,b;c;z) = {}_2\mathrm{F}_1(b,a;c;z)$. One of the most used integral representations is

$${}_2\mathrm{F}_1(a,b;c;z) = \frac{\Gamma(c)}{\Gamma(b)\Gamma(c-b)}\int_0^1 t^{b-1}(1-t)^{c-b-1}(1-zt)^{-a}dt, \quad c > b > 0. \tag{A.69}$$

These functions are solutions of the extremely important second-order differential equation

$$z(1-z)\frac{d^2y}{dz^2} + [c-(a+b+1)z]\frac{dy}{dz} - aby = 0, \tag{A.70}$$

whose regular (fixed) singular points are at $z = 0, 1, \infty$ (e.g., see Bender and Orszag 1978; Ince 1927; Lomen and Mark 1988). As is well known, a large part of elementary and higher transcendental functions can be expressed by using the ${}_2\mathrm{F}_1$ functions for special choices of the parameters, and an impressive number of identities, transformations, recursion relations, and analytical prolongements are knwon (e.g., see chapter 9 in Gradshteyn et al. 2007). Finally, the ${}_2\mathrm{F}_1$ functions are generalized to the ${}_pF_q$ functions, and further to the *Meijer* G-*functions*, and to the even more general family of *Fox* H-*functions* (e.g., Mathai et al. 2009; Prudnikov et al. 1990).

A.2.3 Legendre Functions

As we have seen in Chapter 2, the *Legendre functions* arise naturally when separating the three-dimensional Laplace operator in spherical coordinates (e.g., see Binney and Tremaine 2008; Jackson 1998). In particular, the second-order, linear differential equation relative to the angle ϑ, known as the Legendre (1752–1833) differential equation, is given in its more general form by

$$\frac{1}{\sin\vartheta}\frac{d}{\partial\vartheta}\left(\sin\vartheta\frac{dV}{d\vartheta}\right)+\left[\lambda(\lambda+1)-\frac{\mu^2}{\sin^2\vartheta}\right]V=0,\quad 0\leq\vartheta\leq\pi, \tag{A.71}$$

where λ and μ are parameters that can assume complex values. Equation (A.71) is also encountered in algebraic form, obtained from the substitution $x=\cos\vartheta$

$$\frac{d}{dx}\left[(1-x^2)\frac{dV}{dx}\right]+\left[\lambda(\lambda+1)-\frac{\mu^2}{1-x^2}\right]V=0,\quad -1\leq x\leq 1. \tag{A.72}$$

The literature on the properties of the Legendre equation is immense (e.g., Arfken and Weber 2005; Binney and Tremaine 2008; Erdélyi et al. 1953; Gradshteyn et al. 2007; Ince 1927; Jackson 1998; Lomen and Mark 1988; Prudnikov et al. 1990), so here we only recall what is strictly necessary. Being a linear second-order equation, there are two independent solutions, the P_λ^μ and Q_λ^μ *Legendre associate functions*, and these solutions are special cases of the Gauss hypergeometric function; in some treatise, $\mathrm{P}_\lambda^0\equiv\mathrm{P}_\lambda$ and $\mathrm{Q}_\lambda^0\equiv\mathrm{Q}_\lambda$ are simply named Legendre functions. The Fuchs (1833–1902) theorem proves that there are two *regular singular points* at $x=\pm1$ (i.e., geometrically) along the z-axis. Moreover, the Frobenius (1849–1917) method shows that $\mathrm{Q}_\lambda^\mu(\pm1)$ are singular independently of the values of λ and μ, but P_λ^μ are instead regular at $x=\pm1$ for the special case $\lambda\equiv l=0,1,2,\ldots$, non-negative integer, and $\mu\equiv m$ integer, with $-l\leq m\leq l$. From now on, we restrict our discussion to P_l^m functions (sometimes named *Legendre polynomials*, even though they are true polynomials only for even values of m, with $|m|\leq l$). Remarkably, in the separation problem discussed in Chapter 2, the univocity of the azimuthal function over $0\leq\varphi<2\pi$ selects $\mu=m$, and so with the orientation of the z-axis being arbitrary, only the P_l^m functions with $|m|\leq l$ can be retained in order to describe physical solutions over all of the space. Several recursive formulae are known, and here we report the useful *Rodrigues* identity

$$\mathrm{P}_l^m(x)=(-1)^m(1-x^2)^{\frac{m}{2}}\frac{d^m\mathrm{P}_l(x)}{dx^m}=\frac{(-1)^m}{2^l l!}(1-x^2)^{\frac{m}{2}}\frac{d^{l+m}}{dx^{l+m}}(x^2-1)^l, \tag{A.73}$$

where the first identity holds for $m=0,1,2,\ldots,l$, and the second also allows us to determine the P_l^m functions for $m=-l,-l+1,-l+2,\ldots,-1$. Note that $\mathrm{P}_l^m(x)=0$ for $m=l+1,l+2,\ldots$, and that Eq. (A.73), evaluated for $m=0$, gives the *Legendre polynomials* $\mathrm{P}_l(x)\equiv\mathrm{P}_l^0(x)$; moreover, it can be proved that

$$\mathrm{P}_l^{-m}(x)=(-1)^m\frac{(l-m)!}{(l+m)!}\mathrm{P}_l^m(x),\quad -l\leq m\leq l. \tag{A.74}$$

We note here (as it is quite often source of confusion) that $\mathrm{P}_l^m(x) \neq 0$ for $m = -l - 1$, $-l - 2, \ldots$, as can be proved from eq. (8.752.2) of Gradshteyn et al. (2007), and iterated integral representation extends Eq. (A.73) to values of $m < -l$.

It is not suprising that the Legendre functions obey an impressive number of recurrence relations, and here we simply recall that

$$(2l+1)x\mathrm{P}_l^m(x) = (l-m+1)P_{l+1}^m(x) + (m+l)P_{l-1}^m(x), \tag{A.75}$$

$$\mathrm{P}_l^m(-x) = (-1)^{l+m}\mathrm{P}_l^m(x). \tag{A.76}$$

In particular

$$\mathrm{P}_l^m(0) = \begin{cases} 0, & l+m = \text{odd}, \\ (-1)^{\frac{l-m}{2}} \dfrac{(l+m-1)!!}{(l-m)!!}, & l+m = \text{even}, \end{cases} \tag{A.77}$$

and so from Eqs. (A.44)–(A.46)

$$\mathrm{P}_{2l+1}(0) = 0, \quad \mathrm{P}_{2l}(0) = \frac{(-1)^l\Gamma(l+1/2)}{\sqrt{\pi}\Gamma(l+1)} = \frac{(-1)^l(2l)!}{2^{2l}(l!)^2}, \quad \mathrm{P}_0(0) = 1. \tag{A.78}$$

Finally, $\mathrm{P}_l^m(\pm 1) = 0$ for $m \neq 0$, and

$$\mathrm{P}_l(\pm 1) = (\pm 1)^l. \tag{A.79}$$

It is a fundamental result that Eq. (A.72) obeys the *Sturm–Liouville* theorem[6] at fixed values of m, so that P_l^m at fixed m are a family of *orthogonal* functions of weight 1, and from Eq. (A.73) it can be proved that

$$\int_{-1}^{1} \mathrm{P}_l^m(x)P_{l'}^m(x)dx = \frac{2}{2l+1}\frac{(l+m)!}{(l-m)!}\delta_{ll'}, \quad l, l' \geq |m|, \tag{A.82}$$

so that P_l^m can be easily normalized to an orthonormal set. Some additional work proves that, at fixed m, the $\mathrm{P}_l^m(x)$ are a *complete* set of functions with respect to $l \geq |m|$, and so these functions can be used to expand a function $f(x)$ over the interval $-1 \leq x \leq 1$, or a function $f(\vartheta)$ over the interval $0 \leq \vartheta \leq \pi$, in a *Legendre series*. This property will be used

[6] We recall that a linear second-order ordinary differential equation is in the form of a Sturm–Liouville problem when it is given as

$$\frac{d}{dx}\left[p(x)\frac{dy}{dx}\right] + q(x)y = \lambda w(x)\, y, \tag{A.80}$$

where w is the so-called *weight* and λ is a parameter. Here, we restrict ourselves to the *real* Sturm–Liouville problem. We consider two values λ_1 and λ_2 and two solutions y_1 and y_2 over the interval (a, b). By multiplication of the equation for y_1 with the solution y_2 and integration by parts, interchanging the index in the resulting epression, and subtracting the two identities, one arrives at

$$(\lambda_1 - \lambda_2)\int_a^b y_1(x)y_2(x)w(x)dx = p(x)(y_2y_1' - y_1y_2')|_a^b, \tag{A.81}$$

so that if p vanishes at the integration interval and/or the *Wronskian* vanishes, the two solutions corresponding to different values of λ are orthogonal with weight w.

in Chapter 2 to construct, jointly with a *Fourier series*, the family of *spherical harmonics*, a fundamental tool in potential theory. Finally, again following the discussion in Chapter 2, the generating function of Legendre polynomials can be obtained by equating the Green function for a mass point on the z-axis with the multipole expansion for the same problem, proving that

$$\frac{1}{\sqrt{1-2xt+t^2}} = \sum_{l=0}^{\infty} \mathrm{P}_l(x)\, t^l. \tag{A.83}$$

An analogous formula holds when expanding the function $(1-2xt+t^2)^{-\alpha/2}$ arising in the study of power-law forces, and when the Legendre polynomials are substituted by the more general *Gegenbauer* polynomials $C_n^{\alpha}(x)$ (sometimes known as ultraspherical polynomials; e.g., see Hernquist and Ostriker 1992 for an application to potential theory).

A.2.4 Bessel Functions and Hankel Transforms

The *Bessel functions* arises naturally when separating the three-dimensional Laplace operator in cylindrical coordinates (e.g., see Binney and Tremaine 2008; Jackson 1998), as discussed in Chapter 2. The second-order, linear differential equation relative to radius R, known as the Bessel (1784–1846) differential equation, is given in its more general form by

$$\frac{1}{R}\frac{d}{dR}\left(R\frac{dU}{dR}\right) + \left(k^2 - \frac{\mu^2}{R^2}\right)U = 0, \tag{A.84}$$

where k and μ are parameters that can assume complex values. A remarkable property holds for Eq. (A.84): with the introduction of a scaled radius $x \equiv kR$, Eq. (A.84) becomes

$$\frac{1}{x}\frac{d}{dx}\left(x\frac{dU}{dx}\right) + \left(1 - \frac{\mu^2}{x^2}\right)U = 0, \quad 0 \leq x < \infty. \tag{A.85}$$

In particular, once Eq. (A.85) is solved in terms of $U_\mu(x)$, it is possibile to obtain immediately the solution of Eq. (A.84) as $U_\mu(kR)$, so that the parameter k is actually contained in the argument of the function. As for the Legendre functions, the literature on Bessel functions is immense (e.g., Arfken and Weber 2005; Binney and Tremaine 2008; Erdélyi et al. 1953; Gradshteyn et al. 2007; Ince 1927; Jackson 1998; Lomen and Mark 1988; Prudnikov et al. 1990; Watson 1966), and great attention should be given when consulting specific treatises and tables, because Bessel functions are often known by different names/symbols in different contexts. Here, we adopt the most used nomenclature for the two independent solutions of Eq. (A.85), and we indicate with J_μ and Y_μ the Bessel functions of the *first* and *second* types; needless to say, Bessel functions are special cases of the Gauss hypergeometric functions, and so they obey an impressive number of relevant identities. Concerning our problem originated by separation of variables for the Laplacian in cylindrical coordinates, the Fuchs and Frobenius theorems show that only J_μ functions are well behaved near the origin, and this is only for $\mu = m$ integer. Restricting ourselves to this family of functions, here we simply recall a few fundamental identities of frequent use

$$\mathrm{J}_{m-1}(x)+\mathrm{J}_{m+1}(x)=\frac{2m}{x}\mathrm{J}_m(x),\quad \mathrm{J}_{-m}(x)=(-1)^m\mathrm{J}_m(x), \tag{A.86}$$

and $\mathrm{J}_m(0)=0$ for integer $m\neq 0$, while $\mathrm{J}_0(0)=1$. Moreover

$$\frac{d\mathrm{J}_m(x)}{dx}=\frac{\mathrm{J}_{m-1}(x)-\mathrm{J}_{m+1}(x)}{2},\quad \frac{d\mathrm{J}_0(x)}{dx}=-\mathrm{J}_1(x), \tag{A.87}$$

where the last identity is easily proved from the general one. Finally, by using the Sturm–Liouvelle theorem, a comparison of Eqs. (A.80) and (A.84) shows that the functions $\mathrm{J}_m(kR)$ at fixed m are orthogonal with weight R for different k, and in fact the following *Hankel closure relation* can be proved:

$$\int_0^\infty R\mathrm{J}_m(aR)\mathrm{J}_m(bR)dR=\frac{\delta(a-b)}{a}. \tag{A.88}$$

The case where $a\neq b$ is simply a consequence of orthogonality, but the case where $a=b$ is not trivial, and it is not proved here. The closure relation allows us to introduce an important family of integral transforms, the so-called *Hankel transforms*, based on Bessel functions. In fact, under the assumptions that the closure relation holds and that all of the necessary convergence requirements are satisfied, the student is invited to "prove" by inverting the order of integration that the perfectly symmetric pairs of identities hold at fixed integer m

$$f(x)=\int_0^\infty k\mathrm{J}_m(kx)\hat{f}_m(k)dk,\quad \hat{f}_m(k)=\int_0^\infty x\mathrm{J}_m(kx)f(x)dx. \tag{A.89}$$

We conclude this section by recalling the family of *modified Bessel functions* of the first (I_m) and second (K_m) kinds, which can be obtained from the Bessel functions evaluated for purely imaginary arguments (e.g., see Arfken and Weber 2005) or as solutions of the modified Bessel differential equation

$$\frac{1}{x}\frac{d}{dx}\left(x\frac{dU}{dx}\right)-\left(1+\frac{\mu^2}{x^2}\right)U=0,\quad 0\leq x<\infty. \tag{A.90}$$

Some of the most relevant and used identities are the following:

$$\mathrm{I}_{m-1}(x)-\mathrm{I}_{m+1}(x)=\frac{2m}{x}\mathrm{I}_m(x),\quad \mathrm{I}_{-m}(x)=\mathrm{I}_m(x), \tag{A.91}$$

and $\mathrm{I}_m(0)=0$ for integer $m\neq 0$, while $\mathrm{I}_0(0)=1$. Moreover

$$\frac{d\mathrm{I}_m(x)}{dx}=\frac{\mathrm{I}_{m-1}(x)+\mathrm{I}_{m+1}(x)}{2},\quad \frac{d\mathrm{I}_0(x)}{dx}=\mathrm{I}_1(x). \tag{A.92}$$

Finally

$$\mathrm{K}_{m-1}(x)-\mathrm{K}_{m+1}(x)=-\frac{2m}{x}\mathrm{K}_m(x),\quad \mathrm{K}_{-m}(x)=\mathrm{K}_m(x), \tag{A.93}$$

$$\frac{d\mathrm{K}_m(x)}{dx}=-\frac{\mathrm{K}_{m-1}(x)+\mathrm{K}_{m+1}(x)}{2},\quad \frac{d\mathrm{K}_0(x)}{dx}=-\mathrm{K}_1(x). \tag{A.94}$$

A.2.5 Dirac δ-Distribution

As is well known, the so-called Dirac δ-function is *not* a function in a technical sense, but a *distribution*. The approach in this section is necessarily "phenomenological," but the student is reassured that the theory of distributions is rigorously founded (e.g., see Vladimirov 1979), and all of the results presented here can be formally proved. In $\Re^n$, by using Cartesian coordinates, and for any sufficiently well-behaved function $f(\mathbf{x})$, the δ can be introduced operationally as a mathematical object such that, for an open set Ω and a point $\mathbf{x}_0$

$$\int_{\Omega} f(\mathbf{x})\delta(\mathbf{x}-\mathbf{x}_0)d^n\mathbf{x} = \begin{cases} 0, & \mathbf{x}_0 \notin \Omega, \\ f(\mathbf{x}_0), & \mathbf{x}_0 \in \Omega. \end{cases} \tag{A.95}$$

Note that $\delta(\mathbf{x}-\mathbf{x}_0) = \delta(\mathbf{x}_0-\mathbf{x})$. The meaning of the Dirac δ-function when $\mathbf{x}_0$ belongs to the frontier of Ω would involve some discussion that we avoid here; it is sufficient to recall that in some case (but not always) the integral in Eq. (A.95) can be shown to evaluate to $f(\mathbf{x}_0)/2$ when $\mathbf{x}_0 \in \partial\Omega$. Three immediate consequences derive from Eq. (A.95): the first is obtained by considering $f = 1$, with the formal result being that the value of the integral of the δ-function equals[7] 1 if $\mathbf{x}_0 \in \Omega$ (independently of the specific position of $\mathbf{x}_0$), while it is 0 for external points. The second fact is that when used in physical applications, the δ-function is not a dimensionless object, but has units of the inverse of the n-dimensional volume. Third, if we pretend that the Dirac δ-function obeys the rules for multiple integrals, it is reasonable to expect (and in fact it can be proved) that in Cartesian coordinates

$$\delta(\mathbf{x}-\mathbf{x}_0) = \prod_{i=1}^{n} \delta(x_i - x_{0i}) \tag{A.96}$$

(i.e., the multidimensional δ-function can be factorized as the product of a one-dimensional δ-function, one for each coordinate). When working with the δ-function in curvilinear coordinates, which is a common situation in physics (see Appendix A.8), some care is needed. In fact, it is quite obvious that the defining property of the δ-function in Eq. (A.95) must be independent of the specific coordinates adopted to evaluate the integral, and so the transformation rule of the δ-function from Cartesian to curvilinear coordinates is

$$\delta(\mathbf{x}-\mathbf{x}_0) = \frac{\prod_{i=1}^{n} \delta(q_i - q_{0i})}{|J(\mathbf{q}_0)|}, \quad |J(\mathbf{q}_0)| = \prod_{i=1}^{n} h_i, \tag{A.97}$$

where $J(\mathbf{q})$ is the Jacobian of the coordinate transformation and $\mathbf{q}_0$ are the curvilinear coordinates of the regular point $\mathbf{x}_0$ (i.e., where $J \neq 0$). The last expression in Eq. (A.97) holds for *orthogonal* curvilinear coordinates, and the Lamé coefficients h_i are given in Eq. (A.140). Notice that in Eq. (A.97), the J can indifferently depend on $\mathbf{q}$ or $\mathbf{q}_0$, because

[7] Notice that this property clearly shows that the δ-function is *not* – as sometimes stated – a function equals to 0 everywhere except for a point $\mathbf{x}_0$ where its value is ∞; the volume integral of such an object would be 0 even for $\mathbf{x}_0 \in \Omega$.

when performing integrations J is in any case evaluated at $\mathbf{q}_0$. In particular, in the one-dimensional case

$$\delta[g(x) - y] = \frac{\delta(x - x_0)}{|g'(x_0)|}, \quad g(x_0) = y, \tag{A.98}$$

when x_0 is *not* a critical point for g.

A function strictly related to the δ-function is the Heaviside *step function*

$$\theta(x) = \begin{cases} 1, & x \geq 0, \\ 0, & x < 0, \end{cases} \tag{A.99}$$

where the value for argument 0 is somewhat a matter of definition. For the student, it is a nice exercise to show that, if allowed to perform standard integration by parts, then

$$\int_A^B \frac{d\theta(x-a)}{dx} f(x)dx \tag{A.100}$$

would produce for values of a inside/outside the integration interval, the same result obtained by using the δ-function; in other words, we could formally use

$$\delta(x - a) = \frac{d\theta(x-a)}{dx}. \tag{A.101}$$

This result is at the basis of one of the most useful methods for building the Green function for ordinary differential equations (ODEs; e.g., Bender and Orszag 1978; Jackson 1998; Lomen and Mark 1988; see also Chapter 2 and Appendix A.7).

A.3 Fourier Transforms and Series

The *Fourier* (1768–1830) direct and inverse transforms of a well-behaved function $f(\mathbf{x})$ defined over $\Re^n$ can be written as

$$\hat{f}(\mathbf{k}) = \int_{\Re^n} f(\mathbf{x}) \frac{e^{-i\langle \mathbf{k}, \mathbf{x} \rangle}}{(2\pi)^{\frac{n}{2}}} d^n\mathbf{x}, \quad f(\mathbf{x}) = \int_{\Re^n} \hat{f}(\mathbf{k}) \frac{e^{i\langle \mathbf{k}, \mathbf{x} \rangle}}{(2\pi)^{\frac{n}{2}}} d^n\mathbf{k}, \tag{A.102}$$

where $\mathbf{k} \in \Re^n$ is the so-called *wave vector*. Fourier trasforms are some of the most powerful tools of mathematical physics, and the literature dedicated to them is immense; a very clear introduction to the subject can be found in Arfken and Weber (2005), where other important integral transforms (e.g., the Hankel, Laplace and Mellin transforms, often encountered in stellar dynamics) are also presented. Here, we simply recall that the kernel function appearing in Eq. (A.102) is the spatial part of a plane wave $e^{i(\langle \mathbf{k}, \mathbf{x} \rangle - \omega t)}$ propagating perpendicularly to $\mathbf{k}$, with wavelength $\lambda = 2\pi/\|\mathbf{k}\|$, period $T = 2\pi/\omega$, and *phase velocity* $v_k = \lambda/T = \omega/\|\mathbf{k}\|$. The fact that these functions are eigenfunctions of the Laplace operator in Cartesian coordinates (see Chapter 2 regarding the construction of the Green function for the Laplace operator) is of fundamental importance not only for classical physics, but also in quantum mechanics, where (for example) the kinetic energy of a particle is represented in the nonrelativistic Schroedinger equation as the Laplacian of the wave function (for an illuminating discussion, see Messiah 1967). Mathematically, it can be proved that the Dirac δ-function in Eq. (A.96) can be Fourier transformed, and from a formal integration we have the beautiful integral representation

$$\hat{\delta}(\mathbf{k}) = \frac{e^{-i\langle \mathbf{k}, \mathbf{x}_0 \rangle}}{(2\pi)^{n/2}}, \quad \delta(\mathbf{x}-\mathbf{x}_0) = \frac{1}{(2\pi)^n} \int_{\Re^n} e^{i<\mathbf{k}, \mathbf{x}-\mathbf{x}_0>} d^n\mathbf{k}. \tag{A.103}$$

The student is invited to meditate on the amazing fact that the Dirac δ-function, admittedly the mathematical object nearest to the concept of a "point mass," can also be seen as the sum of infinitely extended waves!

Fourier series are mathematical objects that are strictly related to Fourier transforms. In the one-dimensional case, for any well-behaved function $f(\varphi)$ defined over the interval $0 \leq \varphi < 2\pi$, it is possible to show that, in analogy with Eq. (A.102)

$$f(\varphi) = \sum_{m=-\infty}^{\infty} \hat{f}_m \frac{e^{im\varphi}}{\sqrt{2\pi}}, \quad \hat{f}_m = \int_0^{2\pi} f(\varphi) \frac{e^{-im\varphi}}{\sqrt{2\pi}} d\varphi. \tag{A.104}$$

In practice, this fundamental result tells us that a "generic" function, defined over a finite interval, can be expressed by adding a sufficient number of periodic functions weighted with suitable coefficients; we recall that this family of functions arises naturally when separating variables (e.g., spherical and cylindrical) in order to determine solutions for the Laplace equation, and requiring uniqueness after a whole rotation of 2π around a given direction. In particular, it is possible to expand the Dirac δ-function as follows:

$$\hat{\delta}_m = \frac{e^{-im\varphi_0}}{\sqrt{2\pi}}, \quad \delta(\varphi - \varphi_0) = \frac{1}{2\pi} \sum_{m=-\infty}^{\infty} e^{im(\varphi-\varphi_0)}. \tag{A.105}$$

The mathematical and physical literature on Fourier series and their properties is enormous, and again the student is invited to consult, for example, Arfken and Weber (2005).

A.4 The Gauss and Stokes Theorems

Several excellent presentations of differential calculus in curvilinear coordinates exist. At the level of undergraduate and graduate students, remarkably clear books are Do Carmo (1976) and Edwards (1994), and for the more physically oriented readers, good references are Arfken and Weber (2005), Aris (1989), Becker (1982), Borisenko and Tarapov (1979), and Narashiman (1993). In any case, a reading of the beautiful presentation of the subject in volume 2 of Feynman et al. (1977) is highly recommended.

Two of the most far-reaching results of vector analysis are the Gauss (divergence) and Stokes (circuitation) theorems. These basic results are reported here in Cartesian coordinates and in Appendix A.8 in curvilinear (orthogonal) coordinates. We first introduce the concepts of *gradient*, *circuitation*, and *flux*.

Let $f(\mathbf{x})$ be a well-behaved function from $\Re^n$ to $\Re$, so that (loosely speaking) the identity $f(\mathbf{x}) = c$ determines, for a given value of c, an $n-1$ dimensional "surface" (variety) in $\Re^n$.

Moreover, let $\mathbf{x}(\lambda)$ be a regular *curve* on the surface (i.e., $f[\mathbf{x}(\lambda)] = c$, with $\mathbf{x}(0) = \mathbf{x}_0$). Then, in Cartesian coordinates we can introduce the mathematical object grad f from the obvious identity

$$0 = \frac{df}{d\lambda}\bigg|_{\lambda=0} = \left\langle \text{grad}\, f, \frac{d\mathbf{x}}{d\lambda} \right\rangle\bigg|_{\lambda=0}, \tag{A.106}$$

where adopting the convention of summing over repeated indices

$$\text{grad}\, f = \nabla f, \quad \nabla \equiv \mathbf{e}_i \frac{\partial}{\partial x_i}. \tag{A.107}$$

Being the vector $d\mathbf{x}/d\lambda$ for $\lambda = 0$ tangent to the surface at $\mathbf{x} = \mathbf{x}_0$, it follows that grad f is *perpendicular* to the surface at $\mathbf{x}_0$.

A vector field that can be derived from the gradient of a scalar function is called *exact*, and the scalar function is called a *potential*. The importance of exact fields in physics is immense because they describe, among other things, *conservative* force fields (in the time-independent case[8]), and *irrotational* fluids in simply connected regions.

Now let $\mathbf{g}(\mathbf{x})$ be a well-behaved vector field from $\Re^n$ to $\Re^n$, and let $\gamma(\lambda)$ be some regular closed curve in $\Re^n$ with $a \leq \lambda \leq b$ so that $\mathbf{x}(a) = \mathbf{x}(b)$. The *circuitation* of $\mathbf{g}$ along γ is naturally defined as

$$C(\mathbf{g}, \gamma) \equiv \int_\gamma \langle \mathbf{g}, d\mathbf{x} \rangle = \int_a^b \left\langle \mathbf{g}, \frac{d\mathbf{x}}{d\lambda} \right\rangle d\lambda. \tag{A.109}$$

It follows that $C(\mathbf{g}, \gamma) = 0$ for arbitrary closed curves if $\mathbf{g}$ is exact, and in turn this means that the value of a line integral of $\mathbf{g}$ between two arbitrary points is actually independent of the path adopted to connect the two points. As is well known, physically this means that the work (the line integral) done by a conservative (exact) force field $\mathbf{g}$ depends only on the initial and final positions. Moreover, it can be proved that a field defined over a *connected* region of $\Re^n$ (i.e., qualitatively speaking, a region of space so that two arbitrary points can be connected with a continuous path) *is exact if and only if the circuitation over all closed curves is zero*. Note that this important result does not require for the region to be *simply connected* (i.e., that all of the closed curves can be continuously shrinked to a point).

Finally, the *flux* of the well-behaved vector field $\mathbf{g}(\mathbf{x})$ over some regular $n-1$ dimensional surface of $\Re^n$ is defined as

$$F(\mathbf{g}, \Sigma) \equiv \int_\Sigma \langle \mathbf{g}, \mathbf{n} \rangle d^{n-1}\mathbf{x}, \tag{A.110}$$

[8] Some care is needed here. In fact, notice that a force field could be *exact but time dependent* (i.e., it could be that $\mathbf{g} = -\nabla\phi(\mathbf{x}, t)$). The most famous case is the *restricted three-body problem* expressed in the barycentric inertial (i.e., nonrotating) system: the force field produced by the two massive bodies is exact, but it is not conservative, and along the orbits of the test particle (of vanishingly small mass m)

$$E \equiv T + m\phi \quad \Rightarrow \frac{dE}{dt} = m\frac{\partial\phi}{\partial t}. \tag{A.108}$$

where $\mathbf{n}$ is the unitary normal to Σ at $\mathbf{x}$. In case of $\Re^3$, the surface integral can be calculated by considering the parametrization $\mathbf{x} = \mathbf{x}(u,v)$, and the normal and area of the surface element are given by

$$\mathbf{n} = \frac{\dfrac{\partial \mathbf{x}}{\partial u} \wedge \dfrac{\partial \mathbf{x}}{\partial v}}{\left\| \dfrac{\partial \mathbf{x}}{\partial u} \wedge \dfrac{\partial \mathbf{x}}{\partial v} \right\|}, \quad d^2\mathbf{x} = \left\| \frac{\partial \mathbf{x}}{\partial u} \wedge \frac{\partial \mathbf{x}}{\partial v} \right\| du dv = \sqrt{\left\| \frac{\partial \mathbf{x}}{\partial u} \right\|^2 \left\| \frac{\partial \mathbf{x}}{\partial v} \right\|^2 - \left\langle \frac{\partial \mathbf{x}}{\partial u}, \frac{\partial \mathbf{x}}{\partial v} \right\rangle^2} du dv, \tag{A.111}$$

where the last identity derives from Eqs. (A.5) and (A.14).

The Gauss (or divergence) theorem states that for any well-behaved vector field $\mathbf{g}$ in $\Re^n$ and any well-behaved region of space Ω bounded by the $n-1$ dimensional surface $\partial\Omega$ with outward normal $\mathbf{n}$ at position $\mathbf{x}$, the following striking identity holds:

$$\int_{\partial\Omega} \langle \mathbf{g}, \mathbf{n} \rangle d^{n-1}\mathbf{x} = \int_{\Omega} \operatorname{div} \mathbf{g}\, d^n \mathbf{x}. \tag{A.112}$$

In other words, the integral flux of a vector field equals the volume integral of a mathematical object called the *divergence* of the field. From the identity in Eq. (A.112), we obtain the *coordinate-free* expression of the divergence operator

$$\operatorname{div} \mathbf{g} \equiv \lim_{\Omega \to 0} \frac{1}{V(\Omega)} \int_{\partial\Omega} \langle \mathbf{g}, \mathbf{n} \rangle d^{n-1}\mathbf{x}, \tag{A.113}$$

where $V(\Omega)$ is the volume of the region Ω and the limit must exist independently of how $\Omega \to 0$ at $\mathbf{x}$. In particular, in Cartesian coordinates it can be proved that

$$\operatorname{div} \mathbf{g} = \frac{\partial g_i}{\partial x_i} = \langle \nabla, \mathbf{g} \rangle, \tag{A.114}$$

where the last notation derives from the formal identification of the operator ∇ in Eq. (A.107) with a vector (the student should be aware that such an identification, if applied mechanically, can produce terribly wrong results!). A vector field so that $\operatorname{div} \mathbf{g} = 0$ at all points in its definition (open) domain is called *solenoidal*.

Similarly to the Gauss theorem, the Stokes (or circuitation) theorem says that for any well-behaved vector field in $\Re^3$ and any well-behaved (simple) closed curve $\partial\Sigma$ with a boundary of an (orientable) open surface Σ, the identity

$$\int_{\partial\Sigma} \langle \mathbf{g}, d\mathbf{x} \rangle = \int_{\Sigma} \langle \operatorname{rot} \mathbf{g}, \mathbf{n} \rangle d^2\mathbf{x} \tag{A.115}$$

holds. In other words, the circuitation of a vector field over a closed curve equals the flux of a mathematical object called the rotor (or curl) of the vector field, evaluated over any surface with a boundary given by the closed curve. Recall that in the integral in Eq. (A.115), the integration along $\partial\Sigma$ is in such a sense that it is "seen counterclockwise" when "observed along" $\mathbf{n}$. In analogy with the divergence case, the component of $\operatorname{rot} \mathbf{g}$ along the direction $\mathbf{n}$ at point $\mathbf{x}$ can be obtained by considering

$$\langle \text{rot}\, \mathbf{g}, \mathbf{n} \rangle \equiv \lim_{\Sigma \to 0} \frac{1}{S(\Sigma)} \int_{\partial \Sigma} \langle \mathbf{g}, d\mathbf{x} \rangle, \tag{A.116}$$

provided the limit exists independently of how the surface element Σ with normal $\mathbf{n}$ and area $S(\Sigma)$ shrinks to 0 at $\mathbf{x}$, and in Cartesian coordinates

$$\text{rot}\, \mathbf{g} = \epsilon_{ijk} \mathbf{e}_i \frac{\partial g_j}{\partial x_k} = \nabla \wedge \mathbf{g}. \tag{A.117}$$

A vector field so that rot $\mathbf{g} = 0$ at all points in its definition (open) domain is called *closed* (or *irrotational*). A simple application of the Stokes theorem is the deduction of the *Green theorem* for functions in $\Re^2$ simply by considering a vector field $\mathbf{g} = (g_x(x, y), g_y(x, y), 0)$ and a regular closed curve in the plane $z = 0$.

The integral representation of the operators div and rot is called *coordinate-free* because the operators at the right-hand side are expressed in terms of integral relations, which are of course independent of the specific coordinates adopted. This property is of great importance not only because it allows us to obtain the expressions of the important differential operators in curvilinear coordinates (see Appendix A.8), but also because it makes apparent that an operator is independent of its representation, a concept that should be stressed to students.

From Eqs. (A.107), (A.114), and (A.117), it is a useful exercise to show by direct evaluation that

$$\text{rot (grad}\, f) = 0, \qquad \text{div (rot}\, \mathbf{g}) = 0, \tag{A.118}$$

but the student is encouraged to reach the same conclusions by using the coordinate-free expressions in Eqs. (A.113) and (A.116). The first (i.e., the fact that an exact field is closed) can be obtained by considering the circuitation around arbitrary small surface areas containing the point of interest. The second is from the divergence theorem applied to rot $\mathbf{g}$ over a small volume whose bounding surface is ideally represented by joining two parts $\partial\Omega_1$ and $\partial\Omega_2$ separated by the same closed curve γ. From the Stokes theorem, the total flux is 0, and by shrinking the volume to 0 the statement is proved. Here, we recall that the identities in Eq. (A.118) and the Gauss and Stokes theorems are just special cases of more general results obtained in the framework of *differential forms*.[9] A vector field is called *solenoidal* if div $\mathbf{g}$ over some region of space, *closed* (or *irrotational*) if rot $\mathbf{g} = 0$, and finally *exact* if it can be derived as the gradient of a scalar function (the *potential*). This nomenclature reflects the fundamental importance of these concepts in fluid dynamics, successively extended to Hamiltonian dynamics in phase space and to classical electrodynamics. The two results above are remarkable in the sense that they connect *local* and *global* properties. In particular, the Stokes theorem produces important consequences when applied to *closed* fields in *multiply connected* regions; the student should deduce that circuitation along paths including the nonconnected regions of a closed field leads to results

[9] A differential k-form is called *exact* if it is the external differential ∂ of a $(k-1)$ form and *closed* if its external differential is 0. A theorem from Poincaré states that an exact form f is necessarily closed (i.e., $\partial\partial f = 0$; e.g., Arnold 1978; Edwards 1994).

that are independent of the specific closed path adopted for the integration. The situation is analogous to the case of the residue theorem in complex analysis.

A fundamental second-order linear operator (elliptic) is the *Laplacian* of a scalar function

$$\Delta \equiv \text{div grad } f = \sum \frac{\partial^2 f}{\partial x_i^2}, \tag{A.119}$$

which is of central importance in physics, where the last expression holds in Cartesian coordinates. From the divergence theorem, we have

$$\int_\Omega \Delta f \, d^n\mathbf{x} = \int_{\partial\Omega} \langle \mathbf{n}, \text{grad } f \rangle d^{n-1}\mathbf{x} = \int_{\partial\Omega} \frac{\partial f}{\partial n} d^{n-1}\mathbf{x}, \tag{A.120}$$

where the last expression introduces the so-called *normal derivative* of the function f along the direction $\mathbf{n}$, with $\partial f/\partial n \equiv \langle \nabla f, \mathbf{n} \rangle$. From this expression, we can obtain the Laplacian as a limit in analogy with Eq. (A.113). A scalar field so that $\Delta f = 0$ at all points in its (open) definition domain is called *harmonic*. Therefore, a vector field that is both solenoidal and exact can be expressed as the gradient of a harmonic function.

The divergence theorem can also be used to obtain a very useful and remarkable identity as follows: let us consider the Gauss theorem in $\Re^n$ for the special field $\mathbf{g} = (0, \ldots, g, \ldots, 0)$, where only the i-th component is not 0. Therefore

$$\int_\Omega \frac{\partial g}{\partial x_i} d^n\mathbf{x} = \int_{\partial\Omega} g n_i d^{n-1}\mathbf{x}. \tag{A.121}$$

By multiplication with the basis versors $\mathbf{e}_i$ and with a suitable choice of function g, we can obtain several relevant coordinate-free identities that allow us to derive operators in generic coordinate systems from volume integration. In particular

$$\int_\Omega \text{grad } f \, d^n\mathbf{x} = \int_{\partial\Omega} \mathbf{n} f d^{n-1}\mathbf{x}, \quad \int_\Omega \text{rot } \mathbf{g} \, d^3\mathbf{x} = \int_{\partial\Omega} \mathbf{n} \wedge \mathbf{g} d^2\mathbf{x}, \tag{A.122}$$

and so the corresponding operators can be defined in general coordinate systems by considering the limit of previous identities as in Eq. (A.113). We recall the alternative way to derive Eq. (A.122): applying the divergence theorem to the two vector fields $\mathbf{a} f$ and $\mathbf{a} \wedge \mathbf{g}$, where $\mathbf{a}$ is a constant but otherwise arbitrary vector.

We conclude this section by mentioning that there exist several beautiful identities that can be constructed by using combinations of the differential operators introduced above (e.g., see Arfken and Weber 2005; Aris 1989; Jackson 1998), and the student should spend some time practicing and on inventing problems. Here, we simply recall

$$\text{rot}(\text{rot } \mathbf{g}) = \text{grad}(\text{div } \mathbf{g}) - \Delta \mathbf{g}, \tag{A.123}$$

which is of fundamental importance in electromagnetism, and

$$\text{grad}\langle \mathbf{a}, \mathbf{b} \rangle = \langle \mathbf{a}, \nabla \rangle \mathbf{b} + \langle \mathbf{b}, \nabla \rangle \mathbf{a} + \mathbf{a} \wedge \text{rot } \mathbf{b} + \mathbf{b} \wedge \text{rot } \mathbf{a}, \tag{A.124}$$

which is especially relevant for its use in fluid dynamics, when the two fields are taken to coincide with the fluid velocity **u**, allowing us to rewrite in a convenient form the advective part of the material derivative $\langle \mathbf{u}, \nabla \rangle \mathbf{u}$, and finally

$$\operatorname{rot}(\mathbf{a} \wedge \mathbf{b}) = \mathbf{a} \operatorname{div} \mathbf{b} - \mathbf{b} \operatorname{div} \mathbf{a} - \langle \mathbf{a}, \nabla \rangle \mathbf{b} + \langle \mathbf{b}, \nabla \rangle \mathbf{a}. \tag{A.125}$$

A.5 Green Identities

A fundamental identity, which we used to prove the remarkable unicity theorem for the Poisson equation in Chapter 2, is obtained very easily from the Gauss theorem. We start by considering the special vector field

$$\mathbf{u} = f \nabla g, \tag{A.126}$$

where f and g are two generic, well-behaved scalar functions from $\Re^n$ to $\Re$ of argument **x**. Taking the divergence of **u** and using the Gauss theorem, the *first Green (1793–1841) identity* follows:

$$\int_{\Omega} (f \Delta g + \langle \nabla f, \nabla g \rangle)\, d^n\mathbf{x} = \int_{\partial \Omega} f \frac{\partial g}{\partial n} d^{n-1}\mathbf{x}, \tag{A.127}$$

where $\partial \Omega$ is the surface bounding the arbitrary volume Ω and **n** is the outward normal (unit) vector to $\partial \Omega$. The *second Green identity* can be obtained from the first identity simply by exchanging f and g in Eq. (A.127), and subtracting

$$\int_{\Omega} (f \Delta g - g \Delta f) d^n\mathbf{x} = \int_{\partial \Omega} \left(f \frac{\partial g}{\partial n} - g \frac{\partial f}{\partial n} \right) d^{n-1}\mathbf{x}. \tag{A.128}$$

A third Green identity also holds, but it is not used here, and so it is not reported (e.g., see Jackson 1998).

A.6 The Helmholtz Decomposition Theorem

The following result (which we present in $\Re^3$ but can be generalized in n dimensions in the framework of differential forms) explains *why* in physics we are so interested in the behavior of vector fields under the actions of operator divergence and rotor (just think about the Maxwell equations!). In fact, the Helmholtz (1821–1894) decomposition theorem (e.g., see Arfken and Weber 2005) proves that a generic vector field defined over a simplly connected region of $\Re^3$ is fully (in the sense explained below) specified when its divergence and its rotor are known. For example, in fluid dynamics, this would be related to the knowledge of sources/sinks and vortices, and in electromagnetism to the specification of charges and currents. Technically, the theorem says that all well-behaved vector fields in $\Re^3$ with the same divergence and the same rotor are essentially the same, with the freedom (gauge) to add the gradient of a generic harmonic function. A proof is as follows: suppose we have a field **g** and we search for the most general form of fields $\mathbf{g}_*$ with the same rotor.

From the linearity of rot , the difference of the two fields is closed, and from the assumption that the domain is simply connected, it is also exact, i.e.,

$$\text{rot}\,(\mathbf{g}_* - \mathbf{g}) = 0 \quad \Rightarrow \quad \mathbf{g}_* = \mathbf{g} + \text{grad}\, f \tag{A.129}$$

for generic functions f. In practice, all of the fields with the same rotor as $\mathbf{g}$ are given in Eq. (A.129). We now require that the two fields have the same divergence, so that

$$0 = \text{div}\,(\mathbf{g}_* - \mathbf{g}) = \Delta f, \tag{A.130}$$

so that f is harmonic, and this proves the theorem. Of course, we could start by searching for all fields $\mathbf{g}_*$ with the same divergence of $\mathbf{g}$ and, following a similar approach, conclude that now $\mathbf{g}_* = \mathbf{g} + \text{rot}\,\mathbf{h}$, for arbitrary functions $\mathbf{h}$ (the so-called *vector potential*). We now ask for the same rotor, and we obtain $\text{rot}\,(\text{rot}\,\mathbf{h}) = 0$ over a simply connected region, so that $\text{rot}\,\mathbf{h} = \text{grad}\, f$ for some f, so that again we conclude that f is a harmonic function.

Now, the Green theorem of unicity allows us to restrict the Helmholtz theorem by proving that, in addition to divergence and curl, the field $\mathbf{g}_2$ is also required to coincide with $\mathbf{g}_1$ on the boundary of a volume, then $\mathbf{g}_2 = \mathbf{g}_1$ (i.e., the field is uniquely determined). In fact, with this additional requirement, we have that inside the volume $\Delta\phi = 0$ and on the boundary $\text{grad}\,\phi = 0$. But if $\text{grad}\,\phi = 0$, then $\partial\phi/\partial n = 0$. We now apply the first Green identity with $f = g = \phi$, so that we have a Neumann problem. Following a line analogous to the discussion in Chapter 2, Section 2.1.1, it follows immediately that $\text{grad}\,\phi = 0$ inside the volume, and this proves the theorem.

Note that the unicity result just proved allows us to decompose a given vector field in a unique way. In fact, consider the field $\mathbf{g}$ and compute $\text{div}\,\mathbf{g} = \rho$ and $\text{rot}\,\mathbf{g} = \boldsymbol{\omega}$. Now, it follows that the field $\mathbf{g}$ can be written in a unique way as the sum of a solenoidal $\mathbf{g}_s$ and an irrotational $\mathbf{g}_i$ field. In fact, let $\mathbf{g} = \mathbf{g}_s + \mathbf{g}_i$, with $\text{div}\,\mathbf{g}_s = 0$ and $\text{rot}\,\mathbf{g}_i = 0$; we immediately see that $\text{div}\,\mathbf{g}_i = \rho$ and $\text{rot}\,\mathbf{g}_s = \boldsymbol{\omega}$, so that the two fields have specified divergence and rotor, and from the previous result they are uniquely determined. It is possible to show how to construct $\mathbf{g}$ explicitly by using some suitable gauge condition in Eq. (A.123), but we will not discuss this point here (e.g., see Arfken and Weber 2005; Becker 1982; Jackson 1998).

We conclude by recalling that the Helmholtz decomposition is not the only possible (and remarkable) way to decompose a vector field. Other decompositions can be constructed based on the concept of *complex lamellar* and *Beltrami* fields (e.g., see Aris 1989), defined, respectively, as fields perpendicular and parallel to their rotor, i.e.,

$$\langle \mathbf{g}, \text{rot}\,\mathbf{g} \rangle = 0, \qquad \mathbf{g} \wedge (\text{rot}\,\mathbf{g}) = 0. \tag{A.131}$$

Among several others applications, this characterization plays a fundamental role in the theory of hydrostatic equilibria (e.g., see Tassoul 1978).

A.7 Green Functions

In Chapter 2, we encountered the problem of the construction of the Green function for the Laplacian. Here, we provide general information on this fundamental mathematical tool (e.g., see Arfken and Weber 2005; Bender and Orszag 1978; Jackson 1998; Lomen and Mark 1988). Suppose we look for the solution of the problem

$$\mathcal{D}[f] = h(\mathbf{x}), \tag{A.132}$$

where $\mathcal{D}$ is some linear differential operator,[10] $h(\mathbf{x})$ is a given function, and for simplicity we restrict ourselves to Cartesian coordinates in $\Re^n$. We proceed formally, without entering the important questions of existence, regularity, etc., for a solution f, which we assume exists and has all of the properties needed for the following discussion. In the framework of the Green functions, the idea is not to solve directly Eq. (A.132), but to solve once and for all a problem so that the solutions for different choices of the source function h do not require us to solve different equations, but can be obtained once a Green function for the operator is known. A function $\mathcal{G}(\mathbf{x},\mathbf{y})$ will be called a Green function for the operator $\mathcal{D}$ if

$$\mathcal{D}[\mathcal{G}] = \delta(\mathbf{x}-\mathbf{y}). \tag{A.133}$$

If we know $\mathcal{G}$, than we have solved *all* of the problems in Eq. (A.132)! In fact, consider (formally)

$$f(\mathbf{x}) = \int_{\Re^n} \mathcal{G}(\mathbf{x},\mathbf{y}) h(\mathbf{y})\, d^n\mathbf{y}, \tag{A.134}$$

and apply the operator $\mathcal{D}$ to it. Exchange the operator with the integral (intuitively, this can be done because $\mathcal{D}$ is linear and integrals are just sums) and use Eq. (A.95); this completes the "proof." In practice, the function $\mathcal{G}$ describes the "reaction" of the operator to a very special source term, and the idea is that the solution of the general problem (thanks to the linearity of $\mathcal{D}$) can be obtained as the weighted superposition of these special solutions. How can we construct the function $\mathcal{G}$ for a given $\mathcal{D}$? A consideration is in order: intuitively, we can expect that the Green function is strictly related to the solution of the homogeneous case of Eq. (A.132) – after all, the δ-function in Eq. (A.133) is equal to 0 for all points of space $\mathbf{x} \neq \mathbf{y}$! The idea is to combine the *general* homogeneous solutions $\mathcal{G}_0$ in a clever way so as to obtain the peculiar behavior of the δ-function at $\mathbf{x} = \mathbf{y}$.

We first consider the case of a one-dimensional operator (e.g., see Eq. (6.12))

$$\mathcal{D}[f] = f^{(n)}(x) + a_{n-1}(x) f^{(n-1)}(x) + \cdots + a_1(x) f^{(1)}(x) + a_0(x) f^{(0)}(x), \tag{A.135}$$

where $a_k(x)$ are regular functions and with $f^{(k)}$ we indicate the k-th derivative; as usual $f^{(0)} = f$. Suppose we look for the solution of Eq. (A.133) for our n-th-order, one-dimensional differential operator, and we determined the general solution of the homegeneous problem $\mathcal{D}[f_0] = 0$

$$f_0 = \sum_{i=1}^{n} A_i\, f_{0i}(x), \tag{A.136}$$

[10] We recall that for linear operators $\mathcal{D}[\alpha u + \beta v] = \alpha\mathcal{D}[u] + \beta\mathcal{D}[v]$.

where A_i are n arbitrary constants and $f_{0i}(x)$ are the n linearly independent solutions. The idea is to consider a different set of constants for $x < y$ and $x > y$ and to obtain the sought-after δ-function behavior at $x = y$ by using Eq. (A.101). If we can fix the values of the constants in the two separated regions of space so that $f_0^{(n-1)}$ behaves as the Heaviside θ-function, then $f^{(n)}$ will behave as the Dirac δ-function, and the one-dimensional problem is solved because all of the lower-order derivatives remain finite and do not affect the behavior of the solution at the critical point $x = y$. The choice of the constants is now trivial: if the $n - 1$-th derivative is the θ-function, than the lower-order derivatives are continuous at $x = y$ (integration improves regularity by unity!). In practice, the constants are fixed by requiring that for

$$f_{0,\text{left}}^{(k)}(y) = f_{0,\text{right}}^{(k)}(y), \quad f_{0,\text{left}}^{(n-1)}(y) + 1 = f_{0,\text{right}}^{(n-1)}(y), \tag{A.137}$$

where the first identities must hold for $k = 0, \ldots, n - 2$. Notice that if we look for the solution of the Green problem $\mathcal{D}[\mathcal{G}] = A\delta(x - y)$, the corresponding function is obtained simply by requiring a "jump" of A instead of unity. We can now discuss the full problem in Eq. (A.133): clearly, we now face a new geometric difficulty with respect to the one-dimensional case. In fact, while in the one-dimensional case the singularity of the right-hand side member is geometrically controlled by the fact that the "jump" can only happen along the x-coordinate, in more than one dimension, the jump could happen along arbitrary directions. How we can control this? The solution is provided by the method of separation of variables. In practice, we search for a homogeneous solution of Eq. (A.133) by using the method of the separation of variables. In many cases of interest (see Chapter 2), we obtain (complete) sets of orthogonal functions. Suppose that the separation of variables for our problem produced $n - 1$ families of orthogonal functions: the idea is to *expand* the (unknown) Green function $\mathcal{G}$ on these bases, evaluate $\mathcal{D}[\mathcal{G}]$ for such an expansion, isolate the remaining one-dimensional Green problem, and finally solve it by using the approach described above. Intuitively, the expansion of the orthgonal functions allow us to "control" the jump along the "direction" of the remaining variable. As the student can see in the three explicit examples for the Laplacian in Chapter 2, in practical cases the method is *less* complicated than its description!

A.8 Differential Operators in Orthogonal Curvilinear Coordinates

We recall here the most important properties of (orthogonal) curvilinear coordinates; for simplicity, we will not discuss the strictly related and important subject of *covariant* and *contravariant* bases, and we will further restict ourselves to $\Re^3$. Very clear and rigorous treatments can be found in Arfken and Weber (2005), Do Carmo (1976), and Narashiman (1993), and for the more physically oriented readers, useful references are Becker (1982), Lamb (1945), and Milne-Thomson (1996).

We start by considering a system of curvilinear coordinates q_i, with $i = 1, 2, 3$. With the compact notation $\mathbf{q} = (q_1, q_2, q_3)$ we do *not* indicate a vector, but only a triplet of curvilinear coordinates.[11] Once the $\mathbf{q}$ coordinates are assigned, the relation between Cartesian and curvilinear coordinates is known

$$\mathbf{x} = \mathbf{x}(\mathbf{q}). \tag{A.138}$$

A simple geometric interpretation of coordinates in terms of the intersection of surfaces of constant q_i is extremely useful for the following discussion, and here we simply recall that Cartesian coordinates can be interpreted as the intersection of mutually perpendicular planes, spherical coordinates of spheres, cones, and planes, and cylindrical coordinates of cylinders and two planes.

It follows immediately that by fixing the values of two of the three curvilinear coordinates in Eq. (A.138) and considering the partial derivative of the resulting curve described by $\mathbf{x}$ as a function of the remaining q_i coordinate, we determine the *tangent* to the coordinate curve, which we call

$$\mathbf{x}_i \equiv \frac{\partial \mathbf{x}(\mathbf{q})}{\partial q_i}. \tag{A.139}$$

Curvilinear coordinates such that $\langle \mathbf{x}_i, \mathbf{x}_j \rangle = 0$ for $i \neq j$ are called *orthogonal curvilinear coordinates*, and in the following we restrict ourselves to this special but extremely important family. Of course, the *length* of the tangent

$$h_i(\mathbf{q}) \equiv \|\mathbf{x}_i\| \tag{A.140}$$

in general changes along the coordinate curve, and the functions h_i are known as the *Lamé* (1795–1870) coefficients of the coordinate system. It is natural to define a *local* orthonormal basis as

$$\mathbf{f}_i \equiv \frac{\mathbf{x}_i}{h_i} = \frac{\mathbf{e}_j}{h_i}\frac{\partial x_j}{\partial q_i}, \quad i = 1,2,3, \tag{A.141}$$

where no summation over the index i is intended. Without loss of generality, we can assume that the triplet $\mathbf{q}$ is ordered so that the associated $\mathbf{f}_i$ form a positively oriented local basis, i.e.,

$$\langle \mathbf{f}_i, \mathbf{f}_j \rangle = \delta_{ij}, \quad \mathbf{f}_1 \wedge \mathbf{f}_2 = \mathbf{f}_3, \tag{A.142}$$

where the indices in the last identity can be cyclically permuted.

In Cartesian coordinates, $\mathbf{x} = x\mathbf{e}_x + y\mathbf{e}_y + z\mathbf{e}_z$, and so $h_x = h_y = h_z = 1$. In spherical coordinates (r, ϑ, φ), we have $\mathbf{x} = (r \sin\vartheta \cos\varphi, r \sin\vartheta \sin\varphi, r \cos\vartheta)$, so that

$$\begin{cases} h_r = 1, \quad h_\vartheta = r, \quad h_\varphi = r \sin\vartheta, \\ \mathbf{f}_r = (\sin\vartheta \cos\varphi, \sin\vartheta \sin\varphi, \cos\vartheta), \\ \mathbf{f}_\vartheta = (\cos\vartheta \cos\varphi, \cos\vartheta \sin\varphi, -\sin\vartheta), \\ \mathbf{f}_\varphi = (-\sin\varphi, \cos\varphi, 0), \end{cases} \tag{A.143}$$

[11] It should be unnecessary to remind the reader that not all ordered lists of numbers are vectors, but only those obeying the axioms of vector spaces. As an additional point, also notice that quite often curvilinear coordinates in physical problems have different physical units!

and $\mathbf{x} = r\mathbf{f}_r$. In cylindrical coordinates (R, φ, z), we have $\mathbf{x} = (R\cos\varphi, R\sin\varphi, z)$, so that

$$\begin{cases} h_R = 1, \quad h_\varphi = R, \quad h_z = 1, \\ \mathbf{f}_R = (\cos\varphi, \sin\varphi, 0), \\ \mathbf{f}_\varphi = (-\sin\varphi, \cos\varphi, 0), \\ \mathbf{f}_z = \mathbf{e}_z = (0, 0, 1), \end{cases} \tag{A.144}$$

and $\mathbf{x} = R\mathbf{f}_R + z\mathbf{f}_z$. For the ellipsoidal coordinates (λ, μ, ν) defined for the unitary ellipsoid in Eq. (2.9), some more work is needed. The relevant transformation formulae can be established in a very elegant way starting from the formal factorization

$$0 = \sum_{i=1}^{3} \frac{x_i^2}{a_i^2 + \tau} - 1 = -\frac{(\tau - \lambda)(\tau - \mu)(\tau - \nu)}{\Delta(\tau)}, \tag{A.145}$$

where the function $\Delta(\tau)$ is given in Eq. (2.11). After multiplication of both members of the identity in Eq. (A.145) for $\Delta(\tau)$, the limits for $\tau = -a_1^2, -a_2^2$, and $-a_3^2$ of the resulting identity show that

$$\begin{cases} x^2 = \dfrac{(a_1^2 + \lambda)(a_1^2 + \mu)(a_1^2 + \nu)}{(a_1^2 - a_2^2)(a_1^2 - a_3^2)}, \\[2ex] y^2 = \dfrac{(a_2^2 + \lambda)(a_2^2 + \mu)(a_2^2 + \nu)}{(a_2^2 - a_1^2)(a_2^2 - a_3^2)}, \\[2ex] z^2 = \dfrac{(a_3^2 + \lambda)(a_3^2 + \mu)(a_3^2 + \nu)}{(a_3^2 - a_1^2)(a_3^2 - a_2^2)}, \end{cases} \tag{A.146}$$

where $-a_1^2 \leq \nu \leq -a_2^2 \leq \mu \leq -a_3^2 \leq \lambda$. Surfaces of constant λ are confocal ellipsoids, surfaces of constant μ are confocal hyperboloids of one sheet, and surfaces of constant ν are confocal hyperboloids of two sheets. It is easy to show that the coordinates are locally orthogonal, that the Lamé coefficients are given by[12]

$$\begin{cases} h_\lambda = \dfrac{1}{2}\sqrt{\dfrac{(\lambda - \mu)(\lambda - \nu)}{\Delta(\lambda)}}, \\[2ex] h_\mu = \dfrac{1}{2}\sqrt{\dfrac{(\mu - \lambda)(\mu - \nu)}{\Delta(\mu)}} = \dfrac{1}{2}\sqrt{\dfrac{(\lambda - \mu)(\mu - \nu)}{-\Delta(\mu)}}, \\[2ex] h_\nu = \dfrac{1}{2}\sqrt{\dfrac{(\nu - \lambda)(\nu - \mu)}{\Delta(\nu)}} = \dfrac{1}{2}\sqrt{\dfrac{(\lambda - \nu)(\mu - \nu)}{\Delta(\nu)}}, \end{cases} \tag{A.147}$$

[12] In the expression for h_μ, the positivity of the arguments of the square root has been made apparent (see also Footnote 1 in Chapter 2).

and finally that the positively oriented local basis is $(\mathbf{f}_\lambda, \mathbf{f}_\mu, \mathbf{f}_\nu)$. Note that from Eq. (A.146) it follows that ellipsoidal coordinates are *degenerate*; in other words, at a given triplet (λ, μ, ν), there correspond eight points with reflecting symmetry with respect the three coordinate planes of $\Re^3$, a property with obvious consequences when computing surface or volume integrals in ellipsoidal coordinates (e.g., see Exercise 2.5).

We are now in a position to generalize to the case of orthogonal curvilinear coordinates the basic differential and integral operators introduced in Appendix A.4. As a preparatory consideration, notice that for an arbitrary vector field $\mathbf{g}$, the following obvious geometric relations hold:

$$\mathbf{g} = g_i \mathbf{e}_i = g'_j \mathbf{f}_j, \quad \Rightarrow \quad g'_j = \langle \mathbf{g}, \mathbf{f}_j \rangle = g_i \langle \mathbf{e}_i, \mathbf{f}_j \rangle. \tag{A.148}$$

Therefore, from Eq. (A.107), it follows immediately that in the local $\mathbf{f}_i$ basis (no summing over repeated indices is intended)

$$(\text{grad } f)'_i = \langle \nabla f, \mathbf{f}_i \rangle = \frac{1}{h_i} \frac{\partial f}{\partial q_i}, \quad i = 1, 2, 3. \tag{A.149}$$

We then generalize the path integral for a vector field $\mathbf{g}$, such as that appearing in Eq. (A.109), as follows. We construct a regular curve γ by considering the parameterization $\mathbf{q}(t)$ with $t \in [a,b]$, and then $\mathbf{x}[\mathbf{q}(t)]$, and it follows immediately that

$$\int_\gamma \langle \mathbf{g}, d\mathbf{x} \rangle = \sum_{j=1}^{3} \int_a^b g'_j h_j \frac{dq_j}{dt} dt. \tag{A.150}$$

A special (and relevant) case is for γ coincident with a coordinate curve when, without loss of generality, we can assume $t = q_i$ (say) for some i, and the path integral reduces to $\int_a^b g'_i h_i dq_i$ (no summing over i intended).

In order to generalize flux (and surface) integrals, such as that in Eq. (A.110), we consider the parameterization of a regular surface Σ in $\Re^3$ as $\mathbf{q} = \mathbf{q}(u, v)$ with $u \in [a,b]$ and $v \in [c,d]$, and then $\mathbf{x}[\mathbf{q}(u,v)]$. Therefore, from Eq. (A.111)

$$\int_\Sigma \langle \mathbf{g}, \mathbf{n} \rangle d^2\mathbf{x} = \sum_{i,j=1}^{3} \int_a^b \int_c^d \langle \mathbf{g}, \mathbf{f}_i \wedge \mathbf{f}_j \rangle h_i h_j \frac{\partial q_i}{\partial u} \frac{\partial q_j}{\partial v} du\, dv. \tag{A.151}$$

Notice that from the second identity in Eq. (A.142), for $i \neq j$, the wedge product $\mathbf{f}_i \wedge \mathbf{f}_j = \pm \mathbf{f}_k$, with the sign determined by the parity of the (i, j, k). Again, a special and important case is obtained when the integration surface is coincident with some region of $q_k = const.$ so that for positive orientation

$$\mathbf{n} = \mathbf{f}_k, \quad d^2\mathbf{x} = h_i h_j dq_i dq_j, \tag{A.152}$$

and so

$$\int_S \langle \mathbf{g}, \mathbf{n} \rangle d^2\mathbf{x} = \int_a^b \int_c^d g'_k h_i h_j dq_i dq_j. \tag{A.153}$$

If we now integrate over a small region on the surface and we use the Stokes theorem in Eq. (A.116), it is a nice exercise to show from Eqs. (A.150) and (A.151) that

$$(\operatorname{rot}\mathbf{g})'_1 = \frac{1}{h_2 h_3}\left(\frac{\partial g'_3 h_3}{\partial q_2} - \frac{\partial g'_2 h_2}{\partial q_3}\right) \tag{A.154}$$

(and, correspondingly, making a cyclic exchange of the indices for the directions $\mathbf{f}_2$ and $\mathbf{f}_3$).

We finally consider the volume integral. By changing coordinates from $\mathbf{x}$ to $\mathbf{q}$, the determinant of the Jacobian is given by the triple product of the vectors in Eq. (A.139), so that in our case

$$|\det J| = |\det(\mathbf{x}_1,\mathbf{x}_2,\mathbf{x}_3)| = h_1 h_2 h_3 |\det(\mathbf{f}_1,\mathbf{f}_2,\mathbf{f}_3)| = h_1 h_2 h_3, \tag{A.155}$$

and in particular Eq. (A.97) follows. Considering now Eqs. (A.113) and (A.153) for a small volume limited by the three coordinate surfaces, we obtain (prove it!)

$$\operatorname{div}\mathbf{g} = \frac{1}{h_1 h_2 h_3}\left(\frac{\partial g'_1 h_2 h_3}{\partial q_1} + \frac{\partial g'_2 h_1 h_3}{\partial q_2} + \frac{\partial g'_3 h_1 h_2}{\partial q_3}\right), \tag{A.156}$$

and finally the expression for the Laplace operator is obtained by combining Eqs. (A.149) and (A.156) with $\mathbf{g} = \operatorname{grad} f$ as follows:

$$\Delta f = \frac{1}{h_1 h_2 h_3}\left[\frac{\partial}{\partial q_1}\left(\frac{h_2 h_3}{h_1}\frac{\partial f}{\partial q_1}\right) + \frac{\partial}{\partial q_2}\left(\frac{h_1 h_3}{h_2}\frac{\partial f}{\partial q_2}\right) + \frac{\partial}{\partial q_3}\left(\frac{h_1 h_2}{h_3}\frac{\partial f}{\partial q_3}\right)\right]. \tag{A.157}$$

With this expression, we are now in a position to use the Poisson equation in the coordinate systems adopted in stellar dynamics (see Chapter 2).

A.9 The "Co-Area" Theorem

We used the co-area theorem in Chapter 2 when discussing the gravitational potential produced by a triaxial ellipsoid with density stratified on homologous concentric and coaxial ellipsoidal surfaces described by Eq. (2.8). In particular, we faced the quite unintuitive fact that the spatial density can be interpreted as the sum of surface densities that are *not* constant over each ellipsoidal stratum. The co-area theorem not only explains why this happens, but also provides a general volume integration formula for stratified functions; technically, it can be viewed as a powerful change of the integration variable, and in the following we illustrate the qualitative idea behind the method. In practice (and for simplicity), suppose we have to integrate the product of a generic function $f(\mathbf{x})$ for $\Re^3$ to $\Re$ and a function g stratified on some family of regular surfaces

$$\mathcal{S}(\mathbf{x}) = m, \tag{A.158}$$

in analogy with Eq. (2.8). We search for an alternative expression of the volume integral

$$I(m_1,m_2) \equiv \int_{m_1 \le \mathcal{S}(\mathbf{x}) \le m_2} f(\mathbf{x}) g[\mathcal{S}(\mathbf{x})] d^3\mathbf{x}. \tag{A.159}$$

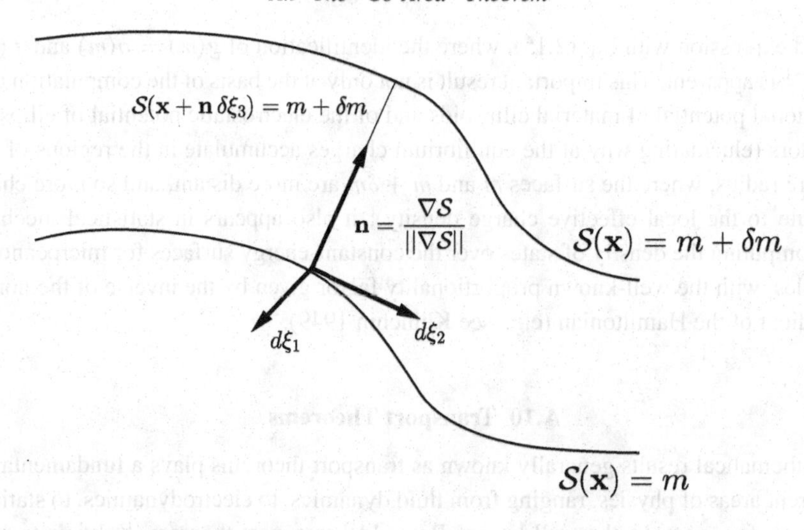

Figure A.2 Qualitative illustration of the geometric idea behind the co-area theorem.

The idea is to split the integration first by conducting a two-dimensional integration over the surfaces of constant m, and then integrating over the range of m. As is shown in Figure A.2, we can imagine first introducing for each surface two coordinates ξ_1 and ξ_2, as well as a coordinate ξ_3 perpendicular to the tangent plane, so that the three coordinates (ξ_1, ξ_2, ξ_3) produce a local coordinate system. We must now substitute the variable ξ_3 with the variable m, and this is accomplished by the requirement that the "length" $\delta\xi_3$ is chosen so that, qualitatively speaking, $\delta\xi_1 \times \delta\xi_2\delta\xi_3$ is the "volume" of the infinitesimal parallelepiped with a vertex at (ξ_1, ξ_2) on the surface $\mathcal{S}(\mathbf{x}) = m$, with lengths of the tangent basis vectors, and with height $\delta\xi_3$ coincident with the distance along the normal to the surface $\mathcal{S}(\mathbf{x}) = m+\delta m$. It is elementary to show that we are looking for a solution of the following identity:

$$\mathcal{S}(\mathbf{x}) = m, \quad \mathcal{S}(\mathbf{x}+\mathbf{n}\delta\xi_3) = m + \delta m, \quad \mathbf{n} = \left.\frac{\nabla\mathcal{S}(\mathbf{x})}{\|\nabla\mathcal{S}(\mathbf{x})\|}\right|_{\mathcal{S}(\mathbf{x})=m}. \tag{A.160}$$

Expanding to the linear order for vanishingly small δm

$$\delta\xi_3 = \frac{\delta m}{\|\nabla\mathcal{S}(\mathbf{x})\|_{\mathcal{S}(\mathbf{x})=m}}, \tag{A.161}$$

so that finally

$$I(m_1, m_2) = \int_{m_1}^{m_2} g(m)dm \int_{\mathcal{S}(\mathbf{x})=m} \frac{f(\mathbf{x})}{\|\nabla\mathcal{S}(\mathbf{x})\|} d^2\mathbf{x}, \tag{A.162}$$

where the appearance of "effective surface density," due to the fact that the two surfaces in Eq. (A.160) are in general not equidistant, is clear.[13] The student should compare the

[13] For the student more interested in rigor, the formal approach would be to determine first the two-dimensional surface element

$$d^2\mathbf{x} = h_1 h_2 \|\mathbf{f}_1 \wedge \mathbf{f}_2\| d\xi_1 d\xi_2, \tag{A.163}$$

obtained expression with Eq. (2.15), where the identification of $g(m) = \rho(m)$ and $f(\mathbf{x}) = \|\mathbf{x}-\mathbf{y}\|^{-1}$ is apparent. This important result is not only at the basis of the computation of the gravitational potential of material ellipsoids and of the electrostatic potential of ellipsoidal conductors (elucidating why at the equilibrium charges accumulate in the regions of short curvature radius, where the surfaces m and $m + \delta m$ are more distant, and so more charges contribute to the local effective charge density); it also appears in statistical mechanics when computing the density of states over the constant-energy surfaces for microcanonical ensembles, with the well-known proportionality factor given by the inverse of the norm of the gradient of the Hamiltonian (e.g., see Khinchin 1949).

A.10 Transport Theorems

The mathematical results generally known as transport theorems plays a fundamental role in different areas of physics, ranging from fluid dynamics, to electrodynamics, to statistical mechanics; for example, the well-known Reynolds' transport theorem, Kelvin's transport theorem, and the transport integrals appearing in the Ampere and Faraday–Maxwell equations in integral form all belong to this family. In particular, in Chapter 9, Reynolds' transport theorem has been applied to derive the collisionless Boltzmann equation in curvilinear coordinates. Transport theorems are important because they describe how some property of a "fluid" (in the broadest meaning) integrated over some region (volume, surface, curve) changes due to the "transport" produced by the solution of the evolutionary differential equation describing the system. The advantages of integral formulation over a differential formulation are obvious because the value of a given integral is independent of the specific coordinates used, and so transport theorems are natural tools to formalize physical principles and to obtain the corresponding differential equations. In this section, we will focus on the main logical steps behind the derivation and meaning of Reynolds' transport theorem (for more extended discussions, see, among others, Aris 1989; Borisenko and Tarapov 1979; Currie 1993; Jackson 1998; Lamb 1945; Meyer 1982; Milne-Thomson 1996; Narashiman 1993).

where no assumptions about the orthogonality of $\mathbf{f}_1$ and $\mathbf{f}_2$ are made. We then ask that the ξ_3 coordinate, measured perpendicular to the surface, is the parameter m; from Eq. (A.158)

$$1 = \left\langle \nabla \mathcal{S}(\mathbf{x}), \frac{\partial \mathbf{x}}{\partial m} \right\rangle = h_m \langle \nabla \mathcal{S}(\mathbf{x}), \mathbf{f}_m \rangle = h_m \|\nabla \mathcal{S}\| \quad \Rightarrow h_m = \frac{1}{\|\nabla \mathcal{S}\|}. \tag{A.164}$$

Therefore, the volume element becomes

$$d^3\mathbf{x} = h_1 h_2 h_m \det(\mathbf{f}_1, \mathbf{f}_2, \mathbf{f}_m) = \frac{d^2\mathbf{x} dm}{\|\nabla \mathcal{S}\|}, \tag{A.165}$$

and this concludes the proof.

A.10.1 Ordinary Differential Equations

In order to introduce the subject, we recall some basic facts about the general properties of ODEs in $\Re^n$ (e.g., see Arnold 1992; Ince 1927; Lomen and Mark 1988; Wintner 1947). As the discussion is quite general, in the following, by $\mathbf{z}$ we mean a list of n variables (not necessarily a vector). For example, in the simplest case $\mathbf{z} = \mathbf{x}$, the Cartesian coordinate vector in $\Re^3$, or for a point moving under the action of a force field in Cartesian coordinates, we could have $\mathbf{z} = (\mathbf{x}, \mathbf{v})$, or $\mathbf{z} = (\mathbf{q}, \mathbf{p})$ when considering the motion in phase space induced by some Hamiltonian function, or $\mathbf{z} = (\mathbf{q}, \dot{\mathbf{q}})$ for the case of curvilinear coordinates and generalized velocities in the Euler–Lagrange framework, or finally $\mathbf{z} = (\mathbf{q}, \boldsymbol{\nu})$ when decomposing the velocities over the local basis as in Chapter 9.

Therefore, let $A \subseteq \Re^n$ be an open set and $\mathbf{W}(\mathbf{z})$ be a list of n functions $A \mapsto \Re^n$. The system

$$\dot{\mathbf{z}} = \frac{d\mathbf{z}}{dt} = \mathbf{W}(\mathbf{z}), \quad \mathbf{z}(0) = \mathbf{z}_0, \tag{A.166}$$

is called an (autonomous) ODE in $\Re^n$ with initial condition $\mathbf{z}_0$; in most physical applications, the parameter t is the time, and the functions $\mathbf{W}$ could also depend explicitly on time, but for simplicity in the following we restrict ourselves to the autonomous case. A function $\Psi : A \times I \mapsto \Re^n$, where I is some (open) interval spanned by the parameter t, so that

$$\dot{\Psi}(\mathbf{z}_0, t) = \mathbf{W}[\Psi(\mathbf{z}_0, t)], \quad \Psi(\mathbf{z}_0, 0) = \mathbf{z}_0, \tag{A.167}$$

is called the *solution* with initial condition $\mathbf{z}_0$ for the ODE.

A central problem of the theory of ODEs is to determine whether a solution exists and what are its properties. For our purposes, the most important results concern the *local existence*, *uniqueness*, and *regularity* of the solution (details and formal proofs can be found in the list of references provided above). Regarding existence, if the components of $\mathbf{W}$ are at least continuous on some open set $U_0 \in A$, then the solution exists at least locally (i.e., a $\delta > 0$ exists for which Eq. (A.167) holds for $|t| < \delta$). Moreover, if $\mathbf{W}$ is Lipschitz on U_0, then the solution not only exists, but it is also unique, with continuous dependence on the initial condition $\mathbf{z}_0$. Finally, if the components of $\mathbf{W}$ are smooth of order $C^r(U_0)$, with $r \geq 1$, then $\Psi(\mathbf{z}_0, t) \in C^r(U_0)$ with respect to $\mathbf{z}_0$ and $\Psi(\mathbf{z}_0, t) \in C^{r+1}(I)$ with respect to t. In summary, for sufficiently regular $\mathbf{W}$, the three results just mentioned imply that the solution is time-invertible; in other words, for $|t| < \delta$ and $\forall \mathbf{z}_0 \in U_0$

$$\mathbf{z} = \Psi(\mathbf{z}_0, t), \quad \mathbf{z}_0 = \Psi[\Psi(\mathbf{z}_0, t), -t]. \tag{A.168}$$

With suggestive notation, we can write[14] $\Psi^{-1}(\mathbf{z}, t) = \Psi(\mathbf{z}, -t)$.

[14] Technically, Ψ is a one-parameter diffeomorphism, with $\Psi[\Psi(\mathbf{z}_0, t_1), t_2] = \Psi(\mathbf{z}_0, t_1 + t_2)$ for arbitrary t_1 and t_2; see Arnold (1992).

In practice, the previous results show that for arbitrary but fixed t, the components of Ψ determine a regular change of variables between $\mathbf{z}$ and $\mathbf{z}_0$. Let $J_0(t)$ be the determinant of the Jacobian matrix of the coordinate transformation $\mathbf{z} = \Psi(\mathbf{z}_0, t)$

$$J_0(t) \equiv \det \left[\frac{\partial \Psi_i(\mathbf{z}_0, t)}{\partial z_{0j}} \right], \quad i, j = 1, \ldots, n. \tag{A.169}$$

From uniqueness, $J_0 \neq 0$ over $U_0 \times I$, so that from the identity $J_0(0) = 1$, it follows that $J_0(t)$ remains positive for $t \in I$. We now prove the fundamental identity

$$\frac{dJ_0}{dt} = J_0 \times \sum_{i=1}^{n} \left. \frac{\partial W_i}{\partial z_i} \right|_{\mathbf{z}=\Psi(\mathbf{z}_0,t)}. \tag{A.170}$$

In fact, from linear algebra, the derivative of the determinant of an $n \times n$ matrix with respect to some parameter equals the sum of n determinants, where in each of them a different row (column) of the matrix is replaced by the derivative of the row (column). Equation (A.170) can be proved as follows: let us compute the first of our determinants associated with dJ_0/dt. For the j-th element in the first row, we have

$$\frac{d}{dt}\frac{\partial \Psi_1}{\partial z_{0j}} = \frac{\partial}{\partial z_{0j}}\frac{d\Psi_1}{dt} = \frac{\partial W_1[\Psi(\mathbf{z}_0,t)]}{\partial z_{0j}} = \frac{\partial W_1}{\partial \Psi_1}\frac{\partial \Psi_1}{\partial z_{0j}} + \sum_{i=2}^{n} \frac{\partial W_1}{\partial \Psi_i}\frac{\partial \Psi_i}{\partial z_{0j}}. \tag{A.171}$$

In practice, the first row of the first determinant is given by the first row of the matrix in Eq. (A.169) multiplied by $\partial W_1/\partial \Psi_1$, plus the linear combination of the rows $i = 2, \ldots, n$ with coefficients $\partial W_1/\partial \Psi_i$. The resulting determinant is independent of such a linear combination, and it sums to $J_0 \partial W_1/\partial \Psi_1$. Equation (A.170) is then proved by repeating the same calculation for the remaining $n - 1$ determinants and adding the results.

A.10.2 Material Derivative

A key concept associated with the solution of the ODE in Eq. (A.167), and central in the following discussion, is built by considering the time evolution of some given function $f = f(\mathbf{z}, t)$ when computed along the solution with initial condition $\mathbf{z}_0$. The quantity

$$f_{\mathcal{L}}(\mathbf{z}_0, t) \equiv f[\Psi(\mathbf{z}_0; t), t] \tag{A.172}$$

is called the *Lagrangian function* associated with the initial condition $\mathbf{z}_0$. From a physical point of view, $f_{\mathcal{L}}$ describes how the "property" f associated with $\mathbf{z}_0$ at $t = 0$, evolves with time.

The *material* (or Lagrangian) derivative of f along the solutions of $\mathbf{W}$ with initial condition $\mathbf{z}_0$ is immediately obtained from the standard differentiation rule of composite functions

$$\frac{df_{\mathcal{L}}(\mathbf{z}_0, t)}{dt} = \left. \frac{Df}{Dt} \right|_{\mathbf{z}=\Psi(\mathbf{z}_0;t)}, \quad \frac{Df}{Dt} \equiv \frac{\partial f}{\partial t} + \sum_{i=1}^{n} W_i \frac{\partial f}{\partial z_i}, \tag{A.173}$$

where the differential operator D/Dt is the material derivative in Eulerian form. Of course, the material derivative obeys the same rules of standard total derivatives when considering sums and products of functions, etc. In Chapters 9 and 10, examples of material derivatives of astrophysical interest are discussed. Notice that a quantity is conserved along a solution if and only if its material derivative vanishes.

A.10.3 Reynolds' Transport Theorem

We are finally in a position to consider an arbitrary region $\Omega(0) \subseteq \Re^n$ at $t = 0$. The solution $\Psi(\mathbf{z}_0; t)$ induced by $\mathbf{W}$ "moves" each point of $\Omega(0)$, transforming $\Omega(0)$ into $\Omega(t)$. The function

$$F(t) \equiv \int_{\Omega(t)} f(\mathbf{z},t) d^n\mathbf{z} = \int_{\Omega(0)} f_{\mathcal{L}}(\mathbf{z}_0,t)\, J_0(t) d^n\mathbf{z}_0 \tag{A.174}$$

is the integral of f over the region $\Omega(t)$. We wish to determine the explicit expression of dF/dt. The method is based on the possibility of changing variables from $\mathbf{z}$ back to $\mathbf{z}_0$ from Eq. (A.168), where $J_0 > 0$ is given by Eq. (A.169) and the integration domain is mapped back to $A(0)$ and so is time-independent. From differentiation under the sign of the integral

$$\begin{aligned}\frac{dF(t)}{dt} &= \int_{\Omega(0)} \left(\frac{Df}{Dt} + f \sum_{i=1}^{n} \frac{\partial W_i}{\partial z_i} \right)_{\mathbf{z}=\Psi(\mathbf{z}_0,t)} J_0(t) d^n\mathbf{z}_0 \\ &= \int_{\Omega(t)} \left(\frac{Df}{Dt} + f \sum_{i=1}^{n} \frac{\partial W_i}{\partial z_i} \right) d^n\mathbf{z},\end{aligned} \tag{A.175}$$

where we used the first identity in Eq. (A.173) and Eq. (A.170), and the last expression[15] is obtained by reverting the coordinates from $\mathbf{z}_0$ to $\mathbf{z}$.

[15] Two other relevant transport theorems (e.g., see Aris 1989; Jackson 1998; and in particular Truesdell 1954) that are often encountered in mechanics, fluid dynamics, and electrodynamics are related to the transport of the flux of a vector field in $\Re^3$ through a time-dependent surface $\Sigma(t)$

$$\begin{aligned}\frac{d}{dt}\int_{\Sigma(t)} \langle \mathbf{g}, \mathbf{n} \rangle d^2\mathbf{x} &= \int_{\Sigma(t)} \left\langle \left[\frac{\partial \mathbf{g}}{\partial t} + \mathbf{W}\,\mathrm{div}\,\mathbf{g} + \mathrm{rot}\,(\mathbf{g} \wedge \mathbf{W}) \right], \mathbf{n} \right\rangle d^2\mathbf{x} \\ &= \int_{\Sigma(t)} \left\langle \left[\frac{D\mathbf{g}}{Dt} + \mathbf{g}\,\mathrm{div}\,\mathbf{W} - \langle \mathbf{g}, \nabla \rangle \mathbf{W} \right], \mathbf{n} \right\rangle d^2\mathbf{x},\end{aligned} \tag{A.176}$$

where the equivalence of the two formulations can be established by using Eq. (A.125), and the second expression is obtained from the result (that we do not prove here)

$$\frac{D}{Dt}(n_i d^2\mathbf{x}) = \left[(\mathrm{div}\,\mathbf{W}) n_i - \frac{\partial W_j}{\partial x_i} n_j \right] d^2\mathbf{x}, \tag{A.177}$$

and to the transport of circuitation of a vector field (sometimes known as Kelvin's theorem when applied to the fluid velocity $\mathbf{u} = \mathbf{W}$) evaluated along a time-dependent curve $\gamma(t)$

$$\frac{d}{dt}\int_{\gamma(t)} \langle \mathbf{g}, d\mathbf{x} \rangle = \int_{\gamma(t)} \left\langle \frac{D\mathbf{g}}{Dt}, d\mathbf{x} \right\rangle + \int_a^b \left\langle \mathbf{g}_{\mathcal{L}}, \frac{d\mathbf{W}_{\mathcal{L}}}{d\lambda} \right\rangle d\lambda. \tag{A.178}$$

In particular, for $\mathbf{g} = \mathbf{u}$, the identity in Eq. (A.178) in the case of an inviscid, barotropic fluid under the action of conservative forces as obtained by specializing Eq. (10.23) shows that the velocity circuitation is conserved along the fluid motion, and so under these assumptions and the Stokes theorem, the fluid vorticity $\boldsymbol{\omega} \equiv \mathrm{rot}\,\mathbf{u}$ cannot be created or destroyed.

A first important application of the previous result is obtained under the *assumption* that $dF(t)/dt = 0$ for *all* possible choices of the control volume $\Omega(0)$, a case that is often encountered in physics. In this, the transport theorem is used to deduce the equation obeyed by the integrand, and we get

$$\frac{Df}{Dt} + f \sum_{i=1}^{n} \frac{\partial W_i}{\partial z_i} = \frac{\partial f}{\partial t} + \sum_{i=1}^{n} \frac{\partial f W_i}{\partial z_i} = 0, \tag{A.179}$$

where the second expression is obtained by using Eq. (A.173). In fact, if the integrand in the last integral of Eq. (A.175) would be different from 0 at some point (and so over some small region around the point for a well-known theorem from calculus), by restricting $\Omega(0)$ to such a region we would obtain a nonzero integral, which goes against the assumption. If in addition **W** is solenoidal, we have

$$\frac{\partial W_i}{\partial z_i} = 0 \quad \Rightarrow \quad \frac{Df}{Dt} = 0. \tag{A.180}$$

A second application of the transport theorem is obtained when we know something about the integrand in Eq. (A.175), and we use this information to deduce the behavior of $dF(t)/dt$. One preeminent example is given by the Liouville (1809–1882) theorem: the fact that *the volume in phase space of a Hamiltonian system is conserved.* In fact, fix $f = 1$ in Eq. (A.174), so that $F(t) = \text{Volume}[\Omega(t)]$, and then use Eq. (A.175). We deduce immediately that the volume contracts in cases of the motion produced by a field **W** having negative divergence and expands if the divergence is positive. As special case is that of solenoidal velocity fields (e.g., Hamiltonian flows in phase space): for all Hamiltonian fields, the phase-space volume is conserved, and so Eq. (A.180) holds. Moreover, the same equation holds for *all* canonical changes of coordinates, as they leave invariant the structure of Hamilton equations, and the Jacobian matrix of the transformation is symplectic, so that its determinant is identically 1 (see Chapter 9).

References

Abraham, R., and Marsden, J. E. 1978. *Foundations of Mechanics*, 2nd ed. Addison-Wesley Publishing Company, Inc.
Abramowitz, M., and Stegun, I. A. 1972. *Handbook of Mathematical Functions*. Dover Publications.
Aguilar, L. A., and Merritt, D. 1990. *ApJ*, **354**, 33.
Alessandrini, E., Lanzoni, B., Miocchi, P., Ciotti, L., and Ferraro, F. R. 2014. *ApJ*, **795**, 169.
Amendt, P., and Cuddeford, P. 1994. *ApJ*, **435**, 93.
Amorisco, N. C., and Bertin, G. 2010. *A&A*, **519**, A47.
An, J., and Evans, N. W. 2019. *MNRAS*, **486**, 3915.
An, J. H., and Evans, N. W. 2006. *ApJ*, **642**, 752.
An, J., Evans, N. W., and Sanders, J. L. 2017. *MNRAS*, **467**, 1281.
An, J., van Hese, E., and Baes, M. 2012. *MNRAS*, **422**, 652.
Appell, P. 1887. *Ann. Math. Lpz.*, **30**, 155.
Arena, S. E., and Bertin, G. 2007. *A&A*, **463**, 921.
Arena, S. E., Bertin, G., Liseikina, T., and Pegoraro, F. 2006. *A&A*, **453**, 9.
Arfken, G. B., and Weber, H. J. 2005. *Mathematical Methods for Physicists*. Elsevier Academic Press.
Aris, R. 1989. *Vectors, Tensors, and the Basic Equations of Fluid Mechanics*. Dover Publications.
Arnold, V. I. 1978. *Mathematical Methods of Classical Mechanics*. Springer.
Arnold, V. I. 1990. *Huygens and Barrow, Newton and Hooke*. Birkhauser Verlag.
Arnold, V. I. 1992. *Ordinary Differential Equations*, 3rd ed., translated from the Russian by Roger Cooke. Springer.
Arnold, V. I. 1997. *Dynamical Systems III*. Springer.
Baes, M., and Ciotti, L. 2019a. *A&A*, **626**, A110.
Baes, M., and Ciotti, L. 2019b. *A&A*, **630**, A113.
Baes, M., and Dejonghe, H. 2002. *A&A*, **393**, 485.
Baes, M., and van Hese, E. 2007. *A&A*, **471**, 419.
Baes, M., and van Hese, E. 2011. *A&A*, **534**, A69.
Barber, J. A., and Zhao, H. 2014. *MNRAS*, **442**, 3533.
Barnabè, M., Ciotti, L., Fraternali, F., and Sancisi, R. 2006. *A&A*, **446**, 61.
Barnes, J. E., and Hernquist, L. 1992. *ARAA*, **30**, 705.
Barrow-Green, J. 1997. *Poincaré and the Three Body Problem*. American Mathematical Society and London Mathematical Society.

Batchelor, G. K. 1967. *An Introduction to Fluid Dynamics*. Cambridge University Press.
Becker, R. 1982. *Electromagnetic Fields and Interactions*. Dover Publications.
Bekenstein, J., and Milgrom, M. 1984. *ApJ*, **286**, 7.
Bender, C. M., and Orszag, S. A. 1978. *Advanced Mathematical Methods for Scientists and Engineers*. McGraw Hill.
Bender, R., Burstein, D., and Faber, S. M. 1992. *ApJ*, **399**, 462.
Bertin, G. 2014. *Dynamics of Galaxies*, 2nd ed. Cambridge University Press.
Bertin, G., and Lin, C. C. 1996. *Spiral Structure in Galaxies. A Density Wave Theory*. MIT Press.
Bertin, G., and Stiavelli, M. 1984. *A&A*, **137**, 26.
Bertin, G., and Trenti, M. 2003. *ApJ*, **584**, 729.
Bertin, G., and Varri, A. L. 2008. *ApJ*, **689**, 1005.
Bertin, G., Ciotti, L., and Del Principe, M. 2002. *A&A*, **386**, 149.
Bertin, G., Liseikina, T., and Pegoraro. 2003. *A&A*, **405**, 73.
Bertin, G., Pegoraro, F., Rubini, F., and Vesperini, E. 1994. *ApJ*, **434**, 94.
Binney, J. 1977. *MNRAS*, **181**, 735.
Binney, J. 1978. *MNRAS*, **183**, 501.
Binney, J. 1980. *MNRAS*, **190**, 873.
Binney, J. 1981. *MNRAS*, **196**, 455.
Binney, J. 1982a. *ARAA*, **20**, 399.
Binney, J. 1982b. *MNRAS*, **200**, 951.
Binney, J. 1985. *MNRAS*, **212**, 767.
Binney, J. 1996. *J. Astrophys. Astr.*, **17**, 81.
Binney, J. 2014. *arXiv e-prints*, arXiv:1411.4937.
Binney, J., and Gerhard, O. 1996. *MNRAS*, **279**, 1005.
Binney, J., and Kumar, S. 1993. *MNRAS*, **261**, 584.
Binney, J., and Mamon, G. A. 1982. *MNRAS*, **200**, 361.
Binney, J., and McMillan, P. J. 2016. *MNRAS*, **456**, 1982.
Binney, J., and Merrifield, M. 1998. *Galactic Astronomy*. Princeton University Press.
Binney, J. J., and Ossipkov, L. P. 2001. Page 317 of: Ossipkov, L. P., and Nikiforov, I. I. (eds.), *Stellar Dynamics: From Classic to Modern*. Saint Petersburg State University.
Binney, J., and Spergel, D. 1982. *ApJ*, **252**, 308.
Binney, J., and Spergel, D. 1984. *MNRAS*, **206**, 159.
Binney, J., and Tremaine, S. 2008. *Galactic Dynamics*, 2nd ed. Princeton University Press.
Bleistein, N., and Handelsman, R. A. 1986. *Asymptotic Expansions of Integrals*. Dover Publications.
Boccaletti, D., and Pucacco, G. 1996. *Theory of Orbits. I: Integrable Systems and Non-Perturbative Methods*. Springer-Verlag.
Boltzmann, L. 1896. *Lectures on Gas Theory*. Dover Publications.
Bontekoe, Tj. R., and van Albada, T. S. 1987. *MNRAS*, **224**, 349.
Borisenko, A. I., and Tarapov, I. E. 1979. *Vector and Tensor Analysis with Applications*. Dover Publications.
Born, M. 1969. *Atomic Physics*. Blackie and Son.
Bovy, J. 2017. *MNRAS*, **468**, L63.
Buchler, J. R., Ipser, J. R., and Williams, C. A. 1988. *Integrability in Dynamical Systems*. New York Academy of Sciences.
Burkert, A. 1995. *ApJL*, **447**, L25.
Byrd, P. F., and Friedman, M. 1971. *Handbook of Elliptic Integrals for Engineers and Physicists*, 2nd ed. Springer-Verlag.

Caon, N., Capaccioli, M., and D'Onofrio, M. 1993. *MNRAS*, **265**, 1013.
Cappellari, M. 2002. *MNRAS*, **333**, 400.
Cappellari, M. 2008. *MNRAS*, **390**, 71.
Cappellari, M. 2016. *ARAA*, **54**, 597.
Cappellari, M. 2020. *MNRAS*, **494**, 4819.
Caravita, C., Ciotti, L., and Pellegrini, S. 2021. arXiv e-prints, arXiv:2102.09440
Carollo, C. M., de Zeeuw, P. T., van der Marel, R. P., Danziger, I. J., and Qian, E. E. 1995a. *ApJL*, **441**, L25.
Carollo, C. M., de Zeeuw, P. T., and van der Marel, R. P. 1995b. *MNRAS*, **276**, 1131.
Casertano, S. 1983. *MNRAS*, **203**, 735.
Cavaliere, A., and Fusco-Femiano, R. 1976. *A&A*, **500**, 95.
Chandrasekhar, S. 1939. *An Introduction to the Study of Stellar Structure*. Dover Publications.
Chandrasekhar, S. 1941. *ApJ*, **93**, 285.
Chandrasekhar, S. 1942. *Principles of Stellar Dynamics*. Dover Publications.
Chandrasekhar, S. 1943a. *ApJ*, **97**, 255.
Chandrasekhar, S. 1943b. *ApJ*, **97**, 263.
Chandrasekhar, S. 1943c. *ApJ*, **98**, 54.
Chandrasekhar, S. 1969. *Ellipsoidal Figures of Equilibrium*. Dover Publications.
Chandrasekhar, S. 1995. *Newton's Principia for the Common Reader*. Clarendon Press.
Chandrasekhar, S., and von Neumann, J. 1942. *ApJ*, **95**, 489.
Chandrasekhar, S., and von Neumann, J. 1943. *ApJ*, **97**, 1.
Cimatti, A., Fraternali, F., and Nipoti, C. 2019. *Introduction to Galaxy Formation and Evolution*. Cambridge University Press.
Ciotti, L. 1991. *A&A*, **249**, 99.
Ciotti, L. 1994. *Celest. Mech. Dyn. Astr.*, **60**, 401.
Ciotti, L. 1996. *ApJ*, **471**, 68.
Ciotti, L. 1999. *ApJ*, **520**, 574.
Ciotti, L. 2000. *Lecture Notes on Stellar Dynamics*. Publications of the Scuola Normale Superiore, Springer.
Ciotti, L. 2009. *Nuovo Cimento*, **32**, 1.
Ciotti, L. 2010. Page 117 of: Bertin, G., de Luca, F., Lodato, G., Pozzoli, R., and Romé, M. (eds.), *Plasmas in the Laboratory and in the Universe: Interactions, Patterns, and Turbulence*. American Institute of Physics Conference Series, Vol. 1242.
Ciotti, L. 2019. *arXiv e-prints*, arXiv:1911.10480.
Ciotti, L. 2020. *arXiv e-prints*, arXiv:2009.06452.
Ciotti, L., and Bertin, G. 1999. *A&A*, **352**, 447.
Ciotti, L., and Bertin, G. 2005. *A&A*, **437**, 419.
Ciotti, L., and Binney, J. 2004. *MNRAS*, **351**, 285.
Ciotti, L., and Dutta, S. N. 1994. *MNRAS*, **270**, 390.
Ciotti, L., and Giampieri, G. 1997. *Celest. Mech. Dyn. Astr.*, **68**, 313.
Ciotti, L., and Giampieri, G. 2007. *MNRAS*, **376**, 1162.
Ciotti, L., and Lanzoni, B. 1997. *A&A*, **321**, 724.
Ciotti, L., and Marinacci, F. 2008. *MNRAS*, **387**, 1117.
Ciotti, L., and Morganti, L. 2009. *MNRAS*, **393**, 179.
Ciotti, L., and Morganti, L. 2010a. *MNRAS*, **401**, 1091.
Ciotti, L., and Morganti, L. 2010b. *MNRAS*, **408**, 1070.
Ciotti, L., and Ostriker, J. P. 2012. *AGN Feedback in Elliptical Galaxies: Numerical Simulations*. Astrophysics and Space Science Library, Vol. 378, p. 83.

Ciotti, L., and Pellegrini, S. 1992. *MNRAS*, **255**, 561.
Ciotti, L., and Pellegrini, S. 1996. *MNRAS*, **279**, 240.
Ciotti, L., and Pellegrini, S. 2004. *MNRAS*, **350**, 609.
Ciotti, L., and Pellegrini, S. 2008. *MNRAS*, **387**, 902.
Ciotti, L., and Pellegrini, S. 2017. *ApJ*, **848**, 29.
Ciotti, L., and Pellegrini, S. 2018. *ApJ*, **868**, 91.
Ciotti, L., and van Albada, T. S. 2001. *ApJL*, **552**, L13. 0
Ciotti, L., and Ziaee Lorzad, A. 2018. *MNRAS*, **473**, 5476.
Ciotti, L., Bertin, G., and Londrillo, P. 2004. Page 322 of: Bertin, G., Farina, D., and Pozzoli, R. (eds.), *Plasmas in the Laboratory and in the Universe: New Insights and New Challenges*. American Institute of Physics Conference Series, Vol. 703.
Ciotti, L., D'Ercole, A., Pellegrini, S., and Renzini, A. 1991. *ApJ*, **376**, 380.
Ciotti, L., Lanzoni, B., and Renzini, A. 1996. *MNRAS*, **282**, 1.
Ciotti, L., Lanzoni, B., and Volonteri, M. 2007. *ApJ*, **658**, 65.
Ciotti, L., Londrillo, P., and Nipoti, C. 2006. *ApJ*, **640**, 741.
Ciotti, L., Mancino, A., and Pellegrini, S. 2019. *MNRAS*, **490**, 2656.
Ciotti, L., Mancino, A., Pellegrini, S., and Ziaee Lorzad, A. 2021. *MNRAS*, **500**, 1054.
Ciotti, L., Morganti, L., and de Zeeuw, P. T. 2009. *MNRAS*, **393**, 491.
Ciotti, L., Stiavelli, M., and Braccesi, A. 1995. *MNRAS*, **276**, 961.
Ciotti, L., Zhao, H., and de Zeeuw, P. T. 2012. *MNRAS*, **422**, 2058.
Clarke, C., and Carswell, B. 2007. *Principles of Astrophysical Fluid Dynamics*. Cambridge University Press.
Contopoulos, G., Spyrou, N. K., and Vlahos, L. 1994. *Galactic Dynamics and N-Body Simulations – Lecture Notes in Physics*, Vol. 433. Springer-Verlag.
Courant, R., and Hilbert, D. 1989. *Methods of Mathematical Physics*. Wiley.
Cuddeford, P. 1991. *MNRAS*, **253**, 414.
Cuddeford, P., and Louis, P. 1995. *MNRAS*, **275**, 1017.
Cullen, C. G. 1990. *Matrices and Linear Transformations*. Dover Publications.
Currie, I. G. 1993. *Fundamental Mechanics of Fluids*. McGraw Hill.
D'Ercole, A., Recchi, S., and Ciotti, L. 2000. *ApJ*, **533**, 799.
D'Onofrio, M., Rampazzo, R., Zaggia, S., Struck, C., Bianchi, L., Poggianti, B. M., Sulentic, J. W., Tully, B. R., Marziani, P., Longair, M. S., Matteucci, F., Ciotti, L., Einasto, J., and Kroupa, P. 2016. Page 509 of: D'Onofrio, M., Rampazzo, R., and Zaggia, S. (eds), *From the Realm of the Nebulae to Populations of Galaxies*. Astrophysics and Space Science Library, Springer, Vol. 435.
Danby, J. 1962. *Fundamentals of Celestial Mechanics*. Macmillan.
de Bruijn, N. G. 1958. *Asymptotic Methods in Analysis*. Dover Publications.
de Bruijne, J. H. J., van der Marel, R. P., and de Zeeuw, P. T. 1996. *MNRAS*, **282**, 909.
de Vaucouleurs, G. 1948. *Ann. d'Astrophys.*, **11**, 247.
de Zeeuw, P. T. 1985a. *MNRAS*, **216**, 273.
de Zeeuw, P. T. 1985b. *MNRAS*, **216**, 599.
de Zeeuw, P. T., and Franx, M. 1991. *ARAA*, **29**, 239.
de Zeeuw, P. T., and Lynden-Bell, D. 1985. *MNRAS*, **215**, 713.
de Zeeuw, P. T., and Pfenniger, D. 1988. *MNRAS*, **235**, 949.
de Zeeuw, P. T., Evans, N. W., and Schwarzschild, M. 1996. *MNRAS*, **280**, 903.
de Zeeuw, P. T., Peletier, R., and Franx, M. 1986. *MNRAS*, **221**, 1001.
Dehnen, W. 1993. *MNRAS*, **265**, 250.
Dehnen, W., and Gerhard, O. E. 1993. *MNRAS*, **261**, 311.
Dehnen, W., and Gerhard, O. E. 1994. *MNRAS*, **268**, 1019.

Dejonghe, H. 1986. *Phys. Rep.*, **133**, 217.

Dejonghe, H. 1987a. Page 495 of: de Zeeuw, P. T. (ed.), *Structure and Dynamics of Elliptical Galaxies*. IAU Symposium, Vol. 127.

Dejonghe, H. 1987b. *MNRAS*, **224**, 13.

Dejonghe, H., and de Zeeuw, T. 1988. *ApJ*, **333**, 90.

Dejonghe, H., and Merritt, D. 1992. *ApJ*, **391**, 531.

Dennery, P., and Krzywicki, A. 1967. *Mathematics for Physicists*. Dover Publications.

Di Cintio, P., and Ciotti, L. 2011. *IJBC*, **21**, 2279.

Di Cintio, P., Ciotti, L., and Nipoti, C. 2013. *MNRAS*, **431**, 3177.

Di Cintio, P., Ciotti, L., and Nipoti, C. 2017. *MNRAS*, **468**, 2222.

Diacu, F., and Holmes, P. 1996. *Celestial Encounters. The Origins of Chaos and Stability*. Princeton University Press.

Djorgovski, S., and Davis, M. 1987. *ApJ*, **313**, 59.

Do Carmo, M. P. 1976. *Differential Geometry of Curves and Surfaces*. Prentice Hall.

Dressler, A., Lynden-Bell, D., Burstein, D., Davies, R. L., Faber, S. M., Terlevich, R., and Wegner, G. 1987. *ApJ*, **313**, 42.

Dubinski, J., and Carlberg, R. G. 1991. *ApJ*, **378**, 496.

Eddington, A. S. 1916. *MNRAS*, **76**, 572.

Edwards, C. H., Jr. 1994. *Advanced Calculus of Several Variables*. Dover Publications.

Einasto, J. 1965. *Trudy Inst. Astrofiz. Alma-Ata*, **5**, 87.

El-Zant, A. A., Hoffman, Y., Primack, J., Combes, F., and Shlosman, I. 2004. *ApJL*, **607**, L75.

Erdélyi, A., Magnus, W., Oberhettinger, F., and Tricomi, F. G. 1953. *Higher Transcendental Functions*. McGraw Hill.

Esposito, L. P. 2014. *Planetary Rings: A Post-Equinox View*. Cambridge University Press.

Evans, N. W. 1990. *PhRvA*, **41**, 5666.

Evans, N. W. 1993. *MNRAS*, **260**, 191.

Evans, N. W. 1994. *MNRAS*, **267**, 333.

Evans, N. W., and Bowden, A. 2014. *MNRAS*, **443**, 2.

Evans, N. W., and de Zeeuw, P. T. 1992. *MNRAS*, **257**, 152.

Evans, N. W., and de Zeeuw, P. T. 1994. *MNRAS*, **271**, 202.

Evans, N. W., and Lynden-Bell, D. 1989. *MNRAS*, **236**, 801.

Evans, N. W., An, J., Bowden, A., and Williams, A. A. 2015. *MNRAS*, **450**, 846.

Evans, N. W., de Zeeuw, P. T., and Lynden-Bell, D. 1990. *MNRAS*, **244**, 111.

Evans, N. W., Hafner, R. M., and de Zeeuw, P. T. 1997. *MNRAS*, **286**, 315.

Faber, S. M., and Jackson, R. E. 1976. *ApJ*, **204**, 668.

Fabricant, D., Lecar, M., and Gorenstein, P. 1980. *ApJ*, **241**, 552.

Ferrarese, L., and Merritt, D. 2000. *ApJL*, **539**, L9.

Ferraro, F. R., Lanzoni, B., Dalessandro, E., Beccari, G., Pasquato, M., Miocchi, P., Rood, R. T., Sigurdsson, S., Sills, A., Vesperini, E., Mapelli, M., Contreras, R., Sanna, N., and Mucciarelli, A. 2012. *Nature*, **492**, 393.

Ferraro, F. R., Lanzoni, B., Raso, S., Nardiello, D., Dalessandro, E., Vesperini, E., Piotto, G., Pallanca, C., Beccari, G., Bellini, A., Libralato, M., Anderson, J., Aparicio, A., Bedin, L. R., Cassisi, S., Milone, A. P., Ortolani, S., Renzini, A., Salaris, M., and van der Marel, R. P. 2018. *ApJ*, **860**, 36.

Ferrers, N. M. 1877. *Quart. J. Pure Appl. Math.*, **14**, 1.

Ferronsky, V. I., Denisik, S. A., and Ferronsky, S. V. 2011. *Jacobi Dynamics*, Vol. 369. Astrophysics and Space Science Library, Springer.

Feynman, R. P., Leighton, R. B., and Sands, M. 1977. *The Feynman Lectures on Physics*. Addison Wesley.
Franx, M. 1988a. *Structure and Kinematics of Elliptical Galaxies*. PhD thesis, Leiden University.
Franx, M. 1988b. *MNRAS*, **231**, 285.
Freeman, K. C. 1966. *MNRAS*, **134**, 15.
Freeman, K. C. 1970. *ApJ*, **160**, 811.
Fricke, W. 1952. *Astron. Nachr.*, **280**, 193.
Fridman, A. M., and Poliachenko, V. L. 1984. *Physics of Gravitating Systems. I – Equilibrium and Stability*. Springer-Verlag.
Gallavotti, G. 2001. *Rend. Mat. Acc. Lincei*, **12**, 125.
Gebhardt, K., Bender, R., Bower, G., Dressler, A., Faber, S. M., Filippenko, A. V., Green, R., Grillmair, C., Ho, L. C., Kormendy, J., Lauer, T. R., Magorrian, J., Pinkney, J., Richstone, D., and Tremaine, S. 2000. *ApJL*, **539**, L13.
Gerhard, O. E., and Binney, J. J. 1996. *MNRAS*, **279**, 993.
Gerhard, O. E. 1991. *MNRAS*, **250**, 812.
Gerhard, O. E. 1993. *MNRAS*, **265**, 213.
Gerhard, O. E, Arnaboldi, M., Freeman, K. C., Kashikawa, N., Okamura, S., and Yasuda, N. 2005. *ApJL*, **621**, L93.
Gidas, B., Ni, W.-M., and Nirenberg, L. 1979. *Comm. Math. Phys.*, **68**, 209.
Gnedin, O. Y., Ostriker, J. P., and Tremaine, S. 2014. *ApJ*, **785**, 71.
Goldreich, P., and Tremaine, S. 1982. *ARAA*, **20**, 249.
Goldstein, H. 1975. *Am. J. Phys.*, **43**, 737.
Goldstein, H., Poole, C., and Safko, J. 2000. *Classical Mechanics*, 3rd ed. Addison Wesley.
González-García, A. C., and van Albada, T. S. 2003. *MNRAS*, **342**, L36.
Gorenflo, R., and Vessella, A. 1991. *Abel Integral Equations*. Springer-Verlag.
Gradshteyn, I. S., Ryzhik, I. M., Jeffrey, A., and Zwillinger, D. 2007. *Table of Integrals, Series, and Products*, 7th ed. Elsevier.
Graham, A. W. 1998. *MNRAS*, **295**, 933.
Graham, A. W., and Colless, M. 1997. *MNRAS*, **287**, 221.
Graham, A. W. 2016. Pp. 263 of Laurikainen, E., Peletier, R., and Gadotti, D. (eds.), *Galactic Bulges*. Astrophysics and Space Science Library book series (ASSL), vol. 418.
Graham, A. W., and Driver, S. P. 2005. *PASA*, **22**, 118.
Graham, A. W., Merritt, D., Moore, B., Diemand , J., and Terzić, B. 2006. *AJ*, **132**, 2701.
Gutzwiller, M. C. 1990. *Chaos in Classical and Quantum Mechanics*. Springer-Verlag.
Hagihara, Y. 1970. *Celestial Mechanics*, Vols. 1–5. MIT Press.
Hansen, S. H., and Moore, B. 2006. *New Astron.*, **11**, 333.
Heggie, D., and Hut, P. 2003. *The Gravitational Million-Body Problem: A Multidisciplinary Approach to Star Cluster Dynamics*. Cambridge University Press.
Hénon, M. 1959. *Ann. d'Astrophys.*, **22**, 126.
Hénon, M. 1960. *Ann. d'Astrophys.*, **23**, 668.
Hernquist, L. 1990. *ApJ*, **356**, 359.
Hernquist, L., and Ostriker, J. P. 1992. *ApJ*, **386**, 375.
Hiotelis, N. 1994. *A&A*, **291**, 725.
Hjorth, J., and Madsen, J. 1995. *ApJ*, **445**, 55.
Hubble, E. P. 1930. *ApJ*, **71**.
Hunter, C. 1963. *MNRAS*, **126**, 299.
Hunter, C. 1975. *AJ*, **80**, 783.

Hunter, C. 1977. *AJ*, **82**, 271.
Hunter, C., and de Zeeuw, P. T. 1992. *ApJ*, **389**, 79.
Hunter, C., and Qian, E. 1993. *MNRAS*, **262**, 401.
Ince, E. L. 1927. *Ordinary Differential Equations*. Dover Publications.
Jackson, J. D. 1998. *Classical Electrodynamics*, 3rd ed. John Wiley & Sons.
Jaffe, W. 1983. *MNRAS*, **202**, 995.
Jeffrey, H., and Jeffrey, B. S. 1950. *Methods of Mathematical Physics*, 2nd ed. Cambridge University Press.
Jorgensen, I., Franx, M., and Kjaergaard, P. 1993. *ApJ*, **411**, 34.
Jorgensen, I., Franx, M., and Kjaergaard, P. 1996. *MNRAS*, **280**, 167.
Kaasalainen, M., and Binney, J. 1994. *MNRAS*, **268**, 1033.
Kahn, P. B. 2004. *Mathematical Methods for Scientists and Engineers*. Dover Publications.
Kalnajs, A. J. 1972. *ApJ*, **175**, 63.
Kalnajs, A. J. 1976a. *ApJ*, **205**, 745.
Kalnajs, A. J. 1976b. *ApJ*, **205**, 751.
Kandrup, H. E. 1980. *Phys. Rep.*, **63**, 1.
Kellogg, O. D. 1953. *Foundations of Potential Theory*. Dover Publications.
Kent, S. M. 1990. *MNRAS*, **247**, 702.
Khinchin, A. I. 1949. *Mathematical Foundations of Statistical Mechanics*. Dover Publications.
Kim, D.-W., and Pellegrini, S. 2012. *Hot Interstellar Matter in Elliptical Galaxies*. Astrophysics and Space Science Library, Springer, Vol. 378.
King, I. R. 1962. *AJ*, **67**, 471.
King, I. R. 1966. *AJ*, **71**, 64.
King, I. R. 1972. *ApJL*, **174**, L123.
Kormendy, J. 1977. *ApJ*, **218**, 333.
Korol, V., Ciotti, L., and Pellegrini, S. 2016. *MNRAS*, **460**, 1188.
Kuzmin, J. G. 1956. *Astron. Zh.*, **33**, 27.
Lamb, H. 1945. *Hydrodynamics*. Dover Publications.
Landau, L. D., and Lifshitz, E. M. 1969. *Mechanics*. Pergamon Press.
Landau, L. D., and Lifshitz, E. M. 1971. *The Classical Theory of Fields*. Pergamon Press.
Landau, L. D., and Lifshitz, E. M. 1986. *Fluid Mechanics*. Pergamon Press.
Lanzoni, B., and Ciotti, L. 2003. *A&A*, **404**, 819.
Lanzoni, B., Ciotti, L., Cappi, A., Tormen, G., and Zamorani, G. 2004. *ApJ*, **600**, 640.
Lee, J., and Suto, Y. 2003. *ApJ*, **585**, 151.
Letelier, P. S. 2007. *MNRAS*, **381**, 1031.
Lewin, L. 1986. *Polylogarithms and Associated Functions*. North Holland.
Lichtenberg, A. J., and Lieberman, M. A. 1992. *Regular and Chaotic Dynamics*. Springer-Verlag.
Łokas, E. L., and Mamon, G. A. 2001. *MNRAS*, **321**, 155.
Lomen, D., and Mark, J. 1988. *Differential Equations*. Prentice Hall.
Long, K., and Murali, C. 1992. *ApJ*, **397**, 44.
Louis, P. D. 1993. *MNRAS*, **261**, 283.
Lynden-Bell, D. 1960. *MNRAS*, **120**, 204.
Lynden-Bell, D. 1962a. *MNRAS*, **123**, 447.
Lynden-Bell, D. 1962b. *MNRAS*, **124**, 95.
Lynden-Bell, D. 1962c. *MNRAS*, **124**, 1.
Lynden-Bell, D. 1967. *MNRAS*, **136**, 101.
Lynden-Bell, D. 2004. *PhRvD*, **70**(10), 105017.

Lynden-Bell, D. 2006. *arXiv e-prints*, astro-ph/0604428.
Lynden-Bell, D., and Eggleton, P. P. 1980. *MNRAS*, **191**, 483.
Lynden-Bell, D., and Lynden-Bell, R. M. 2004. *J. Stat. Phys.*, **117**, 199.
Lynden-Bell, D., and Pineault, S. 1978. *MNRAS*, **185**, 679.
Lynden-Bell, D., and Wood, R. 1968. *MNRAS*, **138**, 495.
Magorrian, J., and Binney, J. 1994. *MNRAS*, **271**, 949.
Magorrian, J., Tremaine, S., Richstone, D., Bender, R., Bower, G., Dressler, A., Faber, S. M., Gebhardt, K., Green, R., Grillmair, C., Kormendy, J., and Lauer, T. 1998. *AJ*, **115**, 2285.
Mathai, A. M., Saxena, R. K., and Haubold, H. J. 2009. *The H-Function: Theory and Applications*. Springer-Verlag.
Mathews, W. G., and Brighenti, F. 2003a. *ARAA*, **41**, 191.
Mathews, W. G., and Brighenti, F. 2003b. *ApJ*, **599**, 992.
McCauley, J. L. 1997. *Classical Mechanics*. Cambridge University Press.
McGill, C., and Binney, J. 1990. *MNRAS*, **244**, 634.
McMillan, W. D. 1958. *The Theory of the Potential*. Dover Publications.
Merrifield, M. R., and Kent, S. M. 1990. *AJ*, **99**, 1548.
Merritt, D. 1985a. *AJ*, **90**, 1027.
Merritt, D. 1985b. *MNRAS*, **214**, 25P.
Merritt, D. 2013. *Dynamics and Evolution of Galactic Nuclei*. Princeton University Press.
Merritt, D., and Ferrarese, L. 2001. *ApJ*, **547**, 140.
Merritt, D., Tremaine, S., and Johnstone, D. 1989. *MNRAS*, **236**, 829.
Messiah, A. 1967. *Quantum Mechanics*, Vols. 1 and 2. North Holland.
Mestel, L. 1963. *MNRAS*, **126**, 553.
Meyer, K. R., and Hall, G. R. 1992. *Introduction to Hamiltonian Dynamical Systems and the N-Body Problem*. Springer-Verlag.
Meyer, R. E. 1982. *Introduction to Mathematical Fluid Dynamics*. Dover Publications.
Michie, R. W. 1963. *MNRAS*, **125**, 127.
Milgrom, M. 1983. *ApJ*, **270**, 365.
Milne-Thomson, L. M. 1996. *Theoretical Hydrodynamics*. Dover Publications.
Miyamoto, M. 1971. *PASJ*, **23**, 21.
Miyamoto, M. 1974. *A&A*, **30**, 441.
Miyamoto, M. 1975. *PASJ*, **27**, 431.
Miyamoto, M., and Nagai, R. 1975. *PASJ*, **27**, 533.
Morse, P. M., and Feshbach, H. 1953. *Methods of Theoretical Physics*. McGraw Hill.
Muccione, V., and Ciotti, L. 2004. *A&A*, **421**, 583.
Nagai, R., and Miyamoto, M. 1976. *PASJ*, **28**, 1.
Narashiman, M. N. L. 1993. *Principles of Continuum Mechanics*. John Wiley & Sons.
Navarro, J. F., Frenk, C. S., and White, S. D. M. 1997. *ApJ*, **490**, 493.
Negri, A., Ciotti, L., and Pellegrini, S. 2014a. *MNRAS*, **439**, 823.
Negri, A., Posacki, S., Pellegrini, S., and Ciotti, L. 2014b. *MNRAS*, **445**, 1351.
Newman, E. T. 1973. *J. Math. Phys.*, **14**, 102.
Nipoti, C., Ciotti, L., Binney, J., and Londrillo, P. 2008. *MNRAS*, **386**, 2194.
Nipoti, C., Ciotti, L., and Londrillo, P. 2011. *MNRAS*, **414**, 3298.
Nipoti, C., Londrillo, P., and Ciotti, L. 2002. *MNRAS*, **332**, 901.
Nipoti, C., Londrillo, P., and Ciotti, L. 2003. *MNRAS*, **342**, 501.
Nipoti, C., Londrillo, P., and Ciotti, L. 2006. *MNRAS*, **370**, 681.
Nipoti, C., Londrillo, P., and Ciotti, L. 2007. *ApJ*, **660**, 256.
Nipoti, C., Treu, T., Ciotti, L., and Stiavelli, M. 2004. *MNRAS*, **355**, 1119.

Ogorodnikov, K. F. 1965. *Dynamics of Stellar Systems*. Pergamon Press.
Oldham, L. J., and Evans, N. W. 2016. *MNRAS*, **462**, 298.
Osipkov, L. P. 1979. *Astron. Lett.*, **5**, 77.
Ostriker, J. P. 1980. *Comments Astrophys.*, **8**, 177.
Ostriker, J. P., and Ciotti, L. 2005. *Philos. Trans. R. Soc.*, **363**, 667.
Ostriker, J. P., and Davidsen, A. F. 1968. *ApJ*, **151**, 679.
Ostriker, J. P., and Peebles, P. J. E. 1973. *ApJ*, **186**, 467.
Paczyńsky, B., and Wiita, P. J. 1980. *A&A*, **500**, 203.
Palmer, P. L. 1994. *Stability of Collisionless Stellar Systems: Mechanisms for the Dynamical Structure of Galaxies*, Vol. 185. Astrophysics and Space Science Library and Kluwer Academic Publishers.
Pellegrini, S., and Ciotti, L. 2006. *MNRAS*, **370**, 1797.
Plummer, H. C. 1911. *MNRAS*, **71**, 460.
Poincaré, H. 1892. *Les Méthodes Nouvelles de la Mécanique Céleste.* Gauthier-Villars.
Polyachenko, V. L., and Shukhman, I. G. 1981. *SvA*, **25**, 533.
Posacki, S., Pellegrini, S., and Ciotti, L. 2013. *MNRAS*, **433**, 2259.
Poveda, A., Iturriaga, R., and Orozco, I. 1960. *Bol. Obs. Tonantzintla y Tacubaya*, **2**, 3.
Pringle, J. E., and King, A. R. 2007. *Astrophysical Flows.* Cambridge University Press.
Prudnikov, A. P., Brychkov, Yu. A., and Marichev, O. I. 1990. *Integrals and Series*. Gordon and Breach.
Prugniel, P., and Simien, F. 1996. *A&A*, **309**, 749.
Prugniel, P., and Simien, F. 1997. *A&A*, **321**, 111.
Qian, E. E., de Zeeuw, P. T., van der Marel, R. P., and Hunter, C. 1995. *MNRAS*, **274**, 602.
Renzini, A., and Ciotti, L. 1993. *ApJL*, **416**, L49.
Reynolds, J. H. 1913. *MNRAS*, **74**, 132.
Richstone, D. O. 1980. *ApJ*, **238**, 103.
Richstone, D. O. 1984. *ApJ*, **281**, 100.
Richstone, D. O., and Tremaine, S. 1984. *ApJ*, **286**, 27.
Riciputi, A., Lanzoni, B., Bonoli, S., and Ciotti, L. 2005. *A&A*, **443**, 133.
Roberts, P. H. 1962. *ApJ*, **136**, 1108.
Rosenbluth, M. N., MacDonald, W. M., and Judd, D. L. 1957. *Phys. Rev.*, **107**, 1.
Routh, E. J. 1922. *A Treatise on Analytical Statics. II.* Cambridge University Press.
Rowley, G. 1988. *ApJ*, **331**, 124.
Roy, A. E. 2005. *Orbital Motion*, 4th ed. Institute of Physics Publishing.
Saglia, R. P., Bender, R., and Dressler, A. 1993. *A&A*, **279**, 75.
Saha, P. 1991. *MNRAS*, **248**, 494.
Sanders, J. L., and Binney, J. 2015. *MNRAS*, **447**, 2479.
Sanders, R. H. 2010. *The Dark Matter Problem: A Historical Perspective*. Cambridge University Press.
Sarazin, C. L. 1988. *X-Ray Emission from Clusters of Galaxies*. Cambridge University Press.
Saslaw, W. C. 1987. *Gravitational Physics of Stellar and Galactic Systems*. Cambridge University Press.
Satoh, C. 1980. *PASJ*, **32**, 41.
Schiaparelli, G. 1926. *Scritti sulla storia dell'astronomia antica.* Zanichelli.
Schulz, E. 2009. *ApJ*, **693**, 1310.
Schulz, E. 2012. *ApJ*, **747**, 106.
Schwarzschild, M. 1954. *AJ*, **59**, 273.
Schwarzschild, M. 1979. *ApJ*, **232**, 236.

Sersic, J. L. 1968. *Atlas de Galaxias Australes*. Observatorio Astronomico, Cordoba, Argentina.
Shu, F. H. 1992. *Physics of Astrophysics. II: Gas Dynamics*. University Science Books.
Shu, F. H. 1999. *ApJ*, **525C**, 347.
Smet, C. O., Posacki, S., and Ciotti, L. 2015. *MNRAS*, **448**, 2921.
Sparke, L. S., and Gallagher, J. S. 2007. *Galaxies in the Universe: An Introduction*. Cambridge University Press.
Spies, G. O., and Nelson, D. B. 1974. *MNRAS*, **17**, 1865.
Spitzer, L. 1987. *Dynamical Evolution of Globular Clusters*. Princeton University Press.
Spitzer, L., Jr. 1942. *ApJ*, **95**, 329.
Stark, A. A. 1977. *ApJ*, **213**, 368.
Stiavelli, M., and Bertin, G. 1985. *MNRAS*, **217**, 735.
Stiavelli, M., and Bertin, G. 1987. *MNRAS*, **229**, 61.
Szebehely, V. 1967. *Theory of Orbits. The Restricted Problem of Three Bodies*. Academic Press.
Tassoul, J.-L. 1978. *Theory of Rotating Stars*. Princeton University Press.
Toomre, A. 1963. *ApJ*, **138**, 385.
Toomre, A. 1964. *ApJ*, **139**, 1217.
Toomre, A. 1977. Page 401 of: Tinsley, B. M., and Larson, R. B. (eds.), *Evolution of Galaxies and Stellar Populations*. Yale University Observatory.
Toomre, A. 1982. *ApJ*, **259**, 535.
Tremaine, S., and Weinberg, M. D. 1984. *MNRAS*, **209**, 729.
Tremaine, S., Henon, M., and Lynden-Bell, D. 1986. *MNRAS*, **219**, 285.
Tremaine, S., Gebhardt, K., Bender, R., Bower, G., Dressler, A., Faber, S. M., Filippenko, A. V., Green, R., Grillmair, C., Ho, L. C., Kormendy, J., Lauer, T. R., Magorrian, J., Pinkney, J., and Richstone, D. 2002. *ApJ*, **574**, 740.
Tremaine, S. D., Ostriker, J. P., and Spitzer, L., Jr. 1975. *ApJ*, **196**, 407.
Tremaine, S., Richstone, D. O., Byun, Y.-I., Dressler, A., Faber, S. M., Grillmair, C., Kormendy, J., and Lauer, T. R. 1994. *AJ*, **107**, 634.
Truesdell, C.-A. 1954. *The Kinematics of Vorticity*. Indiana University Press.
Truesdell, C.-A. 1984. *Rational Thermodynamics*. Springer-Verlag.
Truesdell, C.-A. 1991. *A First Course in Rational Continuum Mechanics: General Concepts*. Academic Press.
Tully, R. B., and Fisher, J. R. 1977. *A&A*, **500**, 105.
Valtonen, M., and Karttunen, H. 2006. *The Three-Body Problem*. Cambridge University Press.
van Albada, T. S. 1982. *MNRAS*, **201**, 939.
van Albada, T. S., and Szomoru, A. 2020. Page 532 of: Bragaglia, A., Davies, M., Sills, A., and Vesperini, E. (eds.), *IAU Symposium*. IAU Symposium, Vol. 351.
van Albada, T. S., Bahcall, J. N., Begeman, K., and Sancisi, R. 1985. *ApJ*, **295**, 305.
van Albada, T. S., Bertin, G., and Stiavelli, M. 1995. *MNRAS*, **276**, 1255.
van de Ven, G., Hunter, C., Verolme, E. K., and de Zeeuw, P. T. 2003. *MNRAS*, **342**, 1056.
van der Marel, R. 1994. *Velocity Profiles and Dynamical Modeling of Galaxies*. PhD thesis, Leiden University.
van der Marel, R. P., and Franx, M. 1993. *ApJ*, **407**, 525.
van der Marel, R. P., Rix, H. W., Carter, D., Franx, M., White, S. D. M., and de Zeeuw, T. 1994. *MNRAS*, **268**, 521.
Van Hese, E., Baes, M., and Dejonghe, H. 2011. *ApJ*, **726**, 80.
Varri, A. L., and Bertin, G. 2009. *ApJ*, **703**, 1911.

Varri, A. L., and Bertin, G. 2012. *A&A*, **540**, A94.
Vladimirov, V. S. 1979. *Generalized Functions in Mathematical Physics – Translated from the Russian by George Yankovsky*. Mir Publishers.
Vogt, D., and Letelier, P. S. 2009a. *MNRAS*, **396**, 1487.
Vogt, D., and Letelier, P. S. 2009b. *MNRAS*, **398**, 1563.
Watson, G. N.. 1966. *A Treatise on the Theory of Bessel Functions*, 2nd ed. Cambridge University Press.
White, M. L. 1949. *ApJ*, **109**, 159.
White, S. D. M. 1976. *MNRAS*, **174**, 19.
Whittaker, E. T. 1917. *A Treatise on the Analytical Dynamics of Particels and Rigid Bodies*. Cambridge University Press.
Williams, A. A., and Evans, N. W. 2017. *MNRAS*, **469**, 4414.
Wilson, C. P. 1975. *AJ*, **80**, 175.
Wintner, A. 1947. *The Analytical Foundations of Celestial Mechanics*. Princeton University Press.
Woltjer, L. 1967. Structure and dynamics of galaxies. Vol. 9, p. 1 of: Ehlers, J. (ed.), *Lectures in Applied Mathematics. Relativity Theory and Astrophysics. 2. Galactic Structure*. American Mathematical Society.
Yoshida, T. 1987. *Eur. J. Phys.*, **8**, 259.
Young, P. J. 1976. *AJ*, **81**, 807.
Zhao, H. 1996. *MNRAS*, **278**, 488.

Index

Printed in the United States
by Baker & Taylor Publisher Services

Printed in the United States
by Baker & Taylor Publisher Services